T0269410

Lecture Notes in Mathematics 2155

More information about this series at http://www.springer.com/series/304

Eric Delabaere

Divergent Series, Summability and Resurgence III

Resurgent Methods and the First Painlevé Equation

Eric Delabaere
Département de Mathématiques
Université d'Angers
Angers, France

ISSN 0075-8434 ISSN 1617-9692 (electronic)
Lecture Notes in Mathematics
ISBN 978-3-319-28999-1 ISBN 978-3-319-29000-3 (eBook)
DOI 10.1007/978-3-319-29000-3

Library of Congress Control Number: 2016940058

Mathematics Subject Classification (2010): 34M30, 34M40, 34M55, 30B40, 30D05

Printed on acid-free paper

This Springer imprint is published by Springer Nature
The registered company is Springer International Publishing AG Switzerland

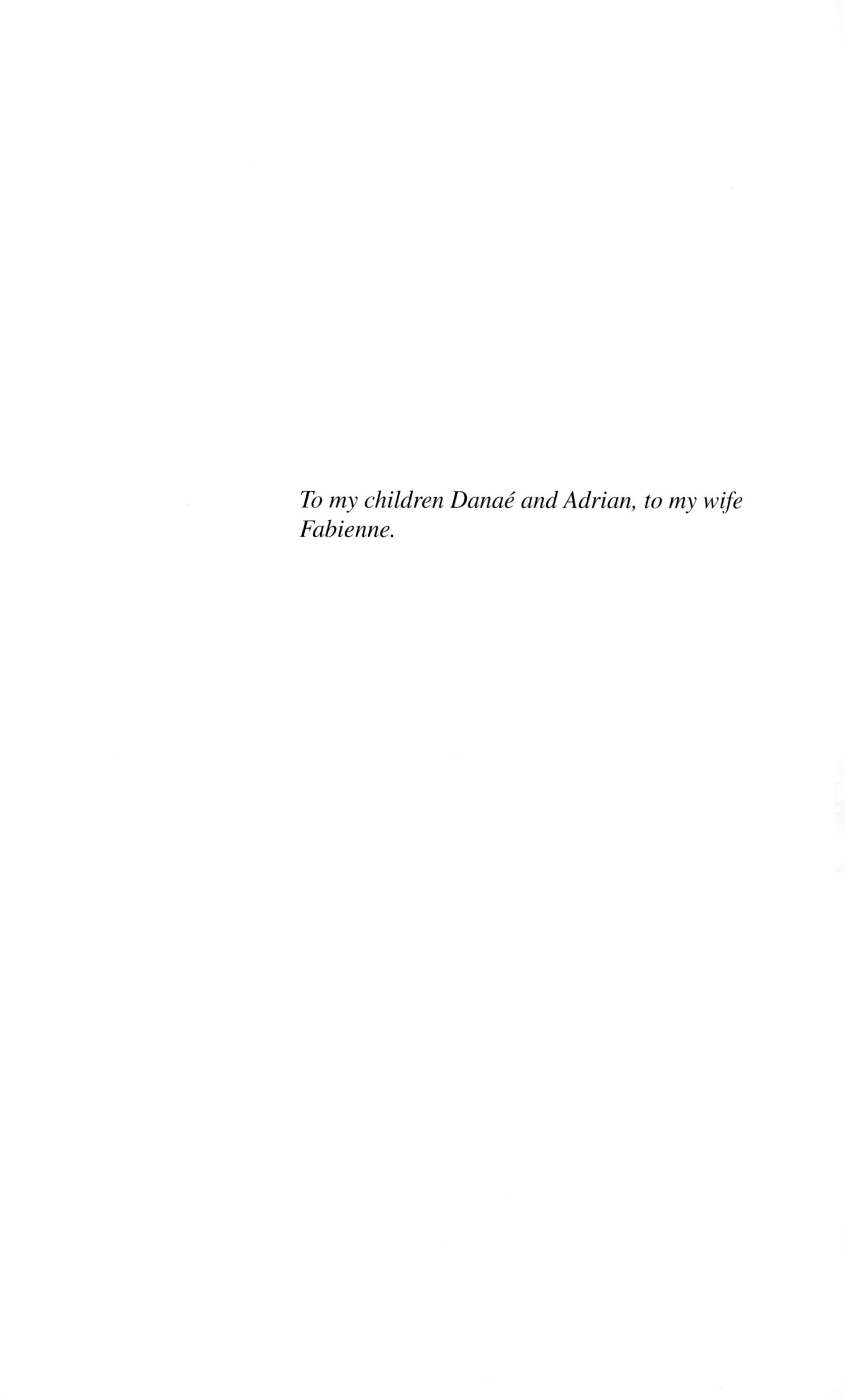

To my children Danaé and Adrian, to my wife Fabienne.

Avant-Propos

Le sujet principal traité dans la série de volumes *Divergent Series, Summability and Resurgence* est la théorie des développements asymptotiques et des séries divergentes appliquée aux équations différentielles ordinaires (EDO) et à certaines équations aux différences dans le champ complexe.

Les équations différentielles dans le champ complexe, et dans le cadre holomorphe, sont un sujet très ancien. La théorie a été très active dans la deuxième moitié du XIX-ème siècle. En ce qui concerne les *équations linéaires*, les mathématiciens de cette époque les ont subdivisées en deux classes. Pour la première, celle des équations *à points singuliers réguliers* (ou *de Fuchs*), généralisant les équations hypergéométriques d'Euler et de Gauss, ils ont enregistré *"des succès aussi décisifs que faciles"* comme l'écrivait René Garnier en 1919. En revanche, pour la seconde, celle des équations dites *à points singuliers irréguliers*, comme l'écrivait aussi Garnier, *"leurs efforts restent impuissants à édifier aucune théorie générale"*. La raison centrale de ce vif contraste est que toute série entière apparaissant dans l'écriture d'une solution d'une équation différentielle de Fuchs est *automatiquement convergente* tandis que pour les équations irrégulières ces séries sont génériquement *divergentes* et que l'on ne savait qu'en faire. La situation a commencé à changer grâce à un travail magistral de Henri Poincaré entrepris juste après sa thèse, dans lequel il "donne un sens" aux solutions divergentes des EDO linéaires irrégulières en introduisant un outil nouveau, et qui était appelé à un grand avenir, la théorie des développements asymptotiques. Il a ensuite utilisé cet outil pour donner un sens aux séries divergentes de la mécanique céleste, et remporté de tels succès que presque tout le monde a oublié l'origine de l'histoire, c'est-à-dire les EDO ! Les travaux de Poincaré ont (un peu...) remis à l'honneur l'étude des séries divergentes, abandonnée par les mathématiciens après Cauchy. L'Académie des Sciences a soumis ce sujet au concours en 1899 ce qui fut à l'origine d'un travail important d'Émile Borel. Celui-ci est la source de nombre des techniques utilisées dans *Divergent Series, Summability and Resurgence*. Pour revenir aux EDO irrégulières, le sujet a fait l'objet de nombreux et importants travaux de G.D. Birkhoff et R. Garnier durant le premier quart du XX-ème siècle. On retrouvera ici de nombreux prolongements des méthodes de Birkhoff. Après 1940, le sujet a étrangement presque disparu, la théorie étant, je ne sais trop pourquoi, considérée comme achevée, tout comme celle des équations de Fuchs. Ces dernières ont réémergé au début des années 1970, avec les travaux de Raymond Gérard, puis un livre de Pierre Deligne. Les équations irrégulières ont suivi avec des travaux de l'école allemande et surtout de l'école française. De nombreuses techniques complètement nouvelles ont été introduites (développements asymptotiques Gevrey, k-sommabilité, multisommabilité, fonctions résurgentes...) permettant en particulier une vaste généralisation du *phénomène de Stokes* et sa mise en relation avec la théorie de Galois différentielle et le problème de Riemann-Hilbert généralisé. Tout ceci a depuis reçu de très nombreuses applications dans des domaines très variés, allant de l'intégrabilité des systèmes hamiltoniens aux problèmes de points tournant pour les EDO singulièrement perturbées ou à divers problèmes de modules. On en trouvera certaines dans *Divergent Series, Summability and Resurgence*, comme l'étude résurgente des germes de difféomorphismes analytiques du plan complexe tangents à l'identité ou celle de l'EDO non-linéaire Painlevé I.

Le sujet restait aujourd'hui difficile d'accès, le lecteur ne disposant pas, mis à part les articles originaux, de présentation accessible couvrant *tous les aspects*. Ainsi *Divergent Series, Summability and Resurgence* comble une lacune. Ces volumes présentent un large panorama des recherches les plus récentes sur un vaste domaine classique et passionnant, en pleine renaissance, on peut même dire en pleine explosion. Ils sont néanmoins accessibles à tout lecteur possédant une bonne familiarité avec les fonctions analytiques d'une variable complexe. Les divers outils sont soigneusement mis en place, progressivement et avec beaucoup d'exemples. C'est une belle réussite.

À Toulouse, le 16 mai 2014,

Jean-Pierre Ramis

Preface to the Three Volumes

This three-volume series arose out of lecture notes for the courses we gave together at a CIMPA[1] school in Lima, Peru, in July 2008. Since then, these notes have been used and developed in graduate courses held at our respective institutions, that is, the universities of Angers, Nantes, Strasbourg (France) and the Scuola Normale Superiore di Pisa (Italy). The original notes have now grown into self-contained introductions to problems raised by analytic continuation and the divergence of power series in one complex variable, especially when related to differential equations.

A classical way of solving an analytic differential equation is the power series method, which substitutes a power series for the unknown function in the equation, then identifies the coefficients. Such a series, if convergent, provides an analytic solution to the equation. This is what happens at an ordinary point, that is, when we have an initial value problem to which the Cauchy-Lipschitz theorem applies. Otherwise, at a singular point, even when the method can be applied the resulting series most often diverges; its connection with "actual" local analytic solutions is not obvious despite its deep link to the equation.

The hidden meaning of divergent formal solutions was already pondered in the nineteenth century, after Cauchy had clarified the notions of convergence and divergence of series. For ordinary *linear* differential equations, it has been known since the beginning of the twentieth century how to determine a full set of linearly independent formal solutions[2] at a singular point in terms of a finite number of complex powers, logarithms, exponentials and power series, either convergent or divergent. These formal solutions completely determine the linear differential equation; hence, they contain all information about the equation itself, especially about its analytic solutions. Extracting this information from the divergent solutions was the underly-

[1] Centre International de Mathématiques Pures et Appliquées, or ICPAM, is a non-profit international organization founded in 1978 in Nice, France. It promotes international cooperation in higher education and research in mathematics and related subjects for the benefit of developing countries. It is supported by UNESCO and IMU, and many national mathematical societies over the world.

[2] One says *a formal fundamental solution.*

ing motivation for the theories of summability and, to some extent, of resurgence. Both theories are concerned with the precise structure of the singularities.

Divergent series may appear in connection with any local analytic object. They either satisfy an equation, or are attached to given objects such as formal first integrals in dynamical systems or formal conjugacy maps in classification problems. Besides linear and non-linear ordinary differential equations, they also arise in partial differential equations, difference equations, q-difference equations, etc. Such series, issued from specific problems, call for suitable theories to extract valuable information from them.

A theory of *summability* is a theory that focuses on a certain class of power series, to which it associates analytic functions. The correspondence should be injective and functorial: one expects for instance a series solution of a given functional equation to be mapped to an analytic solution of the same equation. In general, the relation between the series and the function –the latter is called its sum– is *asymptotic*, and depends on the direction of summation; indeed, with non-convergent series one cannot expect the sums to be analytic in a full neighborhood, but rather in a "sectorial neighborhood" of the point at which the series is considered.

One summation process, commonly known as the Borel-Laplace summation, was already given by Émile Borel in the nineteenth century; it applies to the classical Euler series and, more generally, to solutions of linear differential equations with a single "level", equal to one, although the notion of level was by then not explicitly formulated. It soon appeared that this method does not apply to all formal solutions of differential equations, even linear ones. A first generalization to series solutions of linear differential equations with a single, arbitrary level $k > 0$ was given by Le Roy in 1900 and is called *k-summation*. In the 1980's, new theories were developed, mainly by J.-P. Ramis and Y. Sibuya, to characterize k-summable series, a notion a priori unrelated to equations, but which applies to all solutions of linear differential equations with the single level k. The question of whether any divergent series solution of a linear differential equation is k-summable, known as the *Turrittin problem*, was an open problem until J.-P. Ramis and Y. Sibuya in the early 1980's gave a counterexample. In the late 1980's and in the 1990's *multisummability theories* were developed, in particular by J.-P. Ramis, J. Martinet, Y. Sibuya, B. Malgrange, W. Balser, M. Loday-Richaud and G. Pourcin, which apply to all series solution of linear differential equations with an arbitrary number of levels. They provide a unique sum of a formal fundamental solution on appropriate sectors at a singular point.

It was proved that these theories apply to solutions of non-linear differential equations as well: given a series solution of a non-linear differential equation, the choice of the right theory is determined by the linearized equation along this series. On the other hand, in the case of difference equations, not all solutions are multisummable; new types of summation processes are needed, for instance those introduced by J. Écalle in his theory of resurgence and considered also by G. Immink and B. Braaksma. Solutions of q-difference equations are not all multisummable either; specific processes in this case have been introduced by F. Marotte and C. Zhang in the late 1990's.

Summation sheds new light on the *Stokes phenomenon*. This phenomenon occurs when a divergent series has several sums, with overlapping domains, which correspond to different summability directions and differ from one another by exponentially small quantities. The question then is to describe these quantities. A precise analysis of the Stokes phenomenon is crucial for classification problems. For systems of linear differential equations, the meromorphic classification easily follows from the characterization of the Stokes phenomenon by means of the *Stokes cocycle*. The Stokes cocycle is a 1-cocycle in non-abelian Čech cohomology. It is expressed in terms of finitely many automorphisms of the normal form, the *Stokes automorphisms*, which select and organize the "exponentially small quantities". In practice, the Stokes automorphisms are represented by constant unipotent matrices called the *Stokes matrices*. It turned out that these matrices are precisely the correction factors needed to patch together two contiguous sums, that is, sums taken on the two sides of a singular direction, of a formal fundamental solution.[3]

The theory of *resurgence* was independently developed in the 1980's by J. Écalle, with the goal of providing a theory with a large range of applications, including the summation of divergent solutions of a variety of functional equations, differential, difference, differential-difference, etc. Basically, resurgence theory starts with the Borel-Laplace summation in the case of a single level equal to one, and this is the only situation we consider in these volumes. Let us mention however that there are extensions of the theory based on more general kernels.

The theory focuses on what happens in the Borel plane, that is, after one applies a Borel transform. The results are then pulled back via a Laplace transform to the plane of the initial variable also called the Laplace plane. In the Borel plane one typically gets functions, called *resurgent functions*, which are analytic in a neighborhood of the origin and can be analytically continued along various paths in the Borel plane, yet they are not entire functions: one needs to avoid a certain set Ω of possible singular points and analytic continuation usually gives rise to multiple-valuedness, so that these Borel-transformed functions are best seen as holomorphic functions on a Riemann surface akin to the universal covering of $\mathbb{C}\setminus\{0\}$. Of crucial importance are the singularities[4] which may appear at the points of Ω, and *Écalle's alien operators* are specific tools designed to analyze them.

The development of resurgence theory was aimed at non-linear situations where it reveals its full power, though it can be applied to the formal solutions of linear differential equations (in which case the singular support Ω is finite and the Stokes matrices, hence the local meromorphic classification, determined by the action of finitely many alien operators). The non-linearity is taken into account via the convolution product in the Borel plane. More precisely, we mean here the complex convolution which is turned into pointwise multiplication when returning to the original variable by means of a Laplace transform. Given two resurgent functions, analytic

[3] A less restrictive notion of Stokes matrices exists in the literature, which patch together any two sectorial solutions with same asymptotic expansion, but they are not local meromorphic invariants in general.

[4] The terms *singularity* in Écalle's resurgence theory and *microfunction* in Sato's microlocal analysis have the same meaning.

continuation of their convolution product is possible, but new singularities may appear at the sum of any two singular points of the factors; hence, Ω needs to be stable by addition (in particular, it must be infinite; in practice, one often deals with a lattice in $\mathbb{C}$). All operations in the Laplace plane have an explicit counterpart in the Borel plane: addition and multiplication of two functions of the initial variable, as well as non-linear operations such as multiplicative inversion, substitution into a convergent series, functional composition, functional inversion, which all leave the space of resurgent functions invariant.

To have these tools well defined requires significant work. The reward of setting the foundations of the theory in the Borel plane is greater flexibility, due to the fact that one can work with an *algebra* of resurgent functions, in which the analysis of singularities is performed through *alien derivations*[5].

Écalle's important achievement was to obtain the so-called *bridge equation*[6] in many situations. For a given problem, the bridge equation provides an all-in-one description of the action on the solutions of the alien derivations. It can be viewed as an *infinitesimal version of the Stokes phenomenon*: for instance, for a linear differential system with level one it is possible to prove that the set of Stokes automorphisms in a given formal class naturally has the structure of a unipotent Lie group and the bridge equation gives infinitesimal generators of its Lie algebra.

Summability and resurgence theories have useful interactions with the algebraic and geometrical approaches of linear differential equations such as *differential Galois theory* and the *Riemann-Hilbert problem*. The local differential Galois group of a meromorphic linear differential equation at a singular point is a linear algebraic group, the structure of which reflects many properties of the solutions. At a "regular singular" point [7] for instance, it contains a Zariski-dense subgroup finitely generated by the monodromy. However, at an "irregular singular" point, one needs to introduce further automorphisms, among them the Stokes automorphisms, to generate a Zariski-dense subgroup. For linear differential equations with rational coefficients, when all the singular points are regular, the classical Riemann-Hilbert correspondence associates with each equation a monodromy representation of the fundamental group of the Riemann sphere punctured at the singular points; conversely, from any representation of this fundamental group, one recovers an equation with prescribed regular singular points.[8] In the case of possibly irregular singular points, the monodromy representation alone is insufficient to recover the equation; here too one has to introduce the Stokes automorphisms and to connect them via "analytic continuation" of the divergent solutions, that is, via summation processes.

---

[5] Alien derivations are suitably weighted combinations of alien operators which satisfy the Leibniz rule.

[6] Its original name in French is *équation du pont*.

[7] This means that the formal solutions at that point may contain powers and logarithms but no exponential.

[8] The Riemann-Hilbert problem more specifically requires that the singular points in this restitution be *Fuchsian*, that is, simple poles only, which is not always possible.

These volumes also include an application of resurgence theory to the first Painlevé equation. Painlevé equations are nonlinear second-order differential equations introduced at the turn of the twentieth century to provide new transcendents, that is, functions that can neither be written in terms of the classical functions nor in terms of the special functions of physics. A reasonable request was to ask that all the movable singularities[9] be poles and this constraint led to a classification into six families of equations, now called Painlevé I to VI. Later, these equations appeared as conditions for isomonodromic deformations of Fuchsian equations on the Riemann sphere. They occur in many domains of physics, in chemistry with reaction-diffusion systems and even in biology with the study of competitive species. Painlevé equations are a perfect non-linear example to be explored with the resurgent tools.

We develop here the particular example of Painlevé I and we focus on its now classical truncated solutions. These are characterized by their asymptotics as well as by the fact that they are free of poles within suitable sectors at infinity. We determine them from their asymptotic expansions by means of a Borel-Laplace procedure after some normalization. The non-linearity generates a situation which is more intricate than in the case of linear differential equations. Playing the role of the formal fundamental solution is the so-called *formal integral* given as a series in powers of logarithm-exponentials with power series coefficients. More generally, such expansions are called *transseries* by J. Écalle or *multi-instanton expansions* by physicists. In general, the series are divergent and lead to a Stokes phenomenon. In the case of Painlevé I we prove that they are resurgent. Although the Stokes phenomenon can no longer be described by Stokes matrices, it is still characterized by the alien derivatives at the singular points in the Borel plane (see O. Costin *et al.*). The local meromorphic class of Painlevé I at infinity is the class of all second-order equations locally meromorphically equivalent at infinity to this equation. The characterization of this class requires *all* alien derivatives in all higher sheets of the resurgence surface. These extra invariants are also known as *higher order Stokes coefficients* and they can be given a numerical approximation using the *hyperasymptotic theory* of M. Berry and C. Howls. The complete resurgent structure of Painlevé I is given by its *bridge equation* which we state here, seemingly for the first time.

Recently, in quantum field and string theories, the resurgent structure has been used to describe the instanton effects, in particular for quartic matrix models which yield Painlevé I in specific limits. In the late 1990's, following ideas of A. Voros and J. Écalle, applications of the resurgence theory to problems stemming from quantum mechanics were developed by F. Pham and E. Delabaere. Influenced by M. Sato, this was also the starting point by T. Kawai and Y. Takei of the so-called *exact semi-classical analysis* with applications to Painlevé equations with a large parameter and their hierarchies, based on isomonodromic methods.

[9] The fixed singular points are those appearing on the equation itself; they are singular for the solutions generically. The movable singular points are singular points for solutions only; they "move" from one solution to another. They are a consequence of the non-linearity.

Summability and resurgence theories have been successfully applied to problems in analysis, asymptotics of special functions, classification of local analytic dynamical systems, mechanics, and physics. They also generate interesting numerical methods in situations where the classical methods fail.

In these volumes, we carefully introduce the notions of analytic continuation and monodromy, then the theories of resurgence, k-summability and multisummability, which we illustrate with examples. In particular, we study tangent-to-identity germs of diffeomorphisms in the complex plane both via resurgence and summation, and we present a newly developed resurgent analysis of the first Painlevé equation. We give a short introduction to differential Galois theory and a survey of problems related to differential Galois theory and the Riemann-Hilbert problem. We have included exercises with solutions. Whereas many proofs presented here are adapted from existing ones, some are completely new. Although the volumes are closely related, they have been organized to be read independently. All deal with power series and functions of a complex variable; the words *analytic* and *holomorphic* are used interchangeably, with the same meaning.

This book is aimed at graduate students, mathematicians and theoretical physicists who are interested in the theories of monodromy, summability or resurgence and related problems.

Below is a more detailed description of the contents.

- Volume 1: *Monodromy and Resurgence* by C. Mitschi and D. Sauzin.
  An essential notion for the book and especially for this volume is the notion of analytic continuation "à la Cauchy-Weierstrass". It is used both to define the monodromy of solutions of linear ordinary differential equations in the complex domain and to derive a definition of resurgence.
  Once monodromy is defined, we introduce the Riemann-Hilbert problem and the differential Galois group. We show how the latter is related to analytic continuation by defining a set of automorphisms, including the Stokes automorphisms, which together generate a Zariski-dense subgroup of the differential Galois group. We state the inverse problem in differential Galois theory and give its particular solution over $\mathbb{C}(z)$ due to Tretkoff, based on a solution of the Riemann-Hilbert problem. We introduce the language of vector bundles and connections in which the Riemann-Hilbert problem has been extensively studied and give the proof of Plemelj-Bolibrukh's solution when one of the prescribed monodromy matrices is diagonalizable.
  The second part of the volume begins with an introduction to the 1-summability of series by means of Borel and Laplace transforms (also called Borel or Borel-Laplace summability) and provides non-trivial examples to illustrate this notion. The core of the subject follows, with definitions of resurgent series and resurgent functions, their singularities and their algebraic structure. We show how one can analyse the singularities via the so-called *alien calculus* in resurgent algebras; this includes the *bridge equation* which usefully connects alien and ordinary derivations. The case of tangent-to-identity germs of diffeomorphisms in the complex plane is given a thorough treatment.

- Volume 2: *Simple and Multiple Summability* by M. Loday-Richaud.
The scope of this volume is to thoroughly introduce the various definitions of k-summability and multisummability developed since the 1980's and to illustrate them with examples, mostly but not only, solutions of linear differential equations. For the first time, these theories are brought together in one volume.
We begin with the study of basic tools in Gevrey asymptotics, and we introduce examples which are reconsidered throughout the following sections. We provide the necessary background and framework for some theories of summability, namely the general properties of sheaves and of abelian or non-abelian Čech cohomology. With a view to applying the theories of summability to solutions of differential equations we review fundamental properties of linear ordinary differential equations, including the main asymptotic expansion theorem, the formal and the meromorphic classifications (formal fundamental solution and linear Stokes phenomenon) and a chapter on index theorems and the irregularity of linear differential operators. Four equivalent theories of k-summability and six equivalent theories of multisummability are presented, with a proof of their equivalence and applications. Tangent-to-identity germs of diffeomorphisms are revisited from a new point of view.
- Volume 3: *Resurgent Methods and the First Painlevé equation* by E. Delabaere.
This volume deals with ordinary non-linear differential equations and begins with definitions and phenomena related to the non-linearity. Special attention is paid to the first Painlevé equation, or Painlevé I, and to its tritruncated and truncated solutions. We introduce these solutions by proving the Borel-Laplace summability of transseries solutions of Painlevé I. In this context resonances occur, a case which is scarcely studied. We analyse the effect of these resonances on the formal integral and we provide a normal form. Additional material in resurgence theory is needed to achieve a resurgent analysis of Painlevé I up to its bridge equation.

Acknowledgements. We would like to thank the CIMPA institution for giving us the opportunity of holding a winter school in Lima in July 2008. We warmly thank Michel Waldschmidt and Michel Jambu for their support and advice in preparing the application and solving organizational problems. The school was hosted by IMCA (Instituto de Matemática y Ciencias Afines) in its new building of La Molina, which offered us a perfect physical and human environment, thanks to the colleagues who greeted and supported us there. We thank all institutions that contributed to our financial support: UNI and PUCP (Peru), LAREMA (Angers), IRMA (Strasbourg), IMT (Toulouse), ANR Galois (IMJ Paris), IMPA (Brasil), Universidad de Valladolid (Spain), Ambassade de France au Pérou, the International Mathematical Union, CCCI (France) and CIMPA. Our special thanks go to the students in Lima and in our universities, who attended our classes and helped improve these notes via relevant questions, and to Jorge Mozo Fernandez for his pedagogical assistance.

Angers, Strasbourg, Pisa, November 2015

Éric Delabaere, Michèle Loday-Richaud, Claude Mitschi, David Sauzin

Preface to this Volume

These lecture notes are an extended form of a course given at a CIMPA master class held in LIMA, Peru, in the summer of 2008. The students who attended these lectures were already introduced to linear differential equations, Gevrey asymptotics, k-summability and resurgence by my colleagues Michèle Loday, Claude Mitschi and David Sauzin. The aim was merely to show the resurgent methods acting on an example and along that line, to extend the presentation of the resurgence theory of Jean Écalle provided that the need.

The present lecture notes reflect this plan and this pedagogical point of view. The example that we follow along this course is the First Painlevé differential equation, or Painlevé I for short. Besides its simplicity, various reasons justify this choice. One of them is the non-linearity, which is the field where the resurgence theory reveals its power. Another reason lies on the fact that resonances occur, a case which is scarcely found in the literature. Last but not least, the Painlevé equations and their transcendents appear today to be an inescapable knowledge in analysis for young mathematicians. It was thus certainly worthy to detail the complete resurgent structure for Painlevé I.

We have tried to be as self-contained as possible. Nevertheless, the reader is assumed to have a previous acquaintance with the theories of summability, especially with Borel-Laplace summation and a little background with resurgence theory, amply elaborated in the first two volumes of this book. Since this volume deals with ordinary non-linear differential equations, we begin with definitions and phenomena linked to the non-linearity. Special attention is then brought to Painlevé I and to its so-called tritruncated and truncated solutions. We construct them by proving the Borel-Laplace summability of the transseries solutions. We analyze the formal integral for Painlevé I and, equivalently, the formal transform that brings Painlevé I to its associated normal form. We eventually detail the resurgent structure for Painlevé I via additional material in resurgence theory. As a rule, each chapter ends with some comments on possible extensions for which we provide references to the existing literature.

Acknowledgments. I am indebted to Frédéric Pham who initiated me to Resurgence theory and to many related problems, especially those stemming from theoretical physics. I would like to thank Michèle Loday-Richaud without whom this book would not have been written. Finally I wish to thank my students, Trinh Duc Tai, Jean-Marc Rasoamanana, Yafei Ou and particularly Julie Belpaume, who helped me to work out some parts of this manuscript.

Angers, November 2015 *Éric Delabaere*

Contents

Chapter 1
Some Elements about Ordinary Differential Equations

Abstract This chapter is merely devoted to recalling usual notation and elementary results on ordinary differential equations in the complex domain. We give the fundamental existence theorem for Cauchy problems (Sect. 1.1). We detail the main differences between solutions of linear versus nonlinear ODEs, when the question of their analytic continuation is considered (Sect. 1.2). Finally we provide a short introduction to Painlevé equations (Sect. 1.3).

1.1 Ordinary Differential Equations in the Complex Domain

An ordinary differential equation (ODE) is a functional relation of the type

$$\mathscr{F}\big(x,\mathbf{u}(x),\mathbf{u}'(x),\cdots,\mathbf{u}^{(N)}(x)\big)=0, \qquad \mathbf{u}^{(k)}(x)=\frac{\mathrm{d}^k\mathbf{u}}{\mathrm{d}x^k}(x)\in\mathbb{C}^m. \tag{1.1}$$

We refer to m as the *dimension* of the ODE. The *order* N of the ODE refers to the highest derivative considered in the equation. This ODE of order N is said to be *solved in his highest derivative* if it is written as

$$\mathbf{u}^{(N)}=\mathbf{F}(x,\mathbf{u},\cdots,\mathbf{u}^{(N-1)}). \tag{1.2}$$

1.1.1 The Fundamental Existence Theorem

We recall the fundamental existence theorem for the Cauchy problem, for analytic ODEs (see, e.g. [Inc56, Hil76, Koh99, IY008]). We denote by $D(z,r)\subset\mathbb{C}$ the open disc centred on z and of radius r. For a given domain $U\subset\mathbb{C}^m$ (i.e., U is a connected open set) we denote by $\mathscr{O}(U)$ the complex linear space of functions holomorphic on U.

E. Delabaere, *Divergent Series, Summability and Resurgence III*,
Lecture Notes in Mathematics 2155, DOI 10.1007/978-3-319-29000-3_1

Let $U \subset \mathbb{C}^n$ be an open set and let $f : U \to \mathbb{C}$ be a function. The following statements are equivalent (this is the Osgood lemma):

- f is analytic on U, that is f can be represented by a convergent power series in a neighbourdhood of each $x \in U$;
- f is complex differentiable on U;
- f is weakly holomorphic, that is f is continuous on U and partially differentiable on U with respect to each variable x_i $(x = (x_1, \cdots, x_n))$.

As a matter of fact, it is enough to assume only the holomorphy in each complex variable without the continuity hypothesis (Hartogs theorem).

Theorem 1.1 (Cauchy problem). *Let $U \subset \mathbb{C} \times \mathbb{C}^m$ be a domain and $\mathbf{F} : U \to \mathbb{C}^m$ be a holomorphic vector function. For every $(x_0, \mathbf{u}_0) \in U$, there exist a polydisc $D(x_0, \varepsilon_0) \prod_{1 \le i \le m} D(\mathbf{u}_{0i}, \varepsilon_i) \subset U$ and a solution $\mathbf{u} : D(x_0, \varepsilon_0) \to \prod_{1 \le i \le m} D(\mathbf{u}_{0i}, \varepsilon_i)$ of the analytic ODE of order 1 and dimension m, $\frac{\mathrm{d}\mathbf{u}}{\mathrm{d}x} = \mathbf{F}(x, \mathbf{u})$, which satisfies the initial value condition $\mathbf{u}(x_0) = \mathbf{u}_0$. Moreover this solution is unique, $\mathbf{u}$ belongs to $\mathcal{O}(D(x_0, \varepsilon_0))$ and also depends holomorphically on the initial value $\mathbf{u}_0$.*

In what follows we shall consider essentially *scalar* ODEs, that it ODEs of dimension 1 and of order N. The theorem 1.1 translates to this case as well, since every ODE of order N and of dimension 1, once solved in his highest derivative, is equivalent to an ODE of order 1 and of dimension N :
if $u = v_0$, $u' = v_1$, $\cdots$, $u^{(N-1)} = v_{N-1}$, the following Cauchy problem,

$$\begin{cases} u^{(N)} = F(x, u, \cdots, u^{(N-1)}) \\ \left(u(x_0), \cdots, u^{(N-1)}(x_0)\right) = \left(u_0, \cdots, u_0^{(N-1)}\right), \end{cases}$$

is equivalent to that one:

$$\frac{\mathrm{d}}{\mathrm{d}x}\begin{pmatrix} v_0 \\ \vdots \\ v_{N-2} \\ v_{N-1} \end{pmatrix} = \begin{pmatrix} v_1 \\ \vdots \\ v_{N-1} \\ F(x, v_0, \cdots, v_{N-1}) \end{pmatrix} \quad \text{and} \quad \begin{pmatrix} v_0 \\ \vdots \\ v_{N-1} \end{pmatrix}(x_0) = \begin{pmatrix} u_0 \\ \vdots \\ u_0^{(N-1)} \end{pmatrix}.$$

1.1.2 Some Usual Terminology

The following terminology is commonly used (see, e.g. [Con99-2]):

- The *general solution* of an ODE of order N and of dimension 1 is the set of all solutions determined in application of the Cauchy theorem 1.1. It depends on N arbitrary complex constants.
- A *particular* or *special* solution is a solution derived from the general solution when fixing a particular initial data.
- A *singular* solution is a solution which is not particular.

1.1.3 Algebraic Differential Equations

In a moment we shall concentrate on algebraic differential equations, these we now define.

Let $U \subset \mathbb{C}$ be a domain. We denote by $\mathscr{M}(U)$ the field of meromorphic functions on U. The ODE (1.1) of order N and of dimension 1 is said to be *algebraic on a domain* U if $\mathscr{F} \in \mathscr{M}(U)[u,u',\cdots,u^{(N)}]$ that is, $\mathscr{F}$ is polynomial in $(u,u',\cdots,u^{(N)})$ with meromorphic coefficients in x. An algebraic ODE is *rational* if it is of degree one in the highest derivative $u^{(N)}$, and *linear* (homogeneous) if $\mathscr{F}$ is a linear form in $(u,u',\cdots,u^{(N)})$.

1.2 Solutions of Ordinary Differential Equations and Singularities

1.2.1 Notation

We fix some notation that will be used in a moment.

Definition 1.1. Let $\lambda : [a,b] \subset \mathbb{R} \to \mathbb{C}$ be a path starting at $x_1 = \lambda(a)$ and ending at $x_2 = \lambda(b)$. If u is a (germ of) holomorphic function(s) at x_1 which can be analytically continued along λ, we denote by $\mathrm{cont}_\lambda\, u$ the resulting (germ of) holomorphic function(s) at x_2.

Remark 1.1. Let $\mathscr{O} = \bigsqcup_{x\in\mathbb{C}} \mathscr{O}_x$ be the set of all germs of holomorphic functions. We equip $\mathscr{O}$ with its usual Hausdorff topology, a basis $\mathscr{B} = \{\mathscr{U}(U,\Phi)\}$ of open sets being defined as follows: $\mathscr{U}(U,\Phi) = \{\varphi_x \in \mathscr{O}_x \mid \varphi_x \text{ germ of } \Phi \text{ at } x \in U\}$, where $U \subset \mathbb{C}$ is a domain and $\Phi \in \mathscr{O}(U)$. With the projection $\mathfrak{q} : \begin{array}{c} \mathscr{O} \to \mathbb{C} \\ \varphi_x \in \mathscr{O}_x \mapsto x \in \mathbb{C} \end{array}$ which associates to a germ its support [For81, Ebe007], the (non-connected) topological space $\mathscr{O}$ becomes an étalé space, that is $\mathfrak{q}$ is a local homeomorphism. The analytic continuation of the germ $u \in \mathscr{O}_{x_1}$ along λ, if it exists, is the image of the unique path $\Lambda : [a,b] \to \mathscr{O}$ such that $\Lambda(a) = u$ and whose projection by $\mathfrak{q}$ is λ, $\begin{array}{ccc} & \mathscr{O} & \\ \Lambda \nearrow & & \searrow \mathfrak{q} \\ [a,b] & \xrightarrow[\lambda]{} & \mathbb{C} \end{array}$.

With this notation, $\mathrm{cont}_\lambda\, u = \Lambda(b)$. See the first volume of this book [MS016] for more details.

1.2.2 Problem

We consider an ODE of order N and dimension 1, $\mathscr{F}\left(x,u(x),u'(x),\cdots,u^{(N)}(x)\right)=0$ with $\mathscr{F}:U\to\mathbb{C}$ a holomorphic function on the open domain $U\subset\mathbb{C}\times\mathbb{C}^{N+1}$. Assume that $\left(x_0,u_0,\cdots,u_0^{(N)}\right)\in U$ and that $\begin{cases}\mathscr{F}\left(x_0,u_0,\cdots,u_0^{(N)}\right)=0\\ \partial_{N+2}\mathscr{F}\left(x_0,u_0,\cdots,u_0^{(N)}\right)\neq 0\end{cases}$. By the implicit function theorem, the Cauchy problem

$$\begin{cases}\mathscr{F}\left(x,u(x),u'(x),\cdots,u^{(N)}(x)\right)=0\\ \left(u(x_0),\cdots,u^{(N)}(x_0)\right)=\left(u_0,\cdots,u_0^{(N)}\right)\end{cases}$$

is locally equivalent to a Cauchy problem where the ODE is solved in its highest derivative. Theorem 1.1 thus provides a holomorphic solution u near $x=x_0$. We consider a path $\gamma:[a,b]\to\mathbb{C}$ from x_0 to x_1 in $\mathbb{C}$ and for $s\in[a,b]$ we denote by $\gamma_s:[a,s]\to\mathbb{C}$ the restriction to $[a,s]$ of γ. Assume that u can be analytically continued along the path γ and that for every $s\in[a,b]$, the value at $\gamma(s)$ of the analytic continuation $\mathrm{cont}_{\gamma_s}\left(x,u,u',\cdots,u^{(N)}\right)$ along γ_s belongs to U. Then *the analytic continuation* $\mathrm{cont}_\gamma u$ *along* γ *of the solution* u *still satisfies the differential equation*, thanks to the uniqueness of the analytic continuation.

This property raises the question of describing the singularities of the analytic continuations of solutions of analytic ODEs, for instance for an algebraic differential equation defined on an open domain. As we shall see, appearance of singularities is quite different whether one considers linear or nonlinear ODEs.

1.2.3 Linear Differential Equations

Linear differential equations are studied in the first two volumes of this book [MS016, Lod016], see also, e.g. [Was65, Koh99, IY008, IKSY91]. For linear (homogeneous) ordinary differential equations it results from the Cauchy existence theorem and the Grönwall lemma that the general solution has no other singularities than the so-called *fixed singularities* which arise from the coefficients of the ODE once solved for the highest derivative.

1.2.3.1 Example 1

We start with an equation where $x=0$ is an irregular singular point of Poincaré rank 1,

$$x^2u'+u=0,\qquad u(x)=Ce^{1/x},\qquad C\in\mathbb{C}.$$

Here $x=0$ is a fixed essential singularity for the general solution (but not for the particular solution $u(x)=0$), which arises from the equation itself.

If $u \in \mathscr{O}(D^\star(0,r))$ is holomorphic on the punctured disc $D^\star(0,r) = D(0,r) \setminus \{0\}$, then u can be represented by its Laurent series expansion $\sum_{n\in\mathbb{Z}} a_n x^n$ which converges in $0 < |x| < r$. One says that 0 is an essential singularity if and only if the Laurent series expansion has an infinite number of $n < 0$ such that $a_n \neq 0$ or, equivalently, if u has no limit (finite or infinite) when $x \to 0$. A typical example is provided by the function $e^{1/x}$.

1.2.3.2 Example 2

We consider the Airy equation,

$$u'' - xu = 0, \qquad u(x) = C_1 Ai(x) + C_2 Bi(x), \qquad C_1, C_2 \in \mathbb{C}.$$

Here Ai and Bi are the Airy's special functions of the first and second kind respectively. These are entire functions. When considered on the Riemann sphere $\overline{\mathbb{C}}$ (see the first volume of this book [MS016]), $x = \infty$ appears as a fixed (essential) singularity for the general solution (except again for the particular solution $u(x) = 0$) which arises from the equation : $x = \infty$ is an irregular singular point of Poincaré rank $3/2$.

More generally, for a linear ordinary differential equation

$$\sum_{k=0}^{N} a_k(x) u^{(k)} = 0, \qquad a_k(x) \in \mathscr{O}(U), \tag{1.3}$$

the general solution can be analytically continued as a multivalued function on $U \setminus S$, $S = \{\text{the zeros of } a_N\}$, or more precisely as a single valued holomorphic function once it is considered on a Riemann surface [For81, Ebe007] defined as a covering space, $\begin{array}{c}\mathscr{R}\\ \pi\downarrow\\ U\setminus S\end{array}$. In other words, the general solution is *uniformisable* (or also *stable*) [Con99-2] in the following sense : for any Cauchy data at $x_0 \in U \setminus S$ that determined a unique local solution u of (1.3) on a domain $U_0 \subset U \setminus S$, one can find a domain $\mathscr{U}_0 \subset \mathscr{R}$ such that $\pi|_{\mathscr{U}_0} : \mathscr{U}_0 \to U_0$ is a homeomorphism, and a holomorphic function $\phi : \mathscr{R} \to \mathbb{C}$ so that $\phi|_{\mathscr{U}_0} = u \circ \pi|_{\mathscr{U}_0}$.

Then, for any domain $\mathscr{U}' \subset \mathscr{R}$ so that $\pi|_{\mathscr{U}'} : \mathscr{U}' \to U'$ is a homeomorphism, the function $\phi \circ (\pi|_{\mathscr{U}'})^{-1}$ is still a holomorphic solution of (1.3) on U'.

1.2.4 Nonlinear Differential Equations

When nonlinear ODEs are concerned, beside the possibly fixed singularities arising from the equation, the general solution has as a rule other singularities which depend on the arbitrary coefficients : these are *movable singularities*.

1.2.4.1 Example 1

We consider the following nonlinear ODE,

$$xu' - u^2 = 0, \qquad \begin{array}{l} \text{general solution} : u(x) = \dfrac{1}{C - \log(x)}, \quad C \in \mathbb{C}. \\ \text{singular solution} : u(x) = 0 \end{array}$$

For the general solution, $x = 0$ is a fixed branch point singularity which comes from the equation. The general solution u is uniformisable : considered as a function on the Riemann surface $(\widetilde{\mathbb{C}}, \pi)$ of the logarithm, $\widetilde{\mathbb{C}} = \{x = re^{\mathrm{i}\theta} \mid r > 0,\ \theta \in \mathbb{R}\}$, $\pi : x \in \widetilde{\mathbb{C}} \mapsto \dot{x} = re^{\mathrm{i}\theta} \in \mathbb{C} \setminus \{0\}$, one sees that the general solution u is meromorphic with poles at $\pi^{-1}(\mathrm{e}^C)$: these are movable singularities, depending on the chosen coefficient C.

The logarithmic function is carefully introduced in the first volume of this book [MS016] and distinct notation is introduced there for the principal branch, other branches or the extension to the Riemann surface $\widetilde{\mathbb{C}}$. Throughout this course, we use a single notation log which should be interpreted according to the context.

1.2.4.2 Example 2

The above example is just a special case of a more general rational ODE of order 1, the *Riccati equation*,

$$u' = a_0(x) + a_1(x)u + a_2(x)u^2 \qquad a_i \in \mathscr{M}(U), \tag{1.4}$$

where $U \subset \mathbb{C}$ is a domain. By the change of unknown function $u = -\dfrac{1}{a_2(x)} \dfrac{\mathrm{d}}{\mathrm{d}x} \log v$, equation (1.4) is linearizable into the following linear ODE,

$$v'' + \left(\frac{a_2'(x)}{a_2(x)} - a_1(x) \right) v' + a_2(x)a_0(x)v = 0.$$

The general solution for this linear equation has (fixed) singularities located at the poles of $\dfrac{a_2'(x)}{a_2(x)} - a_1(x)$ and $a_2(x)a_0(x)$. We denote by $S \subset U$ this set of poles. The general solution of the Riccati equation (1.4) is then uniformisable since it can be analytically continued as a meromorphic function on a Riemann surface defined as a covering over $U \setminus S$.
When the a_i belong to $\mathscr{O}(U)$, then the general solution of (1.4) is a meromorphic function on U [Lai93].

1.2.4.3 Example 3

Another well known equation is the following algebraic nonlinear ODE of order 1, of degree 2 in its highest derivative, namely the *elliptic equation*:

$$u'^2 = 4u^3 - g_2 u - g_3, \qquad (g_2, g_3) \in \mathbb{C}. \tag{1.5}$$

A particular solution is provided by the Weierstrass p-function $\wp(x; g_2, g_3)$ which can be obtained as the inverse function of the elliptic integral of the first kind

$$x = \int_\infty^u \frac{dq}{\sqrt{4q^3 - g_2 q - g_3}}, \qquad \left(\frac{\mathrm{d}x}{\mathrm{d}u}\right)^2 = \frac{1}{4u^3 - g_2 u - g_3}.$$

(Just apply the inverse function theorem).
When the discriminant $\mathbb{D} = g_2^3 - 27g_3^2$ satisfies the condition $\mathbb{D} \neq 0$, the polynomial function $4u^3 - g_2 u - g_3 = 4(u - e_1)(u - e_2)(u - e_3)$ has 3 distinct simple roots e_1, e_2, e_3. In that case the elliptic function $\wp(x; g_2, g_3)$ is a doubly periodic meromorphic function with double poles at the period lattice $m\omega_1 + n\omega_2$, $(n,m) \in \mathbb{Z}^2$, $\dfrac{\omega_1}{\omega_2} \notin \mathbb{R}$.
The period lattice can be described as follows : consider the elliptic curve $\mathscr{L} = \{(q,p) \in \mathbb{C}^2, p^2 = 4q^3 - g_2 q - g_3\}$ for $\mathbb{D} \neq 0$. The homology group $H_1(\mathscr{L};\mathbb{Z})$ is a free $\mathbb{Z}$-module of rank 2 and we denote by γ_1 and γ_2 two cycles which generate $H_1(\mathscr{L};\mathbb{Z})$, Fig. 1.1. The period lattice is generated by the period integrals $\omega_1 = \displaystyle\int_{\gamma_1} \frac{dq}{p}$, $\omega_2 = \displaystyle\int_{\gamma_2} \frac{dq}{p}$ (equivalently $\omega_1 = \displaystyle\int_{e_1}^{e_2} \frac{2dq}{\sqrt{4q^3 - g_2 q - g_3}}$, $\omega_2 = \displaystyle\int_{e_1}^{e_3} \frac{2dq}{\sqrt{4q^3 - g_2 q - g_3}}$). The homology group $H_1(\mathscr{L};\mathbb{Z})$ can be seen as a local system on $\mathbb{C}^2 \setminus N(\mathbb{D})$ (that is a locally constant sheaf of $\mathbb{Z}$-modules on $\mathbb{C}^2 \setminus N(\mathbb{D})$), where $N(\mathbb{D})$ is the zero set of $\mathbb{D}$. Viewed as functions of (g_2, g_3), $\omega_{1,2}$ can be analytically continued as "multivalued" analytic functions on $\mathbb{C}^2 \setminus N(\mathbb{D})$. On the discriminant locus $N(\mathbb{D})$, the solutions degenerate into simply periodic solutions, with a string of poles instead of a double array.

Conversely, starting from the period lattice with $\dfrac{\omega_1}{\omega_2} \notin \mathbb{R}$, the Weierstrass $\wp$-function can be obtained by a series,

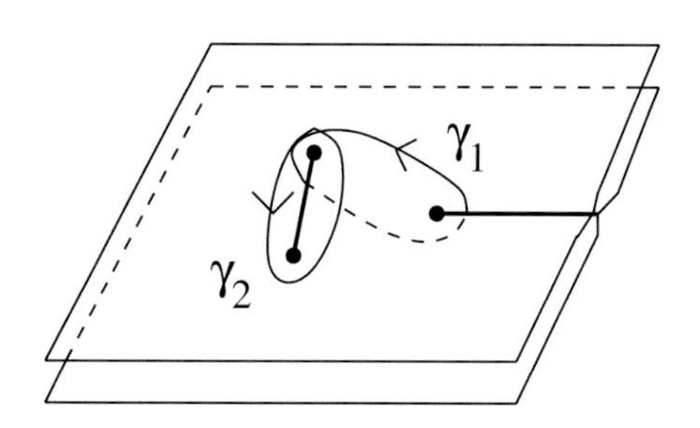

Fig. 1.1 The elliptic curve $\mathscr{L}$ viewed as the Riemann surface of $p = (4u^3 - g_2 u - g_3)^{1/2}$. The homology classes of the cycles γ_1 and γ_2 drawn generate $H_1(\mathscr{L};\mathbb{Z})$

$$\wp(x;g_2,g_3) = x^{-2} + \sum_{\omega\neq 0} \{(x-\omega)^{-2} - \omega^{-2}\} = x^{-2} + g_2\frac{x^2}{20} + g_3\frac{x^4}{28} + \cdots$$

where the first summation extends over all $\omega = m\omega_1 + n\omega_2 \neq 0$, $(n,m) \in \mathbb{Z}^2$ while $g_2 = 60 \sum_{\omega\neq 0} \omega^{-4}$, $g_3 = 140 \sum_{\omega\neq 0} \omega^{-6}$.
The general solution of (1.5) is given by $\wp(x - x_0; g_2, g_3)$, since (1.5) is an autonomous ODE.
To go further on the nice properties of elliptic functions see, e.g. [Sie88].

1.2.4.4 Example 4

The singularities of differential equations may be isolated singularities such as poles, branch points of finite or infinite determinations, or essential singularities. They may be also essential singular lines, or even perfect sets of singular points. For instance, the general solution of the following Chazy equation of class III,

$$u^{(3)} - 2uu^{(2)} + 3u'^2 = 0, \tag{1.6}$$

is defined only inside or outside an open disc whose boundary is a natural movable boundary determined by the initial data [Bur87, CO96].

1.3 The Painlevé Program, Painlevé Property and Painlevé Equations

At the end of the 19th century a list of special transcendental functions was known, most of them being obtained as solutions of linear algebraic differential equations.

> An algebraic function u in one complex variable x is a solution of a polynomial equation $P(x,u) = 0$, $P \in \mathbb{C}[x,u]$. A transcendental function u is a function which is not algebraic.

A challenging problem in analysis was thus to discover new transcendental functions defined by algebraic ODEs which cannot be expressed in term of solutions of linear algebraic ODEs : these new functions should thus be defined by non-linear algebraic differential equations [Con99-2, DL006, IKSY91].

For that purpose a systematic approach needs first to classify the ODEs under convenient criters. This is the goal of the so-called *Painlevé program* (see [Con99-2] and references therein) which consists in classifying all algebraic ODEs of first order, then second order, etc ..., whose general solution can be analytically continued as a *single valued* function[1]. In other words, no branch point is allowed. For instance the elliptic equation (1.5) or the Chazy equation (1.6) are such equations.

According to what we have seen, the Painlevé program splits into two problems:

---

[1] This condition can be weakened by asking the general solution to be only uniformisable.

- absence of fixed branch point for the general solution;
- absence of movable branch point for the general solution : this condition is the so-called *Painlevé property*.

In the literature, the term "Painlevé property" is sometimes used for the stronger property for the general solution of an ODE to be meromorphic, see [Con99-2]

Notice that the Painlevé property for an algebraic ODE $\mathscr{F}(x,u,u',\cdots,u^{(p)})=0$ defined on a domain $U\subset\mathbb{C}$ is preserved by:

- a holomorphic change of variable $x\in U\mapsto X=h(x)$, $h\in\mathscr{O}(U)$;
- a linear fractional change of the unknown with coefficient holomorphic in U (action of the homographic group),

$$u\mapsto v=\frac{a(x)u+b(x)}{c(x)u+d(x)},\quad v\mapsto u=\frac{d(x)v-b(x)}{-c(x)v+a(x)},$$

$a,b,c,d\in\mathscr{O}(U)$, $ad-bc\neq 0$. Therefore, the classification in the Painlevé program is made up to these transformations.

Notice however that other actions preserving the Painlevé property can be considered, see [Con99-2, CM008, IKSY91].

1.3.1 Ordinary Differential Equations of Order One

We consider (nonlinear) ODEs of the form

$$\mathscr{F}\left(x,u,u'\right)=0, \tag{1.7}$$

with $\mathscr{F}\in\mathscr{M}(U)[u,u']$. For that class of ODEs, the Painlevé program can be considered as being achieved and we mainly refer to [Inc56, Hil76, Con99-2, IKSY91] for the classification.

In that case *no essential movable singular point can appear* ([Inc56], Sect. 13.6). Therefore looking for ODEs of type (1.7) with the Painlevé property reduces in asking that the movable singular points are just poles.

When (1.7) is a rational ODE, then the class of ODEs we are looking for is represented only by the Riccati equation (1.4). See [Lai93], in particular the Malmquist-Yosida-Steinmetz type theorems.

The ODEs of type (1.7) of degree ≥ 2 in the highest derivative and satisfying the Painlevé property essentially reduce (up to the transformations mentioned above) to the elliptic equation (1.5). See [Con99-2, Inc56] for more precise statements.

1.3.2 Ordinary Differential Equations of Order Two, Painlevé Equations

In contrast to what happens for algebraic ODEs of order one, essential movable singular points may exist when the order is greater than 2, making the analysis more difficult. Nevertheless, the classification is known for at least algebraic equations of order two

$$\mathscr{F}\left(x,u,u',u''\right)=0, \qquad \mathscr{F}\in\mathscr{M}(U)[u,u',u''] \tag{1.8}$$

which are rational, that is of degree one in u''. Such equations enjoying the Painlevé property reduce (up to transformation) to:

- equations which can be integrated by quadrature,
- or linear equations,
- or one of six ODEs known as the *Painlevé equations*, the first 3 being:

$$\begin{aligned} &(P_I) \quad u''=6u^2+x \\ &(P_{II}) \ u''=2u^3+xu+\alpha \\ &(P_{III}) \ u''=\frac{u'^2}{u}-\frac{u'}{x}+\frac{\alpha u^2+\beta}{x}+\gamma u^3+\frac{\delta}{u} \end{aligned} \tag{1.9}$$

For the complete list see, e.g. [Inc56, Hil76, Con99-2, IKSY91]. In (1.9), $\alpha,\beta,\gamma,\delta$ are arbitrary complex constants. Each Painlevé equation can be derived from the "master equation" P_{VI} by some limit processes [IKSY91].

Painlevé equations have beautiful properties, see e.g. [Con99-1, IKSY91, GLS002]. One of them is the following one:

Theorem 1.2. *The general solution of the Painlevé equation P_J, $J=I,\cdots,VI$ admits no singular points except poles outside the set of fixed singularities.*

Therefore, the Painlevé equations have the Painlevé property but moreover, the general solution is free of movable essential singularities.
Notice that the Painlevé equation should be seen as defined on the Riemann sphere $\overline{\mathbb{C}}$. The set of fixed singular points S_J of P_J is a subset of $\{0,1,\infty\}$. For instance S_I and S_{II} are just $\{\infty\}$, while $S_{III}=\{0,\infty\}$. Theorem 1.2 thus translates as follows : the general solution of P_J can be analytically continued as a meromorphic function on the universal covering of $\overline{\mathbb{C}}\setminus S_J$.

Theorem 1.2 can be proved in various ways. An efficient one uses the relationship between Painlevé equations and monodromy-preserving deformation of some Fuchsian differential equations [JMU81, JM81, Mal83, IKSY91, FIKN].

The general (global) solutions of the Painlevé equations are called the *Painlevé transcendents*. This refers to the fact that, for generic values of the integration constants and of the parameters of the equations, these solutions cannot be written with elementary or classical transcendental functions, a question which has been completely solved only recently with the development of the modern nonlinear differential Galois theory (see [Ume006] and references therein. For an introduction to differential Galois theory, see the first volume of this book [MS016]).

1.3.3 Painlevé Equations and Related Topics

The renewed interest in Painlevé equations mainly came from theoretical physics in the seventies, with the study of PDEs of the soliton type (Boussinesq equation, Korteweg-de Vries KdV and modified Korteweg-de Vries equation mKdV, etc..): when linearized by inverse scattering transform [AC91], these PDEs give rise to ODEs with the Painlevé property. For instance, the Boussinesq equation $u_{tt} - u_{xx} - 6(u^2)_{xx} + u_{xxxx} = 0$ has a self-similar solution of the form $u(x,t) = w(x-t)$ where w is either an elliptic function or satisfies the first Painlevé equation. In the same lines, the (m)KdV hierarchy introduced by Lax in [Lax76] (and already in substance in [Lax68] after the work of Gardner *et al* [GGKM67] on the KdV equation), will later give rise to various *Painlevé hierarchies* which are thought of as higher-order Painlevé equations and much studied since. For instance, the first Painlevé hierarchy is of the form

$$(P_I^{(n)}) \qquad d_{[n+1]}[u] + 4x = 0, \qquad n = 1, 2, \cdots \tag{1.10}$$

where $d_{[n]}[u]$ are differential polynomials recursively determined as follows (see [Shi004] and references therein):

$$\begin{cases} d_{[0]}[u] = 1 \\ \partial d_{[n+1]}[u] = \left(\partial^3 - 8u\partial - 4u'\right) d_{[n]}[u], \quad \partial = \frac{\mathrm{d}}{\mathrm{d}x}, n \in \mathbb{N}. \end{cases} \tag{1.11}$$

(The first Painlevé equation is $(P_I^{(1)})$.) See also [Mor009] and references therein, for an asymptotic study of the Jimbo-Miwa [JM81] and Flaschka-Newell [FN80] second Painlevé hierarchies [GJP006].

For the first and second Painlevé hierarchies, one conjectures that the solutions of each equation are meromorphic, thus satisfy the Painlevé property, but there is no proof up to our knowledge [Lai008].

Discrete (analogues of) *Painlevé equations* are today the matter of an intensive research, after the pioneering work of Bessis *et al* [BIZ80] on the study of partition functions in 2D quantum gravity, yielding what is now known as the first discrete Painlevé equation (dP_I) when the quartic matrix model is considered:

$$(dP_I) \quad w_n\left[w_{n+1} + w_n + w_{n-1}\right] = an + b + cw_n, \quad a, b, c \in \mathbb{C}. \tag{1.12}$$

The first discrete Painlevé equation naturally arises in the context of orthogonal polynomials. Consider the inner product $(f \mid g) = \int_{-\infty}^{+\infty} f(x)g(x)w(x)dx$ with the exponential weight $w(x) = \mathrm{e}^{-NV(x)}$, $V(x) = \frac{\mu}{2}x^2 + \frac{\lambda}{4}x^4$, and look for an orthogonal polynomial sequence $(p_n)_{n\geq 0}$, each p_n being a monic polynomial of degree n. It can be shown that the polynomials p_n are governed by a three-term recurrence equation of the form

$$\begin{cases} xp_n(x) = p_{n+1}(x) + r_n p_{n-1}(x) \\ p_0(x) = 1,\ p_1(x) = x \end{cases} \tag{1.13}$$

where $r_n = \frac{h_n}{h_{n-1}}$ with $(p_n \mid p_m) = h_n \delta_{nm}$ (δ_{nm} is the Kronecker index). This motivates the calculations of the coefficients r_n which themselves satisfy a recurrence relation of the form

$$\lambda r_n \left[r_{n+1} + r_n + r_{n-1} \right] + \mu r_n = \frac{n}{N} \tag{1.14}$$

and we recognize (dP_I). Among remarkable properties, (1.14) has a continuum limit to the first Painlevé equation when the double-scaling limit $n, N \to \infty$, $\frac{n}{N} \to t$ is considered. See for instance [Its011, GKT004, HK007] and references therein.

Non commutative extensions of integrable systems have recently attracted the attention of the specialists, with *non commutative* (analogues of) *Painlevé equations* and their hierarchies as main examples, see e.g. [RR010].

Finally, we could hardly leave untold the important group-theoretic interpretation of Painlevé equations in the line of the work of Okamoto [Oka87], see for instance [DL006] and references therein.

It is not our aim to say more about Painlevé equations in general except for the first Painlevé equation which is used in this course as field of experiments in asymptotic and resurgent analysis, and which is the matter for the next chapter.

References

AC91. M. Ablowitz, P. Clarkson, *Solitons, nonlinear evolution equations and inverse scattering.* London Mathematical Society Lecture Note Series, 149. Cambridge University Press, Cambridge, 1991.

BIZ80. D. Bessis, C. Itzykson, J. B. Zuber, *Quantum field theory techniques in graphical enumeration.* Adv. in Appl. Math. **1** (1980), no. 2, 109-157.

Bur87. F. J. Bureau, *Sur des systèmes différentiels non linéaires du troisième ordre et les équations différentielles non linéaires associées.* Acad. Roy. Belg. Bull. Cl. Sci. (5) **73** (1987), no. 6-9, 335-353.

CO96. P. Clarkson, P. Olver, *Symmetry and the Chazy equation.* J. Differential Equations **124** (1996), no. 1, 225-246.

Con99-1. R. Conte (ed), *The Painlevé property. One century later.* CRM Series in Mathematical Physics. Springer-Verlag, New York, 1999.

Con99-2. R. Conte, *The Painlevé approach to nonlinear ordinary differential equations.* The Painlevé property, 77-180, CRM Ser. Math. Phys., Springer, New York, 1999.

CM008. R. Conte, M. Musette, *The Painlevé handbook.* Springer, Dordrecht, 2008.

DL006. E. Delabaere, M. Loday-Richaud (eds), *Théories asymptotiques et équations de Painlevé. [Asymptotic theories and Painlevé equations]* Séminaires et Congrés, **14**. Société Mathématique de France, Paris, 2006.

Ebe007. W. Ebeling, *Functions of several complex variables and their singularities.* Translated from the 2001 German original by Philip G. Spain. Graduate Studies in Mathematics, 83. American Mathematical Society, Providence, RI, 2007.

FN80. H. Flaschka, A. C. Newell, *Monodromy- and spectrum-preserving deformations. I.* Comm. Math. Phys. **76** (1980), no. 1, 65-116.

FIKN. A.S. Fokas, A.R. Its, A.A. Kapaev, V.Y. Novokshenov, *Painlevé transcendents. The Riemann-Hilbert approach.* Mathematical Surveys and Monographs, 128. American Mathematical Society, Providence, RI, 2006.

For81. O. Forster, *Lectures on Riemann Surfaces*, Graduate texts in mathematics; 81, Springer, New York (1981).

GGKM67. C. S. Gardner, J. M. Greene, M. D. Kruskal, R. M. Miura, *Methods for solving the Korteweg-de Vries equation.* Phys. Rev. Letters **19** (1967), 1095–1097.

GKT004. B. Grammaticos, Y. Kosmann-Schwarzbach, T. Tamizhmani (eds), *Discrete integrable systems.* Lecture Notes in Physics, 644. Springer-Verlag, Berlin, 2004.

GJP006. P. Gordoa, N. Joshi, A. Pickering, *Second and fourth Painlevé hierarchies and Jimbo-Miwa linear problems.* J. Math. Phys. **47** (2006), no. 7, 073504, 16 pp.

GLS002. V. Gromak, I. Laine, S. Shimomura, *Painlevé differential equations in the complex plane.* de Gruyter Studies in Mathematics, 28. Walter de Gruyter & Co., Berlin, 2002.

HK007. R.G. Halburd, R.J. Korhonen, *Meromorphic solutions of difference equations, integrability and the discrete Painlevé equations.* J. Phys. A **40** (2007), no. 6, R1-R38.

Hil76. E. Hille, *Ordinary differential equations in the complex domain.* Pure and Applied Mathematics. Wiley-Interscience [John Wiley & Sons], New York-London-Sydney, 1976.

IY008. Y. Ilyashenko, S. Yakovenko, *Lectures on analytic differential equations.* Graduate Studies in Mathematics, 86. American Mathematical Society, Providence, RI, 2008.

Inc56. E.L. Ince, *Ordinary Differential Equations.* Dover Publications, New York, 1956.

Its011. A. Its, *Discrete Painlevé equations and orthogonal polynomials.* In "Symmetries and integrability of difference equations." Lectures from the Summer School held at the Universit de Montréal, Montréal, QC, June 8-21, 2008. Edited by Decio Levi, Peter Olver, Zora Thomova and Pavel Winternitz. London Mathematical Society Lecture Note Series, **381**. Cambridge University Press, Cambridge, 2011.

IKSY91. K. Iwasaki, H. Kimura, S. Shimomura, M. Yoshida, *From Gauss to Painlevé. A modern theory of special functions.* Aspects of Mathematics, E16. Friedr. Vieweg & Sohn, Braunschweig, 1991.

JM81. M. Jimbo, T. Miwa, *Monodromy preserving deformation of linear ordinary differential equations with rational coefficients. II.* Phys. D **2** (1981), no. 3, 407-448.

JMU81. M. Jimbo, T. Miwa, K. Ueno, *Monodromy preserving deformation of linear ordinary differential equations with rational coefficients. I. General theory and τ-function.* Phys. D **2** (1981), no. 2, 306-352.

Koh99. M. Kohno, *Global analysis in linear differential equations.* Mathematics and its Applications, 471. Kluwer Academic Publishers, Dordrecht, 1999.

Lai93. I. Laine, *Nevanlinna theory and complex differential equations.* de Gruyter Studies in Mathematics, 15. Walter de Gruyter & Co., Berlin, 1993.

Lai008. I. Laine, *Complex differential equations.* Handbook of differential equations: ordinary differential equations. Vol. IV, 269-363, Handb. Differ. Equ., Elsevier/North-Holland, Amsterdam, 2008.

Lax68. P. Lax *Integrals of nonlinear equations of evolution and solitary waves.* Comm. Pure Appl. Math. **21** (1968), 467-490.

Lax76. P. Lax *Almost periodic solutions of the KdV equation.* SIAM Rev. **18** (1976), no. 3, 351-375.

Lod016. M. Loday-Richaud, *Divergent Series, Summability and Resurgence II. Simple and Multiple Summability.* Lecture Notes in Mathematics, **2154**. Springer, Heidelberg, 2016.

Mal83. B. Malgrange, *Sur les déformations isomonodromiques.* Mathematics and physics (Paris, 1979/1982), 401-438, Progr. Math., **37**, Birkhäuser Boston, Boston, MA, 1983.

MS016. C. Mitschi, D. Sauzin, *Divergent Series, Summability and Resurgence I. Monodromy and Resurgence.* Lecture Notes in Mathematics, **2153**. Springer, Heidelberg, 2016.

Mor009. T. Morrison, *Asymptotics of higher-order Painlevé equations.* PhD thesis, The University of Sydney, School of Mathematics and Statistics, June 2009.

Oka87. K. Okamoto, *Studies on the Painlevé equations. I. Sixth Painlevé equation P_{VI}.* Ann. Mat. Pura Appl. (4) **146** (1987), 337-381.

RR010. V. Retakh, V. Rubtsov, *Noncommutative Toda chains, Hankel quasideterminants and the Painlevé II equation.* J. Phys. A **43** (2010), no. 50, 505204, 13 pp.

Shi004. S. Shimomura, *A certain expression of the first Painlevé hierarchy.* Proc. Japan Acad. Ser. A Math. Sci. **80** (2004), no. 6, 105-109.

Sie88. C.L. Siegel, *Topics in complex function theory. Vol. I. Elliptic functions and uniformization theory.* Wiley Classics Library. A Wiley-Interscience Publication. John Wiley & Sons, Inc., New York, 1988.

Ume006. H. Umemura, *Galois theory and Painlevé equations.* In Théories asymptotiques et équations de Painlevé, 299-339, Sémin. Congr., **14**, Soc. Math. France, Paris, 2006.

Was65. W. Wasow, *Asymptotic expansions for ODE.* Reprint of the 1965 edition. Robert E. Krieger Publishing Co., Huntington, N.Y., 1976.

Chapter 2
The First Painlevé Equation

Abstract This chapter aims at introducing the reader to properties of the first Painlevé equation and its general solution. The definition of the first Painlevé equation is recalled (Sect. 2.1). We precise how the Painlevé property translates for the first Painlevé equation (Sect. 2.2), a proof of which being postponed to an appendix. We explain how the first Painlevé equation also arises as a condition of isomonodromic deformations for a linear ODE (Sect. 2.3 and Sect. 2.4). Some symmetry properties are mentioned (Sect. 2.5). We spend some times describing the asymptotic behaviour at infinity of the solutions of the first Painlevé equation and, in particular, we introduce the truncated solutions (Sect. 2.6). We eventually briefly comment the importance of the first Painlevé transcendents for models in physics (Sect. 2.7).

2.1 The First Painlevé Equation

We concentrate on the first Painlevé equation,

$$(P_I)\ u'' = 6u^2 + x. \tag{2.1}$$

We notice that for every $x_0 \in \mathbb{C}$ and every $(u_0, u_0') \in \mathbb{C}^2$, theorem 1.1 ensures the existence of a unique solution of (2.1), holomorphic near x_0, satisfying the initial data $\big(u(x_0), u'(x_0)\big) = (u_0, u_0')$.

2.2 Painlevé Property for the First Painlevé Equation

As already mentioned, the first Painlevé equation satisfies the Painlevé property. The following more precise result holds.

E. Delabaere, *Divergent Series, Summability and Resurgence III*,
Lecture Notes in Mathematics 2155, DOI 10.1007/978-3-319-29000-3_2

Theorem 2.1. *Every solution of the Painlevé equation P_I can be analytically continued as a meromorphic function on $\mathbb{C}$ with only double poles.*

This theorem will be shown in appendix. We add the following result for completeness:

Theorem 2.2. *Every solution of (2.1) is a transcendental meromorphic function on $\mathbb{C}$ with infinitely many poles.*

Proof. We just give an idea of the proof. It is easy to see that every solution u of the first Painlevé equation (2.1) is a transcendental function. Otherwise, since u is meromorphic with double poles, u should be a rational function, $u(x) = \frac{P(x)}{Q(x)^2}$. Reasoning on the degrees of P and Q, one shows that this is impossible. So every solution u is a transcendental meromorphic function. It can be then derived from the Clunie lemma in Nevanlinna theory of meromorphic functions that necessarily u has an infinite set of poles [Lai92, GLS002]. □

The above properties were well-known since Painlevé [Pai00]. The following one was also known by Painlevé, however its complete proof has been given only recently [Nis88], see also [Cas009].

Theorem 2.3. *A solution of P_I cannot be described as any combination of solutions of first order algebraic differential equations and those of linear differential equations on $\mathbb{C}$.*

2.3 First Painlevé Equation and Isomonodromic Deformations Condition

Each Painlevé equation P_J is equivalent to a nonautonomous Hamiltonian system [Oka80-1]. Concerning the first Painlevé equation this Hamiltonian system is given by the following *first Painlevé system*:

$$(\mathcal{H}_I)\begin{cases} \dfrac{du}{dx} = \dfrac{\partial H_I}{\partial \mu} = \mu \\ \\ \dfrac{d\mu}{dx} = -\dfrac{\partial H_I}{\partial u} = 6u^2 + x \end{cases}, \quad H_I(u,\mu,x) = \frac{1}{2}\mu^2 - 2u^3 - xu. \tag{2.2}$$

It is known [Gar12, Oka80-2] that this Hamiltonian system arises as a *condition of isomonodromic deformations* of the following (Schlesinger type) second order linear ODE,

$$(\mathscr{S}\mathscr{L}_I)\begin{cases}\dfrac{\partial^2\Psi}{\partial z^2}=Q_I(z;u,\mu,x)\Psi\\[2ex] Q_I(z;u,\mu,x)=4z^3+2xz+2H_I(u,\mu,x)-\dfrac{\mu}{z-u}+\dfrac{3}{4(z-u)^2}.\end{cases}\tag{2.3}$$

In other words, u is solution of the first Painlevé equation (2.1) if and only if the monodromy data of (2.3) do not depend on x. We explain this point. Equation (2.3) has two fixed singularities $z=u,\infty$, so that any (local) solution of (2.3) can be analytically continued to a Riemann surface which covers $\overline{\mathbb{C}}\setminus\{u,\infty\}$. The point $z=u$ is a regular singular point, and a local analysis easily shows that the monodromy at this point (see the first volume of this book [MS016]) of any fundamental system of solutions of (2.3) does not depend on x. The other singular point $z=\infty$ is an irregular singular point. Thus the only nontrivial monodromy data of (2.3) are given by the Stokes coefficients at $z=\infty$.

The second order linear ODE (2.3) is equivalent to a first order linear ODE in dimension two. Each Stokes matrix is a two by two unipotent matrix (see the first two volumes of this book [Lod016, MS016]), and thus depends on a single complex coefficient called a Stokes coefficient.

In general these Stokes coefficients depend on x, except when Ψ satisfies the following isomonodromic deformation condition:

$$(\mathscr{D}_I)\ \frac{\partial\Psi}{\partial x}=A_I\frac{\partial\Psi}{\partial z}-\frac{1}{2}\frac{\partial A_I}{\partial z}\Psi,\qquad A_I=\frac{1}{2(z-u)}\tag{2.4}$$

The first Painlevé system (2.2) ensures the compatibility between equations (2.3) and (2.4) : solving a Painlevé equation is thus equivalent to solving an inverse monodromy problem (Riemann-Hilbert problem) [MS016, JMK81, JM81, KK93, KT96, Tak000, Kap004, JKT009, FIKN006].

We add another property : we mentioned that the asymptotics of (2.3) at $z=\infty$ are governed by some Stokes coefficients $s_i=s_i(u,\mu,x)$. It can be shown that the space of Stokes coefficients makes a complex manifold $\mathscr{M}_I$ of dimension 2. Also, for any point of $\mathscr{M}_I$ there exists a unique solution of the first Painlevé equation (2.1) for which the monodromy data of equation (2.3) are equal to the corresponding coordinates of this point [KK93].

2.4 Lax Formalism

There is another fruitful alternative to get the Painlevé equations, however related to the previous one, based on the linear representations of integrable systems through the Lax formalism [Lax68]. We exemplify this theory for Painlevé I, for which the so-called *Lax pair* A and B are the matrix operators given as follows [JM81]:

$$A = \begin{pmatrix} v(x) & 4\big(z-u(x)\big) \\ z^2+u(x)z+u(x)^2+x/2 & -v(x) \end{pmatrix}, \quad B = \begin{pmatrix} 0 & 2 \\ z/2+u(x) & 0 \end{pmatrix}.$$

To the matrix operator A one associates a first order ODE in the z variable, whose time evolution (the x variable) is governed by another first order ODE determined by the matrix operator B,

$$\begin{cases} \dfrac{\partial \Psi}{\partial z} = A\Psi \\ \dfrac{\partial \Psi}{\partial x} = B\Psi \end{cases} \tag{2.5}$$

The compatibility condition $\dfrac{\partial^2 \Psi}{\partial z \partial x} = \dfrac{\partial^2 \Psi}{\partial x \partial z}$ provides what is known as the *zero curvature condition* (or Lax equation), namely $\frac{\partial A}{\partial x} - \frac{\partial B}{\partial z} = [B,A]$ where $[B,A] = BA - AB$ stands for the commutator. Expliciting this condition, one recovers the first Painlevé equation under the form $\begin{cases} \dfrac{\mathrm{d}u}{\mathrm{d}x} = v \\ \dfrac{\mathrm{d}v}{\mathrm{d}x} = 6u^2 + x \end{cases}$. From what have been previously seen, the zero curvature condition allows to think of (2.5) as an isomonodromic deformations condition for its first equation.

2.5 Symmetries

We would like to notice here that the cyclic symmetry group of order five acts on the set of solutions (2.1). Indeed, introducing $\omega_k = \mathrm{e}^{\frac{2i\pi}{5}k}$, $k = 0, \cdots, 4$, then any solution u of (2.1) is mapped to another solution u_k through the transformation

$$u_k(x) = \omega_k^2 u\big(\omega_k x\big), \qquad k = 0, \cdots, 4.$$

In general u and u_k will be different solutions, an obvious exception being when u satisfies the initial data $u(0) = u'(0) = 0$.

2.6 Asymptotics at Infinity

Our aim in this section is to describe all the possible behaviors at infinity of the solutions of the first Painlevé equation (2.1).

We first notice that $x = \infty$ is indeed a fixed singularity for P_I : making the change of variable $u(x) = \mathrm{u}(t)$, $t = \dfrac{1}{x}$, equation (2.1) translates into $t^5 \mathrm{u}'' + 2t^4 \mathrm{u}' = 1 + 6t\mathrm{u}^2$, where $t = 0$ appears as a (irregular) singular point.

We mention that, when analysing the asymptotics of solutions of differential equations at singular points, there is a great difference between linear and nonlinear

ODEs. When a linear ODE is concerned, the asymptotics of every solution can be derived from the asymptotics of a fundamental system of solutions. For non linear ODEs some care has to be taken, since as a rule singular solutions may exist, which cannot be deduced from the general solution.

The study of all possible behaviors at infinity was first made by Boutroux [Bou13, Bou14]. Various approaches can be used: a direct asymptotic approach in the line of Boutroux as in [Hil76, Jos99, JK92], or another one based on the relationship between the first Painlevé equation and a convenient Schlesinger type linear ODE as described in Sect. 2.3, see [KK93] (see also [KT96, KKNT004, KT005, Tak000] for an exact semiclassical variant).

2.6.1 Dominant Balance Principle

We only want to give a rough idea of how to get the whole possible asymptotic behaviors and, in the spirit of this course, we follow the viewpoint of asymptotic as in [Hil76, JK92, Jos99]. In this approach, for a given ODE, the first task is to determine the terms in the equation which are dominant and of comparable size when $x \to \infty$ along a path or a inside a sector. The reduced equation obtained by keeping the dominant terms only in the ODE gives the leading behavior.
One usual trick to guess the asymptotics of solutions of ODEs is the *dominant balance principle* [BO78]. A maximal dominant balance corresponds to the case where there is a maximal set of dominant terms of comparable size in the equation. As a rule, this gives rise to the general behavior. The remaining cases are called subdominant balances.

It is useful to introduce the following notation:

- $f \simeq g$ when $x \to \infty$ along a path if $\lim\limits_{x\to\infty} \dfrac{f(x)}{g(x)} = Cte,\ Cte \in \mathbb{C}^\star$.
- $f \ll g$ when $x \to \infty$ along a path if $\lim\limits_{x\to\infty} \dfrac{f(x)}{g(x)} = 0$.

The unique maximal balance for P_I consists in assuming all the three terms in (2.1) of comparable size when $x \to \infty$. In particular u^2 and x have comparable size, so that $u(x) = x^{\frac{1}{2}} O(1)$ when $x \to \infty$. We therefore write $u(x) = x^{\frac{1}{2}} v\big(z(x)\big)$ with $z(x) \to \infty$ and $v\big(z(x)\big) = O(1)$ when $x \to \infty$. If $z(x)$ behaves like a fractional power of x at infinity, then $\dfrac{z'(x)}{z(x)} \simeq \dfrac{z''(x)}{z'(x)} \simeq \dfrac{1}{x}$ and this is what will be assumed. We also make the following remark : if $v(z)$ is an analytic function whose asymptotics at infinity is governed by a (formal, possibly Laurent) series, then $\dfrac{v''(z)}{z^2} \ll \dfrac{v'(z)}{z} \ll v(z)$ at infinity, that is $\dfrac{v''\big(z(x)\big)}{z^2(x)} \ll \dfrac{v'\big(z(x)\big)}{z(x)} \ll v\big(z(x)\big)$ when $x \to \infty$.
Here we will adjust the choice of $z(x)$ by adding the demand:

$$v\big(z(x)\big) \ll z(x)v'\big(z(x)\big) \ll z(x)^2 v''\big(z(x)\big) \quad \text{when} \quad x \to \infty.$$

These assumptions on v and $z(x)$ provide the identities:

$$u'(x) = x^{-\frac{1}{2}} z(x) v'\big(z(x)\big) O(1) + o(1), \qquad u''(x) = x^{-\frac{3}{2}} z^2(x) v''\big(z(x)\big) O(1) + o(1).$$

Thus, if $v\big(z(x)\big) = v'\big(z(x)\big) = v''\big(z(x)\big) = O(1)$ and demanding that u'' and x have comparable size, one gets $z(x) = x^{\frac{5}{4}} O(1)$ as a necessary condition. This suggests with Boutroux [Bou13, Bou14] to make the following transformation,

$$u(x) = \alpha x^{\frac{1}{2}} v(z), \qquad z = \beta x^{\frac{5}{4}}, \tag{2.6}$$

with $\alpha, \beta \neq 0$ some constants, under which equation (2.1) becomes:

$$v'' + \frac{v'}{z} - \frac{4}{25} \frac{v}{z^2} - \frac{96\alpha}{25\beta^2} v^2 - \frac{16}{25\alpha\beta^2} = 0.$$

With the following choice for α and β,

$$\alpha = \frac{\mathrm{e}^{\frac{i\pi}{2}}}{\sqrt{6}}, \qquad \beta = \mathrm{e}^{\frac{5i\pi}{4}} \frac{24^{\frac{5}{4}}}{30}, \tag{2.7}$$

one finally gets:

$$v'' = \frac{1}{2} v^2 - \frac{1}{2} - \frac{v'}{z} + \frac{4}{25} \frac{v}{z^2}. \tag{2.8}$$

We now concentrate on this equation (2.8) and we examine the possible balances.

2.6.2 *Maximal Balance, Elliptic Function-type Behavior*

We consider the maximal balance case, that is we assume that v and its derivatives can be compared to unity. This means that equation (2.8) is asymptotic to the equation $v'' = \frac{1}{2} v^2 - \frac{1}{2}$ whose solutions[1] are the functions $v(z) = 12\wp(z - z_0; \frac{1}{12}, g_3)$ where $\wp$ is the Weierstrass p-function (cf. Sect. 1.2.4), while z_0 and g_3 are two free complex parameters. This indeed provides the general behaviour of the Painlevé transcendents near infinity [Bou13, Bou14, JK92] : for $|z|$ large enough in each open quadrants

$$Q_k = \{z \in \mathbb{C}, k\frac{\pi}{2} < \arg z < (k+1)\frac{\pi}{2}\}, \ k = 0, 1, 2, 3 \mod 4,$$

the generic solution v of (2.8) has an approximate period lattice of poles, Fig. 2.1. In this domain, excluding small neighbourdhoods of poles, the asymptotics of such

[1] Just multiply both sides of the equality by v', then integrate.

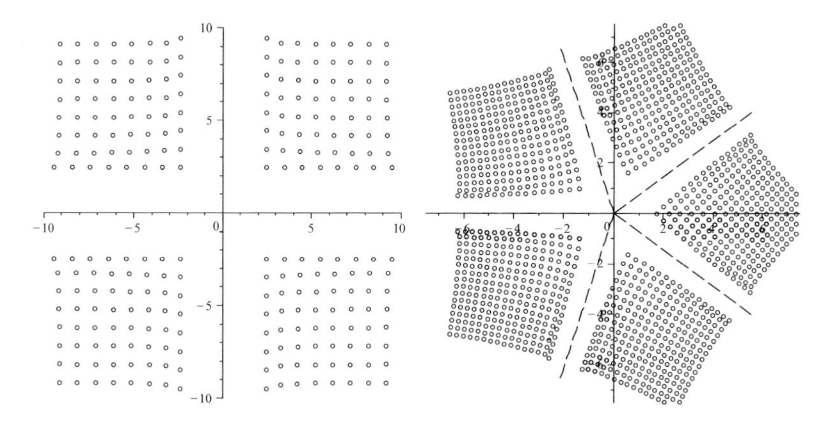

Fig. 2.1 Left hand side : approximate period lattices in each quadrants Q_i of z-plane. Right hand side, their images in the x-plane through the transformation $x \mapsto z$ defined by (2.6)-(2.7)

a generic solution v of (2.8) is governed by Weierstrassian elliptic functions. With Kruskal & Joshi [JK92] one refers to this behavior as an *elliptic function-type* behavior. Through the transformation (2.6)-(2.7), this translates for the Painlevé I transcendents into an asymptotic regime on the sectors:

$$S_k = \{x \in \mathbb{C},\ -\pi + k\frac{2\pi}{5} < \arg x < -\pi + (k+1)\frac{2\pi}{5}\},\ k = 0,1,2,3,4 \mod 5. \quad (2.9)$$

When z approaches the real axis (*resp.* the imaginary axis), $|z|$ large enough and in a small angular strip of width $O\Big((\log|z|)/|z|\Big)$, then the solution v displays a *near oscillatory-type* behaviour with no poles, and $v(z) \to -1$ (*resp.* $v(z) \to +1$) when $|z| \to \infty$, see [JK92]. The five special rays $\arg x = -\pi + k\frac{2\pi}{5}, k = 0, \cdots, 4$ thus play an important role in the asymptotics of the solutions of Painlevé I, the general solutions having lines of poles asymptotic to these rays.

2.6.3 Submaximal Dominant Balances, Truncated Solutions

We now consider submaximal dominant balances, that is when v or one of its derivatives differ from order unity. As shown in [JK92], the single consistent case occurs when $v \simeq 1$ and $v'' \ll 1$. This implies that equation (2.8) is now asymptotic to the equation $\frac{1}{2}v^2 - \frac{1}{2} = 0$, that is $v(z) = \pm 1 + o(1)$. Examining this case leads to the following result:

Theorem 2.4. *The first Painlevé equation (2.1) has:*

- *five complex parameter families of solutions u, the so-called* intégrales tronquées (truncated solutions) *after Boutroux, such that u is free of poles in two adjacent sectors* S_k *and* S_{k+1} *for* $|x|$ *large enough, and* $u(x) = \left(-\frac{x}{6}\right)^{\frac{1}{2}} \left(1 + O(x^{-\frac{5}{2}})\right)$ *at infinity in these sectors (for a convenient determination of the square root).*

- *among these truncated solutions, five special solutions, each of them being free of poles in four adjacent sectors* $S_k, S_{k+1}, S_{k+2}, S_{k+3}$ *for* $|x|$ *large enough, with the above asymptotics at infinity in these sectors. These are the so-called* intégrales tri-tronquées (tritruncated solutions).

This theorem has various proofs, see for instance [JK001, Mas010-1, Mas010-2] for "nonconventional" approaches. We will see in this course how the resurgent analysis can be used to show theorem 2.4.

There are analogues of truncated solutions for each member $(P_I^{(n)})$, $n = 1, 2, \cdots$ in the first Painlevé hierarchy (1.10), with asymptotics at infinity of the form [DZ010]:

$$u(x) = \left((-1)^n \frac{x}{c_n}\right)^{\frac{1}{n+1}} \left(1 + O(x^{-\frac{2n+3}{2n+2}})\right), \quad c_n = \frac{2^{3n+1}\Gamma(n+3/2)}{\Gamma(n+2)\Gamma(1/2)}. \tag{2.10}$$

Similar results occur for the first discrete Painlevé equation (1.12), see [JL015].

2.7 First Painlevé Equation and Physical Models

As already said (Sect. 1.3.3), the Painlevé equations in general and the first Painlevé equation in particular, appear by similarity reductions of integrable PDEs. They play a significant role in others physical models, see e.g. [Kap004] and references therein for the first Painlevé equation. This includes the description of asymptotic regime in transition layers and caustic-type domain. We exemplify this fact with the focusing nonlinear Schrödinger equation $i\varepsilon\Psi_t + \frac{\varepsilon^2}{2}\Psi_{xx} + |\Psi|^2\Psi = 0$ (fNLS). It is shown in [DGK009] that when considering the (so-called) dispersionless limit $\varepsilon \to 0$, the solutions (of convenient Cauchy problems) of (fNLS) are asymptotically governed by a tritruncated solution of the first Painlevé equation. In the same work, theoretical and numerical evidences led the authors to conjecture that the tritruncated solutions of the first Painlevé equation have the following property, shown in [CHT014] under the naming "the Dubrovin conjecture":

Proposition 2.1. *Each tritruncated solution of the first Painlevé equation is holomorphic on a full sector of the form* $\{x \in \mathbb{C} \mid \arg x \in \bar{I}, |x| \geq 0\}$, *where* I *stands for the closure of an open arc* I *of length* $|I| = 8\pi/5$. *Moreover, each tritruncated solution can be analytically continued to a disc* $|x| < r_0$ *with* $r_0 > 0$ *small enough.*

Recently, resurgence theory spectacularly enters the realm of string theory and related models, as an efficient tool for making the connection between perturbative and non-perturbative effects (see, e.g. [Mar014] and references therin). In particular, the first Painlevé equation was particularly adressed in [ASV012] thanks of its physical interpretation in the context of 2D quantum gravity [DGZ95, MSW008, MSW009, GIKM012].

Appendix

The reader only interested in learning applications of resurgence theory may skip this appendix, where we show theorem 2.1 for completeness. We follow the proof given in [Cos99]. See also [Hil76, Inc56] and specially [GLS002] with comments and references therein. We start with two lemmas.

Lemma 2.1. *Let u be any solution of (2.1), holomorphic on a neighbourhood of $x_0 \in \mathbb{C}$. Then the radius R of analyticity at x_0 satisfies $R \geq 1/\rho$ with*

$$\rho = \max\left(\left|u(x_0)\right|^{1/2}, \left|\frac{u'(x_0)}{2}\right|^{1/3}, \left|u^2(x_0)+\frac{x_0}{6}\right|^{1/4}, \left|\frac{u(x_0)u'(x_0)}{2}+\frac{1}{24}\right|^{1/5}\right). \quad (2.11)$$

Proof. If $u(x) = \sum_{k=0}^{\infty} c_k(x-x_0)^k \in \mathbb{C}\{x-x_0\}$ solves (2.1), then

$$\begin{cases} c_0 = u(x_0), & c_1 = u'(x_0) \\ c_2 = 3c_0^2 + \dfrac{x_0}{2}, & c_3 = 2c_0c_1 + \dfrac{1}{6} \\ (k+1)(k+2)c_{k+2} = 6\sum_{m=0}^{k} c_m c_{k-m}, \; k \geq 2 \end{cases} \quad (2.12)$$

Let be $\rho > 0$ given by (2.11) so that for any integer $l \in [0,3]$,

$$|c_l| \leq (l+1)\rho^{l+2}. \quad (2.13)$$

Assume that (2.13) is satisfied for every $0 \leq l \leq k+1$ for a given $k \geq 2$. Then by (2.12),

$$(k+1)(k+2)|c_{k+2}| \leq 6\sum_{m=0}^{k}(m+1)(k-m+1)\rho^{k+4} \leq (k+1)(k+2)(k+3)\rho^{k+4}.$$

The coefficients $\sum_{m=0}^{k}(m+1)(k-m+1)$ are those of the taylor expansions of $(1-x)^{-4}$ at the origin. Indeed, for $|x|<1$, $\dfrac{1}{1-x} = \sum_{k\geq 0} x^k$ so that $\dfrac{1}{(1-x)^2} = \sum_{k\geq 0}(k+1)x^k$. Therefore $\left(\dfrac{1}{(1-x)^2}\right)^2 = \sum_{k\geq 0}\left(\sum_{m=0}^{k}(m+1)(k-m+1)\right)x^k$.

We conclude that (2.13) is satisfied for every $l \geq 0$ and this implies that $R \geq \dfrac{1}{\rho}$, where R is the radius of convergence of the series expansion u. □

Lemma 2.2. *In a neighbourhood of any given point $\widetilde{x} \in \mathbb{C}$, there exists a one-parameter family of meromorphic solutions u of (2.1) having a pole at $\widetilde{x}$. Necessarily $\widetilde{x}$ is a double pole and u is given by the Laurent-series expansions*

$$u(x) = \frac{1}{(x-\widetilde{x})^2} - \frac{\widetilde{x}}{10}(x-\widetilde{x})^2 - \frac{1}{6}(x-\widetilde{x})^3 + c_4(x-\widetilde{x})^4 + \sum_{k\geq 6} c_k(x-\widetilde{x})^k$$

where $c_4 \in \mathbb{C}$ is a free parameter.

Proof. We are looking for a Laurent-series $u(x) = \sum_{k=p}^{\infty} c_k(x-\widetilde{x})^k \in \mathbb{C}\{x-\widetilde{x}\}\left[\frac{1}{x-\widetilde{x}}\right]$ satisfying (2.1). Necessarily $p \geq -2$, $c_{-2} = 1$ or 0, $c_{-1} = 0$. Therefore, either $\widetilde{x}$ is a regular point, or otherwise

$$u(x) = \frac{1}{(x-\widetilde{x})^2} - \frac{\widetilde{x}}{10}(x-\widetilde{x})^2 - \frac{1}{6}(x-\widetilde{x})^3 + c_4(x-\widetilde{x})^4 + \sum_{k\geq 6} c_k(x-\widetilde{x})^k$$

where $c_4 \in \mathbb{C}$ is a free parameter, while for $k \geq 6$ the coefficients are polynomial functions of $(\widetilde{x}, \alpha)$. Indeed, one has $(k-2)(k+5)c_{k+2} = 6\sum_{m=0}^{k} c_m c_{k-m}$, $k \geq 2$. We can define $\rho > 0$ (depending on $(\widetilde{x}, \alpha)$) such that, for $0 \leq l \leq 5$,

$$|c_l| \leq \frac{1}{3}(l+1)\rho^{l+2}. \tag{2.14}$$

Assume that this property is satisfied for every c_l, $0 \leq l \leq k+1$, for a given $k \geq 4$. Then

$$(k-2)(k+5)|c_{k+2}| \leq \frac{2}{3}\sum_{m=0}^{k}(m+1)(k-m+1)\rho^{k+4} \leq \frac{1}{9}(k+1)(k+2)(k+3)\rho^{k+4}$$

and we conclude that $|c_{k+2}| \leq \frac{1}{3}(k+3)\rho^{k+4}$. Therefore (2.14) is true for every $l \geq 0$ and the Laurent series converges in the punctured dics $D(\widetilde{x}, 1/\rho)^{\star}$. □

The following notation will now be used:

- $D_{x_0} \subset \mathbb{C}$ is an open disc, Ω is a discrete subset of D_{x_0} and $x_0 \in D_{x_0} \setminus \Omega$.
- u is a solution of (2.1) defines by some initial data at $x_0 \in D_{x_0} \setminus \Omega$ and u is meromorphic in $D_{x_0} \setminus \Omega$.
- $\lambda(a,b) : [0,1] \to D_{x_0} \setminus \Omega$ denotes a $\mathscr{C}^{\infty}$-smooth path in $D_{x_0} \setminus \Omega$ with endpoints $\lambda(a,b)(0) = a$ and $\lambda(a,b)(1) = b$. When $b \in \partial D_{x_0}$ it is assumed that $\lambda(a,b)$ is a path where b is removed (that is one considers the restriction to $[0,1[$ of $\lambda(a,b)$). Moreover we assume that the length of any subsegment $\lambda(c,d)$ of $\lambda(a,b)$ is less that $2|c-d|$.
  We mention that we use the same notation $\lambda(a,b)$ for the path and its image.
- $\widetilde{x} \in \partial D_{x_0}$ is a singular point for u.

Lemma 2.3. *Assume that* $u(x) = \sum_{k=-2}^{4} a_k(x-\widetilde{x})^k + O(|x-\widetilde{x}|^5)$ *when* $x \to \widetilde{x}$ *along* $\lambda(x_0,\widetilde{x})$*, with* $a_{-2} \neq 0$*. Then* u *is meromorphic at* $\widetilde{x}$ *and* u *is uniquely determined by* $(\widetilde{x}, a_4)$.

Proof. Since u is solution of (2.1) which is analytic at each point of the smooth path $\lambda(x_0,\widetilde{x})$, one has $u''(x) = 6u^2(x) + x = 6\left(\sum_{k=-2}^{4} a_k(x-\widetilde{x})^k + O(|x-\widetilde{x}|^5)\right)^2 + x$ when $x \to \widetilde{x}$ along $\lambda(x_0,\widetilde{x})$. This implies that the asymptotic expansion is differentiable.

This is a consequence of the mean value theorem, $u(x) = u(x_0) + \int_{x_0}^{x} u'(s)\,ds$ along $\lambda(x_0, \widetilde{x})$ which is $\mathscr{C}^\infty$-smooth, and the uniqueness of the asymptotic expansion.

With the same calculus made in the proof of lemma 2.2, we show that

$$u(x) = \frac{1}{(x-\widetilde{x})^2} - \frac{\widetilde{x}}{10}(x-\widetilde{x})^2 - \frac{1}{6}(x-\widetilde{x})^3 + a_4(x-\widetilde{x})^4 + O(|x-\widetilde{x}|^5).$$

We denote by v the meromorphic solution of (2.1) obtained in lemma 2.2 with $c_4 = a_4$. We set

$$w(x) = v(x) - (x-\widetilde{x})^{-2} = O(|x-\widetilde{x}|^2)$$
$$f(x) = u(x) - v(x) = O(|x-\widetilde{x}|^5)$$

and we want to show that $f = 0$. We have $f'' - \frac{12}{(x-\widetilde{x})^2} f = g$ with $g = 12wf + 6f^2$, $g = O(|x-\widetilde{x}|^7)$. Integrating this linear ODE yields:

$$\begin{aligned} f(x) &= C_1(x-\widetilde{x})^{-3} + C_2(x-\widetilde{x})^4 \\ &- 7(x-\widetilde{x})^{-3}\int_{\widetilde{x}}^{x}(s-\widetilde{x})^4 g(s)\,ds + \frac{(x-\widetilde{x})^4}{7}\int_{\widetilde{x}}^{x}(s-\widetilde{x})^{-3}g(s)\,ds. \end{aligned}$$

Since $f(x) = O(|x-\widetilde{x}|^5)$, f is solution of the fixed-point problem $f = \mathsf{N}(f)$ with

$$\mathsf{N}(f)(x) = -7(x-\widetilde{x})^{-3}\int_{\widetilde{x}}^{x}(s-\widetilde{x})^4 g(s)\,ds + \frac{(x-\widetilde{x})^4}{7}\int_{\widetilde{x}}^{x}(s-\widetilde{x})^{-3}g(s)\,ds.$$

For $x_1 \in \lambda(x_0, \widetilde{x})$ we consider the normed vector space $(\mathbb{B}, \|.\|)$,

$$\mathbb{B} = \{f \in \mathscr{C}^0(\lambda(x_1, \widetilde{x})),\ f = O(|x-\widetilde{x}|^5\},\qquad \|f\| = \sup_{x\in\lambda(x_1,\widetilde{x})} |(x-\widetilde{x})^{-5}f(x)|.$$

We show later that $(\mathbb{B}, \|.\|)$ is a Banach space (lemma 2.4). Now for x_1 close enough from $\widetilde{x}$ (see lemma 2.4):

- the mapping N send the unit ball B of $\mathbb{B}$ into itself,
- the mapping $\mathsf{N} : B \to B$ is contractive.

Therefore the fixed-point problem $f = \mathsf{N}(f)$ has a unique solution in B by the contraction mapping theorem. Obviously this solution is $f = 0$ and therefore $u = v$. □

Lemma 2.4. *With the notation of the proof of lemma 2.3:* $(\mathbb{B}, \|.\|)$ *is a Banach space and the mapping* $\mathsf{N} : B \to B$ *is contractive.*

Proof.

$(\mathbb{B}, \|.\|)$ *is a Banach space.* Assume that (f_p) is a Cauchy sequence in $(\mathbb{B}, \|.\|)$,

$$\forall \varepsilon, \exists p_0 : \forall p, q > p_0, \forall x \in \lambda(x_1, \widetilde{x}),\ |(x-\widetilde{x})^{-5}(f_p(x) - f_q(x))| < \varepsilon\,. \tag{2.15}$$

Writing $g_p(x) = (x-\widetilde{x})^{-5} f_p(x)$, condition (2.15) implies that for every $x \in \lambda(x_1, \widetilde{x})$ the sequence $(g_p(x))$ is a Cauchy sequence, hence $g_p(x) \to g(x)$ in $\mathbb{C}$. Now making

$q \to +\infty$ in (2.15) one sees that $g_p \to g$ uniformaly. Therefore $g \in \mathscr{C}^0(\lambda(x_1,\widetilde{x}))$ and is bounded on $\lambda(x_1,\widetilde{x})$. Thus $g = (x-\widetilde{x})^{-5} f$ with $f \in \mathbb{B}$.

The mapping N *is contractive for* x_1 *close enough from* $\widetilde{x}$*.* We set $\mathsf{N}_1(f)(x) = -7(x-\widetilde{x})^{-3} \int_{\widetilde{x}}^{x} (s-\widetilde{x})^4 g(s)\,ds$, $\mathsf{N}_2(f)(x) = \frac{(x-\widetilde{x})^4}{7} \int_{\widetilde{x}}^{x} (s-\widetilde{x})^{-3} g(s)\,ds$ so that $\mathsf{N}(f) = \mathsf{N}_1(f) + \mathsf{N}_1(f)$. One can assume that $|s-\widetilde{x}| \leq |x-\widetilde{x}|$ for $s \in \lambda(x,\widetilde{x})$. Also, there exist $r > 0$ and $a > 0$ such that $|w(x)| \leq a|x-\widetilde{x}|^2$ when $|x-\widetilde{x}| \leq r$. We now assume that $|x_1 - \widetilde{x}| \leq r$. For any $f_1, f_2 \in B$ and $x \in \lambda(x_1,\widetilde{x})$:

$$\begin{aligned}&\left|(x-\widetilde{x})^{-5}\Big(\mathsf{N}_1(f_1) - \mathsf{N}_2(f_2)\Big)\right| \\ &\leq \left| -7(x-\widetilde{x})^{-8} \int_{\widetilde{x}}^{x} (s-\widetilde{x})^4 \Big(12w(s)(f_1(s)-f_2(s)) + 6(f_1^2(s) - f_2^2(s))\Big)\,ds \right|,\end{aligned}$$

thus

$$\begin{aligned}\left|(x-\widetilde{x})^{-5}\Big(\mathsf{N}_1(f_1) - \mathsf{N}_2(f_2)\Big)\right| &\leq 7|x-\widetilde{x}|^{-8}\Big(12a|x-\widetilde{x}|^{11}\|f_1 - f_2\| \\ &\quad + 6|x-\widetilde{x}|^{14}\|f_1-f_2\|\|f_1+f_2\|\Big)\mathrm{Length}(\lambda(x,\widetilde{x})) \\ &\leq 14|x-\widetilde{x}|^4\Big(12a + 12|x-\widetilde{x}|^3\Big)\|f_1 - f_2\|.\end{aligned}$$

The other term of $(x-\widetilde{x})^{-5}\Big(\mathsf{N}_2(f_1) - \mathsf{N}_2(f_2)\Big)$ is worked out in a similar way. Choosing x_1 close enough from $\widetilde{x}$, one obtains the existence of a constant $Cte \in]0,1[$ such that for any $f_1, f_2 \in B$, $\|\mathsf{N}(f_1) - \mathsf{N}(f_2)\| \leq Cte\|f_1 - f_2\|$. □

Lemma 2.5. *When* $x \to \widetilde{x}$ *along* $\lambda(x_0,\widetilde{x})$ *with* $\widetilde{x} \in \partial D_{x_0}$ *a singular point for u:*

1. $|u(x)| + |u'(x)| \to +\infty$,
2. *u is unbounded.*

Proof. 1. Lemma 2.1 implies that $|u(x)|$ or $|u'(x)|$ has to be large for x near $\widetilde{x}$ which is a singular point.
2. Multiplying (2.1) by u' and then integrating yields

$$(u')^2 = 4u^3 + 2xu - 2\int_{x_0}^{x} u(s)\,ds + C \tag{2.16}$$

where $C \in \mathbb{C}$ is a constant. Therefore if u is bounded $x \to \widetilde{x}$ along $\lambda(x_0,\widetilde{x})$ then u' is bounded as well, which contradicts the first property. □

Lemma 2.6. *When* $x \to \widetilde{x}$ *along* $\lambda(x_0,\widetilde{x})$*, with* $\widetilde{x} \in \partial D_{x_0}$ *a singular point for u, then:*

$$u^{-3}(x)\int_{x_0}^{x} u(s)\,ds \to 0, \qquad |u(x)| \to +\infty, \qquad |u'(x)| \to +\infty.$$

Proof. By lemma 2.5, we know that u is unbounded when $x \to \widetilde{x}$ along $\lambda(x_0,\widetilde{x})$, so that $\limsup_{x\to\widetilde{x}} |u(x)| = +\infty$, $\liminf_{x\to\widetilde{x}} |u^{-1}(x)| = 0$.

Reminder: $\limsup_{x\to\widetilde{x}} f(x) = \lim_{\varepsilon\to 0}\Big(\sup\{f(x) \mid x\in\lambda(x_0,\widetilde{x})\cap D(\widetilde{x},\varepsilon)\}\Big)$,
$\liminf_{x\to\widetilde{x}} f(x) = \lim_{\varepsilon\to 0}\Big(\inf\{f(x) \mid x\in\lambda(x_0,\widetilde{x})\cap D(\widetilde{x},\varepsilon)\}\Big)$.

Since $\left|u^{-3}(x)\int_{x_0}^{x} u(s)\,ds\right| \leq |u^{-3}(x)|.\max_{\lambda(x_0,x)}|u|.\,\mathrm{Length}(\lambda(x_0,x))$ for $x\in\lambda(x_0,\widetilde{x})$, it turns out that

$$\liminf_{x\to\widetilde{x}}\left\{\left|u^{-3}(x)\int_{x_0}^{x} u(s)\,ds\right|\right\} \leq \liminf_{x\to\widetilde{x}}\left\{|u^{-3}(x)|.\max_{\lambda(x_0,x)}|u|.\,\mathrm{Length}(\lambda(x_0,x))\right\}.$$

The right hand side term vanishes because u is unbounded when $x\to\widetilde{x}$, thus

$$\liminf_{x\to\widetilde{x}}\left\{\left|u^{-3}(x)\int_{x_0}^{x} u(s)\,ds\right|\right\} = 0. \tag{2.17}$$

In particular, for every $\gamma > 0$, for every $D(\widetilde{x},\varepsilon)$, there exists $x\in\lambda(x_0,\widetilde{x})\cap D(\widetilde{x},\varepsilon)$ so that $\left|u^{-3}(x)\int_{x_0}^{x} u(s)\,ds\right| \leq \gamma$.

We make the following $\boxed{\textit{Assumption}: u^{-3}(x)\int_{x_0}^{x} u(s)\,ds \to 0 \text{ is a false premise.}}$

This assumption translates into the condition : there exists $\gamma > 0$ such that, for every $D(\widetilde{x},\varepsilon)$, there exists $x\in\lambda(x_0,\widetilde{x})\cap D(\widetilde{x},\varepsilon)$ so that $\left|u^{-3}(x)\int_{x_0}^{x} u(s)\,ds\right| \geq \gamma$.

By continuity, we see that for any $\gamma > 0$ small enough, there exists a sequence $x_n\to\widetilde{x}$, $x_n\in\lambda(x_0,\widetilde{x})$, such that

$$\left|\int_{x_0}^{x_n} u(s)\,ds\right| = \gamma\left|u^3(x_n)\right|. \tag{2.18}$$

The arguments used in the proof of lemma 2.5 show that $\limsup_n |u(x_n)| = +\infty$. This means that there exists a subsequence (x_{n_k}) of (x_n) such that $|u(x_{n_k})|\to+\infty$. Therefore we can assume that $\lim_n |u(x_n)| = +\infty$. with the following consequences: from (2.18) we see that

$$\lim_n\left|\int_{x_0}^{x_n} u(s)\,ds\right| = +\infty \tag{2.19}$$

while (2.18) with $\gamma > 0$ chosen small enough and (2.16) imply $\lim_n |u'(x_n)| = +\infty$.

We are going to prove in several steps that the above assumption leads to a contradiction.

First step We consider the solution h_n of the Cauchy problem

$$\begin{cases} (h')^2 = 4h^3 + 2x_n h + \widetilde{C}_n & \text{with } \widetilde{C}_n = C - 2\int_{x_0}^{x_n} u(s)\,ds \\ h(0) = u(x_n), & h'(0) = u'(x_n) \end{cases} \tag{2.20}$$

where C is the constant given in (2.16). Notice by (2.19) that $\lim_n |\widetilde{C}_n| = +\infty$ and by (2.18) then (2.16):

$$\begin{cases} |h_n(0)| = (2\gamma)^{-1/3}|\widetilde{C}_n|^{1/3}(1+o(1)) \\ |h_n'(0)| = \left|2\gamma^{-1}e^{i\phi_n}+1\right|^{1/2}|\widetilde{C}_n|^{1/2}(1+o(1)), \qquad \phi_n \in \mathbb{R}. \end{cases} \tag{2.21}$$

Writing

$$h_n(t) = \widetilde{C}_n^{1/3}H_n(X),\, X = \widetilde{C}_n^{1/6}t, \tag{2.22}$$

the function H_n is solution of the following *elliptic differential equation* (see (1.5)) with a given initial data:

$$\begin{cases} (H')^2 = 4H^3 + 2\theta_n H + 1, \;\text{ with } \theta_n = x_n\widetilde{C}_n^{-2/3} \\ H_n(0) = \widetilde{C}_n^{-1/3}u(x_n), \qquad |H_n(0)| = (2\gamma)^{-1/3}(1+o(1)), \\ H_n'(0) = \widetilde{C}_n^{-1/2}u'(x_n), \qquad |H_n'(0)| = \left|2\gamma^{-1}e^{i\phi_n}+1\right|^{1/2}(1+o(1)). \end{cases} \tag{2.23}$$

From the properties of elliptic functions, H_n can be analytically continued as a doubly periodic meromorphic function with double poles at the period lattice $a_n + m\omega_1(\theta_n) + n\omega_2(\theta_n)$, $(n,m) \in \mathbb{Z}^2$, for some $a_n \in \mathbb{C}$ and $\omega_{1,2}(\theta_n) = Cte_{1,2} + O(\theta_n)$.

Second step Next we consider the function U_n satisfying to the condition:

$$u(x) = \widetilde{C}_n^{1/3}U_n(X),\, X = \widetilde{C}_n^{1/6}(x - x_n). \tag{2.24}$$

From (2.1), U_n is solution of the ODE

$$U'' = 6U^2 + \theta_n + \varepsilon_n X, \;\text{ with } \varepsilon_n = \widetilde{C}_n^{-5/6}, \tag{2.25}$$

and, more precisely from (2.16):

$$\begin{cases} (U')^2 = 4U^3 + 2\theta_n U + 1 + 2\varepsilon_n\left(XU - \int_0^X U(S)\,dS\right) \\ U_n(0) = \widetilde{C}_n^{-1/3}u(x_n), \qquad U_n'(0) = \widetilde{C}_n^{-1/2}u'(x_n) \end{cases} \tag{2.26}$$

Third step We want to show that U_n and H_n are locally holomorphically equivalent: we look for a function G_n holomorphic near 0 such that

$$U_n = H_n \circ G_n \;\text{ with } G_n(X) = X + g_n(X), \qquad g_n(0) = 0,\, g_n'(0) = 0. \tag{2.27}$$

We know from (2.23) that $H_n'' = 6H_n^2 + \theta_n$, hence from (2.25) we deduce that

$$2g_n'H_n'' \circ G_n + (g_n')^2 H_n'' \circ G_n + g_n'' H_n' \circ G_n = \varepsilon_n X$$

or else:

$$2g'_n(H'_n \circ G_n)' + g''_n H'_n \circ G_n = \varepsilon_n X + (g'_n)^2 H''_n \circ G_n.$$

Multiplying both parts of this equality by $H'_n \circ G_n$ and integrating, one gets:

$$\begin{cases} w_n = (H'_n \circ G_n)^{-2} \int_0^X H'_n \circ G_n(S) \left[\varepsilon_n S + w_n^2(S).H''_n \circ G_n(S) \right] dS = \mathsf{N}(w_n) \\ g_n(X) = \int_0^X w_n(S)\, dS, \quad w_n(0) = 0, \quad G_n(X) = X + g_n(X). \end{cases} \tag{2.28}$$

Let $D(0, \frac{|\varepsilon_n|^{-1/4}}{2})$ be the disc centred at 0 of diameter $|\varepsilon_n|^{-1/4}$. We denote by $\widetilde{D}(0, \frac{|\varepsilon_n|^{-1/4}}{2})$ the disc $D(0, \frac{|\varepsilon_n|^{-1/4}}{2})$ deprived from the discs of diameter $d(\gamma)$ around the poles and the zeros of H'_n. We consider a path $\lambda(0, X_0)$ in $\widetilde{D}(0, \frac{|\varepsilon_n|^{-1/4}}{2})$. In (2.28), the integrals $\int_0^X$ are considered along $\lambda(0,X) \subset \lambda(0,X_0)$. We can assume that the length of any subsegment $\lambda(0,X)$ of $\lambda(0,X_0)$ is less that $2|X|$.
Let be $a \in]1/4, 1/2[$ and $(\mathbb{B}, \|.\|)$ be the Banach space $\mathbb{B} = \{f \in \mathscr{C}^0(\lambda(0,X_0))\}$, $\|f\| = \sup_{x \in \lambda(0,X_0)} |f(x)|$. Let B be the ball $B = \{f \in \mathbb{B}, \|f\| \leq |\varepsilon_n|^a\}$. If $w \in B$ and $g(X) = \int_0^X w(S)\, dS$,

$$\|g\| \leq \sup_{X \in \lambda(0,X_0)} \left| \int_0^X w(S)\, dS \right| \leq \|w\|.\text{Length}(\lambda(0,X_0)) \leq |\varepsilon_n|^{a-1/4}.$$

One can assume that $d(\gamma) \geq 3|\varepsilon_n|^{a-1/4}$ so that

$$\|\mathsf{N}(w)\| \leq |\varepsilon_n| Cte_1(\gamma) |\varepsilon_n|^{-1/2} + Cte_2(\gamma) |\varepsilon_n|^{2a-1/4}.$$

Therefore $\|\mathsf{N}(w)\| \leq |\varepsilon_n|^a$ for $|\varepsilon_n|$ small enough. Quite similarly, for $w_1, w_2 \in B$,

$$\|\mathsf{N}(w_1) - \mathsf{N}(w_2)\| = O(|\varepsilon_n|^{a-1/4}) \|w_1 - w_2\|.$$

We conclude by the contraction mapping theorem: N has a unique fixed point in B, for $|\varepsilon_n|$ small enough.

Final step We have seen that for $|\varepsilon_n|$ small enough and $a \in]1/4, 1/2[$,

$$U_n(X) = H_n\big(X + g_n(X)\big), \quad |g(X)| \leq |\varepsilon_n|^{a-1/4}, \quad X \in \widetilde{D}(0, \frac{|\varepsilon_n|^{-1/4}}{2}).$$

Therefore,

$$\sup_{X \in \widetilde{D}(0, \frac{|\varepsilon_n|^{-1/4}}{2})} \left| \widetilde{C}_n^{-1/3} u(x_n + \widetilde{C}_n^{-1/6} X) - H_n(X) \right| = O(|\varepsilon_n|^{a-1/4}). \tag{2.29}$$

Remember that $|\widetilde{C}_n| \to +\infty$ and $|\varepsilon_n| = |\widetilde{C}_n^{-5/6}| \to 0$ when $x_n \to \widetilde{x}$. If $X \in \widetilde{D}(0, \frac{|\varepsilon_n|^{-1/4}}{2})$, then $\widetilde{C}_n^{-1/6} X$ belongs to a disc of radius $|\widetilde{C}_n|^{1/24}$ deprived of some discs of radius

$d(\gamma)|\widetilde{C}_n|^{-1/6}$. Consequently, for n large enough,

$$\forall x \in D_{x_0},\ \exists X \in \widetilde{D}(0, \frac{|\varepsilon_n|^{-1/4}}{2}),\ \left|x - (x_n + \widetilde{C}_n^{-1/6} X)\right| \leq \frac{d(\gamma)}{2} |\widetilde{C}_n|^{-1/6}.$$

Choosing $x = x_0$, we see from (2.29) that u is unbounded near x_0 which is a regular point for u: $\boxed{\text{contradiction.}}$

Therefore, $u^{-3}(x) \int_{x_0}^{x} u(s)\,ds \to 0$ when $x \to \widetilde{x}$ along $\lambda(x_0, \widetilde{x})$. It is now an easy exercice by lemma 2.5 and (2.16) to see that $\min\{|u|, |u'|\} \to +\infty$ necessarily when $x \to \widetilde{x}$. (Just assume that $u^{-1}(x) \to 0$ is false and see that there is a contradiction.)
□

End of the Proof of theorem 2.1. What remains to show is that $\widetilde{x}$ is a second order pole. The substitution $u = 1/v^2$ transforms (2.16) into

$$(v')^2 = 1 + \frac{x}{2} v^4 - \frac{v^6}{2} \int_{x_0}^{x} \frac{ds}{v^2(s)} + \frac{C}{4} v^6. \tag{2.30}$$

We know from lemma 2.6 that $\frac{v^6}{2} \int_{x_0}^{x} \frac{ds}{v^2(s)} ds \to 0$ and $v \to 0$ along a path $\lambda(x_0, \widetilde{x})$ which avoids the poles of u in D_{x_0}. Therefore $(v')^2 = 1 + o(1)$, then $v^2(x) = (x - \widetilde{x})^2 \big(1 + o(1)\big)$. Plugging this last equality in (2.30) yields $(v')^2(x) = 1 + \frac{\widetilde{x}}{2}(x - \widetilde{x})^4 + o((x - \widetilde{x})^4)$, thus $v^2(x) = (x - \widetilde{x})^2 + \frac{\widetilde{x}}{10}(x - \widetilde{x})^6 + o((x - \widetilde{x})^6)$. One uses (2.30) again and eventually concludes with lemma 2.3.

References

ASV012. I. Aniceto, R. Schiappa, M. Vonk, *The resurgence of instantons in string theory.* Commun. Number Theory Phys. **6** (2012), no. 2, 339-496.

BO78. C.M. Bender, S. A. Orszag, *Advanced mathematical methods for scientists and engineers. I. Asymptotic methods and perturbation theory.* Reprint of the 1978 original. Springer-Verlag, New York, 1999.

Bou13. P. Boutroux, *Recherches sur les transcendantes de M. Painlevé et l'étude asymptotique des équations différentielles du second ordre.* Annales scientifiques de l'Ecole Normale Supérieure Sér. 3, **30** (1913), 255-375.

Bou14. P. Boutroux, *Recherches sur les transcendantes de M. Painlevé et l'étude asymptotique des équations différentielles du second ordre (suite).* Annales scientifiques de l'Ecole Normale Supérieure Sér. 3, **31** (1914), 99-159.

Cas009. G. Casale, *Une preuve galoisienne de l'irréductibilité au sens de Nishioka-Umemura de la première équation de Painlevé.* Asterisque **323** (2009), 83-100.

Cos99. O. Costin, *Correlation between pole location and asymptotic behavior for Painlevé I solutions.* Comm. Pure Appl. Math. **52** (1999), no. 4, 461-478.

CHT014. O. Costin, M. Huang, S. Tanveer, *Proof of the Dubrovin conjecture and analysis of the tritronquée solutions of P_I*, Duke Math. J. **163** (2014), no. 4, 665-704.

DZ010. D. Dai, L. Zhang, *On tronquée solutions of the first Painlevé hierarchy.* J. Math. Anal. Appl. **368** (2010), no. 2, 393-399.

DGZ95. P. Di Francesco, P. Ginsparg, J. Zinn-Justin, *2D gravity and random matrices.* Phys. Rep. **254** (1995), no. 1-2, 133 pp.

DGK009. B. Dubrovin, T. Grava, C. Klein, *On universality of critical behavior in the focusing nonlinear Schrödinger equation, elliptic umbilic catastrophe and the tritronquée solution to the Painlevé-I equation.* J. Nonlinear Sci. **19** (2009), no. 1, 57-94.

FIKN006. A.S. Fokas, A.R. Its, A.A. Kapaev, V.Y. Novokshenov, *Painlevé transcendents. The Riemann-Hilbert approach.* Mathematical Surveys and Monographs, 128. American Mathematical Society, Providence, RI, 2006.

Gar12. R. Garnier, *Sur des équations différentielles du troisième ordre dont l'intégrale générale est uniforme et sur une classe d'équations nouvelles d'ordre supérieur dont l'intégrale générale a ses points critiques fixes.* Ann. Sci. Ecole Norm. Sup. (3) **29** (1912), 1-126.

GIKM012. S. Garoufalidis, A. Its, A. Kapaev, M. Mariño, *Asymptotics of the instantons of Painlevé I.* Int. Math. Res. Not. IMRN, 2012, no. 3, 561-606.

GLS002. V. Gromak, I. Laine, S. Shimomura, *Painlevé differential equations in the complex plane.* de Gruyter Studies in Mathematics, 28. Walter de Gruyter & Co., Berlin, 2002.

Hil76. E. Hille, *Ordinary differential equations in the complex domain.* Pure and Applied Mathematics. Wiley-Interscience [John Wiley & Sons], New York-London-Sydney, 1976.

Inc56. E.L. Ince, *Ordinary Differential Equations.* Dover Publications, New York, 1956.

JM81. M. Jimbo, T. Miwa, *Monodromy preserving deformation of linear ordinary differential equations with rational coefficients. II.* Phys. D **2** (1981), no. 3, 407-448.

JMK81. M. Jimbo, T. Miwa, K. Ueno, *Monodromy preserving deformation of linear ordinary differential equations with rational coefficients. I. General theory and τ-function.* Phys. D **2** (1981), no. 2, 306-352.

Jos99. N. Joshi, *Asymptotic studies of the Painlevé equations.* In The Painlevé property, 181-227, CRM Ser. Math. Phys., Springer, New York, 1999.

JK001. N. Joshi, A. V. Kitaev, *On Boutroux's tritronquée solutions of the first Painlevé equation.* Stud. Appl. Math. **107** (2001), no. 3, 253-291.

JKT009. N. Joshi, A. V. Kitaev, P. A Treharne, *On the linearization of the first and second Painlevé equations.* J. Phys. A **42** (2009), no. 5, 055208, 18 pp.

JK92. N. Joshi, M. Kruskal, *The Painlevé connection problem: an asymptotic approach. I.* Stud. Appl. Math. **86** (1992), no. 4, 315-376.

JL015. N. Joshi, C. J. Lustri, *Stokes phenomena in discrete Painlevé I.* Proc. R. Soc. Lond. Ser. A **471** (2015), no. 2177.

Kap004. A.A. Kapaev, *Quasi-linear stokes phenomenon for the Painlevé first equation.* J. Phys. A **37** (2004), no. 46, 11149-11167.

KK93. A.A. Kapaev, A. V. Kitaev, *Connection formulae for the first Painlevé transcendent in the complex domain.* Lett. Math. Phys. **27** (1993), no. 4, 243-252.

KT96. T. Kawai, Y. Takei, *WKB analysis of Painlevé transcendents with a large parameter. I.* Adv. Math. **118** (1996), no. 1, 1-33.

KT005. T. Kawai, Y. Takei, *Algebraic analysis of singular perturbation theory.* Translated from the 1998 Japanese original by Goro Kato. Translations of Mathematical Monographs, 227. Iwanami Series in Modern Mathematics. American Mathematical Society, Providence, RI, 2005.

KKNT004. T. Kawai, Y. T. Koike, Y. Nishikawa, Y. Takei, *On the Stokes geometry of higher order Painlevé equations.* Analyse complexe, systèmes dynamiques, sommabilité des séries divergentes et théories galoisiennes. II. Astérisque No. **297** (2004), 117-166.

Lai92. I. Laine, *Nevanlinna theory and complex differential equations.* de Gruyter Studies in Mathematics, 15. Walter de Gruyter & Co., Berlin, 1993.

Lax68. P. Lax *Integrals of nonlinear equations of evolution and solitary waves.* Comm. Pure Appl. Math. **21** (1968), 467-490.

Lod016. M. Loday-Richaud, *Divergent Series, Summability and Resurgence II. Simple and Multiple Summability.* Lecture Notes in Mathematics, **2154**. Springer, Heidelberg, 2016.

Mar014. M. Mariño, *Lectures on non-perturbative effects in large N gauge theories, matrix models and strings.* Fortschr. Phys. **62** (2014), no. 5-6, 455-540.

MSW008. M. Mariño, R. Schiappa, M. Weiss, *Nonperturbative effects and the large-order behavior of matrix models and topological strings.* Commun. Number Theory Phys. **2** (2008), no. 2, 349-419.

MSW009. M. Mariño, R. Schiappa, M. Weiss, *Multi-instantons and multicuts.* J. Math. Phys. **50** (2009), no. 5, 052301, 31 pp.

Mas010-1. D. Masoero, *Poles of intégrale tritronquée and anharmonic oscillators. A WKB approach.* J. Phys. A: Math. Theor. **43**, no. 9, 095201, (2010).

Mas010-2. D. Masoero, *Poles of intégrale tritronquée and anharmonic oscillators. Asymptotic localization from WKB analysis.* Nonlinearity **23** , no. 10, 2501-2507, (2010).

MS016. C. Mitschi, D. Sauzin, *Divergent Series, Summability and Resurgence I. Monodromy and Resurgence.* Lecture Notes in Mathematics, **2153**. Springer, Heidelberg, 2016.

Nis88. K. Nishioka, *A note on the transcendency of Painlevé's first transcendent.* Nagoya Math. J. **109** (1988), 63-67.

Oka80-1. K. Okamoto, *Polynomial Hamiltonians associated with Painlevé equations. I.* Proc. Japan Acad. Ser. A Math. Sci. **56** (1980), no. 6, 264-268.

Oka80-2. K. Okamoto, *Polynomial Hamiltonians associated with Painlevé equations. II. Differential equations satisfied by polynomial Hamiltonians.* Proc. Japan Acad. Ser. A Math. Sci. **56** (1980), no. 8, 367-371.

Pai00. P. Painlevé, *Mémoire sur les équations différentielles dont l'intégrale générale est uniforme.* Bull. Soc. Math. France **28** (1900), 201-261.

Tak000. Y. Takei, *An explicit description of the connection formula for the first Painlevé equation.* In Toward the exact WKB analysis of differential equations, linear or non-linear (Kyoto, 1998), 204, 271-296, Kyoto Univ. Press, Kyoto, 2000.

Chapter 3
Tritruncated Solutions For The First Painlevé Equation

Abstract This chapter is devoted to the construction of the tritruncated solutions for the first Painlevé equation, the existence of which being announced in Sect. 2.6. This example will introduce the reader to common reasonings in resurgence theory. We construct a prepared form associated with the first Painlevé equation (Sec 3.1). This prepared ODE has a unique formal solution from which we deduce the existence of truncated solutions by application of the "'main asymptotic existence theorem". We then study the Borel-Laplace summability property of the formal solution by various methods (Sect. 3.3). One deduces the existence of the tritruncated solutions for the first Painlevé equation, by Borel-Laplace summation (Sect. 3.4).

3.1 Normalization and Formal Series Solution

Throughout this course, $\mathbb{C}[[z^{-1}]]$ stands for the differential algebra of formal power series of the form $\widetilde{g}(z) = \sum_{n\geq 0} a_n z^{-n}$, while $\mathbb{C}((z^{-1}))$ is the space of formal Laurent series. The space of formal Laurent series is a valuation field with the natural valuation

$$\mathrm{val}: \begin{array}{ccc} \mathbb{C}((z^{-1})) & \to & \mathbb{Z}\cup\infty \\ \sum_{n\in\mathbb{Z}} a_n z^{-n} & \mapsto & \mathrm{val}\,\widetilde{w} = \min\{n\in\mathbb{Z}\,/\,a_n\neq 0\}. \end{array} \tag{3.1}$$

3.1.1 Normalization, Prepared Form

We saw in Sect. 2.6 that the first Painlevé equation is equivalent to the following differential equation,

$$v'' + \frac{v'}{z} = -\frac{1}{2} + \frac{4}{25}\frac{v}{z^2} + \frac{1}{2}v^2, \tag{3.2}$$

E. Delabaere, *Divergent Series, Summability and Resurgence III*, Lecture Notes in Mathematics 2155, DOI 10.1007/978-3-319-29000-3_3

under the Boutroux's transformation: $u(x) = \frac{e^{\frac{i\pi}{2}}}{\sqrt{6}} x^{\frac{1}{2}} v(z)$, $z = e^{\frac{5i\pi}{4}} \frac{24^{\frac{5}{4}}}{30} x^{\frac{5}{4}}$.

The variable z is most often called *critical time* [Cos009].

It is worth mentioning that the symmetries detailed in Sect. 2.5 translate into the fact that any solution v of (3.2) is mapped into another solution v_k through the transformation:

$$v_k(z) = e^{i\pi k} v\left(e^{i\pi k/2} z\right), \qquad k = 0, \cdots, 3. \tag{3.3}$$

We look for a formal solution of (2.8) of the form $\widetilde{v}(z) = \sum_{l=0}^{\infty} b_l z^{-l} \in \mathbb{C}[[z^{-1}]]$. When plugging this formal series in (3.2), one gets the necessary conditions: $b_0^2 = 1, b_1 = 0$ and $b_2 = -\frac{4}{25}$. Thanks to the symmetries (3.3), there is no restriction in assuming $b_0 = 1$. Also, it will be convenient in the sequel to make a new transformation,

$$v(z) = 1 - \frac{4}{25}\frac{1}{z^2} + \frac{1}{z^2} w(z), \tag{3.4}$$

which has the virtue of bringing (3.2) into the following differential equation :

$$w'' - \frac{3}{z} w' - w = \frac{392}{625}\frac{1}{z^2} - \frac{4}{z^2} w + \frac{1}{2z^2} w^2. \tag{3.5}$$

Definition 3.1. The differential equation (3.5), which reads

$$P(\partial)w + \frac{1}{z} Q(\partial) w = F(z,w), \quad \text{with} \quad P(\partial) = \partial^2 - 1, Q(\partial) = -3\partial, \partial = \frac{d}{dz} \tag{3.6}$$

and $F(z,w) = \frac{392}{625}\frac{1}{z^2} - \frac{4}{z^2} w + \frac{1}{2z^2} w^2 = f_0(z) + f_1(z) w + f_2(z) w^2$, is called the *prepared form* equation associated with the first Painlevé equation.

Remark 3.1. For general comments on normalization procedures see, e.g. [Cos009] and exercise 3.1. Notice that the prepared form is not uniquely defined.

3.1.2 Formal Series Solution

Substituting the formal series expansion $\sum_{l=0}^{\infty} a_l z^{-l}$ into equation (3.6) and identifying the powers, yields a quadratic recursion relation, namely:

$$\begin{cases} a_0 = a_1 = 0, \quad a_2 = -\frac{392}{625} \\ a_l = l^2 a_{l-2} - \frac{1}{2}\sum_{p=0}^{l-2} a_{(p)} a_{(l-2-p)}, \; l = 3, 4, \cdots \end{cases} \tag{3.7}$$

The following proposition is a simple exercise.

Proposition 3.1. *There exists a unique formal series solution of (3.6) denoted by:*

$$\widetilde{w}(z) = \sum_{l=0}^{\infty} a_l z^{-l} \in \mathbb{C}[[z^{-1}]]. \tag{3.8}$$

Moreover the series $\widetilde{w}$ *is even,* val $\widetilde{w} = 2$ *and the coefficients* a_l *are all real negative.*

Remark 3.2. 1. One infers from (3.7) that the series $\widetilde{w}$ diverges since obviously $|a_{2m}| \geq (m!)^2 |a_2|$ for $m \geq 1$.

2. The differential equation (3.6) can be written as a fixed point problem, $w = \mathsf{N}(w)$, $\mathsf{N}(w) = -F(z,w) - \frac{3}{z}w' + w''$. On can consider the differential operator N as acting on the ring $\mathbb{C}[[z^{-1}]]$, $\mathsf{N} : \mathbb{C}[[z^{-1}]] \to \mathbb{C}[[z^{-1}]]$. When $\mathbb{C}[[z^{-1}]]$ is seen as a complete metric space (for the so-called Krull topology, see the first volume of this book [MS016]), N appears as a contractive map and the formal solution $\widetilde{w}$ given by lemma 3.1 is the unique solution of the fixed point problem. This way of showing the existence of the formal solution $\widetilde{w}$ is also useful for numerical calculations,

$$\widetilde{w}(z) = -\frac{392}{625} z^{-2} - \frac{6272}{625} z^{-4} - \frac{141196832}{390625} z^{-6} + \cdots$$

In this course, all calculations have been produced that way under Maple 12.0 (released: 2008).

3.1.3 Towards Truncated Solutions

3.1.3.1 Notation

We fix notation (essentially common with the first two volumes of this book [MS016, Lod016]) which will be used in this chapter and throughout the course.

Definition 3.2. We denote by $\mathbb{S}^1$ the circle of directions about 0 of half-lines on $\mathbb{C}$. We usually identify $\mathbb{S}^1$ with $\mathbb{R}/2\pi\mathbb{Z}$. Let $I =]\alpha, \beta[\subset \mathbb{S}^1$ be an open arc. Its *length* is denoted and defined by $|I| = \beta - \alpha$.

Definition 3.3. Let $I \subset \mathbb{S}^1$ be an open arc. For $0 \leq r < R \leq \infty$, we denote by $\dot{\mathcal{S}}_r^R(I)$ the domain defined by $\dot{\mathcal{S}}_r^R(I) = \{\zeta = \xi e^{i\theta} \in \mathbb{C} \mid \theta \in I, r < \xi < R\}$. In particular $\dot{\mathcal{S}}_0^R(I)$ (*resp.* $\dot{\mathcal{S}}_r^\infty(I)$) is an open sector with vertex 0 (*resp.* ∞) and aperture I.

One denotes by $\dot{\bar{\mathcal{S}}}_0^R(I)$ (*resp.* $\dot{\bar{\mathcal{S}}}_r^\infty(I)$) the closure of $\dot{\mathcal{S}}_0^R(I)$ (*resp.* $\dot{\mathcal{S}}_r^\infty(I)$) in $\mathbb{C}^\star = \mathbb{C} \setminus \{0\}$.

We use abridged notation $\dot{\mathcal{S}}_0(I)$, $\dot{\bar{\mathcal{S}}}_0(I)$, $\dot{\mathcal{S}}^\infty(I)$ and $\dot{\bar{\mathcal{S}}}^\infty(I)$ for sectors, when R or r is unspecified.

A sector $\overset{\bullet}{\mathcal{S}}_0(I')$ (*resp.* $\overset{\bullet}{\mathcal{S}}^\infty(I)$) is said to be a proper subsector of $\overset{\bullet}{\mathcal{S}}_0(I)$ (*resp.* $\overset{\bullet}{\mathcal{S}}^\infty(I)$) and one denotes $\overset{\bullet}{\mathcal{S}}_0(I') \Subset \overset{\bullet}{\mathcal{S}}_0(I)$ (*resp.* $\overset{\bullet}{\mathcal{S}}^\infty(I') \Subset \overset{\bullet}{\mathcal{S}}^\infty(I)$) if the closure $\bar{\overset{\bullet}{\mathcal{S}}}_0(I')$ (*resp.* $\bar{\overset{\bullet}{\mathcal{S}}}^\infty(I')$) is included in $\overset{\bullet}{\mathcal{S}}_0(I)$ (*resp.* $\overset{\bullet}{\mathcal{S}}^\infty(I)$).

3.1.3.2 Main Asymptotic Existence Theorem

We have previously seen that the ODE (3.6) is formally solved by a unique formal series $\widetilde{w}(z) \in \mathbb{C}[[z^{-1}]]$.

Question 3.1. Can we associate to $\widetilde{w}$ a holomorphic solution whose Poincaré asymptotics[1] are governed by this formal series ?

This question is the matter of the "main asymptotic existence theorem". This theorem is detailed in the second volume of this book [Lod016] for linear ODEs. It can be formulated to nonlinear equations, see [Was65], theorems 12.1 and 14.1, and [RS89] for extension to Gevrey asymptotics.

Theorem 3.1 (Main asymptotic existence theorem M.A.E.T.). *Let $I \subset \mathbb{S}^1$ be an open arc of length $|I| \le \pi/(q+1)$ where q is a nonnegative integer. Let $\mathbf{F}(z,\mathbf{w})$ be a m-dimensional vector function subject to the following conditions:*

1. *$\mathbf{F}(z,\mathbf{w})$ is holomorphic in $(z,\mathbf{w})$ on the domain of $\overset{\bullet}{\mathcal{S}}^\infty(I) \times B(0,r)$ with $B(0,r) = \{\mathbf{w} \in \mathbb{C}^m, \|\mathbf{w}\| \le r\}$ for some $r > 0$;*
2. *$\mathbf{F}(z,\mathbf{w})$ admits an asymptotic expansion in powers of z^{-1} at infinity in $\overset{\bullet}{\mathcal{S}}^\infty(I)$, uniformaly valid in $\mathbf{w} \in B(0,r)$;*
3. *the equation $z^{-q}\mathbf{w}' = \mathbf{F}(z,\mathbf{w})$ is formally satisfied by a formal power series solution $\widetilde{\mathbf{w}}(z) \in (\mathbb{C}[[z^{-1}]])^m$;*
4. *if $F_j(z,\mathbf{w})$ denotes the components of $\mathbf{F}(z,\mathbf{w})$, the Jacobian matrix*
$$\lim_{z\to\infty, z\in\overset{\bullet}{\mathcal{S}}^\infty(I)} \begin{pmatrix} \frac{\partial F_1}{\partial w_1}(z,0) & \cdots & \frac{\partial F_1}{\partial w_m}(z,0) \\ \cdots & \cdots & \cdots \\ \frac{\partial F_m}{\partial w_1}(z,0) & \cdots & \frac{\partial F_m}{\partial w_m}(z,0) \end{pmatrix}$$
has non zero eigenvalues.

Then there exists a solution $\mathbf{w}$ of the equation $z^{-q}\mathbf{w}' = \mathbf{F}(z,\mathbf{w})$, holomorphic in a domain of the form $\overset{\bullet}{\mathcal{S}}^\infty(I)$, whose asymptotics at infinity in every proper subsector of $\overset{\bullet}{\mathcal{S}}^\infty(I)$ is given by the formal solution $\widetilde{\mathbf{w}}$.

3.1.3.3 Application

Let us transform (3.6) into a one order ODE of dimension 2. We introduce $\mathbf{w} = \begin{pmatrix} w_1 \\ w_2 \end{pmatrix} = \begin{pmatrix} w \\ w' \end{pmatrix}$ and we obtain the companion system:

[1] The reader is referred to the first two volumes of this book [Lod016, MS016] for details on asymptotic expansions.

$$\partial \mathbf{w} = \begin{pmatrix} 0 & 1 \\ 1 & \frac{3}{z} \end{pmatrix} \mathbf{w} + \begin{pmatrix} 0 \\ F(z, w_1) \end{pmatrix} = \begin{pmatrix} F_1(z, \mathbf{w}) \\ F_2(z, \mathbf{w}) \end{pmatrix} = \mathbf{F}(z, \mathbf{w}) \in (\mathbb{C}[z^{-1}, \mathbf{w}])^2. \quad (3.9)$$

We fix an open arc $I \subset \mathbb{S}^1$, arbitrary but of length $|I| \leq \pi$. We also consider a domain of the form $\dot{\mathcal{S}}^\infty(I)$ and we make the following observations:

1. $\mathbf{F}(z, \mathbf{w})$ is polynomial with respect to $\mathbf{w}$, with coefficients belonging to $\mathbb{C}[z^{-1}]$. Therefore $\mathbf{F}(z, \mathbf{w})$ is holomorphic in $(z, \mathbf{w})$ on the domain $\dot{\mathcal{S}}^\infty(I) \times B(0, r)$ with $B(0, r) = \{\mathbf{w} \in \mathbb{C}^2, \|\mathbf{w}\| \leq r\}$ for some $r > 0$;
2. again because $\mathbf{F}(z, \mathbf{w}) \in (\mathbb{C}[z^{-1}, \mathbf{w}])^2$, $\mathbf{F}(z, \mathbf{w})$ admits an asymptotic expansion in powers of z^{-1} at infinity in $\dot{\mathcal{S}}^\infty(I)$, uniformaly valid in $\mathbf{w} \in B(0, r)$;
3. the equation (3.9) is formally satisfied by a formal power series solution $\widetilde{\mathbf{w}}(z) = \begin{pmatrix} \widetilde{w} \\ \widetilde{w}' \end{pmatrix} \in (\mathbb{C}[[z^{-1}]])^2$;
4. the Jacobian matrix $\begin{pmatrix} 0 & 1 \\ 1 & 0 \end{pmatrix} = \begin{pmatrix} \frac{\partial F_1}{\partial w_1}(\infty, 0) & \frac{\partial F_1}{\partial w_2}(\infty, 0) \\ \frac{\partial F_2}{\partial w_1}(\infty, 0) & \frac{\partial F_2}{\partial w_2}(\infty, 0) \end{pmatrix}$ has non zero eigenvalues $\mu_1 = -1$ and $\mu_2 = 1$.

These properties allow to apply the (M.A.E.T.) and this shows the following proposition (see also [JK001]):

Proposition 3.2. *For any open arc $I \subset \mathbb{S}^1$ of length $|I| \leq \pi$, there exists a solution w of (3.6), holomorphic in a domain of the form $\dot{\mathcal{S}}^\infty(I)$, whose Poincaré asymptotics at infinity in every proper subsector of $\dot{\mathcal{S}}^\infty(I)$, is given by the formal solution $\widetilde{w}$ given by proposition 3.1.*

Proposition 3.2 thus describes the minimal opening of sectors on which holomorphic solutions w asymptotic to $\widetilde{w}$ exist. Through the transformations (3.4), (2.6) and (2.7), these solutions w corresponds to holomorphic functions u solutions of the first Painlevé equation, defined on open sectors of aperture $4\pi/5$: we thus get a first insight towards the truncated solutions for the first Painlevé equation (theorem 2.4).

As a matter of fact, from the above informations and the property for any solution of the first Painlevé equation to be a meromorphic function, one can even show the existence of tritruncated solutions [JK001]. However, to get more precise informations, we decide in what follows to turn to the question of the Borel-Laplace summability of $\widetilde{w}$.

3.2 A Reminder

We assume that the reader has a previous acquaintance with Borel-Laplace summation and a little background with resurgence theory, amply elaborated in the first two volumes of this book [MS016, Lod016] to which we refer. For the convenience of the reader, we offer a brief reminder of definitions and results used in this chapter.

Formal Borel transform and convolution product

Definition 3.4. The *formal Borel transform* $\mathscr{B}(z \to \zeta)$ is the linear isomorphism $\mathscr{B} : \mathbb{C}[[z^{-1}]] \to \mathbb{C}\delta \oplus \mathbb{C}[[\zeta]]$ defined by

$$\widetilde{g}(z) = \sum_{l=0}^{\infty} b_l z^{-l} \mapsto b_0\delta + \widehat{g}(\zeta), \qquad \widehat{g}(\zeta) = \sum_{l=1}^{\infty} b_l \frac{\zeta^{l-1}}{\Gamma(l)}.$$

The formal series $\widehat{g}$ is the *minor* of $\widetilde{g}$. The inverse map $\mathscr{L} = \mathscr{B}^{-1}$ is the *formal Laplace transform*.

Definition 3.5. Let $b_0\delta + \widehat{g}(\zeta)$ and $c_0\delta + \widehat{h}(\zeta)$ be two elements of $\mathbb{C}\delta \oplus \mathbb{C}[[\zeta]]$. Their *convolution product* $(b_0\delta + \widehat{g}) * (c_0\delta + \widehat{h})$ is defined by

$$(b_0\delta + \widehat{g}) * (c_0\delta + \widehat{h}) = \mathscr{B}(\widetilde{g}\widetilde{h}), \quad \text{where} \quad \widetilde{g} = \mathscr{L}(b_0\delta + \widehat{g}), \widetilde{h} = \mathscr{L}(c_0\delta + \widehat{h}).$$

When $\widehat{g}(\zeta) = \sum_{n\geq 0} b_n \zeta^n$ and $\widehat{h}(\zeta) = \sum_{n\geq 0} c_n \zeta^n$ are two formal series, their convolution product $\widehat{g} * \widehat{h}$ is given by the Hurwitz product, $\widehat{g} * \widehat{h}(\zeta) = \sum_{k\geq 1} d_k \zeta^k$ with

$$d_k = \sum_{n+m+1=k} \frac{n!m!}{(n+m+1)!} b_n c_m.$$

Proposition 3.3. *The linear map* $\widehat{\partial} : b_0\delta + \widehat{g} \mapsto -\zeta\widehat{g}$ *provides a derivation of* $\mathbb{C}\delta \oplus \mathbb{C}[[\zeta]]$ *and* $\mathscr{B} : \left(\mathbb{C}[[z^{-1}]], \partial\right) \to \left(\mathbb{C}\delta \oplus \mathbb{C}[[\zeta]], \widehat{\partial}\right)$ *is an isomorphism of differential algebras.*

Gevrey series of order 1

Definition 3.6. A formal series $\widetilde{g}(z) = \sum_{n\geq 0} a_n z^{-n} \in \mathbb{C}[[z^{-1}]]$ is *1-Gevrey* when there exist constants $C > 0$, $A > 0$ so that $|a_n| \leq C(n!)A^n$ for all n. The space of 1-Gevrey series is denoted by $\mathbb{C}[[z^{-1}]]_1$.

We recall from [Lod016, MS016] that the space $\mathbb{C}[[z^{-1}]]_1$ of 1-Gevrey series is a differential algebra.

Notice that a formal series $\widetilde{g}$ is 1-Gevrey if and only if its minor $\widehat{g}$ is a convergent power series, thus defines a germ of holomorphic functions (still denoted by $\widehat{g}$). More precisely:

Proposition 3.4. *The restricted linear map* $\mathscr{B}| : \mathbb{C}[[z^{-1}]]_1 \to \mathbb{C}\delta \oplus \mathscr{O}_0$ *is an isomorphism of differential algebras. Also, for any two germs of holomorphic functions* $\widehat{g}, \widehat{h} \in \mathscr{O}_0$, *their convolution product* $\widehat{g} * \widehat{h} \in \mathscr{O}_0$ *has the following integral representation:* $\widehat{g} * \widehat{h}(\zeta) = \int_0^{\zeta} \widehat{g}(\eta)\widehat{h}(\zeta - \eta)d\eta$.

A flavor of resurgence

Definition 3.7. Let Ω be a non-empty closed discrete subset of $\mathbb{C}$ and $\widehat{\varphi} \in \mathscr{O}_0$ be a germ of holomorphic functions at 0. This germ is said to be *Ω-continuable* if there exists $r > 0$ such that $D^\star(0,r) \cap \Omega = \emptyset$ and $\widehat{\varphi}$ can be represented by a function holomorphic on $D(0,r)$ which can be analytically continued along any path of $\mathbb{C} \setminus \Omega$ originating from any point of $D^\star(0,r)$.
The space of all Ω-continuable germs is denoted by $\hat{\mathscr{R}}_\Omega$. The space $\mathbb{C}\delta \oplus \hat{\mathscr{R}}_\Omega$ is called the space of *Ω-resurgent functions*. The space of *Ω-resurgent formal series* is denoted by $\tilde{\mathscr{R}}_\Omega$ and defined by $\tilde{\mathscr{R}}_\Omega = \mathscr{L}\left(\mathbb{C}\delta \oplus \hat{\mathscr{R}}_\Omega\right)$.

Theorem 3.2. *Let Ω_1, Ω_2 be non-empty closed discrete subsets of $\mathbb{C}$. Let $\Omega \subset \mathbb{C}$ be the subset defined by $\Omega = \Omega_1 \cup \Omega_2 \cup (\Omega_1 + \Omega_2)$ where*

$$\Omega_1 + \Omega_2 = \{\omega_1 + \omega_2 \mid \omega_1 \in \Omega_1, \omega_2 \in \Omega_1\}.$$

If Ω is closed and discrete, then $\widehat{\varphi}_1 \in \hat{\mathscr{R}}_{\Omega_1}$ and $\widehat{\varphi}_2 \in \hat{\mathscr{R}}_{\Omega_2}$ imply $\widehat{\varphi}_1 * \widehat{\varphi}_2 \in \hat{\mathscr{R}}_\Omega$.

In particular, the space $\mathbb{C}\delta \oplus \hat{\mathscr{R}}_\mathbb{Z}$ of $\mathbb{Z}$-resurgent functions is stable under convolution product, thus is an algebra with unit δ.

Borel-Laplace summability

Definition 3.8. A formal series $\widetilde{g}(z) = \displaystyle\sum_{n \geq 0} \frac{b_n}{z^n} \in \mathbb{C}[[z^{-1}]]$ is said to be *Borel-Laplace summable* in direction $\theta \in \mathbb{S}^1$ if the following conditions are satisfied:

- the series $\widetilde{g}$ is 1-Gevrey or, equivalently, its minor $\widehat{g}$ is a convergent series whose sum defines a holomorphic function (still denoted by $\widehat{g}$) near the origin ;
- $\widehat{g}$ can be analytically continued to an open sector of the form $\overset{\bullet}{\Delta}{}_0^\infty(I)$ where $I \subset \mathbb{S}^1$ is an open neighbourdhood of θ, with exponential growth of order 1 at infinity.

Under the above conditions, the *Borel-Laplace sum* of $\widetilde{g}$ in direction θ is denoted by $\mathscr{S}^\theta \widetilde{g}$ and defined by $\mathscr{S}^\theta \widetilde{g}(z) = \mathscr{L}^\theta \circ \mathscr{B}\widetilde{g}(z)$ where $\mathscr{L}^\theta$ stands for the *Laplace transform* in direction θ, $\mathscr{L}^\theta(b_0\delta + \widehat{g})(z) = b_0 + \displaystyle\int_0^{\infty e^{i\theta}} e^{-z\zeta}\widehat{g}(\zeta)d\zeta$.

In addition to this definition, we recall that the Borel-Laplace sum $\mathscr{S}^\theta \widetilde{w}$ is holomorphic on a half-plane where its asymptotic behavior is governed by the formal series $\widetilde{g}$. This will be made more precise in a moment.

3.3 Formal Series Solution and Borel-Laplace Summability

We go back to the formal series $\widetilde{w}$ given by proposition 3.1. Since val $\widetilde{w} > 0$, the formal Borel transform of $\widetilde{w}$ just reduces to its minor $\widehat{w}$. Also, $\widetilde{w}(z)$ is the unique solution in $\mathbb{C}[[z^{-1}]]$ of the differential equation (3.6). One easily infers the following result from the general properties of the formal Borel transform.

Proposition 3.5. *The formal series $\widetilde{w}(z) \in \mathbb{C}[[z^{-1}]]$ is solution of (3.6) if and only if its minor $\widehat{w}(\zeta) \in \mathbb{C}[[\zeta]]$ is solution of the following convolution equation:*

$$
\begin{aligned}
&P(\widehat{\partial})\widehat{w} + 1 * \left[Q(\widehat{\partial})\widehat{w}\right] = \widehat{f}_0 + \widehat{f}_1 * \widehat{w} + \widehat{f}_2 * \widehat{w} * \widehat{w}, \\
&P(\widehat{\partial}) = \widehat{\partial}^2 - 1, \quad Q(\widehat{\partial}) = -3\widehat{\partial}, \\
&\widehat{f}_0(\zeta) = \frac{392}{625}\zeta, \quad \widehat{f}_1(\zeta) = -4\zeta, \quad \widehat{f}_2(\zeta) = \frac{1}{2}\zeta.
\end{aligned}
\tag{3.10}
$$

We will see in a moment that $\widetilde{w}$ is 1-Gevrey and even Borel-Laplace summable. In the rest of this chapter, we analyse this Borel-Laplace summability and we offer various approaches.

3.3.1 Formal Series Solution and Borel-Laplace Summability: a Perturbative Approach

We start with a perturbative approach which has the advantage of giving a first insight into the resurgent structure. In practice, we consider (3.10) as a perturbation of the equation $P(\widehat{\partial})\widehat{w} = \widehat{f}_0$ which is quite easy to solve:

- either formally since the map $P(\widehat{\partial}) : \widehat{g} \in \mathbb{C}[[\zeta]] \mapsto (\zeta^2 - 1)\widehat{g} \in \mathbb{C}[[\zeta]]$ is invertible;
- or analytically, in a space of analytic functions, say $\mathscr{O}_0$, because $P(\widehat{\partial}) : \widehat{g} \in \mathscr{O}_0 \mapsto (\zeta^2 - 1)\widehat{g} \in \mathscr{O}_0$ is once again invertible.

To keep on, it is convenient to transform equation (3.10) into the following one parameter family of convolution equations,

$$
P(\widehat{\partial})\widehat{h} = \widehat{f}_0 + \varepsilon\Big(- 1 * \left[Q(\widehat{\partial})\widehat{h}\right] + \widehat{f}_1 * \widehat{h} + \widehat{f}_2 * \widehat{h} * \widehat{h}\Big), \tag{3.11}
$$

and to look for a solution under the form

$$
\widehat{h}(\zeta, \varepsilon) = \sum_{l \geq 0} \widehat{h}_l(\zeta)\varepsilon^l. \tag{3.12}
$$

When plugging (3.12) into (3.11) and identifying the same powers in ε, one obtains a recursive system of convolution equations, namely:

$$
\begin{cases}
P(\widehat{\partial})\widehat{h}_0 = \widehat{f}_0, \\
P(\widehat{\partial})\widehat{h}_1 = -1 * \left[Q(\widehat{\partial})\widehat{h}_0\right] + \widehat{f}_1 * \widehat{h}_0 + \widehat{f}_2 * \widehat{h}_0 * \widehat{h}_0, \\
P(\widehat{\partial})\widehat{h}_n = -1 * \left[Q(\widehat{\partial})\widehat{h}_{n-1}\right] + \widehat{f}_1 * \widehat{h}_{n-1} + \displaystyle\sum_{n_1 + n_2 = n-1} \widehat{f}_2 * \widehat{h}_{n_1} * \widehat{h}_{n_2}, \quad n \geq 1.
\end{cases}
\tag{3.13}
$$

3.3.1.1 Formal Analysis

Lemma 3.1. *The system (3.13) provides a uniquely determined sequence* $(\widehat{h}_l)_{l\geq 0}$ *of formal series. Furthermore* $\widehat{h}_l(\zeta) \in \zeta^{2l+1}\mathbb{C}[[\zeta]]$ *for every* $l \geq 0$.

Proof. Use the fact that the map $P(\widehat{\partial}) : \mathbb{C}[[\zeta]] \to \mathbb{C}[[\zeta]]$ is invertible and the general properties of the convolution product. □

The above lemma has the following consequence:

Proposition 3.6. *The series* $\sum_{l\geq 0} \widehat{h}_l(\zeta)$ *is well defined in* $\mathbb{C}[[\zeta]]$ *and is formally convergent to the unique formal solution* $\widehat{w}(\zeta) \in \mathbb{C}[[\zeta]]$ *of the convolution equation (3.10).*

We mention that proposition 3.6 has a counterpart by formal Laplace transform $\mathscr{L}(\zeta \to z)$. Introducing $\widetilde{h}_l = \mathscr{L}\widehat{h}_l$, one gets from lemma 3.1 that the sequence $(\widetilde{h}_l)_{l\geq 0}$ solves in $\mathbb{C}[[z^{-1}]]$ the following recursive system of linear nonhomogeneous ODEs:

$$\begin{cases} P(\partial)\widetilde{h}_0 = f_0(z) \\ P(\partial)\widetilde{h}_1 = -\dfrac{1}{z}Q(\partial)\widetilde{h}_0 + f_1(z)\widetilde{h}_0 + f_2(z)\widetilde{h}_0^2 \\ P(\partial)\widetilde{h}_n = -\dfrac{1}{z}Q(\partial)\widetilde{h}_{n-1} + f_1(z)\widetilde{h}_{n-1} + f_2(z) \displaystyle\sum_{n_1+n_2=n-1} \widetilde{h}_{n_1}\widetilde{h}_{n_2}, \quad n \geq 1. \end{cases} \tag{3.14}$$

From lemma 3.1 again, one deduces that $\widetilde{h}_l \in z^{-2l-2}\mathbb{C}[[z^{-1}]]$ for every $l \geq 0$, thus:

Proposition 3.7. *The series* $\sum_{l\geq 0} \widetilde{h}_l(z)$ *is well defined in* $\mathbb{C}[[z^{-1}]]$ *and is formally convergent to the unique formal solution* $\widetilde{w}(z) \in \mathbb{C}[[z^{-1}]]$ *of the differential equation (3.6).*

3.3.1.2 Analytic Properties and a Flavor of Resurgence

Instead of working in the space of formal series, one can rather work in a space of analytic functions. The next proposition uses definition 3.7.

Proposition 3.8. *For every* $l \in \mathbb{N}$, *the formal series* $\widehat{h}_l$ *given by (3.13) defines a germ (still denoted by* $\widehat{h}_l$*) of holomorphic functions at* 0, *which can be represented by a function holomorphic on the open disc* $D(0,1)$. *Moreover,* $\widehat{h}_l$ *belongs to the space* $\widehat{\mathscr{R}}_{\Omega_l}$ *of* Ω_l*-resurgent functions, where* $\Omega_l = \{0, \pm 1, \cdots, \pm l, \pm(l+1)\}$. *As a consequence, the germ* $\widehat{h}_l$ *is a* $\mathbb{Z}$*-resurgent function.*

Proof. The proposition is easily shown by induction from (3.13), theorem 3.2 and the following remark : for every $l \in \mathbb{N}$, $\hat{\mathscr{R}}_{\Omega_l} \subset \hat{\mathscr{R}}_{\Omega_{l+1}}$ and the linear map $P(\widehat{\partial}) : \widehat{g} \in \hat{\mathscr{R}}_{\Omega_l} \mapsto (\zeta^2 - 1)\widehat{g} \in \hat{\mathscr{R}}_{\Omega_l}$ is invertible. □

3.3.1.3 Further Preparations

We have previously seen (proposition 3.6) that the minor $\widehat{w}$ of the formal series $\widetilde{w}$ solution of the prepared form equation (3.6), can be written as $\widehat{w}(\zeta) = \sum_{l \geq 0} \widehat{h}_l(\zeta)$ in the space $\mathbb{C}[[\zeta]]$, where the sequence $(\widehat{h}_l)_{l \geq 0}$ solves the recursive system of equations (3.13). To show the Borel-Laplace summability of $\widetilde{w}$, it is thus enough to check the following properties:

- the series of functions $\sum_{l \geq 0} \widehat{h}_l(\zeta)$ converges to a holomorphic function near the origin and can be analytically continued in a convenient sector;
- this function has at most exponential growth of order 1 at infinity in this sector.

We also know by proposition 3.8 that each $\widehat{h}_l(\zeta)$ is a $\mathbb{Z}$-resurgent function. This motivates the following definition, with the notation : $D(a,r)$ is the open disc centred in a with radius r and $\overline{D}(a,r)$ is its closure.

Definition 3.9. One sets $\mathscr{D}_\rho^{(0)} = \bigcup_{\lambda = \pm 1} D(\lambda, \rho)$ for any $\rho \in]0,1[$. We denote by $\overset{\bullet}{\mathscr{R}}{}_\rho^{(0)}$ the star-shaped domain defined by:

$$\overset{\bullet}{\mathscr{R}}{}_\rho^{(0)} = \mathbb{C} \setminus \{ t\zeta \mid t \in [1, +\infty[, \zeta \in \overline{D}(\pm 1, \rho) \} \subset \mathbb{C} \setminus \overline{\mathscr{D}}{}_\rho^{(0)},$$

and $\overset{\bullet}{\mathscr{R}}{}^{(0)} = \bigcup_{0<\rho<1} \overset{\bullet}{\mathscr{R}}{}_\rho^{(0)} = \mathbb{C} \setminus \{\pm[1, +\infty[\}$. (See Fig. 3.1).

Definition 3.10. Let $f(\zeta) = \sum_{l \geq 0} a_l \zeta^l$ be an analytic function on the open disc $D(0,r)$. One denote by $|f|$ the function defines by $|f|(\xi) = \sum_{l \geq 0} |a_l| \xi^l$.

Notice that $|f|$ is also analytic on $D(0,r)$.

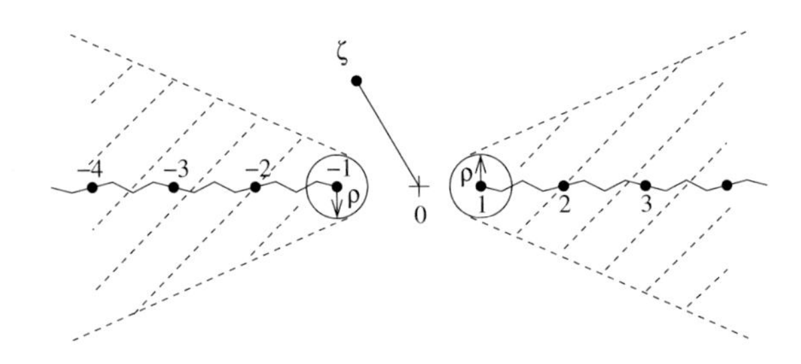

Fig. 3.1 The domain $\overset{\bullet}{\mathscr{R}}{}_\rho^{(0)}$.

Lemma 3.2. *Let be $\rho \in]0,1[$. There exists a constant $M_{\rho,(0)} > 0$ such that for every polynomial $q \in \mathbb{C}[\zeta]$ of degree ≤ 1 and every $\zeta \in \mathbb{C} \setminus \mathscr{D}_\rho^{(0)}$, one has $\left|\frac{q(\zeta)}{P(-\zeta)}\right| \leq M_{\rho,(0)} |q|(1)$. Moreover, on can choose $M_{\rho,(0)} = \frac{1}{\rho}$.*

Proof. By definition of $\mathscr{D}_\rho^{(0)}$, $\frac{1}{|\zeta \pm 1|} \leq \frac{1}{\rho}$ and $\left|\frac{\zeta}{\zeta \pm 1}\right| \leq 1 + \frac{1}{\rho}$ for every $\zeta \in \mathbb{C} \setminus \mathscr{D}_\rho^{(0)}$. Therefore, $\left|\frac{\zeta^p}{P(-\zeta)}\right| \leq \frac{1}{\rho^{2-p}}\left(1+\frac{1}{\rho}\right)^p \leq \frac{1}{\rho^2}(\rho+1)^p \leq \frac{2^p}{\rho^2}$ for $p = 0,1,2$. This means that one can choose $M_{\rho,(0)} = \frac{2}{\rho^2}$ in the lemma. It is possible to be more precise. Suppose for instance that $\Re(\zeta) \geq 0$. Then $|\zeta+1| \geq \max\{1,|\zeta|\}$, thus $\frac{\max\{1,|\zeta|\}}{|P(-\zeta)|} \leq \frac{1}{\rho}$. In a nutshell, one can choose $M_{\rho,(0)} = \frac{1}{\rho}$ in the lemma. □

As a rule, we will combined lemma 3.2 with the following lemma whose proof is left as an exercise (see the first two volumes of this book[MS016, Lod016]):

Lemma 3.3. *Let U be a domain star-shaped from 0. Suppose that $\widehat{f}$ and $\widehat{g}$ are two holomorphic functions on U and satisfy the conditions: for every $\zeta \in U$, $|\widehat{f}(\zeta)| \leq F(|\zeta|)$ and $|\widehat{g}(\zeta)| \leq G(|\zeta|)$ with F, G positive continuous functions on $\mathbb{R}^+$. Then $\widehat{f} * \widehat{g}$ is holomorphic on U and for every $\zeta \in U$, $|\widehat{f} * \widehat{g}(\zeta)| \leq F * G(|\zeta|)$ and $\left|(\zeta\widehat{f}) * \widehat{g}(\zeta)\right| \leq |\zeta|\Big(F * G(|\zeta|)\Big)$.*

3.3.1.4 Majorant Functions

We have in mind to show that the series of functions $\sum_{l\geq 0} \widehat{h}_l(\zeta)$, discussed in propositions 3.6 and 3.8, is uniformaly convergent on any compact subset of $\overset{\bullet}{\mathscr{R}}{}^{(0)}$. We will use majorant functions which we now define.

Definition of the majorant functions We consider, for any $\rho \in]0,1[$, the sequence of functions $(\widehat{H}_l)_{l\geq 0}$ recursively defined by:

$$\begin{cases} \frac{1}{M_{\rho,(0)}}\widehat{H}_0 = |\widehat{f}_0|(\xi), \\ \frac{1}{M_{\rho,(0)}}\widehat{H}_1 = \big(3+|\widehat{f}_1|\big) * \widehat{H}_0 + |\widehat{f}_2| * \widehat{H}_0 * \widehat{H}_0, \\ \frac{1}{M_{\rho,(0)}}\widehat{H}_n = \big(3+|\widehat{f}_1|\big) * \widehat{H}_{n-1} + \sum_{n_1+n_2=n-1} |\widehat{f}_2| * \widehat{H}_{n_1} * \widehat{H}_{n_2}, \quad n \geq 1. \end{cases} \tag{3.15}$$

where $M_{\rho,(0)}$ is given by lemma 3.2 and $|\widehat{f_0}|(\xi) = \frac{392}{625}\xi$, $|\widehat{f_1}|(\xi) = 4\xi$, $|\widehat{f_2}|(\xi) = \frac{1}{2}\xi$. (Compare (3.15) with (3.13).) We claim that for every $l \in \mathbb{N}$, $\widehat{H}_l$ is a majorant function for $\widehat{h}_l$. Precisely:

Lemma 3.4. *For every $\rho \in]0,1[$ and every $l \in \mathbb{N}$, the following properties are satisfied: $\widehat{H}_l(\xi)$ is a polynomial which belongs to $\xi^{l+1}\mathbb{C}[\xi]$; furthermore,*

$$\text{for every } \zeta \in \overset{\bullet}{\mathscr{R}}_\rho^{(0)},\ \left|\widehat{h}_l(\zeta)\right| \leq \widehat{H}_l(\xi) \quad \text{with} \quad \xi = |\zeta|, \tag{3.16}$$

where $(\widehat{h}_l)_{l\geq 0}$ is defined by (3.13).

Proof. The fact that $\widehat{H}_l(\xi) \in \xi^{l+1}\mathbb{C}[\xi]$ is proved by induction from (3.15) and the properties of the convolution product. By (3.13) and lemma 3.2, for every $\zeta \in \overset{\bullet}{\mathscr{R}}_\rho^{(0)}$, $\left|\widehat{h}_0(\zeta)\right| \leq \left|\frac{1}{P(-\zeta)}\right| \left|\widehat{f_0}(\zeta)\right| \leq M_{\rho,(0)}|\widehat{f_0}|(\xi)$ with $\xi = |\zeta|$, so that (3.16) is true for $l = 0$. We now assume that (3.16) is true for $l = 0, \cdots, (n-1)$, for some $n \in \mathbb{N}^\star$. By lemma 3.3 and the induction hypothesis, for every $\zeta \in \overset{\bullet}{\mathscr{R}}_\rho^{(0)}$,

$$\left|\frac{1}{P(-\zeta)}\right| \cdot \left|1 * \left[Q(\widehat{\partial})\widehat{h}_{n-1}\right](\zeta)\right| \leq \left|\frac{1}{P(-\zeta)}\right| |Q|(|\zeta|)\left(1 * \widehat{H}_{n-1}(|\zeta|)\right),$$

where $|Q|(\xi) = 3\xi$. Therefore, by lemma 3.2,

$$\left|\frac{1}{P(-\zeta)}\right| \cdot \left|1 * \left[Q(\widehat{\partial})\widehat{h}_{n-1}\right](\zeta)\right| \leq M_{\rho,(0)}\, |Q|(1)\left(1 * \widehat{H}_{n-1}(\xi)\right)$$

with $\xi = |\zeta|$. More generally, for similar reasons,

$$\frac{1}{M_{\rho,(0)}}|\widehat{h}_n(\zeta)| \leq \left(3 * \widehat{H}_{n-1}(\xi)\right) + |\widehat{f_1}| * \widehat{H}_{n-1}(\xi) + \sum_{n_1+n_2=n-1} |\widehat{f_2}| * \widehat{H}_{n_1} * \widehat{H}_{n_2}(\xi).$$

Thus, for every $\zeta \in \overset{\bullet}{\mathscr{R}}_\rho^{(0)}$, $\left|\widehat{h}_n(\zeta)\right| \leq \widehat{H}_n(\xi)$. This ends the proof. □

Upper bounds for the majorant functions Before keeping on studying the above majorant functions, we state a property which will be useful in the sequel. We first recall some notation.

Definition 3.11. Let $U \subset \mathbb{C}$ be an open set. We denote by $\mathscr{O}(\overline{U})$ the space of functions holomorphic in U and continuous on the closure $\overline{U}$.
For any real positive number $R_0 > 0$, we set $D(\infty, R_0) = \left\{z \in \mathbb{C},\ |z| > \frac{1}{R_0}\right\}$.

Lemma 3.5. *Let be $R_0 > 0$. We suppose $f \in \mathscr{O}\left(\overline{D}(\infty, R_0)\right)$ with $f(z) = O(z^{-m})$ at infinity for a certain $m \in \mathbb{N}$, and let be $M = \sup_{z \in D(\infty,R_0)} |f(z)|$. Then the formal Borel transform $\mathscr{B}f = f_0\delta + \widehat{f}$ of f satisfies the following properties:*

1. $\widehat{f} \in \mathscr{O}(\mathbb{C})$ *and* $|f_0| \leq \dfrac{M}{R_0}$.

2. for every $\zeta \in \mathbb{C}$, $|\widehat{f}(\zeta)| \leq |\widehat{f}|(\xi) \leq \dfrac{M}{R_0} \mathrm{e}^{\frac{\xi}{R_0}}$ *with* $\xi = |\zeta|$ *and, when* $m \geq 2$,

$$|\widehat{f}(\zeta)| \leq \frac{M}{R_0^m} \frac{\xi^{m-2}}{(m-2)!} * \mathrm{e}^{\frac{\xi}{R_0}}, \qquad \xi = |\zeta|.$$

Proof. The Taylor series expansion of f, $\sum_{k \geq m} f_k z^{-k} = z^{-(m-1)} \sum_{l \geq 1} f_{m+l-1} z^{-l}$, converges to f in $D(\infty, R_0)$. By the Cauchy inequalities, $|f_k| \leq \dfrac{M}{R_0^k}$ for any $k \in \mathbb{N}$. The formal Borel transform of f reads $\mathscr{B}f = f_0 \delta + \widehat{f}$ with :

1. $\widehat{f}(\zeta) = \sum_{l \geq 1} f_l \dfrac{\zeta^{l-1}}{(l-1)!}$ as rule,
2. $\widehat{f}(\zeta) = \dfrac{\zeta^{m-2}}{(m-2)!} * \left(\sum_{l \geq 1} f_{m+l-1} \dfrac{\zeta^{l-1}}{(l-1)!} \right)$ when $m \geq 2$.

Also, for every $\zeta \in \mathbb{C}$, $\sum_{l \geq 1} |f_{m+l-1}| \dfrac{|\zeta|^{l-1}}{(l-1)!} \leq \sum_{l \geq 1} \dfrac{M}{R_0^{m+l-1}} \dfrac{\xi^{l-1}}{(l-1)!} \leq \dfrac{M}{R_0^m} \mathrm{e}^{\frac{\xi}{R_0}}$ with $\xi = |\zeta|$. This ensures the uniform convergence on any compact set of $\mathbb{C}$, thus $\widehat{f} \in \mathscr{O}(\mathbb{C})$, and provides the upper bounds. □

We return to the majorant functions defined by (3.15).

Lemma 3.6. *For every* $l \in \mathbb{N}$, *the majorant function* $\widehat{H}_l(\xi)$ *is the formal Borel transform of* $\widetilde{H}_l(z)$ *which has the following properties:* $\widetilde{H}_l(z)$ *belongs to* $\mathbb{C}[z^{-1}]$ *and, for every* $\rho \in]0,1[$, $\widetilde{H}_l(z)$ *is bounded on the domain* $|z| > \frac{8}{\rho}$, *precisely* $\sup_{|z| > \frac{8}{\rho}} |\widetilde{H}_l(z)| \leq \dfrac{1}{2^l}$.

Proof. We introduce the generating function: $\widehat{H}(\xi) = \sum_{l=0}^{\infty} \widehat{H}_l(\xi) \varepsilon^l \in \mathbb{C}[\xi][[\varepsilon]]$. From (3.15), we observe that this generating function formally solves the convolution equation

$$\frac{1}{M_{\rho,(0)}} \widehat{H} = |\widehat{f}_0| + \varepsilon \Big[\big(3 + |\widehat{f}_1|\big) * \widehat{H} + |\widehat{f}_2| * \widehat{H} * \widehat{H} \Big]. \tag{3.17}$$

Therefore, $\widehat{H}$ can be seen as the formal Borel transform of the solution $\widetilde{H}(z, \varepsilon) = \sum_{l=0}^{\infty} \widetilde{H}_l(z) \varepsilon^l \in \mathbb{C}[z^{-1}][[\varepsilon]]$ of the following second order algebraic equation:

$$\begin{gathered} \frac{1}{M_{\rho,(0)}} \widetilde{H} = |f_0|(z) + \varepsilon \Big[\big(\frac{3}{z} + |f_1|\big) \widetilde{H} + |f_2| \widetilde{H}^2 \Big] \\ \text{with } |f_0|(z) = \frac{392}{625} \frac{1}{z^2}, \quad |f_1|(z) = \frac{4}{z^2}, \quad |f_2|(z) = \frac{1}{2z^2}. \end{gathered} \tag{3.18}$$

This equation has two branch solutions and one of them is asymptotic to the equation $\frac{1}{M_{\rho,(0)}}\widetilde{H} = |f_0|$ when ε goes to zero. We are interested in that solution. Instead of using an explicit calculation, we rather use another method which can be generalized. In (3.18) we make the change of variable $t = \frac{1}{z}$ and set $\widetilde{H}(z,\varepsilon) = H(t,\varepsilon)$. The equation (3.18) becomes:

$$\mathscr{F}(t,\varepsilon,H) = 0, \text{ with} \tag{3.19}$$
$$\mathscr{F}(t,\varepsilon,H) = \frac{1}{M_{\rho,(0)}}H - |f_0|(t^{-1}) - \varepsilon\Big[(3t+|f_1|(t^{-1}))H + |f_2|(t^{-1})H^2\Big].$$

Since $\mathscr{F}(0,0,0) = 0$ and $\frac{\partial \mathscr{F}}{\partial H}(0,0,0) = \frac{1}{M_{\rho,(0)}} \neq 0$, the implicit function theorem provides a unique holomorphic solution $H(t,\varepsilon)$ to (3.19), for $|t|$ and $|\varepsilon|$ small enough : there exist $r_1 > 0$, $r_2 > 0$, $r_3 > 0$ and a holomorphic function $H : (t,\varepsilon) \in D(0,r_1) \times D(0,r_2) \mapsto H(t,\varepsilon) \in D(0,r_3)$ such that for every $(t,\varepsilon,H) \in D(0,r_1) \times D(0,r_2) \times D(0,r_3)$, $\Big[\mathscr{F}(t,\varepsilon,H) = 0 \Leftrightarrow H = H(t,\varepsilon)\Big]$.

To get more precise informations, we view the implicit problem (3.19) as a fixed-point problem:

$$H = \mathsf{N}(H), \tag{3.20}$$
$$\mathsf{N}(H) = M_{\rho,(0)}\Big(|f_0|(t^{-1}) + \varepsilon\Big[(3t+|f_1|(t^{-1}))H + |f_2|(t^{-1})H^2\Big]\Big)$$
$$= M_{\rho,(0)}\left(\frac{392}{625}t^2 + \varepsilon\Big[(3t+4t^2)H + \frac{1}{2}t^2H^2\Big]\right).$$

We choose $M_{\rho,(0)} = \frac{1}{\rho}$ (see lemma 3.2) and we introduce the space $\mathscr{O}(\overline{U})$ of functions in (t,ε) which are holomorphic on the polydisc $U = D(0,\frac{\rho}{8}) \times D(0,2)$ and continuous on the closure $\overline{U}$. The space $\big(\mathscr{O}(\overline{U}), \|\ \|\big)$ is a Banach algebra where $\|\ \|$ stands for the maximum norm.

We recall the following theorem [Tre75]: let U be a bounded open subset of $\mathbb{C}^n$, $n \geq 1$, E be a Banach space and $\mathscr{O}(\overline{U})$ be the space of functions $f : x \mapsto f(x) \in E$ which are continuous on $\overline{U}$ and holomorphic on U. With the the maximum norm $\|f\| = \sup_{z\in U}|f(z)|$, $(\mathscr{O}(\overline{U}), \|.\|)$ is a Banach algebra.

For a reason of homogeneity, we introduce the ball $B_\rho = \{H \in \mathscr{O}(\overline{U}),\ \|H\| \leq \rho\}$. For any $H, H_1, H_2 \in B_\rho$, $\|\mathsf{N}(H)\| \leq \frac{1}{\rho}\left(\frac{392}{625}\frac{\rho^2}{64} + 2\Big[\frac{7\rho}{16}\|H\| + \frac{\rho^2}{128}\|H\|^2\Big]\right) \leq \rho$ (remember that $\rho < 1$), while

$$\|\mathsf{N}(H_1) - \mathsf{N}(H_2)\| \leq \frac{2}{\rho}\left(\frac{7\rho}{16}\|H_1 - H_2\| + \frac{\rho^2}{128}\|H_1 - H_2\|\Big(\|H_1\| + \|H_2\|\Big)\right)$$
$$\leq \frac{29}{32}\|H_1 - H_2\|.$$

The mapping $\mathsf{N}_{|B_\rho} : H \in B_\rho \mapsto \mathsf{N}(H) \in B_\rho$ is thus contractive. Since B_ρ is a closed subset of a complete space, $(B_\rho, \|.\|)$ is complete and the contraction mapping theorem can be applied. We deduce the existence of a unique solution H in B_ρ of the fixed-point problem (3.20).
This solution $H(t,\varepsilon)$, thus holomorphic in $U = D(0,\frac{\rho}{8}) \times D(0,2)$, has a Taylor expansion with respect to ε at 0 of the form $H(t,\varepsilon) = \sum_{l=0}^{\infty} H_l(t)\,\varepsilon^l$, where $(H_l)_{l\geq 0}$ is a sequence of holomorphic functions on the disc $D(0,\frac{\rho}{8})$. Moreover, by the Cauchy inequalities and using the fact that $\sup_{(t,\varepsilon)\in U} |H(t,\varepsilon)| \leq \rho$, one gets: for every $l \in \mathbb{N}$, $\sup_{t\in D(0,\frac{\rho}{8})} |H_l(t)| \leq \frac{\rho}{2^l}$. This ends the proof of lemma 3.6. □

Lemma 3.7. *For every $\rho \in]0,1[$ and every $l \in \mathbb{N}$, the majorant function $\widehat{H}_l(\xi)$ is a polynomial which satisfies: for every $\xi \in \mathbb{C}$, $|\widehat{H}_l(\xi)| \leq \frac{8}{2^l} \mathrm{e}^{\frac{8}{\rho}|\xi|}$.*

Proof. This is due to lemmas 3.5 and 3.6. □

3.3.1.5 Formal Series Solution and Borel-Laplace Summability

We are ready to show the following theorem.

Theorem 3.3. *The formal solution $\widetilde{w}$ of the prepared equation (3.6) associated with the first Painlevé equation, is a 1-Gevrey series and satisfies the following properties:*

1. *its minor $\widehat{w}$ is an odd series, convergent to a holomorphic function which can be analytically continued to a function (still denoted by $\widehat{w}$) holomorphic on the cut plane $\overset{\bullet}{\mathscr{R}}{}^{(0)}$;*
2. *$\widehat{w}$ has at most exponential growth of order 1 at infinity along non-horizontal directions. More precisely, for every $\rho \in]0,1[$, there exist $A > 0$ and $\tau > 0$ such that, for every $\zeta \in \overset{\bullet}{\mathscr{R}}{}^{(0)}_\rho$, $|\widehat{w}(\zeta)| \leq A\mathrm{e}^{\tau|\zeta|}$;*
3. *moreover in the above upper bounds one can choose $A = 16$ and $\tau = \frac{8}{\rho}$.*

Proof. Combining lemmas 3.4 and 3.7, we know that, for every $\rho \in]0,1[$ and $l \geq 0$, the function $\widehat{h}_l(\zeta)$, is holomorphic on $\overset{\bullet}{\mathscr{R}}{}^{(0)}_\rho$. Moreover, for every $R > 0$, setting $U_R = D(0,R) \cap \overset{\bullet}{\mathscr{R}}{}^{(0)}_\rho$, $\sum_{l\geq 0} \sup_{\overline{U}_R} |\widehat{h}_l(\zeta)| \leq \sum_{l\geq 0} \widehat{H}_l(R) \leq \sum_{l\geq 0} \frac{8}{2^l} \mathrm{e}^{\frac{8}{\rho}R} \leq 16\mathrm{e}^{\frac{8}{\rho}R}$. This normal convergence ensures the uniform convergence on any compact subset of $\overset{\bullet}{\mathscr{R}}{}^{(0)}$ of the series $\sum_{l\geq 0} \widehat{h}_l(\zeta)$, which thus defines a function holomorphic on $\overset{\bullet}{\mathscr{R}}{}^{(0)}$. However, proposition 3.6 tells use that the series $\sum_{l\geq 0} \widehat{h}_l$ converges to the formal Borel transform $\widehat{w}$ of the formal solution $\widetilde{w}$ of the ODE (3.6). □

Remark 3.3. Better estimates can easily be obtained, see corollary 3.1 and exercise 3.3.

3.3.2 Formal Series Solution and Borel-Laplace Summability: Second Approach

In this second approach, however related to the first one, we introduce a Banach space (following [Cos98, Cos009]), convenient for analyzing the analyticity of the formal Borel transform of the formal series $\widetilde{w}$ solution of the ODE (3.6). We then introduce the reader to a "Grönwall-like lemma" which will give the upper bounds we are looking for.

3.3.2.1 Convolution Algebra and Uniform Norm

Definition 3.12. Let $U = U_R \subset \mathbb{C}$ be an open neighbourdhood of the origin, bounded and star-shaped, $R = \sup_{\zeta \in U} |\zeta|$ the "radius" of U. We denote by $\left(\mathscr{O}(\overline{U}), +, ., *\right)$ the convolution $\mathbb{C}$-algebra (without unit) of functions continuous on $\overline{U}$ and holomorphic on U. We denote by $\mathscr{M}\mathscr{O}(\overline{U})$ the maximal ideal of $\mathscr{O}(\overline{U})$ defined by $\mathscr{M}\mathscr{O}(\overline{U}) = \{f \in \mathscr{O}(\overline{U}), f(0) = 0\}$. We set

$$\widehat{\partial} : f \in \mathscr{O}(\overline{U}) \mapsto \widehat{\partial} f(\zeta) = -\zeta f(\zeta) \in \mathscr{M}\mathscr{O}(\overline{U}).$$

Let be $\nu \geq 0$. The norm $\|.\|_\nu$ is defined as follows: for every $f \in \mathscr{O}(\overline{U})$,

$$\|f\|_\nu = R \sup_{\zeta \in U} \left|\mathrm{e}^{-\nu|\zeta|} f(\zeta)\right|.$$

This norm is extended to $\mathbb{C}\delta \oplus \mathscr{O}(\overline{U})$ by setting: $\|c\delta + f\|_\nu = |c| + \|f\|_\nu$, while $\widehat{\partial}\delta = 0$.

Proposition 3.9. *The space* $\left(\mathbb{C}\delta \oplus \mathscr{O}(\overline{U}), \|.\|_\nu\right)$ *is a Banach algebra. In particular, for every* $f, g \in \mathbb{C}\delta \oplus \mathscr{O}(\overline{U})$, $\|f * g\|_\nu \leq \|f\|_\nu \|g\|_\nu$. *The space* $\mathscr{M}\mathscr{O}(\overline{U})$ *is closed in the normed space* $\left(\mathscr{O}(\overline{U}), \|.\|_\nu\right)$. *Moreover, for* $\nu > 0$*:*

1. *for every* $n \in \mathbb{N}$, *for every* $g \in \mathscr{O}(\overline{U})$, $\|\zeta^n * g\|_\nu \leq \dfrac{n!}{\nu^{n+1}} \|g\|_\nu$, $\|(\zeta \mapsto \zeta^{n+1})\|_\nu \leq \dfrac{n!}{\nu^{n+1}} R$ *and* $\|(\zeta \mapsto 1)\|_\nu = R$.
2. *for every* $f, g \in \mathscr{O}(\overline{U})$, $\|fg\|_\nu \leq \dfrac{1}{R} \|f\|_\nu \|g\|_0$.
3. *for every* $f \in \mathscr{O}(\overline{U})$, $\nu \geq \nu_0 \geq 0 \Rightarrow \|f\|_\nu \leq \|f\|_{\nu_0}$.
4. *for every* $f \in \mathscr{M}\mathscr{O}(\overline{U})$, $\lim_{\nu \to \infty} \|f\|_\nu = 0$.

5. *the derivation* $\widehat{\partial}|_{\mathscr{O}(\overline{U})} : f \in \mathscr{O}(\overline{U}) \mapsto \widehat{\partial} f \in \mathscr{M}\mathscr{O}(\overline{U})$ *is invertible. Its inverse map* $\widehat{\partial}^{-1}$ *satisfies: for every* $f \in \mathscr{O}(\overline{U})$*, for every* $g \in \mathscr{M}\mathscr{O}(\overline{U})$, $\widehat{\partial}^{-1}(f*g) \in \mathscr{M}\mathscr{O}(\overline{U})$ *and* $\|\widehat{\partial}^{-1}(f*g)\|_\nu \leq \dfrac{1}{\nu R}\|f\|_\nu \|\widehat{\partial}^{-1} g\|_0$*. Also, for every* $f \in \mathbb{C}\delta \oplus \mathscr{O}(\overline{U})$*, for every* $g \in \mathscr{M}\mathscr{O}(\overline{U})$, $\widehat{\partial}^{-1}(f*g) \in \mathscr{O}(\overline{U})$ *and* $\|\widehat{\partial}^{-1}(f*g)\|\|_\nu \leq \|f\|_\nu \|\widehat{\partial}^{-1} g\|_\nu$.

Proof. Since $R\mathrm{e}^{-\nu R} \sup\limits_{\zeta \in U} \left|f(\zeta)\right| \leq R \sup\limits_{\zeta \in U} \left|\mathrm{e}^{-\nu|\zeta|} f(\zeta)\right| \leq R \sup\limits_{\zeta \in U} \left|f(\zeta)\right|$, we see that $\|.\|_\nu$ is equivalent to the usual maximum norm on the vector space $\mathscr{O}(\overline{U})$ and this normed vector space is complete. This shows the completeness of $\Big((\mathscr{O}(\overline{U}),+,.),\|.\|_\nu\Big)$ and of $\left(\mathbb{C}\delta \oplus \mathscr{O}(\overline{U}),\|.\|_\nu\right)$ as well.
For $f,g \in \mathscr{O}(\overline{U})$ we have, writing $\zeta = |\zeta|\mathrm{e}^{\mathrm{i}\theta} \in U$,

$$\begin{aligned} R\mathrm{e}^{-\nu|\zeta|} f*g(\zeta) &= R\mathrm{e}^{-\nu|\zeta|} \int_0^{|\zeta|} f(s\mathrm{e}^{\mathrm{i}\theta}) g\big((|\zeta|-s)\mathrm{e}^{\mathrm{i}\theta}\big)\, \mathrm{e}^{\mathrm{i}\theta} ds \\ &= R \int_0^{|\zeta|} f(s\mathrm{e}^{\mathrm{i}\theta})\mathrm{e}^{-\nu s} g\big((|\zeta|-s)\mathrm{e}^{\mathrm{i}\theta}\big)\mathrm{e}^{-\nu(|\zeta|-s)}\, \mathrm{e}^{\mathrm{i}\theta} ds. \end{aligned}$$

Therefore $R|\mathrm{e}^{-\nu|\zeta|} f*g(\zeta)| \leq \|f\|_\nu \|g\|_\nu \displaystyle\int_0^{|\zeta|} \frac{1}{R} ds \leq \|f\|_\nu \|g\|_\nu$. We conclude that for every $f,g \in \mathscr{O}(\overline{U})$, $\|f*g\|_\nu \leq \|f\|_\nu \|g\|_\nu$, hence $\left(\mathscr{O}(\overline{U}),\|.\|_\nu\right)$ is a Banach algebra and $\left(\mathbb{C}\delta \oplus \mathscr{O}(\overline{U}),\|.\|_\nu\right)$ as well.
We now suppose $\nu > 0$.

1. For the particular case $f : \zeta \mapsto \zeta^n$ and $g \in \mathscr{O}(\overline{U})$:

$$\begin{aligned} R\mathrm{e}^{-\nu|\zeta|}\left|(\zeta^n * g)(\zeta)\right| &\leq R \int_0^{|\zeta|} \mathrm{e}^{-\nu s} s^n \left|g\big((|\zeta|-s)\mathrm{e}^{\mathrm{i}\theta}\big)\right| \mathrm{e}^{-\nu(|\zeta|-s)} ds \\ &\leq \|g\|_\nu \int_0^{|\zeta|} \mathrm{e}^{-\nu s} s^n\, ds \\ &\leq \|g\|_\nu \int_0^{\infty} \mathrm{e}^{-\nu s} s^n\, ds. \end{aligned}$$

   This shows that $\|\zeta^n * g\|_\nu \leq \dfrac{n!}{\nu^{n+1}} \|g\|_\nu$. The other properties follow.
2. Obviously, $\|fg\|_\nu \leq \|f\|_\nu \sup\limits_U |g| \leq \dfrac{1}{R}\|f\|_\nu \|g\|_0$, for every $f,g \in \mathscr{O}(\overline{U})$.
3. It is straightforward to see that $\nu \geq \nu_0 \geq 0$ implies $\|f\|_\nu \leq \|f\|_{\nu_0}$ when $f \in \mathscr{O}(\overline{U})$.
4. If $f \in \mathscr{M}\mathscr{O}(\overline{U})$, then $f = \zeta g$ with $g \in \mathscr{O}(\overline{U})$. From the previous property, $\|f\|_\nu \leq \dfrac{1}{R}\|\zeta\|_\nu \|g\|_0 \leq \dfrac{1}{\nu}\|g\|_0$. Thus $\lim_{\nu \to \infty} \|f\|_\nu = 0$.
5. If $f \in \mathbb{C}\delta \oplus \mathscr{O}(\overline{U})$ and $g \in \mathscr{M}\mathscr{O}(\overline{U})$ then $f*g \in \mathscr{M}\mathscr{O}(\overline{U})$. Assume now that $f \in \mathscr{O}(\overline{U})$ and $g \in \mathscr{M}\mathscr{O}(\overline{U})$. Then $\widehat{\partial}^{-1}(f*g)(0) = 0$ and writing $\zeta = |\zeta|\mathrm{e}^{\mathrm{i}\theta} \in U$,

$$Re^{-\nu|\zeta|}f*g(\zeta) = Re^{-\nu|\zeta|}\int_0^{|\zeta|} g(se^{i\theta})f((|\zeta|-s)e^{i\theta}ds \tag{3.21}$$
$$= R\int_0^{|\zeta|} se^{i\theta}(\widehat{\partial}^{-1}g)(se^{i\theta})e^{-\nu s}f((|\zeta|-s)e^{i\theta})e^{-\nu(|\zeta|-s)}e^{i\theta}ds.$$

On the one hand, from (3.21),

$$R|e^{-\nu|\zeta|}f*g(\zeta)| \leq \frac{1}{R}\|f\|_\nu\|\widehat{\partial}^{-1}g\|_\nu\int_0^{|\zeta|} s\,ds \leq \frac{|\zeta|^2}{2R}\|f\|_\nu\|\widehat{\partial}^{-1}g\|_\nu,$$

so that $R|e^{-\nu|\zeta|}\widehat{\partial}^{-1}(f*g)(\zeta)| \leq \frac{|\zeta|}{2R}\|f\|_\nu\|\widehat{\partial}^{-1}g\|_\nu \leq \|f\|_\nu\|\widehat{\partial}^{-1}g\|_\nu$. Thus $\|\widehat{\partial}^{-1}(f*g)\|_\nu \leq \|f\|_\nu\|\widehat{\partial}^{-1}g\|_\nu$. One easily extends this formula to the case $f \in \mathbb{C}\delta \oplus \mathscr{O}(\overline{U})$.

On the other hand, from (3.21),

$$R|e^{-\nu|\zeta|}f*g(\zeta)| \leq \|f\|_\nu \sup_U|\widehat{\partial}^{-1}g|\int_0^{|\zeta|} se^{-\nu s}ds \leq \frac{|\zeta|}{\nu R}\|f\|_\nu\|\widehat{\partial}^{-1}g\|_0,$$

hence $R|e^{-\nu|\zeta|}\widehat{\partial}^{-1}(f*g)(\zeta)| \leq \frac{1}{\nu R}\|f\|_\nu\|\widehat{\partial}^{-1}g\|_0,$. Therefore:

$$\|\widehat{\partial}^{-1}(f*g)\|_\nu \leq \frac{1}{\nu R}\|f\|_\nu\|\widehat{\partial}^{-1}g\|_0.$$

This ends the proof. □

3.3.2.2 A Grönwall-like Lemma

We start with the following observation.

Lemma 3.8. *Let be $a,b,c,d \geq 0$, $N \in \mathbb{N}^\star$ and $(\widehat{F}_n)_{0\leq n\leq N}$ be a sequence of entire functions, real and positive on $\mathbb{R}^+$, with at most exponential growth of order 1 at infinity. Then, the convolution equation*

$$\widehat{W} = d + [a+b\xi]*\widehat{W} + c\left(\widehat{F}_0 + \sum_{n=1}^{N}\widehat{F}_n*\widehat{W}^{*n}\right) \tag{3.22}$$

has a unique solution in $\mathbb{C}[[\xi]]$, whose sum converges to an entire function $\widehat{W}_d(\xi)$ with at most exponential growth of order 1 at infinity. The function $\widehat{W}_d(\xi)$ is real, positive and non-decreasing on $\mathbb{R}^+$ and, for every $\xi \in \mathbb{C}$, the mapping $d \mapsto \widehat{W}_d(\xi)$ is continuous on $\mathbb{R}^+$.

Proof. Obviously, (3.22) has a unique solution $\widehat{W}_d \in \mathbb{R}^+[[\xi]]$. Its formal Laplace transform, $\widetilde{W}_d = \mathscr{L}(\widehat{W}_d) \in \mathbb{R}^+[[z^{-1}]]$, solves the algebraic equation

$$\widetilde{W}(z) = \frac{d}{z} + \left[\frac{a}{z} + \frac{b}{z^2}\right]\widetilde{W}(z) + c\sum_{n=0}^{N} F_n(z)\widetilde{W}^n(z), \tag{3.23}$$

where the $(F_n)_{0\le n\le N}$ is a $(N+1)$-tuple of holomorphic functions on a neighbourhood of infinity with $F_n(z) = O(z^{-1})$. This shows (by a reasoning already done) that $\widetilde{W}_d = O(z^{-1})$ is a holomorphic function in (z,d) for $d \in \mathbb{C}$ and z on a neighbourhood of infinity (independent on d). Therefore, $\widehat{W}_d$ determines a function holomorphic in $(\xi,d) \in \mathbb{C}^2$, with at most exponential growth of order 1 at infinity in ξ. The fact that, for $d \ge 0$, $\widehat{W}_d$ is real, positive and non-decreasing on $\mathbb{R}^+$, is evident. □

Lemma 3.9 (Grönwall lemma). *Let U be a domain star-shaped from 0 and $N \in \mathbb{N}^\star$. Let $(\widehat{f}_n)_{0\le n\le N}$, resp. $(\widehat{F}_n)_{0\le n\le N}$, be a $(N+1)$-tuple of functions in $\mathscr{O}(U)$, resp. of entire functions, real and positive on $\mathbb{R}^+$. We suppose that for every $0 \le n \le N$ and every $\zeta \in U$, $|\widehat{f}_n(\zeta)| \le \widehat{F}_n(\xi)$ with $\xi = |\zeta|$. Let $p,q,r \in \mathbb{C}[\zeta]$ be polynomials such that the function $\zeta \mapsto p(-\zeta)$ is non vanishing on U and the following upper bounds are satisfied: $a = \sup_{\zeta\in U} \frac{|q|(|\zeta|)}{|p(-\zeta)|} < \infty$, $b = \sup_{\zeta\in U} \frac{|r|(|\zeta|)}{|p(-\zeta)|} < \infty$, $c = \sup_{\zeta\in U} \frac{1}{|p(-\zeta)|} < \infty$. We finally assume that $\widehat{w} \in \mathscr{O}(U)$ solves the following convolution equation:*

$$p(\widehat{\partial})\widehat{w} + 1 * [q(\widehat{\partial})\widehat{w}] = \zeta * [r(\widehat{\partial})\widehat{w}] + \widehat{f}_0 + \sum_{n=1}^{N} \widehat{f}_n * \widehat{w}^{*n}. \tag{3.24}$$

Then for every $d \ge 0$, for every $\zeta \in U$, $|\widehat{w}(\zeta)| \le \widehat{W}_d(\xi)$ with $\xi = |\zeta|$, where $\widehat{W}_d \in \mathscr{O}(\mathbb{C})$ is the holomorphic solution of the convolution equation (3.22).

Proof. (Adapted from [LR011]). We assume that $\widehat{w} \in \mathscr{O}(U)$ is a solution of the convolution equation (3.22). We thus have, for every $\zeta \in U$,

$$\begin{aligned} p(\widehat{\partial})\widehat{w}(\zeta) &= \widehat{f}_0(\zeta) - \int_0^\zeta [q(\widehat{\partial})\widehat{w}](\eta)\,d\eta + \int_0^\zeta (\zeta-\eta)[r(\widehat{\partial})\widehat{w}](\eta)\,d\eta \\ &\quad + \sum_{n=1}^{N}\int_0^\zeta \widehat{f}_n(\zeta-\eta)\widehat{w}^{*n}(\eta)\,d\eta \end{aligned}$$

Thus, writing $\xi = |\zeta|$ and $\zeta = \xi e^{i\theta}$,

$$\begin{aligned} |\widehat{w}(\zeta)| &\le \frac{1}{|p(-\zeta)|}\widehat{F}_0(\xi) + \int_0^\xi \left[\frac{|q|(\xi)}{|p(-\zeta)|} + \frac{|r|(\xi)}{|p(-\zeta)|}(\xi - r)\right] |\widehat{w}(re^{i\theta})|\,dr \\ &\quad + \sum_{n=1}^{N}\int_0^\xi \frac{1}{|p(-\zeta)|}\widehat{F}_n(\xi-r)|\widehat{w}^{*n}(re^{i\theta})|\,dr. \end{aligned}$$

Therefore,

$$|\widehat{w}(\zeta)| \le c\widehat{F}_0(\xi) + \int_0^\xi [a + b(\xi-r)]\,|\widehat{w}(re^{i\theta})|\,dr + c\sum_{n=1}^{N}\int_0^\xi \widehat{F}_n(\xi-r)|\widehat{w}^{*n}(re^{i\theta})|\,dr.$$

We notice from (3.24) that $|\widehat{w}(0)| = \left|\frac{\widehat{f}_0(0)}{p(0)}\right|$, while $\widehat{W}_d(0) = c\widehat{F}_0(0) + d$, where $\widehat{W}_d$ solves (3.22). Remark that $|\widehat{w}(0)| \leq c\widehat{F}_0(0)$ by definition of c and by hypothesis on $\widehat{F}_0$.

First case. We assume $\widehat{W}_d(0) > |\widehat{w}(0)|$. We want to show that $|\widehat{w}(\zeta)| < \widehat{W}_d(\xi)$ for ζ on the ray $\zeta = \xi e^{i\theta} \in U$.
Assume on the contrary that there exists $\zeta_1 = \xi_1 e^{i\theta} \in U$ such that $|\widehat{w}(\zeta_1)| \geq \widehat{W}_d(\xi_1)$. Define $\chi = \{\zeta \in [0, \zeta_1] \mid |\widehat{w}(\zeta)| \geq \widehat{W}_d(|\zeta|)\}$. This is a non-empty closed set, bounded from below, and we denote by ζ_2 its infimum.

- If $|\widehat{w}(\zeta)| \geq \widehat{W}_d(|\zeta|)$ for some $\zeta \in]0, \zeta_2[$, then $\zeta \in \chi$ and this contradicts the definition of ζ_2. Thus, for every $\zeta \in [0, \zeta_2[$, $|\widehat{w}(\zeta)| < \widehat{W}_d(|\zeta|)$.
- If $|\widehat{w}(\zeta_2)| > \widehat{W}_d(|\zeta_2|)$ then, by continuity of $\widehat{w}$ and $\widehat{W}_d$, one can find $\alpha > 0$ such that $|\widehat{w}\big((|\zeta_2| - \alpha)e^{i\theta}\big)| > \widehat{W}_d(|\zeta_2| - \alpha)$, but this this contradicts again the definition of ζ_2. Therefore $|\widehat{w}(\zeta_2)| = \widehat{W}_d(|\zeta_2|)$.

Putting things together, one gets with $\xi_2 = |\zeta_2|$:

$$\begin{aligned}
|\widehat{w}(\zeta_2)| \leq c\widehat{F}_0(\xi_2) &+ \int_0^{\xi_2} [a + b(\xi_2 - r)]\, |\widehat{w}(re^{i\theta})|\, dr \\
&+ c\sum_{n=1}^{N} \int_0^{\xi_2} \widehat{F}_n(\xi_2 - r) |\widehat{w}^{*n}(re^{i\theta})|\, dr \\
\leq c\widehat{F}_0(\xi_2) &+ \int_0^{\xi_2} [a + b(\xi_2 - r)]\, \widehat{W}_d(r)\, dr + c\sum_{n=1}^{N} \int_0^{\xi_2} \widehat{F}_n(\xi_2 - r) \widehat{W}_d^{*n}(r)\, dr.
\end{aligned}$$

Therefore $|\widehat{w}(\zeta_2)| \leq \widehat{W}_d(\xi_2) - d$ and we get a contradiction. As a conclusion, for every $d > 0$, for every $\zeta \in U$, $|\widehat{w}(\zeta)| \leq \widehat{W}_d(\xi)$ with $\xi = |\zeta|$.

Second case. The case $\widehat{W}_d(0) = |\widehat{w}(0)|$ (thus, in particular, $d = 0$) is deduced from the above result. Indeed, for a given $\zeta \in U$, one has by $|\widehat{w}(\zeta)| \leq \widehat{W}_d(\xi)$ for every $d > 0$. Since the mapping $d \mapsto \widehat{W}_d(\xi)$ is continuous on $\mathbb{R}^+$ (cf. lemma 3.8), one gets the result by letting $d \to 0$. □

3.3.2.3 Applications

We prove theorem 3.3 with the tools introduced in this section. For $R > 0$ and $\rho > 0$, we introduce the star-shaped domain $U_R = D(0, R) \cap \overset{\bullet}{\mathscr{R}}{}_\rho^{(0)}$. We set $B_r = \{\widehat{v} \in \mathscr{O}(\overline{U}_R), \|\widehat{v}\|_\nu \leq r\}$, $r > 0$ and $\nu > 0$.
We consider the convolution equation (3.10), viewed as a fixed-point problem. Precisely, we consider the mapping

$$\mathsf{N} : \widehat{v} \in B_r \mapsto P(\widehat{\partial})^{-1}\Big[-1 * \big[Q(\widehat{\partial})\widehat{v}\big] + \widehat{f}_0 + \widehat{f}_1 * \widehat{v} + \widehat{f}_2 * \widehat{v} * \widehat{v}\Big].$$

By lemmas 3.2 and proposition 3.9, one first gets:

$$\|\mathsf{N}(\widehat{v})\|_\nu \leq M_{\rho,(0)} \| -1 * \left[Q(\widehat{\partial})\widehat{v}\right] + \widehat{f}_0 + \widehat{f}_1 * \widehat{v} + \widehat{f}_2 * \widehat{v} * \widehat{v}\|_\nu.$$

By proposition 3.9 again, since $Q(\widehat{\partial}) = -3\widehat{\partial}$, one easily obtains:

$$\|1 * \left[Q(\widehat{\partial})\widehat{v}\right]\|_\nu \leq \frac{1}{\nu}\|Q(\widehat{\partial})\widehat{v}\|_\nu \leq \frac{1}{R\nu}\|Q(-\zeta)\|_0 \|\widehat{v}\|_\nu \leq \frac{3}{\nu}\|\widehat{v}\|_\nu.$$

The functions $\widehat{f}_0, \widehat{f}_1, \widehat{f}_2$ belong to $\mathscr{M}\mathscr{O}(\overline{U}_R)$. By proposition 3.9, this implies $\lim_{\nu\to\infty} \|\widehat{f}_i\|_\nu = 0$, $i = 0,1,2$. We then deduce $\|\mathsf{N}(\widehat{v})\|_\nu \leq r$ by choosing $\nu > 0$ large enough.
By the same arguments, one easily sees that $\|\mathsf{N}(\widehat{v}_1) - \mathsf{N}(\widehat{v}_2)\|_\nu \leq k\|\widehat{v}_1 - \widehat{v}_2\|_\nu$ with $k < 1$, for $\widehat{v}_1, \widehat{v}_2 \in B_r$ and for $\nu > 0$ large enough.
This means that N is contractive in the closed set B_r of the Banach space $\left(\mathscr{O}(\overline{U}_R), \|.\|_\nu\right)$, for $\nu > 0$ large enough. The contraction mapping theorem provides a unique solution $\widehat{w} \in B_r$ for the fixed-point problem $\widehat{v} = \mathsf{N}(\widehat{v})$. Since R and ρ can be arbitrarily chosen, we deduce (by uniqueness) that the formal Borel transform $\widehat{w}$ of the unique formal series $\widetilde{w}$ solution of (3.6), defines a holomorphic in $\overset{\bullet}{\mathscr{R}}^{(0)}$.

One turns to the Grönwall lemma to get upper bounds. Working in the star-shaped domain $\overset{\bullet}{\mathscr{R}}{}^{(0)}_\rho$, $\rho \in]0,1[$, one sees by lemma 3.2, lemma 3.3 and the Grönwall lemma 3.9, that for every $\zeta \in \overset{\bullet}{\mathscr{R}}{}^{(0)}_\rho$, $|\widehat{w}(\zeta)| \leq \widehat{W}(\xi)$, $\xi = |\zeta|$, where $\widehat{W}(\xi)$ solves the following convolution equation:

$$\frac{1}{M_{\rho,(0)}}\widehat{W} = |\widehat{f}_0| + \left(3 + |\widehat{f}_1|\right) * \widehat{W} + |\widehat{f}_2| * \widehat{W} * \widehat{W}.$$

This is nothing but (3.17) with $\varepsilon = 1$. We adopt the notation and reasoning made for the proof of lemma 3.6. Let $\widetilde{W}(z)$ be the inverse Borel transform of $\widehat{W}$ and $\widetilde{W}(z) = H(t)$, $t = z^{-1}$. The function H solves the fixed-point problem $H = \mathsf{N}(H)$ with

$$\mathsf{N}(H) = M_{\rho,(0)}\left(\frac{392}{625}t^2 + \left(3t + 4t^2\right)H + \frac{1}{2}t^2H^2\right). \tag{3.25}$$

We set $M_{\rho,(0)} = \frac{1}{\rho}$, $U = D(0, \frac{\rho}{4.22})$, and $B_\rho = \{H \in \mathscr{O}(\overline{U}),\ \|H\| \leq \rho\}$. One easily shows that for any $H, H_1, H_2 \in B_1$,

$$\mathsf{N}(H) \in B_\rho \quad \text{and} \quad \|\mathsf{N}(H_1) - \mathsf{N}(H_2)\| \leq \frac{44150}{44521}\|H_1 - H_2\|.$$

We conclude with the contraction mapping theorem: $\widetilde{W}(z)$ is holomorphic on the domain $|z| > \frac{4.22}{\rho}$ and is bounded by ρ there. Therefore, by lemma 3.5, $\widehat{W}$ is an entire function and satisfies: for every $\xi \in \mathbb{C}$, $|\widehat{W}(\xi)| \leq 4.22 e^{\frac{4.22}{\rho}|\xi|}$. To sum up:

Corollary 3.1. *In theorem 3.3, one can choose* $A = 4.22$ *and* $\tau = \frac{4.22}{\rho}$.

3.4 First Painlevé Equation and Tritruncated Solutions

Theorem 3.3 shows that one can apply the Borel-Laplace summation scheme to the unique formal series expansion $\widetilde{w} \in \mathbb{C}[[z^{-1}]]$ solving equation (3.6). This is what we do in this section which starts with a brief reminder.

3.4.1 Reminder

We complete definitions 3.3 and definition 3.8 with notation essentially common with the first two volumes of this book [MS016, Lod016]. For the convenience of the reader we also recall some results about Borel-Laplace summability and we refer to [MS016, Lod016] for more details.

Definition 3.13. Let $\theta \in \mathbb{S}^1$ be a direction and $I =]\alpha, \beta[\subset \mathbb{S}^1$ be an open arc. We denote by $\breve{\theta} \subset \mathbb{S}^1$ the open arc defined by $\breve{\theta} =]-\frac{\pi}{2} - \theta, -\theta + \frac{\pi}{2}[$, and $\breve{I} = \bigcup_{\theta \in I} \breve{\theta}$. We denote by $\bar{I} = [\alpha, \beta]$ the closure of I and by $I^\star =]-\beta, -\alpha[$ the complex conjugate open arc.

Definition 3.14. For a direction θ and $\tau \in \mathbb{R}$, we denote by $\overset{\bullet}{\Pi}{}^{\theta}_{\tau}$ the following open half-plane, bisected by the half-line $\mathrm{e}^{-\mathrm{i}\theta}\mathbb{R}^+$: $\overset{\bullet}{\Pi}{}^{\theta}_{\tau} = \{z \in \mathbb{C}, \Re(z\mathrm{e}^{\mathrm{i}\theta}) > \tau\}$, of aperture $\breve{\theta}$.
Let $I \subset \mathbb{S}^1$ be an open arc of length $|I| \leq \pi$ and $\gamma : I \to \mathbb{R}$ be a locally bounded function. The domain $\overset{\bullet}{\mathcal{D}}(I, \gamma)$ is defined by $\overset{\bullet}{\mathcal{D}}(I, \gamma) = \bigcup_{\theta \in I} \overset{\bullet}{\Pi}{}^{\theta}_{\gamma(\theta)}$ and is called a *sectorial neighbourhood of infinity*, of aperture $\breve{I}$.

Let $\widetilde{g} = \sum_{n \geq 0} \frac{b_n}{z^n} \in \mathbb{C}[[z^{-1}]]_1$ be a 1-Gevrey series: the minor $\widehat{g}$ thus determines a holomorphic function near the origin (still denoted by $\widehat{g}$). We add the following conditions:

- one can find an open arc $I \subset \mathbb{S}^1$ such that $\widehat{g}$ can be analytically continued to an open sector of the form $\overset{\bullet}{\boldsymbol{\Delta}}{}^{\infty}_{0}(I)$;
- this function (still denoted by) $\widehat{g}$ is of exponential growth of order 1 at infinity: for every proper-subsector $\overset{\bullet}{\boldsymbol{\Delta}}{}^{\infty}(I') \Subset \overset{\bullet}{\boldsymbol{\Delta}}{}^{\infty}_{0}(I)$, there exist $A > 0$ and $\tau > 0$ such that for every $\zeta \in \overset{\bullet}{\boldsymbol{\Delta}}{}^{\infty}(I')$, $|\widehat{g}| \leq A\mathrm{e}^{\tau|\zeta|}$.

Under these conditions, for every direction $\theta \in I'$, the Borel-Laplace sum $\mathscr{S}^{\theta}\widetilde{g}$ is well-defined and holomorphic on the half-plane $\overset{\bullet}{\Pi}{}^{\theta}_{\tau}$. Moreover, for two close directions $\theta_1, \theta_2 \in I'$, the Borel-Laplace sums $\mathscr{S}^{\theta_1}\widetilde{g}$ and $\mathscr{S}^{\theta_2}\widetilde{g}$ coincide on their common domain $\overset{\bullet}{\Pi}{}^{\theta_1}_{\tau} \cap \overset{\bullet}{\Pi}{}^{\theta_2}_{\tau}$, thus can be glued together to give a holomorphic function on $\overset{\bullet}{\Pi}{}^{\theta_1}_{\tau} \cup \overset{\bullet}{\Pi}{}^{\theta_2}_{\tau}$. More generally:

Proposition 3.10. *Let* $\widetilde{g}(z) = \sum_{n\geq 0} \frac{b_n}{z^n} \in \mathbb{C}[[z^{-1}]]_1$ *be a 1-Gevrey series subject to the following conditions:*

- *there exists an open arc* $I \subset \mathbb{S}^1$ *of length* $|I| \leq \pi$ *so that the minor* $\widehat{g}$ *can be analytically continued to the open sector* $\dot{\mathcal{S}}_0^{\infty}(I)$;
- *for every direction* $\theta \in I$, $|\widehat{g}(\xi e^{i\theta})| \leq A(\theta)e^{\gamma(\theta)\xi}$, $\xi > 0$, *where* $A : I \to \mathbb{R}^+$ *and* $\gamma : I \to \mathbb{R}$ *are locally bounded functions.*

Then the family $(\mathscr{S}^{\theta}\widetilde{g})_{\theta\in I}$ *of Borel-Laplace sums determines a holomorphic function on the domain* $\dot{\mathscr{D}}(I,\gamma)$, *denoted by* $\mathscr{S}^I\widetilde{g}$.

Definition 3.15. Under the conditions of proposition 3.10, $\widetilde{g}$ is said to be *Borel-Laplace summable* in the directions of I. The function $\mathscr{S}^I\widetilde{g} \in \mathscr{O}\big(\dot{\mathscr{D}}(I,\gamma)\big)$ is called the Borel-Laplace sum of $\widetilde{g}$ in direction I.

Proposition 3.11. *Let* $\widetilde{g}(z) = \sum_{n\geq 0} \frac{b_n}{z^n} \in \mathbb{C}[[z^{-1}]]$ *be a formal series, Borel-Laplace summable in the directions of* $I \subset \mathbb{S}^1$, *an open arc of length* $|I| \leq \pi$. *Then its Borel-Laplace sum* $\mathscr{S}^I\widetilde{g} \in \mathscr{O}\big(\dot{\mathscr{D}}(I,\gamma)\big)$ *is 1-Gevrey asymptotic to* $\widetilde{g}$ *on* $\dot{\mathscr{D}}(I,\gamma)$: *for any proper-subsector* $\dot{\mathcal{S}}^{\infty} \Subset \dot{\mathscr{D}}(I,\gamma)$, *there exist constants* $C > 0$ *and* $A > 0$ *such that for every* $N \in \mathbb{N}$ *and every* $z \in \dot{\mathcal{S}}^{\infty}$,

$$\left| \mathscr{S}^I\widetilde{g}(z) - \sum_{l=0}^{N-1} \frac{b_l}{z^l} \right| \leq CN!A^N|z|^{-N}. \tag{3.26}$$

In this proposition, the property 3.26 essentially characterizes the Borel-Laplace sum. Indeed, notice that the sectorial neighbourhood of infinity $\dot{\mathscr{D}}(I,\gamma)$ is of aperture $\breve{I}$ which satisfies $\pi < |\breve{I}| \leq 2\pi$, and one can draw the following consequence from the Watson lemma (see the second volume of this book [Lod016]): let $\dot{\mathcal{S}}^{\infty}(I')$ be any sector such that $|I'| > \pi$ and $I' \subset \breve{I}$. Let $f \in \mathscr{O}(\dot{\mathcal{S}}^{\infty}(I'))$ be a holomorphic function which is 1-Gevrey asymptotic to $\widetilde{g}$ on $\dot{\mathcal{S}}^{\infty}(I')$. Then f and $\mathscr{S}^I\widetilde{g}$ coincide on $\dot{\mathcal{S}}^{\infty}(I') \cap \dot{\mathscr{D}}(I,\gamma)$.

We eventually ends this reminder with the following statement:

Proposition 3.12. *Let* $I \subset \mathbb{S}^1$ *be an open arc of length* $|I| \leq \pi$ *and* $\widetilde{f}(z), \widetilde{g}(z) \in \mathbb{C}[[z^{-1}]]$ *be Borel-Laplace summable formal series in the directions of* I. *Then* $\widetilde{f}\widetilde{g}$ *and* $\partial\widetilde{f}$ *are Borel-Laplace summable formal series in the directions of* I *and* $\mathscr{S}^I(\widetilde{f}\widetilde{g}) = (\mathscr{S}^I\widetilde{f})(\mathscr{S}^I\widetilde{g})$, $\mathscr{S}^I(\partial\widetilde{f}) = \partial(\mathscr{S}^I\widetilde{f})$.

3.4.2 *Formal Series Solution and Borel-Laplace Summation*

3.4.2.1 Borel-Laplace Summation

We go back to the formal solution $\widetilde{w}$ of equation (3.6). Theorem 3.3 and corollary 3.1 have the following consequences:

Corollary 3.2. *The Borel transform $\widehat{w} \in \mathscr{O}(\overset{\bullet}{\mathscr{R}}^{(0)})$ of the formal solution $\widetilde{w}$ of equation (3.6) satisfies the following property.*
For every $\delta \in]0, \frac{\pi}{2}[$, there exist $A_\delta > 0$ and $\tau_\delta > 0$ so that

$$\textit{for every } \zeta \in \overset{\bullet}{\Delta}{}_0^\infty(]\delta, \pi - \delta[), \; |\widehat{w}(\zeta)| \leq A_\delta e^{\tau_\delta |\zeta|}. \tag{3.27}$$

Moreover one can choose $A_\delta = 4.22$, $\tau_\delta = \dfrac{4.22}{\sin(\delta)}$.

Proof. One can define $\delta = \sin^{-1}(\rho) = \arcsin(\rho) \in]0, \frac{\pi}{2}[$, for any $\rho \in]0, 1[$. □

From corollary 3.2 and the properties of the Borel-Laplace summation, we see that for every $\delta \in]0, \frac{\pi}{2}[$, the Borel-Laplace sum $\mathscr{S}^\theta \widetilde{w}$ of $\widetilde{w}$ in any direction $\theta \in]\delta, \pi - \delta[$, is well-defined and holomorphic in the half-plane $\overset{\bullet}{\Pi}{}^\theta_{\tau_\delta}$ with $\tau_\delta = \dfrac{4.22}{\sin(\delta)}$. These holomorphic functions glue together to give the Borel-Laplace sum $\mathscr{S}^{]\delta, \pi-\delta[}\widetilde{w}$, holomorphic in the domain $\overset{\bullet}{\mathscr{D}}(]0, \pi[, \tau)$ with

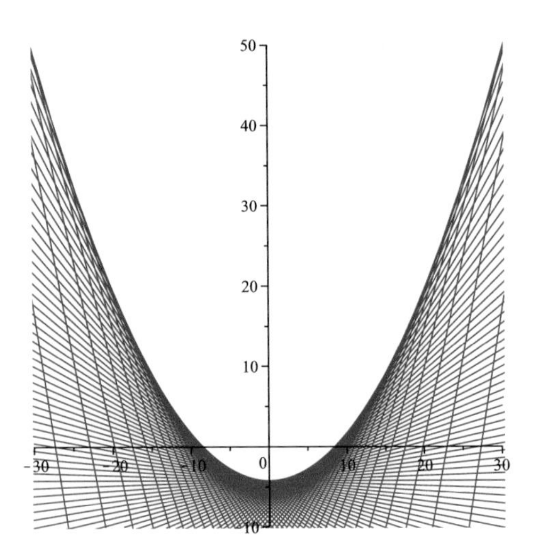

Fig. 3.2 The (shaded) domain $\overset{\bullet}{\mathscr{D}}(]0, \pi[, \tau)$ for $\tau(\theta) = \dfrac{4.22}{\sin(\theta)}$.

$\tau : \theta \in]0,\pi[\mapsto \tau(\theta) = \dfrac{4.22}{\sin(\theta)}$. (See Fig. 3.2 and exercise 3.4).

Moreover, since $\widetilde{w}$ formally solves (3.6), its Borel-Laplace sum $\mathscr{S}^{]0,\pi[}\widetilde{w}$ is a solution of this equation which is 1-Gevrey asymptotic at infinity to $\widetilde{w}$ on $\overset{\bullet}{\mathscr{D}}(]0,\pi[,\tau)$.

Similarly, the formal series $\widetilde{w}$ is Borel-Laplace-summable in the directions of the interval $]\pi,2\pi[$. This provides the Borel-Laplace sum $\mathscr{S}^{]\pi,2\pi[}\widetilde{w}$ which belongs to $\mathscr{O}\big(\overset{\bullet}{\mathscr{D}}(]\pi,2\pi[,\tau)\big)$ and is 1-Gevrey asymptotic to $\widetilde{w}$ on $\overset{\bullet}{\mathscr{D}}(]\pi,2\pi[,\tau)$.

Fine Borel-Laplace summations It is possible to get more precise estimates than those given by (3.26), by appealing to fine Borel-Laplace summations, discussed with much attention in [Lod016, MS016] to whom we refer.

Definition 3.16. We denote by $S_r(\theta)$ the open half-strip $S_r(\theta) = \bigcup\limits_{s\in\mathbb{R}^+} D(se^{i\theta},r)$, for $r > 0$ and θ a direction.

The following proposition is the easy part of a theorem due to Nevanlinna [MS016, Lod016, Mal95, DR007, Sok80].

Proposition 3.13. *Let* $\widetilde{\varphi}(z) = \sum\limits_{n=0}^{+\infty} \dfrac{a_n}{z^n} \in \mathbb{C}[[z^{-1}]]_1$ *be a 1-Gevrey series,* $r > 0$, $A > 0$, $\tau > 0$ *and* θ *a direction. Then property (1) implies property (2) in what follows.*

1. *The minor* $\widehat{\varphi}$ *is analytically continuable on* $S_r(\theta)$ *and for every* $\zeta \in S_r(\theta)$, $|\widehat{\varphi}(\zeta)| \leq Ae^{\tau|\zeta|}$.
2. *The Borel-Laplace sum* $\mathscr{S}^{\theta}\widetilde{\varphi}(z)$ *is holomorphic in* $\overset{\bullet}{\Pi}{}^{\theta}_{\tau}$ *and for every* $p \geq 0$, $N \geq 0$ *and* $z \in \overset{\bullet}{\Pi}{}^{\theta}_{\tau}$ *:*

$$\left|\frac{d^p \mathscr{S}^{\theta}\widetilde{\varphi}}{dz^p}(z) - \sum_{k=p}^{N} (-1)^p a_{k-p} \frac{(k-p)\cdots(k-1)}{z^k}\right| \leq R_{as}(r,A,\tau,N,ze^{i\theta};p) \tag{3.28}$$

where

$$R_{as}(r,A,\tau,N,z;p) = A\frac{N!e^{\tau r}}{r^N|z|^N}\frac{p!}{(\Re(z)-\tau)^{p+1}}\sum_{l=0}^{p}\frac{\big(r(\Re(z)-\tau)\big)^l}{l!} \tag{3.29}$$

Applications We return to theorem 3.3 and corollary 3.1. We consider a direction $\theta \in]0,\pi[$ and we choose $r > 0$ and $0 < \rho < 1$ such that $\sin(\theta) = r+\rho$. This ensures that the half-strip $S_r(\theta)$ is a subset of the domain $\overset{\bullet}{\mathscr{R}}{}^{(0)}_{\rho}$ and, by theorem 3.3, there exist $A > 0$ and $\tau > 0$ such that for every $\zeta \in S_r(\theta)$, $|\widehat{w}(\zeta)| \leq Ae^{\tau|\zeta|}$ with $\sin(\theta) = r+\rho$. Also, from corollary 3.1, one can choose $A = 4.22$, $\tau = \dfrac{4.22}{\rho}$. As a consequence, proposition 3.13 can be applied. The reader will easily adapt the previous considerations when the directions $\theta \in]\pi,2\pi[$ are considered.
We summarize what have been obtained.

Proposition 3.14. *The 1-Gevrey series $\widetilde{w} \in \mathbb{C}[[z^{-1}]]_1$, solution of the prepared equation (3.6) associated with the first Painlevé equation, is Borel-Laplace summable in the directions of the arc $I_0 =]0,\pi[$, resp. $I_1 =]\pi,2\pi[$. The Borel-Laplace sum $w_{tri,0} = \mathscr{S}^{]0,\pi[}\widetilde{w}$, resp $w_{tri,1} = \mathscr{S}^{]\pi,2\pi[}\widetilde{w}$. is a holomorphic solution of the differential equation (3.6) and $w_{tri,0}, w_{tri,1}$ satisfy the following properties. For every $\theta \in I_0$, resp. $\theta \in I_1$, for every $r > 0$ and $\rho > 0$ so that $|\sin(\theta)| = r + \rho$, there exist $\tau > 0$ and $A > 0$ such that :*

- $w_{tri,j} \in \mathscr{O}(\overset{\bullet}{\Pi}{}_{\tau}^{\theta})$, *$j = 0$ resp. $j = 1$;*
- *for every $z \in \overset{\bullet}{\Pi}{}_{\tau}^{\theta}$, for every $N \in \mathbb{N}$, for $j = 0$ resp. $j = 1$,*

$$\left| w_{tri,j}(z) - \sum_{k=0}^{N} \frac{a_k}{z^k} \right| \leq A \frac{N! \mathrm{e}^{\tau r}}{r^N |z|^N} \frac{1}{\Re(z\mathrm{e}^{\mathrm{i}\theta}) - \tau}; \tag{3.30}$$

$$\left| \frac{dw_{tri,j}}{dz}(z) + \sum_{k=1}^{N} \frac{(k-1)a_{(k-1)}}{z^k} \right| \leq A \frac{N! \mathrm{e}^{\tau r}}{r^N |z|^N} \frac{1 + r(\Re(z\mathrm{e}^{\mathrm{i}\theta}) - \tau)}{\left(\Re(z\mathrm{e}^{\mathrm{i}\theta}) - \tau\right)^2} \tag{3.31}$$

  where the coefficients a_k are given by (3.7);
- *morover one can choose $A = 4.22$, $\tau = \dfrac{4.22}{\rho}$. In particular $w_{tri,0}$, resp. $w_{tri,1}$, is holomorphic in $\overset{\bullet}{\mathscr{D}}(I_0,\tau)$, resp. in $\overset{\bullet}{\mathscr{D}}(I_1,\tau)$, with $\tau(\theta) = \dfrac{4.22}{|\sin(\theta)|}$.*

3.4.2.2 A Link with 1-summability Theory

We assume that the reader has a previous acquaintance with 1-summability theory, introduced and much discussed in the second volume of this book [Lod016], to which we refer. We only fix some notation, these are classical [Lod016, Mal95] but for the fact that we consider asymptotics at infinity, and we recall some properties.

Definition 3.17. Let $I \subset \mathbb{S}^1$ be an open arc and $\overset{\bullet}{\Delta}{}^{\infty} = \overset{\bullet}{\Delta}{}^{\infty}(I)$ a sector.

1. $\overline{\mathscr{A}}(\overset{\bullet}{\Delta}{}^{\infty})$, *resp.* $\overline{\mathscr{A}}(I)$, is the differential algebra of holomorphic functions on the sector $\overset{\bullet}{\Delta}{}^{\infty}$ admitting Poincaré asymptotics at infinity in this sector, *resp.* asymptotics germs at infinity over I.
   The linear map $T : \overline{\mathscr{A}}(\overset{\bullet}{\Delta}{}^{\infty}) \to \mathbb{C}[[z^{-1}]]$, *resp.* $T : \overline{\mathscr{A}}(I) \to \mathbb{C}[[z^{-1}]]$, which assigns to each $f \in \overline{\mathscr{A}}(\overset{\bullet}{\Delta}{}^{\infty})$, *resp.* $f \in \overline{\mathscr{A}}(I)$, its asymptotic expansion at infinity, is called the Taylor map.

   The Taylor map T is a morphism of differential algebras and this map is onto (Borel-Ritt theorem).

2. $\overline{\mathscr{A}}_1(\overset{\bullet}{\Delta}{}^{\infty})$, *resp.* $\overline{\mathscr{A}}_1(I)$, is the differential algebra of holomorphic functions on the sector $\overset{\bullet}{\Delta}{}^{\infty}$ with 1-Gevrey asymptotics at infinity in this sector, *resp.* 1-Gevrey

asymptotics germs at infinity over I.
On denotes by $T_1 : \overline{\mathscr{A}}_1(\overset{\bullet}{\Delta}^\infty) \to \mathbb{C}[[z^{-1}]]_1$, *resp.* $T_1 : \overline{\mathscr{A}}_1(I) \to \mathbb{C}[[z^{-1}]]_1$, the Taylor map restricted $\overline{\mathscr{A}}_1(\overset{\bullet}{\Delta}^\infty)$, *resp.* $\overline{\mathscr{A}}_1(I)$, called the 1-Gevrey Taylor map.

The 1-Gevrey Taylor map T_1 is morphism of differential algebras. This map is onto when $|I| \leq \pi$ (Borel-Ritt theorem). This map is injective when $|I| > \pi$ (Watson lemma).

3. $\overline{\mathscr{A}}^{<0}(\overset{\bullet}{\Delta}^\infty)$, *resp.* $\overline{\mathscr{A}}^{<0}(I)$, is the space of flat functions on $\overset{\bullet}{\Delta}^\infty$, *resp.* flat germs at infinity over I.

   $\overline{\mathscr{A}}^{<0}(\overset{\bullet}{\Delta}^\infty)$ is thus the kernel of the Taylor map $T : \overline{\mathscr{A}}(\overset{\bullet}{\Delta}^\infty) \to \mathbb{C}[[z^{-1}]]$

4. $\overline{\mathscr{A}}^{\leq -1}(\overset{\bullet}{\Delta}^\infty)$, *resp.* $\overline{\mathscr{A}}^{\leq -1}(I)$, is the space of 1-exponentially flat functions on $\overset{\bullet}{\Delta}^\infty$, *resp.* 1-exponentially flat germs at infinity over I.

   $\overline{\mathscr{A}}^{\leq -1}(\overset{\bullet}{\Delta}^\infty)$ is the kernel of the 1-Gevrey Taylor map $T_1 : \overline{\mathscr{A}}_1(\overset{\bullet}{\Delta}^\infty) \to \mathbb{C}[[z^{-1}]]_1$.

5. $\mathscr{A}$ is the sheaf over $\mathbb{S}^1$ of asymptotic functions at infinity associated with the presheaf $\overline{\mathscr{A}}$. We denote by $\mathscr{A}_1$ the sheaf over $\mathbb{S}^1$ of 1-Gevrey asymptotic functions at infinity associated with the presheaf $\overline{\mathscr{A}}_1$. We denote by $\mathscr{A}^{<0}$ the sheaf over $\mathbb{S}^1$ of flat germs at infinity associated with the presheaf $\overline{\mathscr{A}}^{<0}$. Finally $\mathscr{A}^{\leq -1}$ stands for the sheaf over $\mathbb{S}^1$ of 1-Gevrey flat germs at infinity associated with the presheaf $\overline{\mathscr{A}}^{\leq -1}$.

Theorem 3.4 (Borel-Ritt). *The quotient sheaf $\mathscr{A}/\mathscr{A}^{<0}$, resp. $\mathscr{A}_1/\mathscr{A}^{\leq -1}$, is isomorphic via the Taylor map T, resp. the 1-Gevrey Taylor map T_1, to the constant sheaf, $\mathbb{C}[[z^{-1}]]$ resp. $\mathbb{C}[[z^{-1}]]_1$*

We now go back to proposition 3.14. On the one hand, The domain $\overset{\bullet}{\mathscr{D}}(I_0,\tau)$ is a sectorial neighbourhood of ∞ with aperture $\breve{I}_0 =]-\frac{3}{2}\pi, +\frac{1}{2}\pi[$. On the other hand, while $\overset{\bullet}{\mathscr{D}}(I_1,\tau) = \mathrm{e}^{-\mathrm{i}\pi}\overset{\bullet}{\mathscr{D}}(I_0,\tau)$ is a sectorial neighbourhood of ∞ with aperture $\breve{I}_1 =]-\frac{5}{2}\pi, -\frac{1}{2}\pi[$. These two open arcs provide a good covering $\{\breve{I}_0, \breve{I}_1\}$ of the circle of directions $\mathbb{S}^1$. Let $J_0 =]-\frac{1}{2}\pi, \frac{1}{2}\pi[$ and $J_1 =]-\frac{3}{2}\pi, -\frac{1}{2}\pi[$ be the two intersection arcs. Both $w_{tri,0}$ and $w_{tri,1}$ can be considered as defining sections of $\mathscr{A}_1$, namely $w_{tri,0} \in \Gamma(\breve{I}_0, \mathscr{A}_1)$ and $w_{tri,1} \in \Gamma(\breve{I}_1, \mathscr{A}_1)$, and are asymptotic to the same 1-Gevrey formal series $\widetilde{w}$. The pair $(w_{tri,0}, w_{tri,1})$ defines a 0-cochain in the sense of Čech cohomology, and the 1-coboundary $(w_{tri,0} - w_{tri,1}, w_{tri,1} - w_{tri,0})$ belongs to $\Gamma(J_0, \mathscr{A}^{\leq -1}) \times \Gamma(J_1, \mathscr{A}^{\leq -1})$.

3.4.2.3 Miscellaneous Properties

We discuss various properties for the Borel-Laplace sums $w_{tri,j}$.

For any $j \in \mathbb{Z}$ and $I_j = I_0 + j\pi =]0,\pi| + j\pi$, one can of course consider the Borel-Laplace sum $w_{tri,j} = \mathscr{S}^{I_j}\widetilde{w}$, which defines a holomorphic function on the domain $\overset{\bullet}{\mathscr{D}}(I_j,\tau)$, a sectorial neighbourhood of ∞ with aperture $\breve{I}_j = \breve{I}_0 - j\pi$, $\breve{I}_j =]-\frac{3}{2}\pi, +\frac{1}{2}\pi[-j\pi$. Morever, for every $j \in \mathbb{Z}$,

$$w_{tri,j+2}(z) = w_{tri,j}(z) \text{ for } z \in \overset{\bullet}{\mathscr{D}}(I_j,\tau) \tag{3.32}$$

because $\widetilde{w} \in \mathbb{C}[[z^{-1}]]_1$.
We mentioned in proposition 3.1 that the formal series $\widetilde{w}(z)$ is even. One deduces that for any $\theta \in]0,\pi[$, for every $z \in \overset{\bullet}{\Pi}{}_\tau^{\pi-\theta}$

$$\mathscr{S}^{\pi-\theta}\,\widetilde{w}(z) = \mathscr{S}^{-\theta}\,\widetilde{w}(-z).$$

Therefore, for every $j \in \mathbb{Z}$,

$$\text{for every } z \in \overset{\bullet}{\mathscr{D}}(I_j,\tau),\ w_{tri,j}(z) = w_{tri,j+1}(-z). \tag{3.33}$$

We know by proposition 3.1 that $\widetilde{w}(z)$ belongs to $\mathbb{R}[[z^{-1}]]$. This has the following consequence : for any $\theta \in]0,\pi[$, for $z \in \overset{\bullet}{\Pi}{}_\tau^{\theta}$, $\overline{\mathscr{S}^{\theta}\,\widetilde{w}(z)} = \mathscr{S}^{-\theta}\,\widetilde{w}(\overline{z})$ (where $\overline{a}$ stands for the complex conjugate of $a \in \mathbb{C}$). In other words, for any $j \in \mathbb{Z}$, the two functions $w_{tri,j}$ and $w_{tri,j+1}$ are complex conjugate,

$$\text{for every } z \in \overset{\bullet}{\mathscr{D}}(I_j,\tau),\ \overline{w_{tri,j}}(z) = w_{tri,j+1}(\overline{z}). \tag{3.34}$$

However, neither $w_{tri,0}$ nor $w_{tri,1}$ are real analytic functions, since this would mean that the 1-coboundary $w_{tri,0} - w_{tri,1}$ is zero which is not as we shall see later on.
The properties (3.33) and (3.34) have the following consequences: for every $j \in \mathbb{Z}$, $w_{tri,j}$ is " $\mathscr{P}\mathscr{T}$-symmetric" [DP98, DP99, DT000], in the sense that for every $z \in \overset{\bullet}{\mathscr{D}}(I_j,\tau)$,

$$w_{tri,j}(z) = \overline{w_{tri,j}}(-\overline{z}). \tag{3.35}$$

In particular, for $r > 0$ large enough,

$$w_{tri,0}(r\mathrm{e}^{-\mathrm{i}\pi/2}) \in \mathbb{R}, \qquad w'_{tri,0}(r\mathrm{e}^{-\mathrm{i}\pi/2}) \in i\mathbb{R}. \tag{3.36}$$

3.4.2.4 Asymptotics and Approximations

By Stirling formula one has $N! \sim \sqrt{2\pi}N^{N+\frac{1}{2}}\mathrm{e}^{-N}$ for large N. Since for a given $z \neq 0$ the function $N \mapsto \dfrac{N^N\mathrm{e}^{-N}}{(r|z|)^N}$ reaches its minimal value at $n = r|z|$, it turns out from formula (3.30) that one can estimate the value of $w_{tri,0}$ or $w_{tri,1}$ from the truncated series expansion $\sum_{k=0}^{N} \frac{a_k}{z^k}$ with $N = \left[r|z|\right]$ where $\left[.\right]$ is the entire part. This gives rise to the *summation to the least term.*

Along this state of mind, there are many ways of computing Borel-Laplace sums approximately in practice (see, e.g., [Jen004, CMRSJ007]). Among them, one may quote the so-called *hyperasymptotic* methods [BH91] of Berry & Howls which have strong links with resurgence theory. These methods, originally arising from (and extending to) geometrical considerations on (multiple) singular integrals [Pha005, DH002, Del005, Del010, PK001], can be applied to a wide class of problems stemming from applied mathematics and physics, see [Old96, Old97, Old005, CHKO007] and references therein. Other ways are available, for instance those based on the use of conformal mappings [BLS002] with realistic upper bounds. It is also theoretically possible to calculate a Borel-Laplace sum exactly by means of factorial series expansions [Mal95, DR007].

3.4.3 Tritruncated Solutions

3.4.3.1 Tritruncated Solutions

One can easily translate proposition 3.14 into properties for the first Painlevé equation (2.1). However, to use the Boutroux's transformations (2.6), (2.7) properly, it is worth to work on the Riemann surface of the logarithm and we thus fix some notation.

Definition 3.18. We denote by $\widetilde{\mathbb{C}}$ the Riemann surface of the logarithm,

$$\widetilde{\mathbb{C}} = \{z = re^{\mathrm{i}\theta} \mid r > 0,\ \theta \in \mathbb{R}\}, \qquad \pi : z \in \widetilde{\mathbb{C}} \mapsto \overset{\bullet}{z} = re^{\mathrm{i}\theta} \in \mathbb{C}^{\star}.$$

For any $z = re^{\mathrm{i}\theta} \in \widetilde{\mathbb{C}}$, we refer to θ as to its argument, denoted by $\theta = \arg z$.
We denote by $\widetilde{\mathbb{S}}^1$ (usually identified with $\mathbb{R}$) the set of directions of half-lines about 0 on $\widetilde{\mathbb{C}}$. We (still) denote by $\pi : \widetilde{\mathbb{S}}^1 \to \mathbb{S}^1$ the natural projection which makes $\widetilde{\mathbb{S}}^1$ an étalé space on $\mathbb{S}^1$ (and even a universal covering).

Definition 3.19. Let $\theta \in \widetilde{\mathbb{S}}^1$ be a direction and $\tau \in \mathbb{R}$. We set

$$\Pi_{\tau}^{\theta} = \{z = re^{\mathrm{i}\alpha} \in \widetilde{\mathbb{C}} \mid \alpha \in \breve{\theta} \text{ and } \pi(z) \in \overset{\bullet}{\Pi}{}_{\tau}^{\theta}\}.$$

Let $I \subset \widetilde{\mathbb{S}}^1$ be an open arc and $\gamma : I \to \mathbb{R}$ be a locally bounded function. We set $\mathscr{D}(I,\gamma) = \bigcup_{\theta \in I} \Pi_{\gamma(\theta)}^{\theta} \subset \widetilde{\mathbb{C}}$. One calls $\mathscr{D}(I,\gamma)$ a *sectorial neighbourhood of infinity* on $\widetilde{\mathbb{C}}$.

In order to define the transformations (2.6) and (2.7) safely, we introduce a biholomorphic mapping.

Definition 3.20. The biholomorphic mapping $\mathscr{T}$ is defined by:

$$\widetilde{\mathbb{C}} \xrightarrow{\mathscr{T}} \widetilde{\mathbb{C}}, \qquad z \mapsto x = \mathscr{T}(z) = \frac{30^{4/5}}{24} \mathrm{e}^{-\mathrm{i}\pi} z^{4/5}. \tag{3.37}$$

For $I \subset \widetilde{\mathbb{S}}^1$ an open arc and $\gamma : I \to \mathbb{R}$ locally bounded, the domain $\mathscr{D}(I,\gamma)$ is sent onto $\mathscr{T}\big(\mathscr{D}(I_j,\tau)\big) \subset \widetilde{\mathbb{C}}$ through the mapping $\mathscr{T}$, and we set

$$\mathcal{S}(I,\gamma) = \mathscr{T}\big(\mathscr{D}(I,\gamma)\big), \qquad \dot{\mathcal{S}}(I,\gamma) = \pi\Big(\mathscr{S}(I,\gamma)\Big). \tag{3.38}$$

We will consider the domains $\mathscr{D}(I_j,\tau)$, $j \in \mathbb{Z}$, for $I_j = I_0 + j\pi =]0,\pi[+ j\pi$ and $\tau(\theta) = \dfrac{4.22}{|\sin(\theta)|}$. Notice that $\mathscr{D}(I_{j+1},\tau) = \mathrm{e}^{-\mathrm{i}\pi}\mathscr{D}(I_j,\tau)$ for any $j \in \mathbb{Z}$.
The domain $\mathcal{S}(I_j,\tau)$ (see Fig. 3.3 and Fig. 3.4) is a sectorial neighbourhood of infinity of aperture $K_j =]-\frac{11}{5}\pi, -\frac{3}{5}\pi[-\frac{4}{5}j\pi$ and we may notice that, for any $j \in \mathbb{Z}$, $\mathcal{S}(I_{j+1},\tau) = \mathrm{e}^{-4\mathrm{i}\pi/5}\mathscr{S}(I_j,\tau)$. In particular, $\dot{\mathcal{S}}(I_{j+5},\tau) = \dot{\mathcal{S}}(I_j,\tau)$.
We now think of $w_{tri,j} = \mathscr{S}^{I_j}\widetilde{w}$ as a holomorphic function on $\mathscr{D}(I_j,\tau)$. By (3.33) and (3.35), these functions satisfy some relationships: for any $j \in \mathbb{Z}$, for every $z \in \mathscr{D}(I_j,\tau)$,

$$\begin{aligned} w_{tri,j}(z) &= w_{tri,j+1}(z\mathrm{e}^{-\mathrm{i}\pi}), \\ \overline{w_{tri,j}}(z) &= w_{tri,j}(\overline{z}\mathrm{e}^{-(2j+1)\mathrm{i}\pi}), \end{aligned} \tag{3.39}$$

with the convention $\overline{z} = r\mathrm{e}^{-\mathrm{i}\alpha} \in \widetilde{\mathbb{C}}$ for $z = r\mathrm{e}^{\mathrm{i}\alpha} \in \widetilde{\mathbb{C}}$.

This gives sense without ambiguity to (3.4), (2.6) and (2.7), with the transform

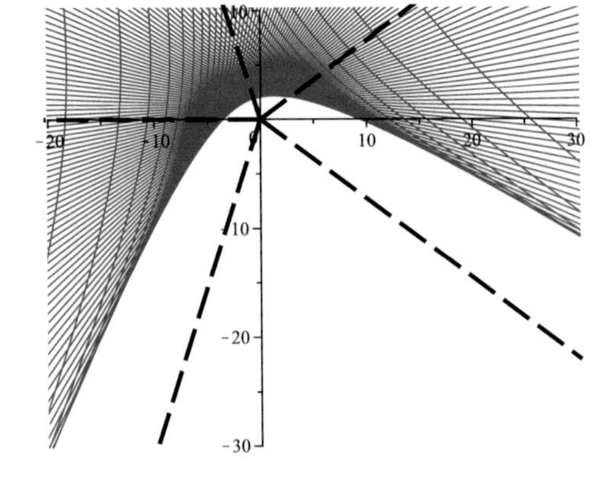

Fig. 3.3 The shaded domain is the projection of $\mathcal{S}(I_0,\tau)$, image by the transformation (3.37), of the domain $\mathscr{D}(I_0,\tau)$ drawn on Fig. 3.2 for $\tau(\theta) = \dfrac{4.22}{|\sin(\theta)|}$. The dash lines recall the sectors (2.9).

$$z \in \mathscr{D}(I_j, \tau) \leftrightarrow x \in \mathscr{S}(I_j, \tau) \tag{3.40}$$
$$w_{tri,j}(z) \leftrightarrow u_{tri,j}(x) = \frac{e^{i\pi/2}}{\sqrt{6}} x^{1/2} \left(1 - \frac{4}{25\left(\mathscr{T}^{-1}(x)\right)^2} + \frac{w_{tri,j}\left(\mathscr{T}^{-1}(x)\right)}{\left(\mathscr{T}^{-1}(x)\right)^2} \right).$$

The functions $u_{tri,j}$ are solutions of the first Painlevé equation (2.1) and by (3.39) and (3.40), they satisfy the following relationships: for any $j \in \mathbb{Z}$, for every $x \in \mathscr{S}(I_j, \tau)$,

$$\begin{aligned} u_{tri,j}(x) &= e^{2i\pi/5} u_{tri,j+1}(xe^{-4i\pi/5}), \\ \overline{u_{tri,j}}(x) &= e^{\frac{2}{5}(2j+1)i\pi} u_{tri,j}(\overline{x}e^{-\frac{2}{5}(4j+7)i\pi}), \end{aligned} \tag{3.41}$$

We recover here the symmetries discussed in Sect. 2.5.

By projection, $u_{tri,j}$ becomes a function holomorphic on the domain $\overset{\bullet}{\mathscr{S}}(I_j, \tau)$. This provides five distinct functions $u_{tri,j}(x)$, $j = 0, \cdots, 4$, the so-called *tri-truncated solutions*.

We now use notions developed in the second volume of this book [Lod016] to which the reader is referred. Since $w_{tri,j}$ is a section on $\breve{I}_j$ of $\mathscr{A}_1$, we deduce that the tritruncated solution $u_{tri,j}(x)$ belongs to the space of holomorphic functions with Gevrey asymptotic expansion of order $4/5$ at infinity in $\overset{\bullet}{\mathscr{S}}(I_j, \tau)$. One can thus recover $u_{tri,j}(x)$ by its asymptotics through $5/4$-summability.

It is also worth mentioning that $u_{tri,2}(x)$ is a real analytic function, as a consequence of property (3.41).

Proposition 3.15. *Let be* $\overset{\bullet}{\mathscr{S}}(I_0, \tau) = \pi\Big(\mathscr{T}\big(\mathscr{D}(I_0, \tau)\big)\Big)$ *with* $\tau(\theta) = \dfrac{4.22}{|\sin(\theta)|}$ *and, for* $j = 0, \cdots, 4$, $\overset{\bullet}{\mathscr{S}}(I_j, \tau) = \omega_j^2 \overset{\bullet}{\mathscr{S}}(I_0, \tau)$, $\omega_j = e^{-\frac{2i\pi}{5}j}$. *The first Painlevé equation (2.1) has 5 tri-truncated solutions* $u_{tri,j}(x)$, $j = 0, \cdots 4$. *The tri-truncated solution* $u_{tri,j}(x)$ *is holomorphic in* $\overset{\bullet}{\mathscr{S}}(I_j, \tau)$, *a sectorial neighbourhood of infinity of aperture* $K_j =]-\frac{11}{5}\pi, -\frac{3}{5}\pi[- \frac{4}{5} j\pi$, *and has in* $\overset{\bullet}{\mathscr{S}}(I_j, \tau)$ *a Gevrey asymptotic expansion of order* $4/5$ *which determined* $u_{tri,j}(x)$ *uniquely. Moreover, for every* $x \in \overset{\bullet}{\mathscr{S}}(I_j, \tau)$, $u_{tri,j}(x) = \omega_j u_{tri,0}(\omega_j^{-2} x)$, $\omega_j = e^{-\frac{2i\pi}{5}j}$, $j = 0, \cdots, 4$, *and* $u_{tri,2}$ *is a real analytic function.*

Remark 3.4. It is shown in exercise 3.3 that for any $j = 0, \cdots, 4$, the tri-truncated solution $u_{tri,j}$ can be analytically continued to the domain $\overset{\bullet}{\mathscr{S}}(I_j, \tau)$ with $\tau(\theta) = \frac{1.4}{|\sin(\theta)|}$. We will see later on that each tri-truncated solution $u_{tri,j}$ can be analytically continued to a wider domain than $\overset{\bullet}{\mathscr{S}}(I_j, \tau)$.

Exercices

3.1. We consider an ordinary differential equation of the form

$$P(\partial)w = G(z,w,w',\dots,w^{(n-1)}) \tag{3.42}$$
$$P(\partial) = \sum_{m=0}^{n} \alpha_{n-m}\partial^m \in \mathbb{C}[\partial],\ \alpha_0 \neq 0,\ \alpha_n \neq 0$$

where $G(z,\mathbf{y})$ is holomorphic in a neighbourhood of $(z,\mathbf{y}) = (\infty,\mathbf{0}) \in \mathbb{C}\times\mathbb{C}^n$, $n \in \mathbb{N}^\star$. We furthermore suppose that $G(z,\mathbf{0}) = O(z^{-1})$ and $\dfrac{\partial^{|\mathbf{l}|}G(z,\mathbf{0})}{\partial \mathbf{y}^{\mathbf{l}}} = O(z^{-1})$ when $|\mathbf{l}| = 1$.

1. Show that for every $M \in \mathbb{N}$ and up to making transformations of the type
$$w = \sum_{k=1}^{M} a_k z^{-k} + v, \tag{3.43}$$
one can instead assume that $G(z,\mathbf{0}) = O(z^{-M-1})$.
2. We suppose that for some $M \in \mathbb{N}^\star$, $G(z,\mathbf{y})$ satisfies $G(z,\mathbf{0}) = O(z^{-M-1})$. Show that, up to making a (so called) shearing transformation of the form
$$w = z^{-M}v, \tag{3.44}$$
one can rather assume that $G(z,\mathbf{0}) = O(z^{-1})$, $\dfrac{\partial^{|\mathbf{l}|}G(z,\mathbf{0})}{\partial \mathbf{y}^{\mathbf{l}}} = O(z^{-1})$ when $|\mathbf{l}| = 1$ and $\dfrac{\partial^{|\mathbf{l}|}G(z,\mathbf{0})}{\partial \mathbf{y}^{\mathbf{l}}} = O(z^{-M(|\mathbf{l}|-1)})$ when $|\mathbf{l}| \geq 2$.
3. Deduce that, through transformations of the type (3.43) and (3.44), one can bring equation (3.42) under the prepared form:
$$P(\partial)w + \frac{1}{z}Q(\partial)w = F(z,w,w',\dots,w^{(n-1)}) \tag{3.45}$$
$$P(\partial) = \sum_{m=0}^{n} \alpha_{n-m}\partial^m \in \mathbb{C}[\partial]\ ,\ Q(\partial) = \sum_{m=0}^{n-1} \beta_{n-m}\partial^m \in \mathbb{C}[\partial]$$
where $F(z,\mathbf{y})$ is holomorphic in a neighbourhood of $(z,\mathbf{y}) = (\infty,\mathbf{0}) \in \mathbb{C}\times\mathbb{C}^n$ such that $F(z,\mathbf{0}) = O(z^{-2-M_0})$, $M_0 \in \mathbb{N}$, $\dfrac{\partial^{|\mathbf{l}|}F(z,\mathbf{0})}{\partial \mathbf{y}^{\mathbf{l}}} = O(z^{-2})$ when $|\mathbf{l}| = 1$ and $\dfrac{\partial^{|\mathbf{l}|}F(z,\mathbf{0})}{\partial \mathbf{y}^{\mathbf{l}}} = O(z^{-2-M_{|\mathbf{l}|}})$, $M_{|\mathbf{l}|} \in \mathbb{N}$, when $|\mathbf{l}| \geq 2$.
4. Show that the shearing transform $w = z^{-M}v$, $M \in \mathbb{N}^\star$, brings equation (3.45) into an equation of the form $P(\partial)v + \frac{1}{z}\left(Q(\partial) - MP'(\partial)\right)v = g(z,v,v',\cdots,v^{(n-1)})$.

3.2. We consider the ODE (3.10) and its unique solution $\widehat{w} \in \mathscr{O}(\overset{\bullet}{\mathscr{R}}{}^{(0)})$.

1. Show that, for any $\rho \in]0,1[$, for any $\zeta = \xi e^{i\theta} \in \overset{\bullet}{\mathscr{R}}_\rho^{(0)}$, $\xi = |\zeta|$,

$$\rho|\widehat{w}(\zeta)| \leq \frac{392}{625} + 7\int_0^\xi |\widehat{w}(re^{i\theta})|\,dr + \frac{1}{2}\int_0^\xi |\widehat{w}^{*2}(re^{i\theta})|\,dr.$$

2. Let be $\rho \in]0,1[$. We consider the (unique) entire function $\widehat{W}$ solution of the convolution equation $\rho\widehat{W}(\xi) = \frac{392}{625} + 7*\widehat{W}(\xi) + \frac{1}{2}*\widehat{W}*\widehat{W}(\xi)$. We denote by $\widetilde{W}(z)$ the inverse Borel transform of $\widehat{W}$.
   Show that $\widetilde{W}(z) \in \mathscr{O}\left(\left\{|z| > \frac{203}{25\rho}\right\}\right)$ (consider the discriminant locus). Show that for $|z| > \frac{203}{25\rho}$, $\widetilde{W}(z) = \frac{784}{625}\left((\rho z - 7) + \left((\rho z-7)^2 - \frac{784}{625}\right)^{1/2}\right)^{-1}$, $\widetilde{W}(z) = O(z^{-1})$ at infinity, and $|\widetilde{W}(z)| \leq \frac{784}{625}\frac{1}{|\rho z - 7|} \leq \frac{28}{25}$.
3. Show that $|\widehat{W}(\xi)| \leq \frac{5684}{625\rho} e^{\frac{203}{25\rho}|\xi|}$ for every $\xi \in \mathbb{C}$.
4. Deduce that for every $\rho \in]0,1[$ and every $\zeta \in \overset{\bullet}{\mathscr{R}}_\rho^{(0)}$, $|\widehat{w}(\zeta)| \leq \frac{5684}{625\rho} e^{\frac{203}{25\rho}|\zeta|}$.

3.3. We consider the ODE

$$y'' + \frac{y'}{z} - y = \frac{392}{625} z^{-4} + \frac{1}{2} y^2. \tag{3.46}$$

deduced from (3.2) by the transformation $v(z) = 1 - \dfrac{4}{25z^2} + y(z)$ or, from (3.6) through the transformation $y(z) = z^{-2}w(z)$. In particular there exists a unique formal series $\widetilde{y}(z) = z^{-2}\widetilde{w}(z) \in \mathbb{C}[[z^{-1}]]$ solution of (3.46). We thus know that the formal Borel transform $\widehat{y}$ belongs to $\mathscr{M}\mathscr{O}(\overset{\bullet}{\mathscr{R}}^{(0)})$ and satisfies the convolution equation associated with (3.46) by formal Borel transformation:

$$(\zeta^2 - 1)\widehat{y} - 1*(\zeta\widehat{y}) = \frac{392}{625}\frac{\zeta^3}{\Gamma(4)} + \frac{1}{2}\widehat{y}*\widehat{y}. \tag{3.47}$$

1. Let $f \in \mathscr{O}_0$ be a germ such that $f(0) = 0$. Show that the solutions $g \in \mathscr{O}_0$ of the convolution equation $(\zeta^2 - 1)g - 1*(\zeta g) = f$ are given by

$$g(\zeta) = \frac{C}{(1-\zeta^2)^{1/2}} - \frac{f(\zeta)}{1-\zeta^2} + \frac{1}{(1-\zeta^2)^{1/2}}\int_0^\zeta \frac{\eta}{(1-\eta^2)^{3/2}} f(\eta)\,d\eta, \quad C \in \mathbb{C}.$$

   (Hint : set $g(\zeta) = \frac{G(\zeta)}{1-\zeta^2}$, differentiate the convolution equation to obtain a non-homogeneous linear differential equation of order 1, and solve this equation).
2. Show that $\widehat{y}$ satisfies the convolution equation (3.47) in $\mathscr{M}\mathscr{O}(\overset{\bullet}{\mathscr{R}}^{(0)})$ if and only if $\widehat{y}$ satisfies the following fixed-point problem:

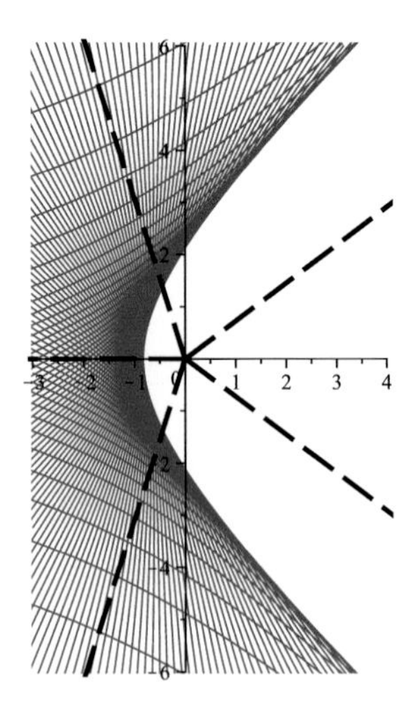

Fig. 3.4 The shaded domain is the projection of $\mathcal{S}_2(I_2,\tau)$, image of the domain $\mathscr{D}(I_2,\tau)$ by the conformal mapping (3.37), for $\tau(\theta)=\dfrac{1.4}{|\sin(\theta)|}$. The dash lines recall the sectors (2.9).

$$\begin{aligned}\widehat{y}&=\mathscr{P}\left(\frac{392}{625}\frac{\zeta^3}{\Gamma(4)}\right)+\frac{1}{2}\mathscr{P}\big(\widehat{y}*\widehat{y}\big)\qquad\text{with}\\ (\mathscr{P}g)(\zeta)&=-\frac{g(\zeta)}{1-\zeta^2}+\frac{1}{(1-\zeta^2)^{1/2}}\int_0^\zeta\frac{\eta}{(1-\eta^2)^{3/2}}g(\eta)\,d\eta,\end{aligned}\tag{3.48}$$

3. Show that for any $\rho\in]0,1[$ and any $\zeta\in\overset{\bullet}{\mathscr{R}}{}^{(0)}_\rho$, and $\left|\frac{\zeta}{(1-\zeta^2)^{3/2}}\right|\le\frac{1}{\rho^{3/2}}$.
4. Show that for any $\rho\in]0,1[$ and any $\zeta\in\overset{\bullet}{\mathscr{R}}{}^{(0)}_\rho$, $|\widehat{y}(\zeta)|\le\widehat{Y}(\xi)$ with $\xi=|\zeta|$, where $\widehat{Y}$ is an entire function which solves the fixed-point problem:

$$\begin{aligned}\widehat{Y}&=\mathscr{Q}\left(\frac{392}{625}\frac{\xi^3}{\Gamma(4)}\right)+\frac{1}{2}\mathscr{Q}\big(\widehat{Y}*\widehat{Y}\big)\\ (\mathscr{Q}G)(\xi)&=\frac{G(\xi)}{\rho}+\frac{1}{\rho^2}\Big(1*G\Big)(\xi)\end{aligned}\tag{3.49}$$

5. For any $\rho\in]0,1[$ we denote by $\widetilde{Y}(z)$ the inverse Borel transform of $\widehat{Y}$. Show that $\widetilde{Y}(z)$ satisfies the algebraic equation

$$\rho\widetilde{Y}=\left(\frac{392}{625}\frac{1}{z^4}+\frac{1}{2}\widetilde{Y}^2\right)\left(1+\frac{1}{\rho z}\right),\quad \widetilde{Y}(z)=\frac{392}{625}\frac{1}{\rho z^4}+O(z^{-5}).\tag{3.50}$$

6. We set $U=D(\infty,\frac{\rho}{1.4})$. Show that the fixed-point problem (3.50) has a unique solution in $B_{\rho^3/2}=\{H\in\mathscr{O}(\overline{U}),\ \|H\|\le\frac{\rho^3}{2}\}$, for
7. Deduce that the minor $\widehat{y}$ of the formal series $\widetilde{y}$ solution of equation (3.46) can be analytically continued is on $\overset{\bullet}{\mathscr{R}}{}^{(0)}$ and that, for any $\rho\in]0,1[$ and any $\zeta\in\overset{\bullet}{\mathscr{R}}{}^{(0)}_\rho$:

$$|\widehat{y}(\zeta)| \leq 0.7\rho^2 e^{\frac{1.4}{\rho}|\zeta|}. \tag{3.51}$$

8. We set $I_j = I_0 + j\pi =]0,\pi| + j\pi$, $j \in \mathbb{Z}$. Show that the Borel-Laplace sum $y_{tri,j} = \mathscr{S}^{I_j}\widetilde{y}$ defines a function holomorphic on $\overset{\bullet}{\mathscr{D}}(I_j,\tau)$ with $\tau(\theta) = \frac{1.4}{|\sin(\theta)|}$.
9. Deduce that the tri-truncated solution $u_{tri,j}$, $j \in \mathbb{Z}$, is holomorphic on the domain $\mathcal{S}(I_j,\tau) = \mathscr{T}\big(\mathscr{D}(I_j,\tau)\big)$ with $\tau(\theta) = \frac{1.4}{|\sin(\theta)|}$. See Fig. 3.4.

3.4. We consider the domain $\overset{\bullet}{\mathscr{D}}(]0,\pi[,\tau)$ for $\tau(\theta) = \frac{\lambda}{\sin(\theta)}$, $\lambda > 0$. We want to describe the boundary $\partial\overset{\bullet}{\mathscr{D}}(]0,\pi[,\tau)$ of this domain.

1. show that $\partial\overset{\bullet}{\mathscr{D}}(]0,\pi[,\tau)$ is the envelope of the following family of line curves: $z = x + iy$, $x\cos(\theta) - y\sin(\theta) = \frac{\lambda}{\sin(\theta)}$, $\theta \in]0,\pi[$.
2. Deduce that $\partial\overset{\bullet}{\mathscr{D}}(]0,\pi[,\tau)$ is the parabolic curve of equation $y = \frac{x^2}{4\lambda} - \lambda$.

References

BH91. M.V. Berry, C.J. Howls, *Hyperasymptotics for integrals with saddles.* Proc. Roy. Soc. Lond. A **434** (1991), 657-675.

BLS002. W. Balser, D.A. Lutz, R. Schäfke, *On the convergence of Borel approximants.* J. Dynam. Control Systems **8** (2002), no. 1, 65-92

CMRSJ007. E. Caliceti, M. Meyer-Hermann, P. Ribeca, A. Surzhykov, U.D. Jentschura, U. D. *From useful algorithms for slowly convergent series to physical predictions based on divergent perturbative expansions.* Phys. Rep. **446** (2007), no. 1-3, 1-96.

CNP93-1. B. Candelpergher, C. Nosmas, F. Pham, *Approche de la résurgence*, Actualités mathématiques, Hermann, Paris (1993).

CNP93-2. B. Candelpergher, C. Nosmas, F. Pham, *Premiers pas en calcul étranger*, Ann. Inst. Fourier (Grenoble) **43** (1993) 201-224.

CHKO007. S. J. Chapman, C. J Howls, J. R. King, A. B. Olde Daalhuis, *Why is a shock not a caustic? The higher-order Stokes phenomenon and smoothed shock formation.* Nonlinearity **20** (2007), no. 10, 2425-2452.

Cos98. O. Costin, *On Borel summation and Stokes phenomena for rank-*1 *nonlinear systems of ordinary differential equations.* Duke Math. J. **93** (1998), no. 2, 289-344.

Cos009. O. Costin, *Asymptotics and Borel summability*, Chapman & Hall/CRC Monographs and Surveys in Pure and Applied Mathematics, 141. CRC Press, Boca Raton, FL, 2009.

Del005. E. Delabaere, *Addendum to the hyperasymptotics for multidimensional Laplace integrals.* Analyzable functions and applications, 177-190, Contemp. Math., 373, Amer. Math. Soc., Providence, RI, 2005.

Del010. E. Delabaere, *Singular integrals and the stationary phase methods. Algebraic approach to differential equations*, 136-209, World Sci. Publ., Hackensack, NJ, 2010.

DH002. E. Delabaere, C. J. Howls, *Global asymptotics for multiple integrals with boundaries.* Duke Math. J. **112** (2002), 2, 199-264.

DP98. E. Delabaere, F. Pham, *Eigenvalues of complex Hamiltonians with PT-symmetry*, Phys. Lett. A **250** (1998), no. 1-3, 25-32.

DP99. E. Delabaere, F. Pham, *Resurgent methods in semi-classical asymptotics*, Ann. Inst. Henri Poincaré, Sect. A **71** (1999), no 1, 1-94.

DR007. E. Delabaere, J.-M. Rasoamanana, *Sommation effective d'une somme de Borel par séries de factorielles.* Ann. Inst. Fourier **57** (2007), no. 2, 421-456.

DT000. E. Delabaere, D.T. Trinh, *Spectral analysis of the complex cubic oscillator,* J.Phys. A: Math. Gen. **33** (2000), no. 48, 8771-8796.

Jen004. U. Jentschura, Habilitation Thesis, Dresden University of Technology, 3rd edition (2004).

JK001. N. Joshi, A. V. Kitaev, *On Boutroux's tritronquée solutions of the first Painlevé equation.* Stud. Appl. Math. **107** (2001), no. 3, 253-291.

Lod016. M. Loday-Richaud, *Divergent Series, Summability and Resurgence II. Simple and Multiple Summability.* Lecture Notes in Mathematics, **2154**. Springer, Heidelberg, 2016.

LR011. M. Loday-Richaud, P. Remy, *Resurgence, Stokes phenomenon and alien derivatives for level-one linear differential systems.* J. Differential Equations **250** (2011), no. 3, 1591-1630.

Mal95. B. Malgrange, *Sommation des séries divergentes.* Expositiones Mathematicae **13**, 163-222 (1995).

PK001. R. B. Paris, D. Kaminski, *Asymptotics and Mellin-Barnes integrals.* Encyclopedia of Mathematics and its Applications, **85**. Cambridge University Press, Cambridge, 2001.

MS016. C. Mitschi, D. Sauzin, *Divergent Series, Summability and Resurgence I. Monodromy and Resurgence.* Lecture Notes in Mathematics, **2153**. Springer, Heidelberg, 2016.

Old96. A.B. Olde Daalhuis, *Hyperterminants I.* J. Comput. Appl. Math. **76** (1996), 255-264.

Old97. A.B. Olde Daalhuis, *Hyperterminants II.* J. Comput. Appl. Math. **89** (1997), 87-95.

Old005. A.B. Olde Daalhuis, *Hyperasymptotics for nonlinear ODEs. II. The first Painlevé equation and a second-order Riccati equation.* Proc. R. Soc. Lond. Ser. A Math. Phys. Eng. Sci. **461** (2005), no. 2062, 3005-3021.

Pha005. F. Pham, *Intégrales singulières.* Savoirs Actuels (Les Ulis). EDP Sciences, Les Ulis; CNRS Editions, Paris, 2005

RS89. J.-P. Ramis, Y. Sibuya, *Hukuhara domains and fundamental existence and uniqueness theorems for asymptotic solutions of Gevrey type.* Asymptotic Anal. **2** (1989), no. 1, 39-94.

Sok80. A.D. Sokal, *An improvement of Watson's theorem on Borel summability.* J. Math. Phys. **21** (1980), no. 2, 261-263.

Tre75. F. Trèves, *Basic linear partial differential equations.*, Pure and Applied Mathematics, Vol. 62. Academic Press, New York-London, 1975.

Was65. W. Wasow, *Asymptotic expansions for ODE.* Reprint of the 1965 edition. Robert E. Krieger Publishing Co., Huntington, N.Y., 1976.

Chapter 4
A Step Beyond Borel-Laplace Summability

Abstract We previously showed that the minor $\widehat{w}$ of the unique formal series solution $\widetilde{w}$ of the prepared ODE associated with the first Painlevé equation, defines a function holomorphic on a cut plane. We further analyze the analytic properties of $\widehat{w}$. We show in Sect. 4.5 how $\widehat{w}$ can be analytically continued to a domain of a Riemann surface, defined in Sect. 4.2, and we draw some consequences. This question is related to the problem of mastering the analytic continuations of convolution products and, as a byproduct, of getting qualitative estimates on any compact set. This is what we will partly do in Sect. 4.3 and Sect. 4.4, using only elementary geometrical arguments. We end with some supplements in Sect. 4.6.

4.1 Introduction

We previously analyzed the Borel-Laplace summability of $\widetilde{w}(z) \in \mathbb{C}[[z^{-1}]]$, the unique formal solution of the prepared ODE (3.6) associated with the first Painlevé equation. This was done by two approaches. In one of them, we defined a sequence $(\widehat{h}_l)_{l\in\mathbb{N}}$ of $\mathbb{Z}$-resurgent functions (proposition 3.8) and we showed that the minor $\widehat{w}$ of $\widetilde{w}$ can be represented as the sum of the series $\sum_{l\geq 0} \widehat{h}_l$ which converges to a holomorphic function on the cut plane $\overset{\bullet}{\mathscr{R}}{}^{(0)} = \mathbb{C} \setminus \{\pm[1,+\infty[\}$. The key issue is:

Question 4.1. Does $\widehat{w}$ belong to the space of $\mathbb{Z}$-resurgent functions or, in other words, is $\widetilde{w}$ a $\mathbb{Z}$-resurgent formal series ?

The answer is "yes" and this will allow an in-depth examination of the (so-called) non-linear Stokes phenomenon for the first Painlevé equation, in the spirit of the various examples already handled in the first volume of this book [MS016]. However, this question requires further tools and we postpone the complete answer to the last chapter of this course. One of these tools consists in sharpening our understanding of Ω-resurgent functions, at least when $\Omega = \mathbb{Z}$. This is our aim in this chapter.

E. Delabaere, *Divergent Series, Summability and Resurgence III*,
Lecture Notes in Mathematics 2155, DOI 10.1007/978-3-319-29000-3_4

The Ω-resurgent functions have been recalled in definition 3.7. This can be rephrased as follows for $\Omega = \mathbb{Z}$:

> the germ $\widehat{\varphi} \in \mathscr{O}_0$ is a $\mathbb{Z}$-resurgent function if and only if $\widehat{\varphi}$ can be represented by a function Φ holomorphic on $U_0 = D(0,r)$, $0 < r \leq 1$, and for any given $\zeta_0 \in U_0^\star = U \setminus \{0\}$, this function can be analytically continued along any path γ of $\mathbb{C} \setminus \mathbb{Z}$ originating from ζ_0.

Notice in this rephrasing that U_0 could have been replaced by any connected and simply connected neighbourdhood of the origin, for instance $\overset{\bullet}{\mathscr{R}}^{(0)}$. (Exercise: why?)

We would like to characterize $\mathbb{Z}$-resurgent functions by means of Riemann surfaces. Let $\widehat{\varphi} \in \mathscr{O}_0$ be a germ of holomorphic functions at 0 and $(\mathscr{O}, \mathfrak{q})$ be the étalé space associated with the sheaf $\mathscr{O}$ (cf. remark 1.1). We denote by $\mathscr{R}(\widehat{\varphi})$ the connected component of $\mathscr{O}$ containing $\widehat{\varphi}$. Endowed with the restricted projection $\mathfrak{q}' = \mathfrak{q}|_{\mathscr{R}(\widehat{\varphi})}$, $\mathscr{R}(\widehat{\varphi})$ is the Riemann surface of $\widehat{\varphi}$.

> We recall that a Riemann surface is a connected one-dimensional complex manifold [MS016, For81, Ebe007]. Notice that $\mathscr{R}(\widehat{\varphi})$ is not necessarily simply connected. (Exercise: why?)

We now assume that $\widehat{\varphi}$ is a $\mathbb{Z}$-resurgent, determined by a function Φ holomorphic on $U_0 \subset \mathbb{C}$, a connected and simply connected neighbourdhood of the origin. Let us draw some conclusions about $\mathscr{R}(\widehat{\varphi})$ from this hypothesis.
In the first place by the very construction of $\mathscr{R}(\widehat{\varphi})$, one can find a neighbourhood $\mathscr{U}_0 \subset \mathscr{R}(\widehat{\varphi})$ of $\widehat{\varphi}$ such that $\mathfrak{q}'(\mathscr{U}_0) = U$ and the mapping $\mathfrak{q}'|_{\mathscr{U}_0} : \mathscr{U}_0 \to U_0$ is a homeomorphism. In particular, $\mathscr{U}_0$ is connected and simply connected.
In addition, let be $\zeta_0 \in U$ and denote by $\widehat{\varphi}_0 = \mathfrak{q}'|_{\mathscr{U}_0}^{-1}(\zeta_0) \in \mathscr{U}_0$ the germ of holomorphic functions at ζ_0 determined by Φ. Since $\widehat{\varphi}$ is $\mathbb{Z}$-resurgent, $\widehat{\varphi}_0$ can be analytically continued along any path γ of $\mathbb{C} \setminus \mathbb{Z}$ originating from ζ_0. In other words, any such path γ can be lifted to $\mathscr{R}(\widehat{\varphi})$ from $\widehat{\varphi}_0$ with respect to $\mathfrak{q}'$, and this lifting is unique by uniqueness of lifting [For81]. We denote by Γ this lifting, $\gamma = \mathfrak{q}' \circ \Gamma$. Now assume that γ is a loop homotopic in $\mathbb{C} \setminus \mathbb{Z}$ to a loop γ' in $U_0^\star$. Then $\mathrm{cont}_\gamma \widehat{\varphi}_0 = \widehat{\varphi}_0$ because cont_γ only depends on the homotopy class of γ in $\mathbb{C} \setminus \mathbb{Z}$, meanwhile $\widehat{\varphi}_0$ is represented by $\Phi \in \mathscr{O}(U_0)$ on U_0. In regard, lifting the homotopy, Γ is homotopic to a loop in $\mathscr{U}_0$, thus null-homotopic since $\mathscr{U}_0$ is simply connected.

This being said, we raise the following question:

Question 4.2. Can we determine a simply connected Riemann surface $\mathscr{R}_\mathbb{Z}$ on which *any* $\mathbb{Z}$-resurgent function can be analytically continued ?

We answer to this question in Sect. 4.2, through an explicit construction of $\mathscr{R}_\mathbb{Z}$. We also describe there various sheets of this Riemann surface which will be usefull for later purposes.

Next we turn to the convolution product. We already know by theorem 3.2 that the space of $\mathbb{Z}$-resurgent functions is stable under convolution product. In other words, if the germs $\widehat{\varphi}, \widehat{\psi} \in \mathscr{O}_0$ can be analytically continued to the Riemann surface $\mathscr{R}_\mathbb{Z}$, then it is the same for their convolution product $\widehat{\varphi} * \widehat{\psi}$. But what about the question of upper bounds ? In the previous chapter, the answer was essentially the matter of lemma 3.3 and the new issue is:

Question 4.3. Can we formulate an analogue of lemma 3.3 for holomorphic functions defined on the Riemann surface $\mathscr{R}_{\mathbb{Z}}$?

The main result of this chapter, namely theorem 4.1 and its corollaries detailed in Sect. 4.4, gives a partial to this question. Its proof relies on the use of shortest symmetrically contractile paths which we describe in Sect. 4.3. We then apply our results to the first Painlevé equation in Sect. 4.5, to get theorem 4.2. A theoretical supplement ends this chapter.

4.2 Resurgent Functions and Riemann Surface

This section is devoted to defining the Riemann surface $\mathscr{R} = \mathscr{R}_{\mathbb{Z}}$ and some of its sheets. We first recall usual notation.

4.2.1 Notation

In this course, a *path* (or a parametrized curve) λ in a topological space X is any continuous function $\lambda : [a, a+l] \to X$, where $[a, a+l] \subset \mathbb{R}$ is a (compact) interval possibly reduced to $\{a\}$.
One denotes by λ^{-1} the inverse path, that is $\lambda^{-1} : t \in [a, a+l] \mapsto \lambda(2a+l-t)$
We often work with *standard paths*, that is paths defined on $[0,1]$. The path $\underline{\underline{\lambda}} : t \in [0,1] \mapsto \lambda(a+tl)$ is the *standardized path* of λ.
For two paths $\lambda_1 : [a, a+l] \to X$, $\lambda_2 : [b, b+k] \to X$ so that $\lambda_1(a+l) = \lambda_2(b)$, one denotes by $\lambda_1\lambda_2$ their *product (or also concatenation)*,

$$\lambda_1\lambda_2 : t \in [a, a+l+k] \mapsto \begin{cases} \lambda_1(t), t \in [a, a+l] \\ \lambda_2(t-a-l+b), t \in [a+l, a+l+k] \end{cases}$$

We denote by $\sim_X$ the equivalence relation of homotopy of paths with fixed extremities in X : $\lambda_1 \sim_X \lambda_2$ if the two paths λ_1, λ_2 in X have same extremities and there exists a continuous map $H : [0,1] \times [0,1] \to X$ that realizes a homotopy between the standardized paths $\underline{\underline{\lambda}}_1$ and $\underline{\underline{\lambda}}_2$.
When X has a (finite $\mathbb{R}$-dimensional and $\mathscr{C}^\infty$) differential structure, one can define smooth paths. We recall that any path can be uniformaly approached by $\mathscr{C}^\infty$-paths. Typically in this course, $X = \mathbb{C}$ with its 2-dimensional real differential structure. For a piecewise $\mathscr{C}^1$-path $\lambda : I \to \mathbb{C}$, its length is denoted by length(λ) where length$(\lambda) = \sum_{k=1}^{n} \int_{t_{k-1}}^{t_k} |\underline{\underline{\lambda}}'(t)| dt$, for any partition $0 = t_0 < t_1 < \cdots < t_n = 1$ of $[0,1]$ for which $\underline{\underline{\lambda}}$ has a continuous derivative on each interval $[t_{k-1}, t_k]$.

4.2.2 The Riemann Surface of $\mathbb{Z}$-Resurgent Functions

4.2.2.1 The Space $\mathscr{R}_{\mathbb{Z},\zeta_0}$

Definition 4.1. Let U_0 be a connected and simply connected neighbourdhood of the origin in $\mathbb{C}$ and $\zeta_0 \in U_0^\star = U_0 \setminus \{0\}$. We denote by $\mathfrak{A}_{\zeta_0}$ (*resp.* $\mathfrak{B}_{\zeta_0}$) the set of paths in U_0 (*resp.* $\mathbb{C}\setminus\mathbb{Z}$) originating from ζ_0, endowed with the equivalence relation $\sim_{U_0}$ (*resp.* $\sim_{\mathbb{C}\setminus\mathbb{Z}}$) of homotopy of paths with fixed extremities.

We set $\mathfrak{R}_{\zeta_0} = \mathfrak{A}_{\zeta_0} \cup \mathfrak{B}_{\zeta_0}$ and denote by $\overset{\zeta_0}{\sim}$ the relation on $\mathfrak{R}_{\zeta_0}$ defined as follows. For any two $\gamma_1, \gamma_2 \in \mathfrak{R}_{\zeta_0}$, $\gamma_1 \overset{\zeta_0}{\sim} \gamma_2$ when one of the following conditions is satisfied:

- either $\gamma_1 \sim_{U_0} \gamma_2$ or $\gamma_1 \sim_{\mathbb{C}\setminus\mathbb{Z}} \gamma_2$
- or else there exists $\gamma_3 \in \mathfrak{A}_{\zeta_0} \cap \mathfrak{B}_{\zeta_0}$ such that $\begin{cases} \gamma_1 \sim_{U_0} \gamma_3 \\ \gamma_2 \sim_{\mathbb{C}\setminus\mathbb{Z}} \gamma_3 \end{cases}$ or $\begin{cases} \gamma_1 \sim_{\mathbb{C}\setminus\mathbb{Z}} \gamma_3 \\ \gamma_2 \sim_{U_0} \gamma_3 \end{cases}$.

Exercise 4.1. Show that $\overset{\zeta_0}{\sim}$ is an equivalence relation on $\mathfrak{R}_{\zeta_0}$.

Definition 4.2. Let γ be an element of $\mathfrak{R}_{\zeta_0}$. We denote by $\mathrm{cl}_{\zeta_0}(\gamma)$ its equivalence class for the relation $\overset{\zeta_0}{\sim}$. We set $\overset{\bullet}{\mathscr{R}} = \mathbb{C}\setminus\mathbb{Z}^\star$ and we define:

$$\mathscr{R}_{\mathbb{Z},\zeta_0} = \{\mathrm{cl}_{\zeta_0}(\gamma) \mid \gamma \in \mathfrak{R}_{\zeta_0}\} \quad \text{and} \quad \mathfrak{p}_{\zeta_0} : \mathrm{cl}_{\zeta_0}(\gamma) \mapsto \underline{\underline{\gamma}}(1) \in \overset{\bullet}{\mathscr{R}}. \tag{4.1}$$

Notice that $\mathfrak{p}_{\zeta_0}^{-1}(0)$ is reduced to a single distinguished point, the equivalence class of any $\gamma \in \mathfrak{A}_{\zeta_0}$ ending at the origin, because U_0 is simply connected.

Definition 4.3. One denotes by $\underline{0} \in \mathscr{R}_{\mathbb{Z},\zeta_0}$ the unique pre-image of 0 by $\mathfrak{p}_{\zeta_0}$. Let be $\zeta \in \overset{\bullet}{\mathscr{R}}$, one denotes by $\underline{\zeta} \in \mathscr{R}_{\mathbb{Z},\zeta_0}$ one of its pre-image if exists. For any $\zeta \in \mathscr{R}_{\mathbb{Z},\zeta_0}$, one denotes by $\overset{\bullet}{\zeta} = \mathfrak{p}_{\zeta_0}(\zeta)$ its projection by $\mathfrak{p}_{\zeta_0}$.

4.2.2.2 The Riemann Surface $\mathscr{R}_{\mathbb{Z},\zeta_0}$

The topological space $\mathscr{R}_{\mathbb{Z},\zeta_0}$ We endow $\mathscr{R}_{\mathbb{Z},\zeta_0}$ with a topology, a basis $\mathscr{B} = \{\mathscr{U}\}$ of open sets being given as follows. Let $\underline{\zeta}$ be an element of $\mathscr{R}_{\mathbb{Z},\zeta_0}$ and set $\zeta = \mathfrak{p}_{\zeta_0}(\underline{\zeta})$.

- Assume that $\underline{\zeta} = \mathrm{cl}_{\zeta_0}(\gamma)$ with $\gamma \in \mathfrak{A}_{\zeta_0}$ (thus $\zeta \in U_0$). Let $U \subset U_0$ be any connected and simply connected open neighbourhood of ζ. To $(U, \underline{\zeta})$ we associate the set $\mathscr{U} \subset \mathscr{R}_{\mathbb{Z},\zeta_0}$ made of all $\xi = \mathrm{cl}_{\zeta_0}(\gamma_1\gamma_2)$ where γ_1 satisfies $\gamma_1 \sim_{U_0} \gamma$ while γ_2 is any path in U originating from ζ.
- Assume that $\underline{\zeta} = \mathrm{cl}_{\zeta_0}(\gamma)$ with $\gamma \in \mathfrak{B}_{\zeta_0}$ (in particular $\zeta \neq 0$). Let $U \subset \mathbb{C}\setminus\mathbb{Z}$ be any connected and simply connected open neighbourhood of ζ. To $(U, \underline{\zeta})$ we associate the set $\mathscr{U} \subset \mathscr{R}_{\mathbb{Z},\zeta_0}$ made of all $\xi = \mathrm{cl}_{\zeta_0}(\gamma_1\gamma_2)$ where γ_1 satisfies $\gamma_1 \sim_{\mathbb{C}\setminus\mathbb{Z}} \gamma$ and γ_2 is any path in U originating from ζ.

Exercise 4.2. Show the following properties (hint : see the classical construction of the universal covering of $\mathbb{C}\setminus\mathbb{Z}$ [For81, Ebe007, MS016] and adapt the arguments):

1. $\mathscr{B}=\{\mathscr{U}\}$ provides a Hausdorff topology on $\mathscr{R}_{\mathbb{Z},\zeta_0}$;
2. the projection $\mathfrak{p}_{\zeta_0}$ is a continuous mapping and even, a local homeomorphism : for every $\mathscr{U}\in\mathscr{B}$, the mapping $\mathfrak{p}_{\zeta_0}|_{\mathscr{U}}\to U=\mathfrak{p}_{\zeta_0}(\mathscr{U})$ is a homeomorphism.
3. $\mathscr{R}_{\mathbb{Z},\zeta_0}$ is arc-connected and simply connected.

The Riemann surface $\mathscr{R}_{\mathbb{Z},\zeta_0}$ The following proposition is a direct consequence of the properties detailed in exercise 4.2.

Proposition 4.1. *The space $\mathscr{R}_{\mathbb{Z},\zeta_0}$ is a topologically separated space, arc-connected and simply connected The projection $\mathfrak{p}_{\zeta_0}$ makes $\mathscr{R}_{\mathbb{Z},\zeta_0}$ an étalé space on $\overset{\bullet}{\mathscr{R}}$. By pulling back by $\mathfrak{p}_{\zeta_0}$ the complex structure of $\mathbb{C}$, the space $\mathscr{R}_{\mathbb{Z},\zeta_0}$ becomes a Riemann surface with a uniquely defined distinguished point $\underline{0}=\mathfrak{p}_{\zeta_0}^{-1}(0)$.*

Notice that $\mathfrak{p}_{\zeta_0}$ is not a covering map since the curve lifting property [For81, Ebe007] is not satisfied. For instance, as a rule, a path starting from and ending at 0 cannot be lifted from $\underline{0}$ on $\mathscr{R}_{\mathbb{Z},\zeta_0}$ with respect to $\mathfrak{p}_{\zeta_0}$.
We precise the "pull back" of the complex structure. If $\mathscr{U}_1,\mathscr{U}_2$, $\mathscr{U}_1\cap\mathscr{U}_2\neq\emptyset$ are two open sets of $\mathscr{R}_{\mathbb{Z},\zeta_0}$ such that the mappings $\mathfrak{p}_{\zeta_0}|_{\mathscr{U}_1}:\mathscr{U}_1\to\mathfrak{p}_{\zeta_0}(\mathscr{U}_1)$ and $\mathfrak{p}_{\zeta_0}|_{\mathscr{U}_2}:\mathscr{U}_2\to\mathfrak{p}_{\zeta_0}(\mathscr{U}_2)$ are two homeomorphisms, then the chart transition $\mathfrak{p}_{\zeta_0}|_{\mathscr{U}_2}\circ\mathfrak{p}_{\zeta_0}|_{\mathscr{U}_1}^{-1}:\mathfrak{p}_{\zeta_0}(\mathscr{U}_1\cap\mathscr{U}_2)\to\mathfrak{p}_{\zeta_0}(\mathscr{U}_1\cap\mathscr{U}_2)$ is nothing but the identity map, thus is biholomorphic. This makes $\mathscr{R}_{\mathbb{Z},\zeta_0}$ a Riemann surface.

Exercise 4.3. Let U_0,ζ_0 be as in definition 4.1. Let $U_1\subset\mathbb{C}\setminus\mathbb{Z}$ be a connected and simply connected open neighbourhood of ζ_0 such that $U_0\cap U_1$ is connected. We denote by $\mathscr{U}_0\subset\mathscr{R}_{\mathbb{Z},\zeta_0}$ the uniquely defined open set such that $\mathfrak{p}_{\zeta_0}|_{\mathscr{U}_0}:\mathscr{U}_0\to U_0$ is a homeomorphism and we set $\underline{\zeta}_0=\mathfrak{p}_{\zeta_0}|_{\mathscr{U}_0}^{-1}(\zeta_0)$. We denote by $\mathscr{U}_1\subset\mathscr{R}_{\mathbb{Z},\zeta_0}$ the uniquely defined neighbourdhood of $\underline{\zeta}_0$ such that $\mathfrak{p}_{\zeta_0}|_{\mathscr{U}_1}:\mathscr{U}_1\to U_1$ is a homeomorphism.

1. Show that $U=U_0\cup U_1$ is simply connected.
2. We set $\mathscr{U}=\mathscr{U}_0\cup\mathscr{U}_1$. Show that $\mathfrak{p}_{\zeta_0}|_{\mathscr{U}}$ is a homeomorphism between $\mathscr{U}$ and U.

4.2.2.3 The Riemann Surface $\mathscr{R}_{\mathbb{Z}}$

Up to now, the Riemann surface $(\mathscr{R}_{\mathbb{Z},\zeta_0},\mathfrak{p}_{\zeta_0})$ depends on the given of U_0, a connected and simply connected neighbourdhood of the origin, and of $\zeta_0\in U_0$.

Lemma 4.1. *Let U_0 (resp. U_1) be a connected and simply connected neighbourdhood of the origin in $\mathbb{C}$ and $\zeta_0\in U_0^\star$ (resp. $\zeta_1\in U_1^\star$). Then there exists a fiber preserving homeomorphism $\tau:\mathscr{R}_{\mathbb{Z},\zeta_0}\to\mathscr{R}_{\mathbb{Z},\zeta_1}$ between the Riemann surfaces $(\mathscr{R}_{\mathbb{Z},\zeta_0},\mathfrak{p}_{\zeta_0},\underline{0})$ and $(\mathscr{R}_{\mathbb{Z},\zeta_1},\mathfrak{p}_{\zeta_1},\underline{0})$.*

Proof. Left as an exercise to the reader.

Definition 4.4. The class of isomorphisms of the Riemann surfaces $(\mathscr{R}_{\mathbb{Z},\zeta_0}, \mathfrak{p}_{\zeta_0}, \underline{0})$ is denoted by $(\mathscr{R}_{\mathbb{Z}}, \mathfrak{p}, \underline{0})$. In this course we often use abridged notation $\mathscr{R}$.

Proposition 4.2. *Let $\widehat{\varphi}_0 \in \mathscr{O}_0$ be a germ of holomorphic functions at the origin and let $(\mathscr{R}(\widehat{\varphi}_0), \mathfrak{q}, \widehat{\varphi}_0)$ be its Riemann surface. Then $\widehat{\varphi}_0$ is a $\mathbb{Z}$-resurgent function if and only if $(\mathscr{R}, \mathfrak{p}, \underline{0})$ is contained in $(\mathscr{R}(\widehat{\varphi}_0), \mathfrak{q}, \widehat{\varphi}_0)$, that is there exists a fiber preserving continuous map $\tau : \mathscr{R} \to \mathscr{R}(\widehat{\varphi}_0)$, $\mathfrak{q} \circ \tau = \mathfrak{p}$ and $\tau(\underline{0}) = \widehat{\varphi}_0$.*

Proof. Assume that $\widehat{\varphi}_0$ is a $\mathbb{Z}$-resurgent function. We set $U_0 = D(0,1)$ and we pick a point $\zeta_0 \in U_0^\star$. On the one hand, there is a uniquely determined domain $\mathscr{U}_0 \subset \mathscr{R}$ homeomorphic to U_0 by $\mathfrak{p}|_{\mathscr{U}_0}$ and we set $\underline{\zeta}_0 = \mathfrak{p}|_{\mathscr{U}_0}^{-1}(\zeta_0)$. On the other hand, there is a uniquely determined domain $\mathscr{U}_0' \subset \mathscr{R}(\widehat{\varphi}_0)$ homeomorphic to U_0 by $\mathfrak{q}|_{\mathscr{U}_0'}$ and we set $\underline{\zeta}_0' = \mathfrak{q}|_{\mathscr{U}_0'}^{-1}(\zeta_0)$. We get this way a natural fiber preserving homeomorphism $\tau|_{\mathscr{U}_0} : \underline{\zeta} \in \mathscr{U}_0 \mapsto \underline{\zeta}' \in \mathscr{U}_0'$. We now extend $\tau|_{\mathscr{U}_0}$ as follows: pick any path γ in $\mathbb{C} \setminus \mathbb{Z}$, originating from ζ_0, let Γ be its lifting from $\underline{\zeta}_0$ on $\mathscr{R}$ with respect to $\mathfrak{p}$ and set $\underline{\zeta} = \underline{\underline{\Gamma}}(1) \in \mathscr{R}$. The path γ can be lifted as well on $\mathscr{R}(\widehat{\varphi}_0)$ with respect to $\mathfrak{q}$ from $\underline{\zeta}_0'$ into a path Γ', because $\widehat{\varphi}_0$ is $\mathbb{Z}$-resurgent. We set $\underline{\underline{\Gamma}}'(1) = \underline{\zeta}'$. The extended mapping $\tau : \underline{\zeta} \in \mathscr{R} \mapsto \mathscr{R}(\widehat{\varphi}_0)$ thus (well)-defined is injective by uniqueness of lifting [For81], continuous because we work with étalé spaces, and preserves fibers.
The converse of the proposition is left to the reader as an exercise. □

In other words, $\widehat{\varphi}_0 \in \mathscr{O}_0$ is a $\mathbb{Z}$-resurgent function if and only $\widehat{\varphi}_0$ can be analytically continued to the Riemann surface $\mathscr{R}_{\mathbb{Z}}$. This means that one can identify the space $\hat{\mathscr{R}}$ with the space $\mathscr{O}(\mathscr{R})$ of functions holomorphic on the Riemann surface $\mathscr{R}$.

Definition 4.5. The Riemann surface $(\mathscr{R}_{\mathbb{Z}}, \mathfrak{p}, \underline{0})$ is called the *Riemann surface of $\mathbb{Z}$-resurgent functions.*

4.2.3 Riemann Surface and Sheets

We introduce various sheets and domains on $\mathscr{R}_{\mathbb{Z}}$. At first sight artificially complicated, these constructions will be needed to state one of main results of this chapter, namely theorem 4.1 and its consequences.

4.2.3.1 Principal Sheet

By the very construction of the Riemann surface $\mathscr{R}$, there exists a unique domain $\mathscr{R}^{(0)}$ of $\mathscr{R}$ so that $\mathfrak{p}|_{\mathscr{R}^{(0)}}$ realizes a homeomorphism between $\mathscr{R}^{(0)}$ and the simply connected domain $\overset{\bullet}{\mathscr{R}}{}^{(0)}$. The domain $\mathscr{R}^{(0)}$ is made of endpoints $\underline{\zeta}$ of paths deduced from any segment $[0,\zeta] \subset \overset{\bullet}{\mathscr{R}}{}^{(0)}$, by lifting from $\underline{0}$ with respect to $\mathfrak{p}$.

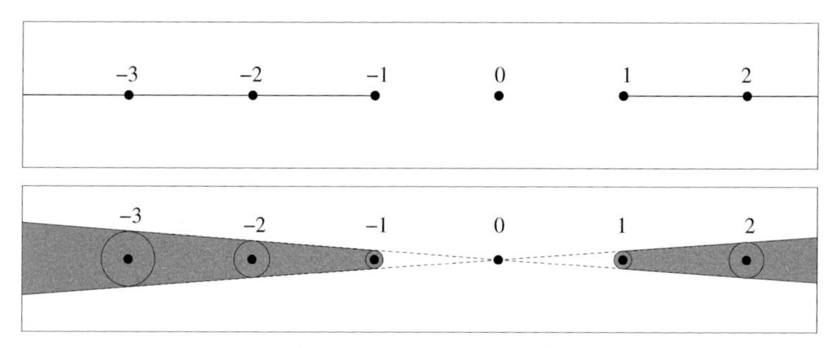

Fig. 4.1 Above, the domain $\overset{\bullet}{\mathscr{R}}^{(0)}$. Below, the domain $\overset{\bullet}{\mathscr{R}}_\rho^{(0)}$.

Definition 4.6. One refers to $\mathscr{R}^{(0)}$ as to the *principal sheet* of the pointed Riemann surface $(\mathscr{R},\underline{0})$. For every $\rho \in]0,1[$, one denotes by $\mathscr{R}_\rho^{(0)}$ the unique open subset of $\mathscr{R}^{(0)}$ such that $\mathfrak{p}(\mathscr{R}_\rho^{(0)}) = \overset{\bullet}{\mathscr{R}}_\rho^{(0)}$. (See Fig. 4.1).

4.2.3.2 Other Sheets

Definition 4.7. Let be $m \in \mathbb{N}^\star$, $\varepsilon = (\varepsilon_1,\cdots,\varepsilon_{m-1}) \in \{+,-\}^{m-1}$ a $(m-1)$-tuple of signs and $\mathbf{n} = (n_1,\cdots,n_{m-1}) \in (\mathbb{N}^\star)^{m-1}$ a $(m-1)$-tuple of positive integers. Let $\theta_1 \in \{0,\pi\} \subset \mathbb{S}^1$ be a direction. Let γ be a path in $\mathbb{C}$ originating from 0.
When $m=1$, one says that the path γ is of type $\gamma_{()}^{\theta_1}$ when γ closely follows the segment $e^{i\theta_1}]0,1[=]0,\omega_1[$ toward $\omega_1 = e^{i\theta_1}$.
Otherwise, for $m \geq 2$, on says that the γ is of type type $\gamma_{\varepsilon^{\mathbf{n}}}^{\theta_1}$ if γ connects the segment $]0,\omega_1[$ to the segment $]\omega_{m-1},\omega_m[$, $\omega_m - \omega_{m-1} = e^{i\theta_m}$, through the following steps:

- γ closely follows the segment $]0,\omega_1[$ toward the direction θ_1, makes n_1 half-turns around the point ω_1, anti clockwise when $\varepsilon_1 = +$, clockwise when $\varepsilon_1 = -1$, and finally closely follows the segment $]\omega_1,\omega_2[$, $\omega_2 - \omega_1 = e^{i\theta_2}$, toward the direction $\theta_2 = \theta_1 + \varepsilon_1(n_1-1)\pi$;
- then, successively for $k = 2,\cdots,m-1$, γ makes n_k half-turns around the point ω_k, anti clockwise when $\varepsilon_k = +$, clockwise when $\varepsilon_k = -1$, and eventually closely follows the segment $]\omega_k,\omega_{k+1}[$, $\omega_{k+1} - \omega_k = e^{i\theta_{k+1}}$, toward the direction $\theta_{k+1} = \theta_k + \varepsilon_k(n_k-1)\pi$.

When $\mathbf{n} = (1,\cdots,1) \in \{1\}^{m-1}$, we simply say that γ is of type $\gamma_\varepsilon^{\theta_1}$. (See Fig. 4.2 and Fig. 4.3).

For instance, if γ is of type $\gamma_\varepsilon^\theta$, then someone standing at $0 \in \mathbb{C}$ and looking in the direction of the half-line $]0,e^{i\theta}\infty[$ will see the path γ avoiding the point $\omega_n = ne^{i\theta} \in \mathbb{C}^\star$ by swerving in the direction of his right hand when $\varepsilon_n = +$, of his left hand when $\varepsilon_n = -$.

Definition 4.8. Let be $m \in \mathbb{N}^\star$, $\varepsilon \in \{+,-\}^m$, $\mathbf{n} \in (\mathbb{N}^\star)^m$ and $\theta \in \{0,\pi\}$. We denote by $\mathscr{R}^{\varepsilon^{\mathbf{n}},\theta} \subset \mathscr{R}$ the domain made of endpoints $\zeta = \underline{\underline{\Gamma}}(1)$ where Γ is the lift from $\underline{0}$

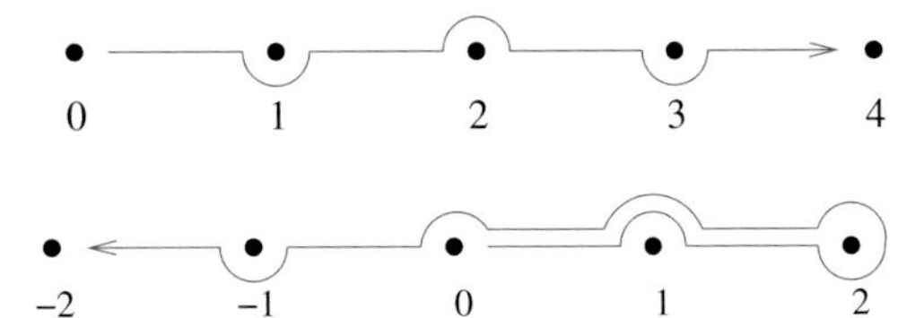

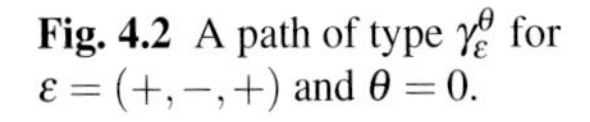

Fig. 4.2 A path of type $\gamma_{\varepsilon}^{\theta}$ for $\varepsilon = (+,-,+)$ and $\theta = 0$.

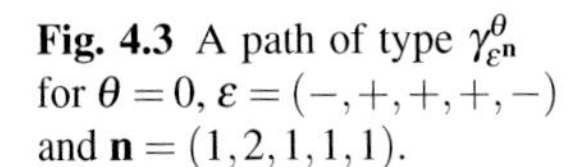

Fig. 4.3 A path of type $\gamma_{\varepsilon^{\mathbf{n}}}^{\theta}$ for $\theta = 0$, $\varepsilon = (-,+,+,+,-)$ and $\mathbf{n} = (1,2,1,1,1)$.

with respect to $\mathfrak{p}$ of any path γ of the form $\gamma = \gamma_1\gamma_2$ with the conditions : γ_1 is a path of type $\gamma_{\varepsilon^{\mathbf{n}}}^{\theta}$ ending at $\overset{\bullet}{\xi} \in]p,(p+1)[=]\omega_m,\omega_{m+1}[$, γ_2 is a path starting from $\overset{\bullet}{\xi}$, and contained in the simply connected domain $\mathbb{C} \setminus \{]-\infty,p] \cup [p+1,+\infty[\}$, star-shaped from $\overset{\bullet}{\xi}$. When $\mathbf{n} = (1,\cdots,1) \in \{1\}^m$, we simply write $\mathscr{R}^{\varepsilon,\theta} = \mathscr{R}^{\varepsilon^{\mathbf{n}},\theta}$.

The collection of sheets $\{\mathscr{R}^{(0)},\mathscr{R}^{\varepsilon^{\mathbf{n}},\theta}\}$ provides an open covering of $\mathscr{R}$, with the following property: the restriction $\mathfrak{p}|_{\mathscr{R}^{\varepsilon^{\mathbf{n}},\theta}}$ is a homeomorphism between $\mathscr{R}^{\varepsilon^{\mathbf{n}},\theta}$ and the simply connected domain $\mathbb{C} \setminus \{]-\infty,p] \cup [p+1,+\infty[\}$ where $]p,(p+1)[=]\omega_m,\omega_{m+1}[$, with ω_m,ω_{m+1} as given by definition 4.7.

Remark that for every $\theta \in \{0,\pi\}$, for every $m \in \mathbb{N}^\star$ and for every ε either in $\{+\}^m$ or in $\{-\}^m$, $\mathscr{R}^{(0)}$ and $\mathscr{R}^{\varepsilon,\theta}$ have a non-empty intersection (a half-plane on projection). This justifies the following definitions.

Definition 4.9. Let be $m \in \mathbb{N}^\star$. We set $(+)_{m-1} = (+,\cdots,+) \in \{+\}^{m-1}$ and $(-)_{m-1} = (-,\cdots,-) \in \{-\}^{m-1}$. We denote by $(\pm)_{m-1}$ any $(m-1)$-tuple of the form $(\pm,\cdots,\pm) \in \{+,-\}^{m-1}$. Also, $(+)_0 = (-)_0 = (\pm)_0 = ()$ is the 0-tuple.

Thus the set of all $(\pm)_m$ is made of 2^m elements.

Definition 4.10. The domain $\mathscr{R}^{\varepsilon,\theta}$ is called a $\mathscr{R}^{(0)}$*-nearby sheet* if $\varepsilon \in \bigcup_{m\in\mathbb{N}^\star} \{(+)_m,(-)_m\}$. One denotes by $\mathscr{R}^{(1)} \subset \mathscr{R}$ the union of the principal sheet and of all nearby sheets: $\mathscr{R}^{(1)} = \mathscr{R}^{(0)} \bigcup_{\theta\in\{0,\pi\},m\in\mathbb{N}^\star} \mathscr{R}^{(+)_m,\theta} \cup \mathscr{R}^{(-)_m,\theta}$.

More generally, for any $k \in \mathbb{N}^\star$, one defines:

$$\mathscr{R}^{(k+1)} = \mathscr{R}^{(k)} \bigcup_{\substack{\theta\in\{0,\pi\},m\in\mathbb{N}^\star \\ \mathbf{n}\in(\mathbb{N}^\star)^k}} \mathscr{R}^{((\pm)_k^{\mathbf{n}},(+)_{m-1}),\theta} \cup \mathscr{R}^{((\pm)_k^{\mathbf{n}},(-)_{m-1}),\theta}.$$

Remark 4.1. Notice that $\mathfrak{p}(\mathscr{R}^{(+)_m,\theta}) = \mathfrak{p}(\mathscr{R}^{(-)_m,\theta}) = \mathbb{C} \setminus e^{i\theta}\{]-\infty,m] \cup [m+1,+\infty[\}$ and $\bigcup_k \mathscr{R}^{(k)} = \mathscr{R}$.

For every integer $k \in \mathbb{N}$, the domain $\mathscr{R}^{(k)}$ inherits from $\mathscr{R}$ the structure of complex manifold, thus is is a Riemann surface.

4.2.4 Nearby Domains

Our aim is to introduce various domains of the Riemann surface $\mathscr{R}$ which will be convenient for later purposes.

We start with the following remark: for $\rho \in]0,1[$ and $m \in \mathbb{N}^\star$, the closed discs $\overline{D}(m, m\rho)$ and $\overline{D}(m+1,(m+1)\rho)$ are disjoint as soon as $m < \frac{\rho^{-1}-1}{2}$. Thus, now assuming that $\rho \in]0, \frac{1}{5}[$ and introducing the integer part $\mathscr{M}(\rho)+1 = \lfloor \frac{\rho^{-1}-1}{2} \rfloor \geq 2$ ($\lfloor . \rfloor$ is the floor function), one observes that the discs $\overline{D}(m, |m|\rho)$ do not overlap when $|m| \leq \mathscr{M}(\rho)+1$.

Definition 4.11. Let be $\rho \in]0, \frac{1}{5}[$. We denote by $\mathscr{M}(\rho) \in \mathbb{N}^\star$ the positive integer defined by $\mathscr{M}(\rho) = \lfloor \frac{\rho^{-1}-1}{2} \rfloor - 1$. For any integer $m \in \mathbb{Z}^\star$ such that $|m| \leq \mathscr{M}(\rho)+1$, we denote by $\overline{D}_m = \overline{D}(m, |m|\rho)$ the closed disc centered at m with radius $|m|\rho$, and $\overline{D}_0 = \{0\}$. For any $\theta \in \{0, \pi\}$, we denote by $\overline{\mathscr{D}}_\rho^\theta \subset \mathbb{C}$ the closed subset defined by

$$\overline{\mathscr{D}}_\rho^\theta = \left\{ t\zeta \mid t \in [1,+\infty[, \, \zeta \in \overline{D}_{\mathrm{e}^{\mathrm{i}\theta}(M+1)} \right\} \bigcup_{0 \leq m \leq \mathscr{M}(\rho)} \overline{D}_{\mathrm{e}^{\mathrm{i}\theta} m}.$$

We set $\overset{\bullet}{\mathscr{P}}{}_\rho^\theta = \mathbb{C} \setminus \overline{\mathscr{D}}_\rho^\theta$. We denote by $\overset{\bullet}{\mathscr{R}}_\rho$ the domain defined by $\overset{\bullet}{\mathscr{R}}_\rho = \left(\overset{\bullet}{\mathscr{P}}{}_\rho^0 \cap \overset{\bullet}{\mathscr{P}}{}_\rho^\pi \right) \cup \{0\}$ and by $\overline{\overset{\bullet}{\mathscr{R}}}_\rho$ its closure. (See Fig. 4.4).

Notice that $\overset{\bullet}{\mathscr{R}} = \bigcup_{0<\rho<1/5} \overset{\bullet}{\mathscr{R}}_\rho$. The domains $\overset{\bullet}{\mathscr{P}}{}_\rho^\theta$ satisfy the following property, the proof of which being left as an exercise :

Lemma 4.2. *Let be ζ be an element of $\overset{\bullet}{\mathscr{P}}{}_\rho^\theta$. For every $n \in [1, \mathscr{M}(\rho)]$, the closed set $\zeta - \overline{D}_{\mathrm{e}^{\mathrm{i}\theta} n} = \{\zeta - \xi \mid \xi \in \overline{D}_{\mathrm{e}^{\mathrm{i}\theta} n}\}$ is a subset of $\overset{\bullet}{\mathscr{P}}{}_\rho^\theta$.*

Definition 4.12. Under the hypotheses of definition 4.11, for any integer $m \in [0, \mathscr{M}(\rho)]$ and $\theta \in \{0, \pi\}$, we define:

$$\overline{\mathscr{E}}_\rho^{m,\theta} = \bigcup_{(\zeta,\xi) \in \overline{D}_{\mathrm{e}^{\mathrm{i}\theta} m} \times \overline{D}_{\mathrm{e}^{\mathrm{i}\theta}(m+1)}} \left\{ \xi + t(\xi - \zeta), \zeta + t(\zeta - \xi) \mid t \in [0,+\infty[\right\}$$

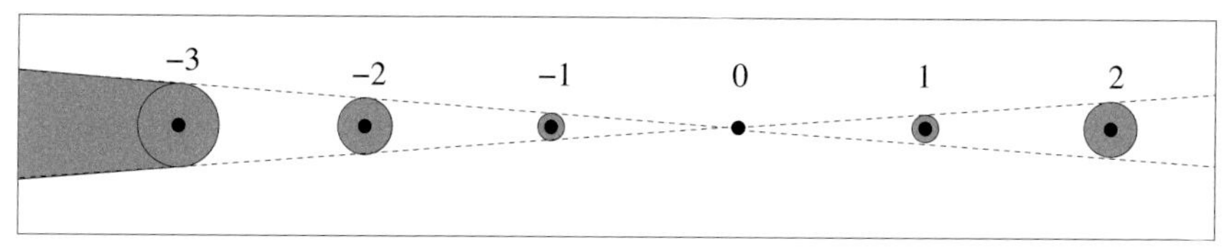

Fig. 4.4 The domain $\overset{\bullet}{\mathscr{R}}_\rho$ when $\frac{1}{9} < \rho \leq \frac{1}{7}$ (the scale is not correct).

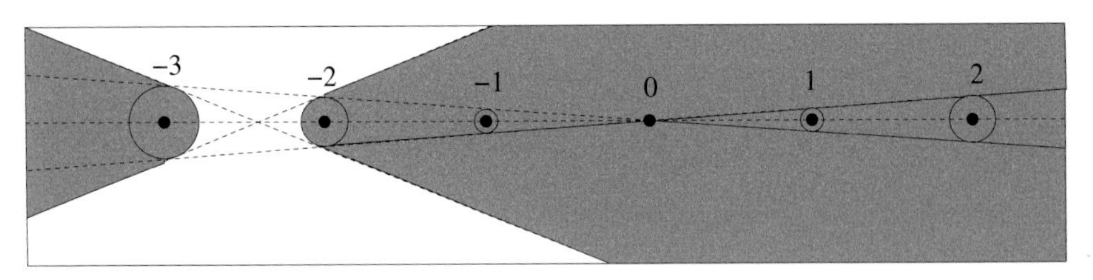

Fig. 4.5 The domain $\overset{\bullet}{\mathscr{Q}}_\rho^{2,\pi}$. The set $\overset{\bullet}{\mathscr{Q}}_\rho^{(-)_{2},\pi}$ lies below the real axis, $\overset{\bullet}{\mathscr{Q}}_\rho^{(+)_{2},\pi}$ lies above the real axis.

and $\overset{\bullet}{\mathscr{Q}}_\rho^{m,\theta} = \mathbb{C} \setminus \overline{\mathscr{E}}^{\theta}_{\rho,m}$. for any integer $m > \mathscr{M}(\rho)$, we set $\overset{\bullet}{\mathscr{Q}}_\rho^{m,\theta} = \emptyset$. For any positive integer $m \geq 1$ and $\varepsilon = \pm$, we set $\overset{\bullet}{\mathscr{Q}}_\rho^{(\varepsilon)_m,\theta} = \overset{\bullet}{\mathscr{Q}}_\rho^{m,\theta} \cap \{\zeta \mid \varepsilon e^{i\theta}(\Im\zeta) \leq 0\}$. See Fig. 4.5.

The domains $\overset{\bullet}{\mathscr{Q}}_\rho^{m,\theta}$ have been defined so as to enjoy the following property :

Lemma 4.3. *Let be $\zeta \in \overset{\bullet}{\mathscr{Q}}_\rho^{m,\theta}$ for some integer $m \in [1, \mathscr{M}(\rho)]$ and some $\theta \in \{0, \pi\}$. Then, for every integer $n \in [1, m]$, $\zeta - \overline{D}_{e^{i\theta}n}$ is a subset of $\overset{\bullet}{\mathscr{Q}}_\rho^{m-n,\theta}$.*

Proof. We only consider the case $\theta = 0$ and we suppose $\zeta \in \overset{\bullet}{\mathscr{Q}}_\rho^{m,0}$. Le be $n \in [1, m]$. Assume the existence of $\zeta_n \in \overline{D}_n$ such that $\zeta - \zeta_n \notin \overset{\bullet}{\mathscr{Q}}_\rho^{m-n,0}$, thus $\zeta - \zeta_n \in \overline{\mathscr{E}}_\rho^{m-n,0}$ (see definition 4.12). Therefore, there exist $\zeta_{m-n} \in \overline{D}_{m-n}$, $\zeta_{m-n+1} \in \overline{D}_{m-n+1}$ and $t \in [0, +\infty[$ such that

$$\zeta - \zeta_n = \zeta_{m-n} + t(\zeta_{m-n} - \zeta_{m-n+1}) \quad \text{or} \quad \zeta - \zeta_n = \zeta_{m-n+1} + t(\zeta_{m-n+1} - \zeta_{m-n}).$$

We look only at the first case, which we write as follows:

$$\zeta = (\zeta_{m-n} + \zeta_n) + t\Big((\zeta_{m-n} + \zeta_n) - (\zeta_{m-n+1} + \zeta_n)\Big).$$

We observe that $\zeta_{m-n} + \zeta_n \in \overline{D}_m$ while $\zeta_{m-n+1} + \zeta_n \in \overline{D}_{m+1}$. Therefore $\zeta \in \overline{\mathscr{E}}_\rho^{m,0}$ and this contradicts the assumption $\zeta \in \overset{\bullet}{\mathscr{Q}}_\rho^{m,0}$. □

Definition 4.13. Under the hypotheses of definition 4.11, for any integer $m \in [1, \mathscr{M}(\rho)]$ and any $\theta \in \{0, \pi\}$, we denote by $\overline{\mathscr{D}}_\rho^{m,\theta} \subset \mathbb{C}$ the closed subset defined by

$$\overline{\mathscr{D}}_\rho^{m,\theta} = \Big\{t\zeta \mid t \in]-\infty, 1],\, \zeta \in \overline{D}_{e^{i\theta}m}\Big\} \cup \Big\{t\zeta \mid t \in [1, +\infty[,\, \zeta \in \overline{D}_{e^{i\theta}(m+1)}\Big\}.$$

We set $\overset{\bullet}{\mathscr{P}}_\rho^{m,\theta} = \mathbb{C} \setminus \overline{\mathscr{D}}_\rho^{m,\theta}$ and $\overset{\bullet}{\mathscr{P}}_\rho^{0,\theta} = \overset{\bullet}{\mathscr{Q}}_\rho^{0,0}$. For any integer $m > \mathscr{M}(\rho)$, we set $\overset{\bullet}{\mathscr{P}}_\rho^{m,\theta} = \emptyset$.

For $\varepsilon = \pm$ we denote by $\overset{\bullet}{\mathscr{P}}_\rho^{(\varepsilon)_m,\theta}$ the domain $\overset{\bullet}{\mathscr{P}}_\rho^{(\varepsilon)_m,\theta} = \overset{\bullet}{\mathscr{P}}_\rho^{m,\theta} \cap \{\zeta \mid \varepsilon e^{i\theta}(\Im\zeta) \leq 0\}$. (See Fig. 4.6).

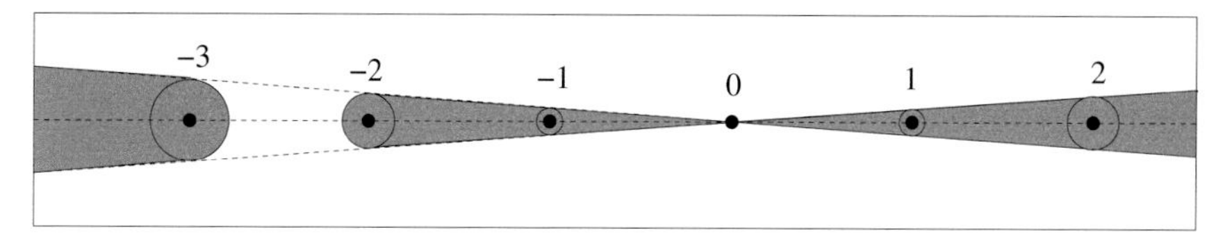

Fig. 4.6 The domain $\overset{\bullet}{\mathscr{P}}_\rho^{2,\pi}$. The set $\overset{\bullet}{\mathscr{P}}_\rho^{(-)2,\pi}$ lies below the real axis, the set $\overset{\bullet}{\mathscr{P}}_\rho^{(+)2,\pi}$ lies above the real axis.

Definition 4.14. Under the hypotheses of definition 4.11, for any $\theta \in \{0,\pi\}$, $\varepsilon = \pm$ and $m \in \mathbb{N}$, we denote by $\overset{\bullet}{\mathscr{R}}_\rho^{(\varepsilon)m,\theta}$ the domain $\overset{\bullet}{\mathscr{R}}_\rho^{(\varepsilon)m,\theta} = \overset{\bullet}{\mathscr{P}}_\rho^{(\varepsilon)m,\theta} \cup \overset{\bullet}{\mathscr{Q}}_\rho^{(-\varepsilon)m,\theta}$ (see Fig. 4.7), and we set:

$$\overset{\bullet}{\mathscr{R}}^{m,\theta} = \bigcup_{0<\rho\leq 1/5} \overset{\bullet}{\mathscr{R}}_\rho^{(+)m,\theta} = \bigcup_{0<\rho\leq 1/5} \overset{\bullet}{\mathscr{R}}_\rho^{(-)m,\theta} = \mathbb{C} \setminus \mathrm{e}^{\mathrm{i}\theta}\{]-\infty,m] \cup [m+1,+\infty[\}.$$

We have already noticed that for $\theta \in \{0,\pi\}$ and $m \in \mathbb{N}^\star$, the restriction $\mathfrak{p}|_{\mathscr{R}^{(+)m,\theta}}$ and $\mathfrak{p}|_{\mathscr{R}^{(-)m,\theta}}$ respectively, realises a homeomorphism between the nearby sheet $\mathscr{R}^{(+)m,\theta}$ and $\mathscr{R}^{(-)m,\theta}$ respectively, and the simply connected domain

$$\mathfrak{p}(\mathscr{R}^{(+)m,\theta}) = \mathfrak{p}(\mathscr{R}^{(-)m,\theta}) = \overset{\bullet}{\mathscr{R}}^{m,\theta}.$$

This justifies the following definition.

Definition 4.15. With the above notation, with $\varepsilon = \pm$ and $m \in [1, \mathscr{M}(\rho)]$ an integer, one sets $\mathscr{R}_\rho^{(\varepsilon)m,\theta} = \mathfrak{p}|^{-1}_{\mathscr{R}^{(\varepsilon)m,\theta}}\big(\overset{\bullet}{\mathscr{R}}_\rho^{(\varepsilon)m,\theta}\big)$. The domain $\mathscr{R}_\rho^{(\varepsilon)m,\theta}$ is called a $\mathscr{R}_\rho^{(0)}$*-nearby domains* .The connected and simply connected domain $\mathscr{R}_\rho^{(1)} \subset \mathscr{R}^{(1)}$ is defined by $\mathscr{R}_\rho^{(1)} = \mathscr{R}_\rho^{(0)} \bigcup_{\substack{1\leq m\leq \mathscr{M}(\rho)\\ \theta\in\{0,\pi\},\varepsilon=\pm}} \mathscr{R}_\rho^{(\varepsilon)m,\theta}$. We denote by $\overline{\mathscr{R}}_\rho^{(1)}$ the closure of $\mathscr{R}_\rho^{(1)}$ in $\mathscr{R}^{(1)}$.

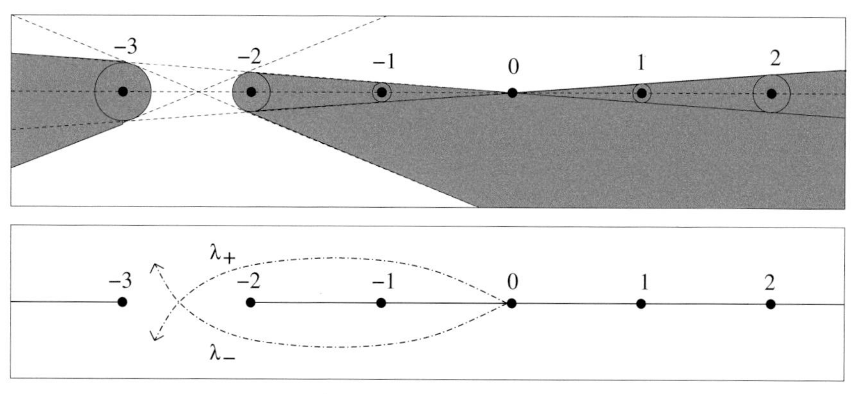

Fig. 4.7 Figure above, the domain $\overset{\bullet}{\mathscr{R}}_\rho^{(+)2,\pi}$. Figure below, the domain $\overset{\bullet}{\mathscr{R}}^{2,\pi}$.

Observe that $\mathfrak{p}\left(\mathscr{R}_\rho^{(1)}\right) = \overset{\bullet}{\mathscr{R}}_\rho$. In the same line, the following lemma is a consequence of lemmas 4.2 and 4.3.

Lemma 4.4. *Let be $m \in [1, \mathscr{M}(\rho)]$, $\theta \in \{0, \pi\}$, $\varepsilon = \pm$ and let $\mathscr{V}$ be the closure of $\mathscr{R}_\rho^{(\varepsilon)_{m},\theta} \setminus \mathscr{R}_\rho^{(0)}$. For every $\zeta \in \mathscr{V}$, and every integer $n \in [1, m]$, $\overset{\bullet}{\zeta} - D_{e^{i\theta}n}$ is a subset of $\overset{\bullet}{\mathscr{R}}_\rho$ and there exists an open set $\mathscr{U} \subset \mathscr{R}_\rho^{(\varepsilon)_{m-n},\theta}$ such that $\mathscr{U}$ and $\overset{\bullet}{\zeta} - D_{e^{i\theta}n}$ are $\mathfrak{p}$-homeomorphic.*

4.2.5 Geodesics

The closed space $\overline{\overset{\bullet}{\mathscr{R}}}_\rho \subset \overset{\bullet}{\mathscr{R}}$ (definition 4.11) can be thought of as a complete real 2-dimensional Riemannian manifold with smooth ($\mathscr{C}^1$) boundary embedded in the 2-dimensional euclidean space. The following lemma thus makes sense.

Lemma 4.5. *Let $X \subseteq \overline{\overset{\bullet}{\mathscr{R}}}_\rho$ be any closed space with smooth($\mathscr{C}^1$) boundary. For every two points $\zeta_1, \zeta_2 \in X$, there exists a geodesic in every homotopy class of curves from ζ_1 to ζ_2 in X, and this geodesic may be chosen as a shortest path in the homotopy class.*

In this lemma, a geodesic means a locally shortest path for the euclidean metric. Lemma 4.5 can be seen as a corollary of the Hopf-Rinow theorem [Jos95]. As a matter of fact, the situation is quite simple here : inside X, a geodesic is nothing but a straight line, otherwise one just follows the smooth boundary ∂X. (See [AB91] and references therein for more general cases.)

The Riemann surface $(\mathscr{R}, \mathfrak{p}, \underline{0})$ of $\mathbb{Z}$-resurgent functions can also be thought of as a real 2-dimensional Riemannian manifold, by pulling-back by $\mathfrak{p}$ the standard euclidean metric on the complex plane. It follows from its very construction that $\overline{\mathscr{R}}_\rho^{(1)}$ (definition 4.15) meets the requirement:

Lemma 4.6. *The closed, connected and simply connected space with smooth $\mathscr{C}^1$-boundary $\overline{\mathscr{R}}_\rho^{(1)} \subset \mathscr{R}$ is a complete real 2-dimensional Riemannian manifold.*

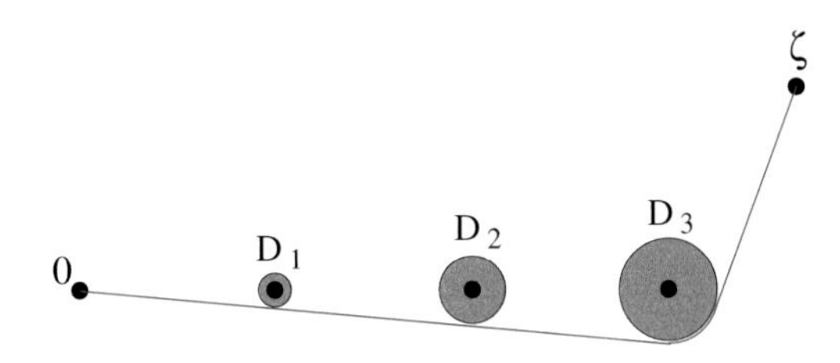

Fig. 4.8 Shortest curve for ζ in $\overline{\mathscr{R}}_\rho^{(+)_3,0} \setminus \overline{\mathscr{R}}_\rho^{(0)}$.

Pick a point $\zeta \in \overline{\mathscr{R}}_\rho^{(1)}$. Up to homotopy, there exists a unique path Λ joining $\underline{0}$ to ζ in $\overline{\mathscr{R}}_\rho^{(1)}$, because $\overline{\mathscr{R}}_\rho^{(1)}$ is (path-)connected and simply connected. Moreover, from the Hopf-Rinow theorem, Λ can be chosen as a shortest ($\mathscr{C}^1$-)path in this homotopy class, and is uniquely determined when parametrized by arc-length. To sum up:

Lemma 4.7. *For every $\zeta \in \overline{\mathscr{R}}_\rho^{(1)}$, there exists a unique path Λ in $\overline{\mathscr{R}}_\rho^{(1)}$, originating from $\underline{0}$ and ending at ζ, such that Λ is a shortest path in its homotopy class and is parametrized by arc-length.*

Remark 4.2. It is easy to construct Λ by hand.

First case: ζ belongs to $\overline{\mathscr{R}}_\rho^{(0)}$. Consider the curve λ, with its arc-length parametrization, starting from 0 which follows the segment $[0,\overset{\bullet}{\zeta}] \subset \overset{\bullet}{\overline{\mathscr{R}}}{}_\rho^{(0)}$. The path Λ is obtained by lifting λ from $\underline{0}$ with respect to $\mathfrak{p}$ on $\overline{\mathscr{R}}_\rho^{(1)}$.

Second case: ζ belongs to $\overline{\mathscr{R}}_\rho^{(\varepsilon)m,\theta} \setminus \overline{\mathscr{R}}_\rho^{(0)}$ for some $\theta \in \{0,\pi\}$, $\varepsilon = \pm$ and some $m \in [1,\mathscr{M}(\rho)]$. Consider the path $\lambda = \gamma_0\delta_0\delta_1$ where $\gamma_0,\delta_0,\delta_1$ stands for the following geodesics with their arc-length parametrizations (see Fig. 4.8) :

- γ_0 follows the segment $[0,\overset{\bullet}{\zeta}_0] \subset \partial\big(\overset{\bullet}{\overline{\mathscr{P}}}{}_\rho^{(\varepsilon)m,\theta} \cap \overset{\bullet}{\overline{\mathscr{R}}}{}_\rho^{(0)}\big)$ that circumvents the segment $e^{i\theta}[1,m]$ to the right when $\varepsilon = +$ and to the left when $\varepsilon = -$;
- δ_0 is the arc-curve from $\overset{\bullet}{\zeta}_0$ to $\overset{\bullet}{\zeta}_1$ that follows in $\overset{\bullet}{\overline{\mathscr{R}}}{}_\rho^{(\varepsilon)m,\theta}$ the boundary $\partial D_{e^{i\theta}m}$;
- δ_1 follows the segment $[\overset{\bullet}{\zeta}_1,\overset{\bullet}{\zeta}]$ in $\overset{\bullet}{\overline{\mathscr{R}}}{}_\rho^{(\varepsilon)m,\theta}$ (possibly reduced to the point $\overset{\bullet}{\zeta}$).

Once again, one deduces Λ from λ by lifting.

Definition 4.16. Let ζ be an element of $\zeta \in \overline{\mathscr{R}}_\rho^{(1)}$. The unique $\mathscr{C}^1$-path Λ in $\overline{\mathscr{R}}_\rho^{(1)}$ given by lemma 4.7 is called the shortest path from $\underline{0}$ to ζ.

4.3 Shortest Symmetric $(\mathbb{Z},\rho)$-Homotopy

4.3.1 Symmetric $\mathbb{Z}$-Homotopy

The notion of *symmetric Ω-homotopy* has been introduced in the first volume of this book [MS016] and used there for analyzing the convolution product of resurgent functions, see also [Eca81-1, Ou012, Sau013, Sau015]. For the convenience of the reader, we recall it here for $\Omega = \mathbb{Z}$.

Definition 4.17. A continuous map $H : I \times I \to \mathbb{C}$, $I = [0,1]$, is called a *symmetric $\mathbb{Z}$-homotopy* if, for each $t \in I$, the path $H_t : s \in I \mapsto H(s,t)$ satisfies:

1. $H_t(0) = 0$ and H_t can be lifted on the Riemann surface $\mathscr{R}_{\mathbb{Z}}$ with respect to $\mathfrak{p}$ from $\underline{0}$;

2. $H_t(1) - H_t(s) = H_t^{-1}(s)$ for every $s \in I$.

The path H_0 (*resp.* H_1) is called the *initial path of* H (*resp.* final path) and the path $t \in I \mapsto H_t(1)$ is called the *endpoint path of* H.
A path λ in $\mathbb{C}$ is called a *symmetrically* $\mathbb{Z}$*-contractile path* if its standardized path $\underline{\underline{\lambda}}$ is the final path of a symmetric $\mathbb{Z}$-homotopy whose initial path follows a segment $[0, \zeta]$ of $\overset{\bullet}{\mathscr{R}}$ in the forward direction.

Let ζ be any point of the Riemann surface $\mathscr{R}_{\mathbb{Z}}$ and pick a path joining $\underline{0}$ to ζ, thus uniquely defined up to homotopy. It is known that one can find a path Λ in this homotopy class with the further condition : its projection $\lambda = \mathfrak{p} \circ \Lambda$ is a symmetrically $\mathbb{Z}$-contractile path. This is a key result to analyze the convolution product, as detailed in the first volume of this book [MS016].

However, there are plenty of paths with the above properties and we raise the question:

Question 4.4. In the homotopy class of these paths, is it possible to find a shortest curve ?

This question is meaningless because $\mathscr{R}_{\mathbb{Z}}$ is not a complete Riemannian manifold, but makes sense on $\overline{\mathscr{R}}_{\rho}^{(1)}$ which is our frame in what follows.

4.3.2 Shortest Symmetric $(\mathbb{Z}, \rho)$-Homotopy

Definition 4.18. Let Λ be a path in $\overline{\mathscr{R}}_{\rho}^{(1)}$ originating from $\underline{0}$ and let $\lambda = \mathfrak{p} \circ \Lambda$ be its projection. The path Λ is said to be *symmetric* if λ satisfies the condition: $\underline{\underline{\lambda}}(1) - \underline{\underline{\lambda}}(s) = \underline{\underline{\lambda}}^{-1}(s)$for every $s \in [0,1]$. A symmetric path Λ in $\overline{\mathscr{R}}_{\rho}^{(1)}$ is said to be *shortest-symmetric* when Λ is a shortest ($\mathscr{C}^1$-)path among the symmetric paths belonging to the same homotopy class in $\overline{\mathscr{R}}_{\rho}^{(1)}$.

For instance, pick a point ζ in $\overline{\mathscr{R}}_{\rho}^{(0)}$ and let λ be the smooth path which follows the segment $[0, \overset{\bullet}{\zeta}] \subset \overset{\bullet}{\overline{\mathscr{R}}}_{\rho}^{(0)}$ in the forward direction with a constant velocity. The path Λ in $\overline{\mathscr{R}}_{\rho}^{(0)}$ deduced from λ by lifting from $\underline{0}$ with respect to $\mathfrak{p}$, is shortest-symmetric.

Proposition 4.3. *Let ζ be any given point in $\overline{\mathscr{R}}_{\rho}^{(1)}$. There exists a unique continuous map $\mathscr{H} : (s,t) \in I \times I \mapsto \mathscr{H}(s,t) \in \overline{\mathscr{R}}_{\rho}^{(1)}$, $I = [0,1]$, which satisfies the following conditions:*

1. *for each $t \in I$, the path $\mathscr{H}_t : s \in I \mapsto \mathscr{H}(s,t)$ is shortest symmetric;*
2. *the projection $H_0 = \mathfrak{p} \circ \mathscr{H}_0$ of the initial path $\mathscr{H}_0$ follows a segment in $\overset{\bullet}{\overline{\mathscr{R}}}_{\rho}^{(0)}$;*
3. *denoting by Γ the endpoint path $t \in I \mapsto \mathscr{H}_t(1)$, the product $\mathscr{H}_0\Gamma$, when reparametrized by arc-length parametrization, coincides with the shortest path from $\underline{0}$ to ζ.*

Proof. Let ζ be a point in $\overline{\mathscr{R}}_\rho^{(1)}$ and Λ be the shortest path from $\underline{0}$ to ζ. We denote by $\lambda = \mathfrak{p}\circ\Lambda$ its projection.

First case: Either ζ belongs to $\overline{\mathscr{R}}_\rho^{(0)}$. Then λ follows the segment $[0,\dot{\zeta}] \subset \dot{\overline{\mathscr{R}}}_\rho^{(0)}$. We set $H : (s,t) \in I \times I \mapsto H(s,t) = s\,\dot{\zeta} \in \dot{\overline{\mathscr{R}}}_\rho^{(0)}$. For each $t \in I$, the path $H_t : s \in I \mapsto H(s,t)$ can be lifted uniquely on $\overline{\mathscr{R}}_\rho^{(0)}$ from $\underline{0}$ with respect to $\mathfrak{p}$ into a path $\mathscr{H}_t : I \to \overline{\mathscr{R}}_\rho^{(0)}$. From the lifting theorem for homotopies [For81, Ebe007], the mapping $\mathscr{H} : (s,t) \in I\times I \mapsto \mathscr{H}_t(s)$ is continuous and matches the other conditions.

Second case: Or else ζ belongs to $\overline{\mathscr{R}}_\rho^{(\varepsilon)_m,\theta} \setminus \overline{\mathscr{R}}_\rho^{(0)}$ for some $\theta \in \{0,\pi\}$, $\varepsilon = \pm$ and some $m \in [1,\mathscr{M}(\rho)]$. For simplicity, we suppose $\theta = 0$ and $\varepsilon = +$. The path λ, *resp.* Λ, can be written as a product $\lambda = \gamma_0\lambda_1$, *resp.* $\Lambda = \Gamma_0\Lambda_1$, where $\gamma_0 = \mathfrak{p}\circ\Gamma_0$ and $\lambda_1 = \delta_0\delta_1 = \mathfrak{p}\circ\Lambda_1$ are the geodesics described in remark 4.2 with their arc-length parametrizations.
We set $H_0 = \underline{\underline{\gamma_0}}$, *resp.* $\mathscr{H}_0 = \underline{\underline{\Gamma_0}}$ the standardized path deduced from γ_0, *resp.* Γ_0, which follows a segment in $\dot{\overline{\mathscr{R}}}_\rho^{(0)}$, *resp.* $\overline{\mathscr{R}}_\rho^{(0)}$. This path can be lifted from $\underline{0}$ with respect to $\mathfrak{p}$ into a unique path $\mathscr{H}_0$ whose endpoint is denoted by $\zeta_0 = \mathscr{H}_0(1)$. By its very construction, the point ζ_0 belongs to $\mathscr{V}$, the closure of $\mathscr{R}_\rho^{(+)_m,0} \setminus \mathscr{R}_\rho^{(0)}$, and we can apply lemma 4.4. Therefore H_0 can be thought of as a geodesic in $X_{\zeta_0} = \dot{\overline{\mathscr{R}}}_\rho \setminus \bigcup_{1\le n\le m} \{\dot{\zeta}_0 - D_n\}$ and is a shortest path in its homotopy class, by application of lemma 4.5.

According to lemma 4.4 again, the space $X_\xi = \dot{\overline{\mathscr{R}}}_\rho \setminus \bigcup_{1\le n\le m} \{\dot{\xi} - D_n\}$ remains in the field of application of lemma 4.5 for every $\xi \in \mathscr{V}$. One gets this way a local system $(X_\xi)_{\xi\in\mathscr{V}}$ of Riemannian manifolds with smooth boundary.
Let $\ell_1 > 0$ be the length of Λ_1 and $T : t \in [0,1] \mapsto t\ell_1 \in [0,\ell_1]$. For any $t \in [0,1]$, the point $\zeta_t = \Lambda_1 \circ T(t)$ belongs to $\mathscr{V}$ by construction. To the path $\Lambda_1 \circ T$ is associated a section $t \in [0,1] \mapsto X_{\zeta_t}$, thus a map $t \in [0,1] \mapsto [\gamma_t]$ which allows to follow the continuous deformation of the homotopy class $[\gamma_0]$ of γ_0, the extremities 0 and ζ_0 being kept fixed. In the homotopy class $[\gamma_t]$ we choose, for any $t \in [0,1]$, a shortest path γ_t

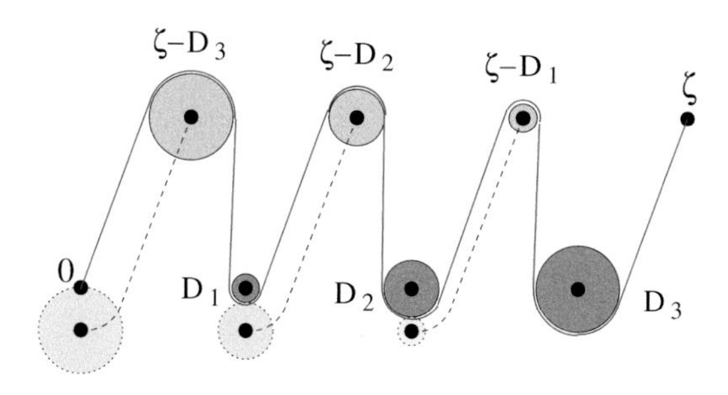

Fig. 4.9 The shortest symmetrically contractile path for ζ in $\overline{\mathscr{R}}_\rho^{(+)_3,0} \setminus \overline{\mathscr{R}}_\rho^{(0)}$

in X_{ζ_t} with its arc-length parametrization. Let $K_t = \gamma_t \lambda_1|_{[0,T^{-1}(t)]}$ be the (minimal) geodesic defined as the product of γ_t with the restriction of λ_1 to $[0, T^{-1}(t)]$. We denote by H_t the path deduced from K_t by standardization and we eventually obtain a continuous mapping $H : (s,t) \in I \times I \mapsto H(s,t) = H_t(s)$, $I = [0,1]$. See Fig. 4.9. For any $t \in [0,1]$, the path H_t can be lifted to $\overline{\mathscr{R}}_\rho^{(1)}$ from $\underline{0}$ with respect to $\mathfrak{p}$. This gives a path, denoted by $\mathscr{H}_t$, which is shortest symmetric by construction and the mapping $\mathscr{H} : (s,t) \in I \times I \mapsto \mathscr{H}(s,t) = \mathscr{H}_t(s) \in \overline{\mathscr{R}}_\rho^{(1)}$ is continuous by the lifting theorem for homotopies [For81, Ebe007]. The reader is encouraged to check the remaining properties. □

Definition 4.19. Let ζ be any given point in $\overline{\mathscr{R}}_\rho^{(1)}$. The uniquely determined continuous map $\mathscr{H}$ given by proposition 4.3 is called the *shortest symmetric* $(\mathbb{Z}, \rho)$-*homotopy* associated with ζ. The path $\mathscr{H}_1 : s \in [0,1] \mapsto \mathscr{H}(s,1)$ is called the *shortest symmetrically contractile path* associated with ζ in $\overline{\mathscr{R}}_\rho^{(1)}$ and its length is denoted by $\text{leng}_\rho(\zeta)$

Remark 4.3. Let $\mathscr{H}$ be a shortest symmetric $(\mathbb{Z}, \rho)$-homotopy. Consider the path $\mathscr{H}_t : s \in [0,1] \mapsto \mathscr{H}(s,t)$ for any given $t \in [0,1]$. Then $\mathscr{H}_t$ is the shortest symmetrically contractile path associated the endpoint $\mathscr{H}_t(1)$ in $\overline{\mathscr{R}}_\rho^{(1)}$.

The next two statements are left as exercises.

Lemma 4.8. *The mapping* $\zeta \in \overline{\mathscr{R}}_\rho^{(1)} \mapsto \text{leng}_\rho(\zeta) \in \mathbb{R}^+$ *is continuous.*

Lemma 4.9. *Let be* $\zeta \in \overline{\mathscr{R}}_\rho^{(1)}$ *and* $\mathscr{H}$ *be its associated shortest symmetric* $(\mathbb{Z}, \rho)$-*homotopy. Then for every* $t \in [0,1]$ *and every* $s \in [0,1]$*:*

- $\text{leng}_\rho\big(\mathscr{H}_t(s)\big) \leq \text{length}\big(\mathscr{H}_t|_{[0,s]}\big)$;
- $\text{leng}_\rho\big(\mathscr{H}_t^{-1}(s)\big) \leq \text{leng}_\rho\big(\mathscr{H}_t(1)\big) - \text{length}\big(\mathscr{H}_t|_{[0,s]}\big)$.

We finally state a result drawn from [GS001], which gives an upper bound for the length of the shortest symmetrically contractile path we work with.

Lemma 4.10. *Let be* $\zeta \in \overline{\mathscr{R}}_\rho^{(1)}$. *Either* $\zeta \in \overline{\mathscr{R}}_\rho^{(0)}$ *and then* $\text{leng}_\rho(\zeta) = |\overset{\bullet}{\zeta}|$, *or* $|\overset{\bullet}{\zeta}| \leq \text{leng}_\rho(\zeta) \leq \frac{1}{\rho}|\overset{\bullet}{\zeta}| + \frac{1}{\rho}\left(\frac{1}{\rho} - 2\right)$.

Proof. The first case is obvious. The second case means that $\zeta \in \overline{\mathscr{R}}_\rho^{(\varepsilon)_m,\theta} \setminus \overline{\mathscr{R}}_\rho^{(0)}$ for some $\theta \in \{0, \pi\}$, $\varepsilon = \pm$ and some $m \in [1, \mathscr{M}(\rho)]$. Let us assume that $\theta = 0$ and $\varepsilon = +$ for simplicity. We return to the construction of the shortest symmetrically contractile path $\mathscr{H}_1$ associated with ζ (see also Fig. 4.9) and we denote by $H_1 = \mathfrak{p} \circ \mathscr{H}_1$ its projection. The path H_1 is made of :

- $m+1$ segments between $\partial\overline{D}_n$ and $\partial\big(\overset{\bullet}{\zeta} - \overline{D}_{m-n}\big)$, $n \in [0,m]$. Each of these segments has length less than $|\overset{\bullet}{\zeta} - m| + m\rho$.

- m segments between $\partial\big(\overset{\bullet}{\zeta} - \overline{D}_{m-n}\big)$ and $\partial\overline{D}_{n+1}$, $n\in[1,m]$. Each of these segments has length less than $|\overset{\bullet}{\zeta} - (m+1)| + (m+1)\rho$.
- $2m$ arcs of circle, the total length of which being less than $2(1+\cdots+m)2\pi\rho$.

Putting things together:

$$\mathrm{leng}_\rho(\zeta) \leq (2m+1)|\overset{\bullet}{\zeta}| + 2m(m+1)(1+\rho) + 2m(m+1)\pi\rho.$$

Since $\rho\leq\frac{1}{5}$, one has $|\overset{\bullet}{\zeta}| \leq \mathrm{leng}_\rho(\zeta) \leq (2m+1)|\overset{\bullet}{\zeta}| + 4m(m+1)$. Remember that $\mathscr{M}(\rho)+1 = \lfloor\frac{\rho^{-1}-1}{2}\rfloor$, thus $m\leq\mathscr{M}(\rho)\leq\frac{1}{2\rho}-1$ and one concludes. □

4.4 Convolution Product and Related Properties

It has been recalled that the space $\hat{\mathscr{R}} = \hat{\mathscr{R}}_{\mathbb{Z}}$ of $\mathbb{Z}$-resurgent germs is a convolution algebra without unit (definition 3.7 and theorem 3.2).

Question 4.5. Is it possible to give quantitave estimates for the convolution product of two $\mathbb{Z}$-resurgent functions ?

The answer is “yes”, as detailed (without proof) in the first volume of this book [MS016] (see also [CNP93-1, Ou012]). Even, quantitave estimates can be obtained for iterated convolutions and this allows non-linear operations in the frame of resurgent functions [Sau015].

Nevertheless, these results are difficult to use in our context. This is why we follow another strategy in the sequel.

4.4.1 A New Convolution Algebra

Definition 4.20. Let $k\in\mathbb{N}^\star$ be a positive integer. We denote by $\hat{\mathscr{R}}^{(k)}$ the space of germs of holomorphic functions at the origin which can be analytically continued to the Riemann surface $\mathscr{R}^{(k)}$.

In other words, the germ $\widehat{\varphi}\in\mathscr{O}_0$ belongs to $\hat{\mathscr{R}}^{(k)}$ when there exists a function $\Phi\in\mathscr{O}(\mathscr{R}^{(k)})$ holomorphic on $\mathscr{R}^{(k)}$ whose germ $\phi_{\underline{0}}\in\mathscr{O}_{\underline{0}}$ at $\underline{0}$ satisfies $\phi_{\underline{0}} = \widehat{\varphi}\circ\mathfrak{p}$. Notice that the linear map $\widehat{\partial}: \widehat{g}\in\hat{\mathscr{R}}^{(k)} \mapsto -\zeta\widehat{g}$ still provides a derivation of $\hat{\mathscr{R}}^{(k)}$.

Theorem 4.1. *1. The space $\hat{\mathscr{R}}^{(1)}$ is a convolution algebra (without unit).*
2. Let $\widehat{\varphi},\widehat{\psi}\in\hat{\mathscr{R}}^{(1)}$ be two germs and let $\Phi,\Psi\in\mathscr{O}(\mathscr{R}^{(1)})$ be their associated holomorphic functions on $\mathscr{R}^{(1)}$. Assume that the following properties hold : for every $\zeta\in\overline{\mathscr{R}}^{(1)}_\rho$, $|\Phi(\zeta)|\leq F\big(\mathrm{leng}_\rho(\zeta)\big)$ and $|\Psi(\zeta)|\leq G\big(\mathrm{leng}_\rho(\zeta)\big)$, where F,G are two positive, non-decreasing and continuous functions on $\mathbb{R}^+$. Then the

convolution product $\widehat{\varphi} * \widehat{\psi}$*, resp.* $(\widehat{\partial \varphi}) * \widehat{\psi}$*, can be analytically continued to* $\mathscr{R}^{(1)}$ *and the corresponding function* $\chi \in \mathscr{O}(\mathscr{R}^{(1)})$*, resp.* $\Upsilon \in \mathscr{O}(\mathscr{R}^{(1)})$*, satisfies the following properties: for every* $\zeta \in \overline{\mathscr{R}}_\rho^{(1)}$*,* $|\chi(\zeta)| \leq F * G(\mathrm{leng}_\rho(\zeta))$*, resp.* $|\Upsilon(\zeta)| \leq \mathrm{leng}_\rho(\zeta) \left(F * G(\mathrm{leng}_\rho(\zeta))\right)$.

Proof. The standard proof for proving that $\hat{\mathscr{R}}$ is a convolution algebra [MS016, Ou012] can be copied as it stands for $\hat{\mathscr{R}}^{(1)}$. We sketch it here, essentially so as to fix some notation that will be used later on, more details can be found in the first volume of this book [MS016].
Let be $\widehat{\varphi}, \widehat{\psi} \in \hat{\mathscr{R}}^{(1)}$ and let $\Phi, \Psi \in \mathscr{O}(\mathscr{R}^{(1)})$ be their associated holomorphic functions on $\mathscr{R}^{(1)}$.
The convolution product $\widehat{\varphi} * \widehat{\psi}(\overset{\bullet}{\zeta})$ is well-defined for every $\zeta \in \mathscr{R}^{(0)}$ and we set $\chi(\zeta) = \widehat{\varphi} * \widehat{\psi}(\overset{\bullet}{\zeta})$. For every $\overset{\bullet}{\zeta}_0 \in \overline{\overset{\bullet}{\mathscr{R}}}_\rho^{(0)}$ and $\overset{\bullet}{\xi} \in \mathbb{C}$ such that $|\overset{\bullet}{\xi}| < \frac{\rho}{2}$, the point $\overset{\bullet}{\zeta}_0 + \overset{\bullet}{\xi}$ belongs to $\overset{\bullet}{\mathscr{R}}^{(0)}$, thus there exists a uniquely determined point denoted by $\zeta_0 + \overset{\bullet}{\xi} \in \mathscr{R}^{(0)}$ such that $\mathfrak{p}(\zeta_0 + \overset{\bullet}{\xi}) = \overset{\bullet}{\zeta}_0 + \overset{\bullet}{\xi}$. Therefore, the convolution product $\chi(\zeta_0 + \overset{\bullet}{\xi}) = \widehat{\varphi} * \widehat{\psi}(\overset{\bullet}{\zeta}_0 + \overset{\bullet}{\xi})$ reads :

$$\begin{aligned}
\chi(\zeta_0 + \overset{\bullet}{\xi}) &= \int_0^{\overset{\bullet}{\zeta}_0 + \overset{\bullet}{\xi}} \widehat{\varphi}(\eta)\widehat{\psi}(\overset{\bullet}{\zeta}_0 + \overset{\bullet}{\xi} - \eta)d\eta \\
&= \int_0^{\overset{\bullet}{\zeta}_0} \widehat{\varphi}(\eta)\widehat{\psi}(\overset{\bullet}{\zeta}_0 + \overset{\bullet}{\xi} - \eta)d\eta + \int_0^{\overset{\bullet}{\xi}} \widehat{\varphi}(\overset{\bullet}{\zeta}_0 + \eta)\widehat{\psi}(\overset{\bullet}{\xi} - \eta)d\eta.
\end{aligned}$$

Let now ζ be any given point in $\overline{\mathscr{R}}_\rho^{(1)}$. We denote by $\mathscr{H}$ the associated shortest symmetric $(\mathbb{Z}, \rho)$-homotopy given by proposition 4.3. We want to construct the analytic continuation of χ at ζ. We therefore assume that ζ_0 is the endpoint $\mathscr{H}_0(1)$ of the intial path $\mathscr{H}_0$. The above equality yields :

$$\begin{aligned}
\chi\left(\mathscr{H}_0(1) + \overset{\bullet}{\xi}\right) = &\int_0^1 \Phi(\mathscr{H}_0(s))\Psi(\mathscr{H}_0^{-1}(s) + \overset{\bullet}{\xi})H_0'(s)ds \\
&+ \overset{\bullet}{\xi} \int_0^1 \Phi(\mathscr{H}_0(1) + \overset{\bullet}{\xi}s)\psi\left(\overset{\bullet}{\xi}(1-s)\right)ds.
\end{aligned}$$

where $H_0 = \mathfrak{p} \circ \mathscr{H}_0$ stands for the projection of $\mathscr{H}_0$. The analytic continuation of χ from $\mathscr{H}_0(1)$ along the path $t \in [0,1] \mapsto \mathscr{H}_t(1) \in \overline{\mathscr{R}}_\rho^{(1)}$ is thus given by (see the arguments in [MS016]):

$$\begin{aligned}
\chi\left(\mathscr{H}_t(1) + \overset{\bullet}{\xi}\right) = &\int_0^1 \Phi(\mathscr{H}_t(s))\Psi(\mathscr{H}_t^{-1}(s) + \overset{\bullet}{\xi})H_t'(s)ds \\
&+ \overset{\bullet}{\xi} \int_0^1 \Phi(\mathscr{H}_t(1) + \overset{\bullet}{\xi}s)\psi\left(\overset{\bullet}{\xi}(1-s)\right)ds.
\end{aligned}$$

In particular when $\zeta = \mathscr{H}_1(1)$, $\chi(\zeta) = \int_0^1 \Phi(\mathscr{H}_1(s))\Psi(\mathscr{H}_1^{-1}(s))\, H_1'(s)ds$ where $\mathscr{H}_1$ is the shortest symmetrically contractile path associated with $\overset{\bullet}{\zeta}$.

Notice that the germ $\chi(\zeta+\xi)$ of holomorphic functions at ζ thus obtained does not depend on the chosen path $\mathscr{H}_1$ since $\mathscr{R}^{(1)}$ is simply connected.

We turn to estimates. Let T be the homothety $s \in [0,1] \mapsto s.\mathrm{leng}_\rho(\zeta)$ so that $\chi(\zeta) = \int_0^{\mathrm{leng}_\rho(\zeta)} \Phi(\mathscr{H}_1 \circ T^{-1}(\ell))\Psi(\mathscr{H}_1^{-1} \circ T^{-1}(\ell))\, d\ell$. We then use lemma 4.9 to get:

$$\begin{aligned}|\chi(\zeta)| &\leq \int_0^{\mathrm{leng}_\rho(\zeta)} F(\ell)G(\mathrm{leng}_\rho(\zeta)-\ell)d\ell \\ &\leq F * G(\mathrm{leng}_\rho(\zeta)).\end{aligned}$$

The proof for the last assertion is left as an exercise. □

4.4.2 Convolution Space and Uniform Norm

The following definition makes sense by lemma 4.8 and lemma 4.9.

Definition 4.21. Let $L > 0$ be a real positive number and $\rho \in]0, \frac{1}{5}]$. One denotes by $\mathscr{U}_{\rho,L}$ the open subset of $\mathscr{R}_\rho^{(1)}$ defined by: $\mathscr{U}_{\rho,L} = \{\zeta \in \mathscr{R}_\rho^{(1)} \mid \mathrm{leng}_\rho(\zeta) < L\}$. An element of $\mathscr{U}_{\rho,L}$ is called a *L-point.*
We denote by $\mathscr{O}(\overline{\mathscr{U}}_{\rho,L})$ the space of functions holomorphic on $\mathscr{U}_{\rho,L}$ and continuous on $\overline{\mathscr{U}}_{\rho,L}$. For any two elements $f, g \in \mathscr{O}(\overline{\mathscr{U}}_{\rho,L})$, one sets

$$f * g(\zeta) = \int_0^{\mathrm{leng}_\rho(\zeta)} f(\mathscr{H}_1 \circ T_\zeta^{-1}(\ell))g(\mathscr{H}_1^{-1} \circ T_\zeta^{-1}(\ell))\, d\ell \tag{4.2}$$

where $\mathscr{H}_1$ stands for the shortest symmetrically contractile path associated with $\zeta \in \overline{\mathscr{U}}_{\rho,L}$, while T_ζ is the homothety $T_\zeta : s \in [0,1] \mapsto s.\mathrm{leng}_\rho(\zeta)$.
The function $\zeta \in \overline{\mathscr{U}}_{\rho,L} \mapsto f * g(\zeta)$ is called the convolution product of f and g.

Proposition 4.4. *For any two elements $f, g \in \mathscr{O}(\overline{\mathscr{U}}_{\rho,L})$, their convolution product belongs to $\mathscr{O}(\overline{\mathscr{U}}_{\rho,L})$. In other words, $\mathbb{C}\delta \oplus \mathscr{O}(\overline{\mathscr{U}}_{\rho,L})$ is a convolution algebra.*

Proof. Use lemma 4.9 and adapt the proof of theorem 4.1. □

The following definition is similar to definition 3.12.

Definition 4.22. Let $\mathscr{U} = \mathscr{U}_{\rho,L}$ be an open set of L-points. We denote by $\mathscr{M}\mathscr{O}(\overline{\mathscr{U}})$ the maximal ideal of $\mathscr{O}(\overline{\mathscr{U}}))$ defined by $\mathscr{M}\mathscr{O}(\overline{\mathscr{U}}) = \{f \in \mathscr{O}(\overline{\mathscr{U}}), f(\underline{0}) = 0\}$.
Let $\nu \geq 0$ be a nonnegative real number. The norm $\|.\|_\nu$ on $\mathscr{O}(\overline{\mathscr{U}}))$ is defined by

$\|f\|_\nu = L \sup_{\zeta\in\mathscr{U}} \left|\mathrm{e}^{-\nu \mathrm{leng}_\rho(\zeta)} f(\zeta)\right|$. We extend this norm to $\mathbb{C}\delta \oplus \mathscr{O}(\overline{\mathscr{U}}_{\rho,L})$ by setting $\|c\delta + f\|_\nu = |c| + \|f\|_\nu$ for every $f \in \mathscr{O}(\overline{\mathscr{U}})$ and every $c \in \mathbb{C}$.

We now state an analogue of proposition 3.9.

Proposition 4.5. *The normed space $\left(\mathbb{C}\delta \oplus \mathscr{O}(\overline{\mathscr{U}}), \|.\|_\nu\right)$ is a Banach algebra. In particular, for every $f,g \in \mathbb{C}\delta \oplus \mathscr{O}(\overline{\mathscr{U}}$, $\|f*g\|_\nu \leq \|f\|_\nu \|g\|_\nu$. The space $\mathscr{M}\mathscr{O}(\overline{\mathscr{U}})$ is closed in $\left(\mathscr{O}(\overline{\mathscr{U}}), \|.\|_\nu\right)$. Moreover, for $\nu > 0$:*

1. *for every $n \in \mathbb{N}$, for every $g \in \mathscr{O}(\overline{\mathscr{U}})$, $\|(\zeta \mapsto \overset{\bullet}{\zeta}{}^n) * g\|_\nu \leq \dfrac{n!}{\nu^{n+1}}\|g\|_\nu$, $\|(\zeta \mapsto \overset{\bullet}{\zeta}{}^n)\|_\nu \leq \dfrac{n!}{\nu^{n+1}} L$ and $\|(\zeta \mapsto 1)\|_\nu = L$.*
2. *for every $f,g \in \mathscr{O}(\overline{\mathscr{U}})$, $\|fg\|_\nu \leq \dfrac{1}{L}\|f\|_\nu \|g\|_0$.*
3. *for every $f \in \mathscr{O}(\overline{\mathscr{U}})$, $\nu \geq \nu_0 \geq 0 \Rightarrow \|f\|_\nu \leq \|f\|_{\nu_0}$.*
4. *for every $f \in \mathscr{M}\mathscr{O}(\overline{\mathscr{U}})$, $\lim_{\nu\to\infty} \|f\|_\nu = 0$.*
5. *the derivation $\widehat{\partial}_{|\mathscr{O}(\overline{\mathscr{U}})} : f \in \mathscr{O}(\overline{\mathscr{U}}) \mapsto -\overset{\bullet}{\zeta} f \in \mathscr{M}\mathscr{O}(\overline{\mathscr{U}})$ is invertible and the inverse map $\widehat{\partial}^{-1}$ satisfies: for every $f \in \mathscr{O}(\overline{\mathscr{U}})$, for every $g \in \mathscr{M}\mathscr{O}(\overline{\mathscr{U}})$, $\widehat{\partial}^{-1}(f*g)$ belongs to $\mathscr{M}\mathscr{O}(\overline{\mathscr{U}})$ and $\|\widehat{\partial}^{-1}(f*g)\|_\nu \leq \dfrac{1}{\nu L}\|f\|_\nu \|\widehat{\partial}^{-1} g\|_0$.*
   For every $\mathbb{C}\delta \oplus \mathscr{O}(\overline{\mathscr{U}})$, for every $g \in \mathscr{M}\mathscr{O}(\overline{\mathscr{U}})$, $\widehat{\partial}^{-1}(f*g)$ belongs to $\mathscr{O}(\overline{\mathscr{U}})$ and $\|\widehat{\partial}^{-1}(f*g)\|\|_\nu \leq \|f\|_\nu \|\widehat{\partial}^{-1} g\|_\nu$.

Proof. The norm $\|.\|_\nu$ is obviously equivalent to the maximum norm on the vector space $\mathscr{O}(\overline{\mathscr{U}})$. This shows the completeness of $\left(\mathscr{O}(\overline{\mathscr{U}}), \|.\|_\nu\right)$ and of $\left(\mathbb{C}\delta \oplus \mathscr{O}(\overline{\mathscr{U}}), \|.\|_\nu\right)$ as well.
Pick a point $\zeta \in \mathscr{U}$. For any $f,g \in \mathscr{O}(\overline{\mathscr{U}})$,

$$f*g(\zeta) = \int_0^{\mathrm{leng}_\rho(\zeta)} d\ell\, \mathrm{e}^{\nu\left[\mathrm{leng}_\rho\left(\mathscr{H}_1\circ T_\zeta^{-1}(l)\right)+\mathrm{leng}_\rho\left(\mathscr{H}_1^{-1}\circ T_\zeta^{-1}(l)\right)\right]} \times f\left(\mathscr{H}_1\circ T_\zeta^{-1}(\ell)\right)\mathrm{e}^{-\nu \mathrm{leng}_\rho\left(\mathscr{H}_1\circ T_\zeta^{-1}(l)\right)} g\left(\mathscr{H}_1^{-1}\circ T_\zeta^{-1}(\ell)\right)\mathrm{e}^{-\nu \mathrm{leng}_\rho\left(\mathscr{H}_1^{-1}\circ T_\zeta^{-1}(l)\right)}. \tag{4.3}$$

We know from lemma 4.9 that $\mathrm{leng}_\rho(\mathscr{H}_1(s)) + \mathrm{leng}_\rho(\mathscr{H}_1^{-1}(s)) \leq \mathrm{leng}_\rho(\zeta)$ for any $s \in [0,1]$. Therefore $L\mathrm{e}^{-\nu \mathrm{leng}_\rho(\zeta)}|f*g(\zeta)| \leq \|f\|_\nu \|g\|_\nu \int_0^{\mathrm{leng}_\rho(\zeta)} \frac{1}{L} d\ell \leq \|f\|_\nu \|g\|_\nu$.
This shows that for every $f,g \in \mathscr{O}(\overline{\mathscr{U}})$, $\|f*g\|_\nu \leq \|f\|_\nu \|g\|_\nu$, hence $\left(\mathscr{O}(\overline{\mathscr{U}}), \|.\|_\nu\right)$ is a Banach algebra and $\left(\mathbb{C}\delta \oplus \mathscr{O}(\overline{\mathscr{U}}), \|.\|_\nu\right)$ as well. We encourage the reader to show the other properties. $\square$

Remark 4.4. We have already noticed that the space $\hat{\mathscr{R}}^{(1)}$ can be identified with the space $\mathscr{O}(\mathscr{R}^{(1)})$ of holomorphic functions on the Riemann surface $\mathscr{R}^{(1)}$. Since

$\mathscr{O}(\mathscr{R}^{(1)}) = \bigcap_{\substack{L>0 \\ 0<\rho\leq 1/5}} \mathscr{O}(\mathscr{U}_{\rho,L})$, formula (4.2) provides the convolution product on $\mathscr{O}(\mathscr{R}^{(1)})$.

4.4.3 An Extended Grönwall-like Lemma

The following statement is similar to lemma 3.9.

Lemma 4.11 (Extended Grönwall lemma). *Let $N \in \mathbb{N}^\star$ be a positive integer. Let and $(\widehat{f}_n)_{0\leq n\leq N}$, resp. $(\widehat{F}_n)_{0\leq n\leq N}$, be a $(N+1)$-tuple of functions in $\mathscr{O}(\mathscr{R}^{(1)})$, resp. of entire functions, real, positive and non-decreasing on $\mathbb{R}^+$, with at most exponential growth of order 1 at infinity. We suppose that for every $0\leq n\leq N$ and every $\zeta\in\overline{\mathscr{R}}_\rho^{(1)}$, $|\widehat{f}_n(\zeta)|\leq \widehat{F}_n\big(\mathrm{leng}_\rho(\zeta)\big)$. Otherwise, let p,q,r be polynomial functions such that the function $\zeta\mapsto p(-\overset{\bullet}{\zeta})$ does not vanish on $\overline{\mathscr{R}}_\rho^{(1)}$ and we assume that the following upper bounds are valid:*

$$a=\sup_{\zeta\in\overline{\mathscr{R}}_\rho^{(1)}}\frac{|q|(\mathrm{leng}_\rho(\zeta))}{|p(-\overset{\bullet}{\zeta})|}<\infty,\ b=\sup_{\zeta\in\overline{\mathscr{R}}_\rho^{(1)}}\frac{|r|(\mathrm{leng}_\rho(\zeta))}{|p(-\overset{\bullet}{\zeta})|}<\infty,\ c=\sup_{\zeta\in\overline{\mathscr{R}}_\rho^{(1)}}\frac{1}{|p(-\overset{\bullet}{\zeta})|}<\infty.$$

We furthermore assume that $\widehat{w}\in\mathscr{O}(\overline{\mathscr{R}}_\rho^{(1)})$ solves the following convolution equation:

$$p(\widehat{\partial})\widehat{w}+1*[q(\widehat{\partial})\widehat{w}]=\zeta*[r(\widehat{\partial})\widehat{w}]+\widehat{f}_0+\sum_{n=1}^{N}\widehat{f}_n*\widehat{w}^{*n}. \tag{4.4}$$

Then, for every $d\geq 0$ and every $\zeta\in\overline{\mathscr{R}}_\rho^{(1)}$, $|\widehat{w}(\zeta)|\leq\widehat{W}_d\big(\mathrm{leng}_\rho(\zeta)\big)$, where $\widehat{W}_d\in\mathscr{O}(\mathbb{C})$ stands for the holomorphic solution of the following convolution equation:

$$\widehat{W}=d+[a+b\xi]*\widehat{W}+c\left(\widehat{F}_0+\sum_{n=0}^{N}\widehat{F}_n*\widehat{W}^{*n}\right). \tag{4.5}$$

Proof. Let $\widehat{w}\in\mathscr{O}(\overline{\mathscr{R}}_\rho^{(1)})$ be a solution of convolution equation (4.4). This means that for every $\zeta\in\overline{\mathscr{R}}_\rho^{(1)}$:

$$\begin{aligned}
p(\widehat{\partial})\widehat{w}(\zeta) &= \widehat{f}_0(\zeta)-\int_0^{\mathrm{leng}_\rho(\zeta)} q\big(\mathscr{H}_1\circ T_\zeta^{-1}(\ell)\big)\widehat{w}\big(\mathscr{H}_1\circ T_\zeta^{-1}(\ell)\big)\,d\ell\\
&+\int_0^{\mathrm{leng}_\rho(\zeta)}\mathscr{H}_1^{-1}\circ T_\zeta^{-1}(\ell)r\big(\mathscr{H}_1\circ T_\zeta^{-1}(\ell)\big)\widehat{w}\big(\mathscr{H}_1\circ T_\zeta^{-1}(\ell)\big)\,d\ell\\
&+\sum_{n=1}^{N}\int_0^{\mathrm{leng}_\rho(\zeta)}\widehat{f}_n\big(\mathscr{H}_1^{-1}\circ T_\zeta^{-1}(\ell)\big)\widehat{w}^{*n}\big(\mathscr{H}_1\circ T_\zeta^{-1}(\ell)\big)\,d\ell.
\end{aligned}$$

where $\mathscr{H}_1$ stands for the shortest symmetrically contractile path associated with ζ. From lemma 4.9 and the hypotheses, one obtains with $\xi = \text{leng}_\rho(\zeta)$:

$$|\widehat{w}(\zeta)| \leq \frac{1}{|p(-\overset{\bullet}{\zeta})|}\widehat{F}_0(\xi) + \int_0^\xi \left[\frac{|q|(\ell)}{|p(-\overset{\bullet}{\zeta})|} + \frac{|r|(\ell)}{|p(-\overset{\bullet}{\zeta})|}(\xi - \ell)\right] |\widehat{w}(\mathscr{H}_1 \circ T_\zeta^{-1}(\ell))| \, d\ell$$
$$+ \sum_{n=1}^N \int_0^\xi \frac{1}{|p(-\overset{\bullet}{\zeta})|}\widehat{F}_n(\xi - \ell)|\widehat{w}^{*n}(\mathscr{H}_1 \circ T_\zeta^{-1}(\ell))| \, d\ell.$$

Therefore

$$|\widehat{w}(\zeta)| \leq c\widehat{F}_0(\xi) + \int_0^\xi [a + b(\xi - s)] \, |\widehat{w}(\mathscr{H}_1 \circ T_\zeta^{-1}(\ell))| \, d\ell \tag{4.6}$$
$$+ c\sum_{n=1}^N \int_0^\xi \widehat{F}_n(\xi - \ell)|\widehat{w}^{*n}(\mathscr{H}_1 \circ T_\zeta^{-1}(\ell))| \, d\ell.$$

The existence and the properties of $\widehat{W}_d$, solution of (4.5), have been given in lemma 3.8. We adapt the proof of lemma 3.9. We first notice that $|\widehat{w}(\underline{0})| \leq c\widehat{F}_0(0)$ by definition of c and by hypothesis on $\widehat{F}_0$. Since $\widehat{W}(0) = d + c\widehat{F}_0(0)$, we have $|\widehat{w}(\underline{0})| \leq \widehat{W}(0)$.

First case. We first assume $|\widehat{w}(\underline{0})| < \widehat{W}(0)$. One considers, for $L > 0$, the open set $\mathscr{U}_{\rho,L}$ of L-points. We remark that, once $L_0 > 0$ is chosen small enough, then for every $0 < L \leq L_0$, for very $d > 0$, for every $\zeta \in \overline{\mathscr{U}}_{\rho,L}$, $|\widehat{w}(\zeta)| < \widehat{W}_d(\xi)$ with $\xi = \text{leng}_\rho(\zeta)$. This is just a consequence of lemma 3.9. (For $L > 0$ small enough, $\text{leng}_\rho(\zeta) = |\zeta|$).

We now assume that there exist $L_1 > 0$ and $\zeta_1 \in \overline{\mathscr{U}}_{\rho,L_1}$ such that $|\widehat{w}(\zeta_1)| \geq \widehat{W}_d(\xi_1)$, $\xi_1 = \text{leng}_\rho(\zeta_1)$. We recall that the mapping $\zeta \in \mathscr{R}_\rho^{(1)} \mapsto \text{leng}_\rho(\zeta)$ is continuous and we define $\chi = \{L \in [L_0, L_1] \mid \text{there exists } \zeta \in \overline{\mathscr{U}}_{\rho,L}, \, |\widehat{w}(\zeta)| \geq \widehat{W}_d(\text{leng}_\rho(\zeta))\}$. This is a closed set bounded from below and we denote by $L_2 \in]L_0, L_1]$ its infimum. This implies that:

- for every $\zeta \in \overline{\mathscr{U}}_{\rho,L_2}$, $|\widehat{w}(\zeta)| \leq \widehat{W}_d(\text{leng}_\rho(\zeta))$;
- there exists $\zeta_2 \in \overline{\mathscr{U}}_{\rho,L_2}$ such that $|\widehat{w}(\zeta_2)| = \widehat{W}_d(\xi_2)$ and $\xi_2 = \text{leng}_\rho(\zeta_2) = L_2$.

We pick such a $\zeta_2 \in \overline{\mathscr{U}}_{\rho,L_2}$. By (4.6),

$$|\widehat{w}(\zeta_2)| \leq c\widehat{F}_0(\xi_2) + \int_0^{\xi_2} [a + b(\xi_2 - \ell)] \, |\widehat{w}(\mathscr{H}_1 \circ T_{\zeta_2}^{-1}(\ell))| \, d\ell$$
$$+ c\sum_{n=1}^N \int_0^{\xi_2} \widehat{F}_n(\xi_2 - \ell)|\widehat{w}^{*n}(\mathscr{H}_1 \circ T_{\zeta_2}^{-1}(\ell))| \, d\ell$$

where $\mathscr{H}_1$ is the shortest symmetrically contractile path associated with ζ_2. We know by lemma 4.9 that $\text{leng}_\rho(\mathscr{H}_1 \circ T_{\zeta_2}^{-1}(\ell)) \leq \ell$ for every $\ell \in [0, \xi_2]$, while $\widehat{W}_d$ is real, positive and non-decreasing on $\mathbb{R}^+$. Therefore,

$$|\widehat{w}(\zeta_2)| \le c\widehat{F}_0(\xi_2) + \int_0^{\xi_2} [a + b(\xi_2 - \ell)]\,\widehat{W}_d(\ell)\,d\ell + c\sum_{n=1}^{N}\int_0^{\xi_2} \widehat{F}_n(\xi_2 - \ell)\widehat{W}_d^{*n}(\ell)\,d\ell.$$

This shows that $|\widehat{w}(\zeta_2)| \le \widehat{W}_d(\xi_2) - d$ and we get a contradiction.

Second case The case $|\widehat{w}(0)| = \widehat{W}(0)$ (thus $d = 0$) is done by an argument of continuity already used in the proof of lemma 3.9. □

4.5 Application to the First Painlevé Equation

4.5.1 A Step Beyond Borel-Laplace Summability

We come back to the minor $\widehat{w}$ of the formal series solution of the prepared equation (3.6). We already know that $\widehat{w}$ can be analytically continued to the star-shaped domain $\dot{\mathscr{R}}^{(0)} \subset \mathbb{C}$, with at most exponential growth of order 1 at infinity there (theorem 3.3). Said in other words, $\widehat{w}$ can be interpreted as a holomorphic function on the principal sheet $\mathscr{R}^{(0)}$ of the Riemann surface $\mathscr{R}$. We claim that this function can be analytically continued to every $\mathscr{R}^{(0)}$-nearby sheets: this is the matter of the next theorem.

Theorem 4.2. *The formal solution $\widetilde{w}$ of the prepared equation (3.6) associated with the first Painlevé equation satisfies the following properties:*

1. *its minor $\widehat{w}$ can be analytically continued to the Riemann surface $\mathscr{R}^{(1)}$. This provides a function in $\mathscr{O}(\mathscr{R}^{(1)})$ still denoted by $\widehat{w}$;*
2. *this function $\widehat{w}$ has at most exponential growth of order 1 at infinity on $\mathscr{R}^{(1)}$. More precisely, for every $\rho \in]0, \frac{1}{5}]$, there exist real positive constants $A = A(\rho) > 0$ and $\tau = \tau(\rho) > 0$ such that for every $\zeta \in \overline{\mathscr{R}}_\rho^{(1)}$, $|\widehat{w}(\zeta)| \le Ae^{\tau\xi}$ with $\xi = \mathrm{leng}_\rho(\zeta)$;*
3. *moreover $\mathrm{leng}_\rho(\zeta) \le \frac{1}{\rho}|\dot{\zeta}| + \frac{1}{\rho}\left(\frac{1}{\rho} - 2\right)$ and one can choose $A = 4$ and $\tau = \frac{4}{\rho^3}$ in the above estimates.*

Proof. We begin this proof with a preliminary result which should be compared with lemma 3.2.

Lemma 4.12. *There exists a real positive number $M_{\rho,(1)} > 0$ such that for every polynomial q of degree ≤ 1, for every $\zeta \in \overline{\mathscr{R}}_\rho^{(1)}$, $\frac{|q|(\mathrm{leng}_\rho(\zeta))}{|P(-\dot{\zeta})|}| \le M_{\rho,(1)}|q|(1)$. Moreover one can choose $M_{\rho,(1)} = \frac{6}{5\rho^3}$*

Proof. From lemma 3.2 and lemma 4.10, one sees that for every $\zeta \in \overline{\mathscr{R}}_\rho^{(1)}$, $\frac{\mathrm{leng}_\rho(\zeta)}{|P(-\dot{\zeta})|} \le \left[\frac{1}{\rho} + \frac{1}{\rho}\left(\frac{1}{\rho} - 2\right)\right] M_{\rho,(0)}$. Then use the fact that $\rho \in]0, \frac{1}{5}]$. □

Holomorphy of $\widehat{w}$ on $\mathscr{R}^{(1)}$ Let $r > 0$ and $\nu > 0$ be positive real numbers, $\mathscr{U} = \mathscr{U}_{\rho,L} \subset \mathscr{R}^{(1)}$ be the open set of L-points, $L > 0$, and $B_r = \{\widehat{v} \in \mathscr{O}(\overline{\mathscr{U}}), \|\widehat{v}\|_\nu \leq r\}$. The convolution equation (3.10) can be viewed as a fixed-point problem on B_r and we set:

$$\mathsf{N} : \widehat{v} \in B_r \mapsto P^{-1}(\widehat{\partial})\Big[-1 * \big[Q(\widehat{\partial})\widehat{v}\big] + \widehat{f}_0 + \widehat{f}_1 * \widehat{v} + \widehat{f}_2 * \widehat{v} * \widehat{v}\Big].$$

By lemmas 4.12 and proposition 4.5,

$$\|\mathsf{N}(\widehat{v})\|_\nu \leq M_{\rho,(1)} \| -1 * \big[Q(\widehat{\partial})\widehat{v}\big] + \widehat{f}_0 + \widehat{f}_1 * \widehat{v} + \widehat{f}_2 * \widehat{v} * \widehat{v}\|_\nu.$$

By proposition 4.5, since $Q(\widehat{\partial}) = -3\widehat{\partial}$:

$$\|1 * \big[Q(\widehat{\partial})\widehat{v}\big]\|_\nu \leq \frac{1}{\nu}\|Q(\widehat{\partial})\widehat{v}\|_\nu \leq \frac{1}{L\nu}\|(\zeta \mapsto Q(-\overset{\bullet}{\zeta}))\|_0 \|\widehat{v}\|_\nu \leq \frac{3L}{\nu}\|\widehat{w}\|_\nu.$$

The functions $\widehat{f}_0, \widehat{f}_1, \widehat{f}_2$ belong to $\mathscr{M}\mathscr{O}(\overline{\mathscr{U}})$, therefore by proposition 4.5: $\lim\limits_{\nu \to \infty} \|\widehat{f}_i\|_\nu = 0$, $i = 0, 1, 2$. Hence, $\|\mathsf{N}(\widehat{v})\|_\nu \leq r$ for $\nu > 0$ large enough.
The same arguments shows that $\|\mathsf{N}(\widehat{v}_1) - \mathsf{N}(\widehat{v}_2)\|_\nu \leq k\|\widehat{v}_1 - \widehat{v}_2\|_\nu$ with $k < 1$, for $\widehat{v}_1, \widehat{v}_2 \in B_r$ and for $\nu > 0$ large enough.
The mapping N is thus contractive on the closed subset B_r of the Banach space $\big(\mathscr{O}(\overline{\mathscr{U}}), \|.\|_\nu\big)$, for $\nu > 0$ large enough. The contraction mapping theorem ensures the existence of a unique solution $\widehat{w} \in B_r$ for the fixed-point problem $\widehat{v} = \mathsf{N}(\widehat{v})$. Since L and ρ can be arbitrarily chosen, we deduce (by uniqueness) that the minor $\widehat{w}$ of the unique formal series $\widetilde{w}$ solution of (3.6) is a germ of holomorphic functions which can be analytically continued to $\mathscr{R}^{(1)}$.

Upper bounds We use the Grönwall lemma 4.11 (with $d = 0$), which tells us that for every $\zeta \in \overline{\mathscr{R}}_\rho^{(1)}$, $|\widehat{w}(\zeta)| \leq \widehat{W}_d(\xi)$, $\xi = \mathrm{leng}_\rho(\zeta)$, where $\widehat{W}(\xi)$ solves the following convolution equation $\frac{1}{M_{\rho,(1)}}\widehat{W} = |\widehat{f}_0| + (3 + |\widehat{f}_1|) * \widehat{W} + |\widehat{f}_2| * \widehat{W} * \widehat{W}$ (just use lemma 4.12). Moreover, one can choose $M_{\rho,(1)} = \frac{6}{5\rho^3}$. We would like to get explicit estimates. We consider $\widehat{W}$ as the Borel transform of the function $\widetilde{W}$, solution of the second order algebraic equation,

$$\frac{1}{M_{\rho,(1)}}\widetilde{W} = |f_0|(z) + \big(\frac{3}{z} + |f_1|\big)\widetilde{W} + |f_2|\widetilde{W}^2, \tag{4.7}$$

holomorphic at infinity and asymptotic to $|f_0|(z)$ there. Remember that $|f_0|(z) = \frac{392}{625}\frac{1}{z^2}$, $|f_1|(z) = \frac{4}{z^2}$, $|f_2|(z) = \frac{1}{2z^2}$. Setting $\widetilde{W}(z) = H(t)$, $t = z^{-1}$, the above problem reads as a fixed-point problem,

$$H = \mathsf{N}(H), \;\; \mathsf{N}(H) = M_{\rho,(1)}\Big(|f_0|(t^{-1}) + \big(3t + |f_1|(t^{-1})\big)H + |f_2|(t^{-1})H^2\Big). \tag{4.8}$$

From homogeneity reasons, we introduce $U = D(0,\rho^3/4)$, we consider the Banach algebra $\left(\mathcal{O}(\overline{U}), \|\ \|\right)$ where $\|\ \|$ stands for the maximum norm, and we set $B_{\rho^3} = \{H \in \mathcal{O}(\overline{U}),\ \|H\| \leq \rho^3\}$. It is easy to show that the mapping $\mathsf{N}_{|B_{\rho^3}} : H \in B_{\rho^3} \mapsto \mathsf{N}(H) \in B_{\rho^3}$ is contractive (remember: $\rho \in]0,1/5]$). Therefore, from contraction mapping theorem, the fixed-point problem (4.8) has a unique solution H in B_{ρ^3}. In return we deduce that $\widehat{W}$ is an entire function and $|\widehat{W}(\xi)| \leq 4e^{\frac{4}{\rho^3}|\xi|}$ for every $\xi \in \mathbb{C}$ (see lemma 3.5). One ends with lemma 4.10.

4.5.2 Concluding Remarks

The following comments rely on notions introduced in the first volume of this book [MS016] to which the reader is referred.

It turns out from theorem 4.2 that the minor $\widehat{w}$ can be analytically continued along any path of type $\gamma^{\theta}_{(+)_{m-1}}$ and $\gamma^{\theta}_{(-)_{m-1}}$ (definitions 4.7 and 4.9), for any $m \in \mathbb{N}^\star$ and any direction $\theta \in \{0,\pi\} \subset \mathbb{S}^1$.
To fix our mind, we consider a path γ of type $\gamma^{0}_{(+)_{m-1}}$. The analytic continuation of $\widehat{w}$ along γ gives a germ $\mathrm{cont}_\gamma \widehat{w}$ which can be represented by a function holomorphic on the open disc $D(\frac{2m-1}{2},\frac{1}{2})$ adherent to m. Writing $f_m(\zeta) = \mathrm{cont}_\gamma \widehat{w}(m+\zeta)$, we get a function f_m holomorphic on $D = D(-\frac{1}{2},\frac{1}{2})$. However, theorem 4.2 translates into the fact that f_m can be analytically continued to a wider domain as a "multi-valued function". Precisely, pick a point $\underline{\zeta}_0 = \frac{1}{2}e^{i\theta_0} \in \pi^{-1}(-\frac{1}{2})$ above $-\frac{1}{2}$ on the Riemann surface $(\widetilde{\mathbb{C}},\pi)$ of the logarithm. Let $\widetilde{D} = \widetilde{D}(\underline{\zeta}_0,\frac{1}{2}) \subset \widetilde{\mathbb{C}}$ be the neighbourhood of $\underline{\zeta}_0$ which is π-homeomorphic to D. One obtains a function $\overset{\vee}{f}_m = f_m \circ \pi$, $\overset{\vee}{f} \in \mathscr{O}(\widetilde{D})$. This function can be holomorphically extended to a function (still denoted by) $\overset{\vee}{f}_m \in \mathscr{O}(\mathcal{S}_0^{1/2})$, where $\mathcal{S}_0^{1/2} \subset \widetilde{\mathbb{C}}$ is the open sector defined by: $\mathcal{S}_0^{1/2} = \{\underline{\zeta} = \xi e^{i\theta} \in \widetilde{\mathbb{C}} \mid \theta \in]-\pi+\theta_0, \theta_0+2\pi[, \xi \in]0,1/2[\}$.

Question 4.6. Can we analytically continue each $\overset{\vee}{f}_m$ into an element of ANA ?

The answer is "yes" but requires further effort and supplements to resurgence theory, given in chapter 7. Taking this for granted, to $\overset{\vee}{f}_m$ thus corresponds a singularity $\overset{\triangledown}{f}_m \in$ SING deduced from $\widehat{w}$ through the action of the alien operator denoted by $\mathscr{A}_m^{\gamma,\underline{\zeta}_0}$ in the first volume of this book [MS016], or more precisely to Δ_m^+.

Question 4.7. Can we describe more precisely the singularities $\overset{\triangledown}{f}_m$?

This is of course the key-question for describing the Stokes phenomenon. Partly, the reply relies on the formal integral associated with equation (3.6), which is the matter of the next chapter 5. The final answer will be given in the last chapter 8 of this course, with the use of the alien derivations. In the same spirit:

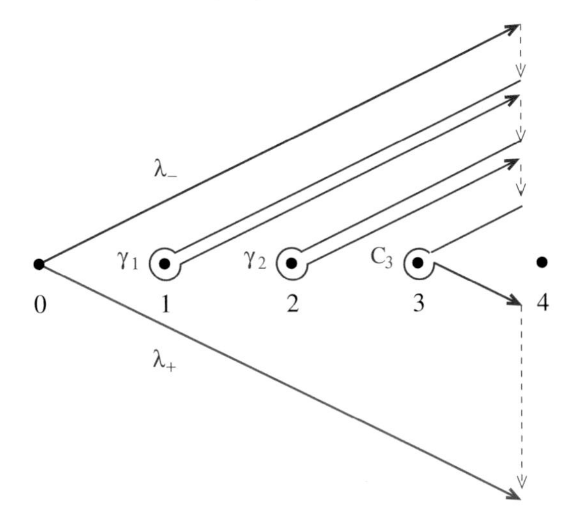

Fig. 4.10 Comparison of right and left Borel-Laplace summation for the direction $\theta = 0$.

Question 4.8. At this stage, can we compare the sums $\mathscr{S}^{]-\pi,0[}\widetilde{w} \in \mathscr{O}(\overset{\bullet}{\mathscr{D}}(]-\pi,0[,\tau))$ and $\mathscr{S}^{]0,\pi[}\widetilde{w} \in \mathscr{O}(\overset{\bullet}{\mathscr{D}}(]0,\pi[,\tau))$? (See proposition 3.14.) In other words, are we able to analyze the Stokes phenomenon ?

Formulated another way, we would like to compare the right Borel-Laplace summation $\mathscr{S}^{0^+}\widetilde{w}$ and the left one $\mathscr{S}^{0^-}\widetilde{w}$. Look at Fig. 4.10 : we have chosen two directions $\theta^+ \in]-\pi,0[$ and $\theta^- \in]-\pi,0[$ closed to zero, so that the Borel-Laplace sums $\mathscr{S}^{\theta^+}\widetilde{w}(z) = \int_{\lambda_+} e^{-z\zeta}\widehat{w}(\zeta)d\zeta$ and $\mathscr{S}^{\theta^-}\widetilde{w}(z) = \int_{\lambda_-} e^{-z\zeta}\widehat{w}(\zeta)d\zeta$ can be compared on their nonempty common domain of definition $\overset{\bullet}{\Pi}{}^{\theta^+}_{\tau(\theta^+)} \cap \overset{\bullet}{\Pi}{}^{\theta^-}_{\tau(\theta^-)}$. The curve $\lambda_-^{-1}\lambda_+$ can be seen as a chain on the Riemann sphere $\overline{\mathbb{C}} = \mathbb{C} \cup \{\infty\}$ running from ∞ to ∞ and avoiding the points $\mathbb{Z}^\star \subset \mathbb{C}$. In other words, $\lambda_-^{-1}\lambda_+$ represents a cycle for the 1-homology group $H_1(\overset{\bullet}{\mathscr{R}} \cup \{\infty\},\infty)$ [DH002, Del010] which is homologous to the sum $\sum_{k=1}^m \gamma_k + C_{m+1}$ (with $m = 2$ on Fig. 4.10). Interpreting this result on $\mathscr{R}^{(1)}$ where $\widehat{w}$ is holomorphic with at most exponential growth of order 1 at infinity (theorem 4.2), we get: for every $z \in \overset{\bullet}{\Pi}{}^{\theta^+}_{\tau(\theta^+)} \cap \overset{\bullet}{\Pi}{}^{\theta^-}_{\tau(\theta^-)}$ with $|z|$ large enough,

$$\mathscr{S}^{\theta^+}\widetilde{w}(z) = \mathscr{S}^{\theta^-}\widetilde{w}(z) + \sum_{k=1}^{m}\int_{\gamma_k} e^{-z\zeta}\widehat{w}(\zeta)d\zeta + \int_{C_{m+1}} e^{-z\zeta}\widehat{w}(\zeta)d\zeta. \tag{4.9}$$

One recognizes in this equation the very construction of the Stokes automorphism, detailed in the first volume of this book [MS016], see also [DP99, DH002, Del005, Del010]. The asymptotics at infinity of the integrals $\int_{\gamma_k} e^{-z\zeta}\widehat{w}(\zeta)d\zeta$ are of the form $e^{-kz}\widetilde{W}_k(z)$ where $\widetilde{W}_k$ stands for a formal series which only depends on the still unknown singularity $\overset{\triangledown}{f}_k$. The remaining integral $\int_{C_{m+1}} e^{-z\zeta}\widehat{w}(\zeta)d\zeta$ provides an exponentially smaller vanishing behaviour. It will be shown in this course that the right-hand side of the equality (4.9) when letting $m \to \infty$, is nothing but the Borel-Laplace sum of a "transseries" introduced and studied in chapters 5 and 6.

4.6 Some Supplements

We end this chapter with some supplements which will be used later on.

Definition 4.23. Let be $\theta \in \{0, \pi\}$, $\alpha \in]0, \pi/2]$ and $L > 0$ be a real positive number. we denote by $\mathfrak{R}^{(\theta,\alpha)}(L)$ the set of paths λ in $\dot{\mathscr{R}}$ originating from 0, piecewise $\mathscr{C}^1$, with length$(\lambda) < L+1$, with the conditions:

- either λ stays in the open disc $D(0,1)$;
- or else, for every $t \in [0,1]$, the right and left derivatives $\underline{\underline{\lambda}}'(t)$ do not vanish and $\arg \underline{\underline{\lambda}}'(t) \in]-\alpha+\theta, \theta+\alpha[$.

We denote by $\mathscr{R}^{(\theta,\alpha)}(L)$ the subset of the Riemann surface $\mathscr{R}$ defined as follows:

$$\mathscr{R}^{(\theta,\alpha)}(L) = \{\underline{\underline{\Lambda}}(1) \in \mathscr{R} \mid \lambda = \mathfrak{p} \circ \Lambda \text{ belongs to } \mathfrak{R}^{(\theta,\alpha)}(L) \text{ and } \underline{\underline{\Lambda}}(0) = \underline{0}\}.$$

We should note in passing that every path $\lambda \in \mathfrak{R}^{(\theta,\alpha)}(L)$ can be lifted on $\mathscr{R}$ from $\underline{0}$ with respect to $\mathfrak{p}$.
The following assertion is left as an exercise.

Lemma 4.13. *The set $\mathscr{R}^{(\theta,\alpha)}(L)$ is an open and connected neighbourhood of $\underline{0}$ in $\mathscr{R}$ and $\mathscr{R}^{(\theta,\alpha)}(L) \subset \mathscr{R}^{(0)} \bigcup_{1 \le j \le m} \mathscr{R}^{(\pm)_j,\theta}$ with $m = \lceil L \rceil$. Also, for any $m \in \mathbb{N}^\star$, for any path γ of type $\gamma_\varepsilon^\theta$ with $\varepsilon \in \{+,-\}^j$ and $1 \le j \le m$, the endpoint $\underline{\underline{\Gamma}}(1)$ belongs to $\mathscr{R}^{(\theta,\alpha)}(m)$, where Γ is the lift of γ from $\underline{0}$ with respect to $\mathfrak{p}$ on $\mathscr{R}$.*

In the above lemma, $\lceil . \rceil$ is the ceiling function.

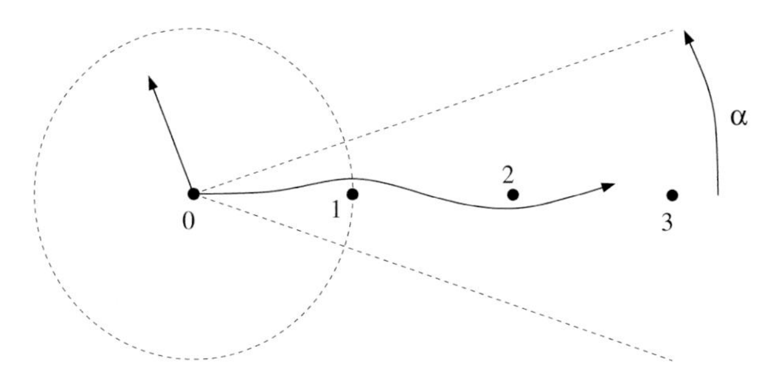

Fig. 4.11 Two paths belonging to $\mathfrak{R}^{(\theta,\alpha)}(L)$ for $\theta = 0$ and $L \ge 2$.

Remark 4.5. Notice that $\mathscr{R}^{(\theta,\alpha)}(L_1) \subset \mathscr{R}^{(\theta,\alpha)}(L_2)$ when $L_1 < L_2$. Also, since $\mathscr{R}^{(\theta,\alpha)}(L)$ is open and connected in $\mathscr{R}$, $\mathscr{R}^{(\theta,\alpha)}(L)$ inherits from $\mathscr{R}$ the structure of complex manifold, thus is a Riemann surface.

Definition 4.24. We denote by $\hat{\mathscr{R}}^{(\theta,\alpha)}(L) \supset \hat{\mathscr{R}}$ the space of germs of holomorphic functions at the origin which can be analytically continued to the Riemann surface $\mathscr{R}^{(\theta,\alpha)}(L)$.

Example 4.1. The formal solution $\widetilde{w}$ of the prepared equation (3.6) has a minor $\widehat{w}$ which belongs to $\hat{\mathscr{R}}^{(\theta,\pi/2)}(L)$, for any direction $\theta \in \{0,\pi\}$ and any $L \in]0,1]$: this is a consequence of theorem 4.2.

Proposition 4.6. *The space $\hat{\mathscr{R}}^{(\theta,\alpha)}(L)$ is a (non unitary) convolution algebra.*

Proof. We have to prove the stability by convolution product. It is shown in the first volume of this book [MS016] that for any smooth path $\gamma : I = [0,1] \to \mathbb{C} \setminus \mathbb{Z}$ such that $|\gamma(0)| < 1$, one can find a symmetric $\mathbb{Z}$-homotopy $H : (s,t) \in I \times I \mapsto H(s,t) = H_t(s)$ whose initial path H_0 is $H_0 : s \in I \mapsto s\gamma(0)$ and whose endpoint path $t \in I \mapsto H_t(1)$ coincides with γ. Lifting every path H_t from $\underline{0}$ on $\mathscr{R}$ with respect to $\mathfrak{p}$, one gets the mapping $\mathscr{H} : (s,t) \in I \times I \mapsto \mathscr{H}(s,t) = \mathscr{H}_t(s) \in \mathscr{R}$, which is continuous by the lifting theorem for homotopies, and the following diagram commutes:

$$\begin{array}{ccc} & & \mathscr{R} \\ & \mathscr{H} \nearrow & \downarrow \mathfrak{p} \\ & & \bullet \\ I \times I & \longrightarrow & \mathscr{R}. \\ & H & \end{array} \tag{4.10}$$

We recall from [MS016] how this symmetric $\mathbb{Z}$-homotopy can be constructed. Pick a $\mathscr{C}^1$ function $\eta : \mathbb{C} \to [0,1]$ satisfying $\{\zeta \in \mathbb{C} \mid \eta(\zeta) = 0\} = \mathbb{Z}$ and consider the non-autonomous vector field $X(\zeta,t) = \dfrac{\eta(\zeta)}{\eta(\zeta) + \eta(\gamma(t) - \zeta)} \gamma'(t)$. The path H_t is obtained by deformation of the initial path H_0 through the flow of the vector field $g : (t_0,t,\zeta) \in [0,1]^2 \times \mathbb{C} \mapsto g^{t_0,t}(\zeta) \in \mathbb{C}$ of X, precisely $H_t(s) = g^{0,t}\big(H_0(s)\big)$.

Let ζ be any point in $\mathscr{R}^{(\theta,\alpha)}(L)$. This point is the endpoint of a path Λ in $\mathscr{R}$ originating from $\underline{0}$ and whose projection $\lambda = \mathfrak{p} \circ \Lambda$ belongs to $\mathfrak{R}^{(\theta,\alpha)}(L)$. We forget the case where λ stays in $D(0,1)$ and, without loss of generality, we can assume that $\lambda = \lambda_0 \gamma$ with $\lambda_0 : s \in [0,1] \mapsto s\gamma(0)$. Let us analyze the above symmetric $\mathbb{Z}$-homotopy H constructed from γ and $H_0 = \lambda_0$, and the associated mapping $\mathscr{H}$. We would like to show that $\mathscr{H}(s,t) \in \mathscr{R}^{(\theta,\alpha)}(L)$ for every $(s,t) \in I \times I$. For this purpose we introduce the path $H^s : t \in I \mapsto H^s(t) = H(s,t)$. We notice that $H^0 \equiv 0$ while for any $s \in]0,1]$:

1. $H^s(0) = H_0(s)$,
2. $\frac{\mathrm{d}H^s(t)}{\mathrm{d}t} = X\big(H^s(t),t\big)$, thus $0 < \left|\frac{\mathrm{d}H^s(t)}{\mathrm{d}t}\right| \leq |\gamma'(t)|$ and $\arg \frac{\mathrm{d}H^s(t)}{\mathrm{d}t} \in]-\alpha+\theta, \theta+\alpha[$.

Denoting by $H_0|_{[0,s]} : s' \in I \mapsto H_0(s's)$ the restriction path, we see that the product of paths $F^s = H_0|_{[0,s]} H^s$ has the following properties, for any $s \in]0,1]$:

1. F^s is a path in $\overset{\bullet}{\mathscr{R}}$ originating from 0 and is piecewise $\mathscr{C}^1$;
2. $\text{length}(F^s) \leq \text{length}(H_0|_{[0,s]}) + \text{length}(H^s) \leq \text{length}(\lambda) \leq L+1$,
3. for every $t \in [0,1]$, the right and left derivatives $(F^s)'(t)$ do not vanish and $\arg(F^s)'(t) \in]-\alpha+\theta, \theta+\alpha[$.

Therefore F^s belongs to $\mathfrak{R}^{(\theta,\alpha)}(L)$ and as a consequence, $\mathscr{H}(s,t)$ belongs to $\mathscr{R}^{(\theta,\alpha)}(L)$ for every $(s,t) \in I \times I$. We end the proof with the arguments used in the proof of theorem 4.1. □

4.7 Comments

As a rule in resurgence theory, one has to deal with *endlessly continuable* functions. This notion is defined in [CNP93-1], a more general definition of which being given by Ecalle in [Eca85, Eca93-1]. The key point is the construction of *endless Riemann surfaces* [CNP93-1, Ou012]). For such an endless Riemann surface, one can define "nearby sheets" in the way we did in Sect. 4.2 and analogues of theorem 4.1 and proposition 4.6 can be stated.

References

AB91. F. Albrecht, I.D. Berg, *Geodesics in Euclidean space with analytic obstacle.* Proc. Amer. Math. Soc. **113** (1991), no. 1, 201-207.

CNP93-1. B. Candelpergher, C. Nosmas, F. Pham, *Approche de la résurgence*, Actualités mathématiques, Hermann, Paris (1993).

Del005. E. Delabaere, *Addendum to the hyperasymptotics for multidimensional Laplace integrals.* Analyzable functions and applications, 177-190, Contemp. Math., 373, Amer. Math. Soc., Providence, RI, 2005.

Del010. E. Delabaere, *Singular integrals and the stationary phase methods.* Algebraic approach to differential equations, 136-209, World Sci. Publ., Hackensack, NJ, 2010.

DH002. E. Delabaere, C. J. Howls, *Global asymptotics for multiple integrals with boundaries.* Duke Math. J. **112** (2002), 2, 199-264.

DP99. E. Delabaere, F. Pham, *Resurgent methods in semi-classical asymptotics*, Ann. Inst. Henri Poincaré, Sect. A **71** (1999), no 1, 1-94.

Ebe007. W. Ebeling, *Functions of several complex variables and their singularities.* Translated from the 2001 German original by Philip G. Spain. Graduate Studies in Mathematics, 83. American Mathematical Society, Providence, RI, 2007.

Eca81-1. J. Écalle, *Les algèbres de fonctions résurgentes*, Publ. Math. d'Orsay, Université Paris-Sud, 1981.05 (1981).

Eca85. J. Ecalle, *Les fonctions résurgentes. Tome III : l'équation du pont et la classification analytique des objets locaux.* Publ. Math. d'Orsay, Université Paris-Sud, 1985.05 (1985).

Eca93-1. J. Écalle, *Six lectures on transseries, Analysable functions and the Constructive proof of Dulac's conjecture*, Bifurcations and periodic orbits of vector fields (Montreal, PQ, 1992), 75-184, NATO Adv. Sci. Inst. Ser. C Math. Phys. Sci., 408, Kluwer Acad. Publ., Dordrecht, 1993.

For81. O. Forster, *Lectures on Riemann Surfaces*, Graduate texts in mathematics; 81, Springer, New York (1981).

GS001. V. Gelfreich, D. Sauzin, *Borel summation and splitting of separatrices for the Hénon map*, Ann. Inst. Fourier (Grenoble) **51** (2001), no 2, 513-567.

Jos95. J. Jost, *Riemannian geometry and geometric analysis.* Universitext. Springer-Verlag, Berlin, 1995.

MS016. C. Mitschi, D. Sauzin, *Divergent Series, Summability and Resurgence I. Monodromy and Resurgence.* Lecture Notes in Mathematics, **2153**. Springer, Heidelberg, 2016.

Ou010. Y. Ou, *On the stability by convolution product of a resurgent algebra.* Ann. Fac. Sci. Toulouse Math. (6) **19** (2010), no. 3-4, 687-705.

Ou012. Y. Ou, *Sur la stabilité par produit de convolution d'algèbres de résurgence.* PhD thesis, Université d'Angers (2012).

Sau013. D. Sauzin, *On the stability under convolution of resurgent functions.* Funkcial. Ekvac. **56** (2013), no. 3, 397-413.

Sau015. D. Sauzin, *Nonlinear analysis with resurgent functions.* Ann. Sci. Ecole Norm. Sup. (6) **48** (2015), no. 3, 667-702.

Chapter 5
Transseries And Formal Integral For The First Painlevé Equation

Abstract This chapter has two purposes. Our first goal is to construct the so-called "formal transseries solutions" for the prepared form associated with the first Painlevé equation, which will be used later on to get the truncated solutions : this is done in Sect. 5.3, after some preliminaries in Sect. 5.1 and Sect. 5.2. Our second goal is to build the formal integral for the first Painlevé equation and, equivalently, the canonical normal form equation to which the first Painlevé equation is formally conjugated. This is what we do in Sect. 5.4. These informations will be used in a next chapter to investigate the resurgent structure for the first Painlevé equation.

5.1 Introduction

We have seen in chapter 3 that the prepared equation (3.6) has a unique formal solution, from now on denoted by $\widetilde{w}_{(0,0)}$. This solution is 1-Gevrey and even Borel-Laplace summable in every directions apart from the directions $k\pi$, $k \in \mathbb{Z}$ (theorem 3.3 and proposition 3.14). To each interval $I_j =]0,\pi[+j\pi$, $j \in \mathbb{Z}$, one associates the Borel-Laplace sum $w_{tri,j}(z) = \mathscr{S}^{I_j}\widetilde{w}_{(0,0)}(z) \in \mathscr{O}\big(\overset{\bullet}{\mathscr{D}}(I_j,\tau)\big)$ where $\overset{\bullet}{\mathscr{D}}(I_j,\tau)$ is a sectorial neighbourhoods of ∞ with aperture $\breve{I}_j =]-\frac{3}{2}\pi, +\frac{1}{2}\pi[-j\pi$. As said in Sect. 3.4.2.2, each $w_{tri,j}$ can be thought of as a section over $\breve{I}_j$ of the sheaf $\mathscr{A}_1$ of 1-Gevrey asymptotic functions, $w_{tri,j} \in \Gamma(\breve{I}_j, \mathscr{A}_1)$. These sections are asymptotic to the same 1-Gevrey series $\widetilde{w}_{(0,0)}$. Therefore the 1-coboundary $w_{tri,1} - w_{tri,0}$ belongs to $\Gamma(\breve{I}_1 \cap \breve{I}_0, \mathscr{A}^{\leq -1})$, while $w_{tri,2} - w_{tri,1}$ belongs to $\Gamma(\breve{I}_2 \cap \breve{I}_1, \mathscr{A}^{\leq -1})$, where $\mathscr{A}^{\leq -1}$ is the sheaf of 1-Gevrey flat germs. In other words, the 1-coboundaries

$$\begin{aligned} \mathscr{W}_{1,0}(z) &= w_{tri,1}(z) - w_{tri,0}(z), \quad -\frac{3}{2}\pi < \arg(z) < -\frac{1}{2}\pi, \quad |z| \text{ large enough}, \\ \mathscr{W}_{2,1}(z) &= w_{tri,2}(z) - w_{tri,1}(z), \quad -\frac{5}{2}\pi < \arg(z) < -\frac{3}{2}\pi, \quad |z| \text{ large enough}, \end{aligned} \tag{5.1}$$

E. Delabaere, *Divergent Series, Summability and Resurgence III*,
Lecture Notes in Mathematics 2155, DOI 10.1007/978-3-319-29000-3_5

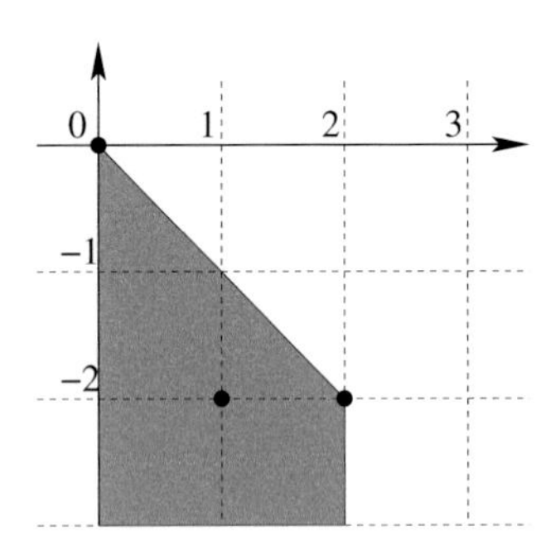

Fig. 5.1 The Newton polygon at infinity $\mathcal{N}_\infty(\mathfrak{P}_0)$ associated with the linear operator (5.3).

are exponentially flat functions of order 1 at infinity, and we deduce from equation (3.6) that $\mathscr{W}_{(j+1),j}$, $j = 0,1$, satisfies the linear ODE:

$$P(\partial)\mathscr{W}_{(j+1),j} + \frac{1}{z}Q(\partial)\mathscr{W}_{(j+1),j} = f_1(z)\mathscr{W}_{(j+1),j} + f_2(z)(w_{tri,j+1} + w_{tri,j})\mathscr{W}_{(j+1),j}. \tag{5.2}$$

Question 5.1. Can we get more informations about $\mathscr{W}_{(j+1),j}$? In other words, are we able to analyze the Stokes phenomenon ?

We have already made some advances in reply to this query in Sect. 4.5.2, as an application of theorem 4.2. Here we return to the asymptotics. Denoting by T_1 the 1-Gevrey Taylor map, we set $\widetilde{\mathscr{W}}_{(j+1),j} = T_1(\breve{I}_{j+1} \cap \breve{I}_j)\mathscr{W}_{(j+1),j}$. We have $\widetilde{\mathscr{W}}_{(j+1),j} = 0$ by construction but, more interestingly for our purpose and since T_1 is a morphism of differential algebras, we deduce from (5.2) that $\widetilde{\mathscr{W}}_{(j+1),j}$ solves the problem $\mathfrak{P}_0\widetilde{\mathscr{W}} = 0$, where $\mathfrak{P}_0$ stands for the second order linear differential operator deduced from the operator $P(\partial) + \frac{1}{z}Q(\partial) - F(z,\cdot)$ by linearisation at $\widetilde{w}_{(0,0)}$:

$$\begin{aligned} \mathfrak{P}_0 &= P(\partial) + \frac{1}{z}Q(\partial) - \frac{\partial F(z,\widetilde{w}_{(0,0)})}{\partial w} \\ &= P(\partial) + \frac{1}{z}Q(\partial) - f_1(z) - 2\widetilde{w}_{(0,0)}(z)f_2(z) \\ &= (\partial^2 - 1) - \frac{3}{z}\partial + O(z^{-2}). \end{aligned} \tag{5.3}$$

For a moment, let us think of $\mathfrak{P}_0\widetilde{\mathscr{W}} = 0$ as a linear ODE with holomorphic coefficients (thus we think of $\widetilde{w}_{(0,0)}$ as a convergent series at infinity). The formal invariants for this this equation are governed by the Newton polygon at infinity $\mathcal{N}_\infty(\mathfrak{P}_0)$, drawn on Fig. 5.1.

The definition and properties of the Newton polygon are amply elaborated in the second volume of this book [Lod016], to which the reader is referred. We only mention that the valuation v_∞ defined there is the opposite of our valuation val defined by (3.1).

The polygon $\mathcal{N}_\infty(\mathfrak{P}_0)$ has a single non-vertical side of slope -1: this corresponds to the fact that $\mathcal{W}_{(j+1),j}$, $j=0,1$, are exponentially flat functions of order 1 at infinity. The characteristic equation associated with this side is nothing but the equation

$$P(\mu)=0, \qquad P(\mu)=\mu^2-1.$$

The polynomial $P(\mu)$ has two simple roots, $\mu_1=-1$ and $\mu_2=1$. Therefore, from the theory of linear ODE [Was65, Lod016], we expect for $\mathcal{W}_{1,0}$ to behave like $e^{\mu_2 z}z^{-\tau_2}O(1)$ at infinity, and for $\mathcal{W}_{2,1}$ to behave like $e^{\mu_1 z}z^{-\tau_1}O(1)$ at infinity.
Pursuing in that direction, the coefficients τ_1,τ_2 can be easily found : $\widetilde{\mathcal{W}}=e^{\mu z}z^{-\tau}\widetilde{w}_\mu(z)$ solves the ODE (5.3) with $P(\mu)=0$ and $\widetilde{w}_\lambda\in\mathbb{C}[[z^{-1}]]$, only under the condition

$$\tau=\frac{Q(\mu)}{P'(\mu)}=-\frac{3}{2}.$$

As a matter of fact, these behaviours are direct consequences of the analytic properties of the minor $\widehat{w}_{(0,0)}$ of $\widetilde{w}_{(0,0)}$. In particular, $\lambda_1=-\mu_1$ and $\lambda_2=-\mu_2$ are precisely the so-called seen singularities of $\widehat{w}_{(0,0)}$, cf. theorem 4.2.

The differential equation $\mathfrak{P}_0\widetilde{\mathcal{W}}=0$ has thus its general formal solution under the form $\widetilde{\mathcal{W}}=U_1e^{\mu_1 z}z^{-\tau_1}\widetilde{w}_{\mu_1}+U_2e^{\mu_2 z}z^{-\tau_2}\widetilde{w}_{\mu_2}$ and, as we will see later on, both $\widetilde{w}_{\mu_1}$ and $\widetilde{w}_{\mu_2}$ are 1-Gevery series whose minors have the same properties than $\widehat{w}_{(0,0)}$.

However, the expectation that $\mathcal{W}_{1,0}$ could be obtained from $U_1e^{\mu_1 z}z^{-\tau_1}\widetilde{w}_{\mu_1}$ by Borel-Laplace summation for some well-chosen $U_1\in\mathbb{C}$ is wrong. Indeed, this would mean that $w_{tri,1}=\mathscr{S}^{I_1}\left(\widetilde{w}_{(0,0)}+U_1e^{\mu_1 z}z^{-\tau_1}\widetilde{w}_{\mu_1}\right)$, thus $\widetilde{w}_{(0,0)}+U_1e^{\mu_1 z}z^{-\tau_1}\widetilde{w}_{\mu_1}$ is a formal solution of (3.6). This is not the case because of the nonlinearity of (3.6) and to the very nature of the Riemann surface $\mathscr{R}^{(1)}$ on which $\widehat{w}_{(0,0)}$ can be analytically continued (theorem 4.2). This raises the question:

Question 5.2. can we define an analogue of the general formal solution for the non-linear equation (3.6) ?

The answer is given by the notion of "formal integral" which we now introduce.

5.2 Formal Integral : Setting

5.2.1 Notation

It will be useful in the sequel to fix customary notation.

Definition 5.1. We suppose $n\in\mathbb{N}^\star$, $\mathbf{k},\mathbf{h}\in\mathbb{N}^n$, $\mathbf{a},\mathbf{b}\in\mathbb{C}^n$.

- If $\mathbf{k}=(k_1,\cdots,k_n)$, then $|\mathbf{k}|=k_1+\cdots+k_n$.
- If $\mathbf{a}=(a_1,\cdots,a_n)$ or $\mathbf{a}={}^t(a_1,\cdots,a_n)$, then $\mathbf{a}^{\mathbf{k}}=a_1^{k_1}\cdots a_n^{k_n}$.
- If $\mathbf{b}=(b_1,\cdots,b_n)$, then $\mathbf{a}.\mathbf{b}=a_1b_1+\cdots+a_nb_n$.
- We denote by $\mathbf{e}_j$ the j-th unit vector of $\mathbb{C}^n$.

5.2.2 Setting

5.2.2.1 Single Level 1 ODE

To introduce the reader to the notion of Ecalle's *formal integral* [Eca85], it will be useful to skip a moment from the ODE (3.6) to a more general one[1] with the same kind of properties. Namely we introduce

$$P(\partial)w + \frac{1}{z}Q(\partial)w = F(z,w) \tag{5.4}$$

$$P(\partial) = \sum_{m=0}^{n} \alpha_{n-m}\partial^m \in \mathbb{C}[\partial] \ , \ Q(\partial) = \sum_{m=0}^{n-1} \beta_{n-m}\partial^m \in \mathbb{C}[\partial]$$

with $n \in \mathbb{N}^\star$. We assume that P is a polynomial of degree n, that is $\alpha_0 \neq 0$, and that $F(z,w)$ is holomorphic in a neighbourhood of $(\infty,0)$ in $\mathbb{C}^2$ with the condition $\frac{\partial^m F}{\partial w^m}(z,0) = O(z^{-2})$, $m \in \mathbb{N}$. (See exercise 3.1). We will add other assumptions to guarantee that the ODE (5.4) has a single level 1 at infinity.

When assuming furthermore that $\alpha_n \neq 0$, what have been said in Sect. 5.1 can be applied as well for (5.4). The equation (5.4) has a unique formal solution $\widetilde{w}_0 \in \mathbb{C}[[z^{-1}]]$ and val $\widetilde{w}_0 \geq 2$. The Newton polygon at infinity $\mathcal{N}_\infty(\mathfrak{P}_\mathbf{0})$ associated with the linear differential operator $\mathfrak{P}_\mathbf{0} = P(\partial) + \frac{1}{z}Q(\partial) - \frac{\partial F}{\partial w}(z,\widetilde{w}_0)$ deduced from the operator $P(\partial) + \frac{1}{z}Q(\partial) - F(z,\cdot)$ by linearisation at $\widetilde{w}_0$, has still a single non-vertical side of slope -1 and the characteristic equation associated with this single side remains the equation $P(\mu) = 0$.

Since $\alpha_n \neq 0$, the roots of the characteristic equation do not vanish. We will also assume that the polynomial

$$\mu \mapsto P(\mu) = \sum_{m=0}^{n} \alpha_{n-m}\mu^m = \alpha_0(\mu-\mu_1)\cdots(\mu-\mu_n)$$

has only simple roots $\mu = \mu_i$, $i = 1,\cdots,n$. The following definitions are adapted from [Arn83, Eca85].

Definition 5.2. Let $\{\mu_i\}$ be the set of the roots of the polynomial $P(\mu)$, and we set $\lambda_i = -\mu_i$, $i = 1,\cdots,n$. The complex numbers $\lambda_1,\cdots,\lambda_n$ are called the *multipliers* of the ODE (5.4).
The ODE (5.4) is said to have a *single level* 1 *at infinity* when the multipliers are all nonzero.

[1] Though far from the more general. For instance in (5.4) one could replace $F(z,w)$ by $F(z,w,\partial w,\cdots,\partial^{n-1}w)$, with F holomorphic in a neighbourhood of $(\infty,\mathbf{0})$ in $\mathbb{C}\times\mathbb{C}^{n-1}$, see exercise 3.1. We refrain of doing that only for a matter of simplicity. See [Eca85] for more general results.

One says that the multipliers are *non-resonant* when they are rationally independent, that is linearly independent over $\mathbb{Z}$. The multipliers are *positively resonant* when there exists $\mathbf{k}_{\text{reson}} = (k_1, \cdots, k_n) \in \mathbb{N}^n \setminus \{\mathbf{0}\}$ so that $\boldsymbol{\lambda}.\mathbf{k}_{\text{reson}} = 0$, where $\boldsymbol{\lambda} = (\lambda_1, \cdots, \lambda_n) \in (\mathbb{C}^\star)^n$ The number $|\mathbf{k}_{\text{reson}}| + 1$ is the order of the resonance, since the positive resonance brings *semi-positively resonances*, that is relationships of the type $\boldsymbol{\lambda}.(\mathbf{k}_{\text{reson}} + \mathbf{e}_j) = \lambda_j$ for any $j \in [1, n]$.

We mention that the following constants are properly defined, since P has only simple roots:

$$\tau_i = \frac{Q(-\lambda_i)}{P'(-\lambda_i)}, \qquad i = 1, \cdots, n. \tag{5.5}$$

From the theory of linear differential equations (see the first two volumes of this book [MS016, Lod016], see also [Bal000, Was65]), we notice that the linear equation
$P(\partial)w + \frac{1}{z}Q(\partial)w = 0$ has a formal general solution under the form

$$w(z) = \sum_{i=1}^{n} v_i(z) y_i(z). \tag{5.6}$$

In (5.6), $y_i(z) = U_i \mathrm{e}^{-\lambda_i z} z^{-\tau_i}$, $U_i \in \mathbb{C}$, stands for the general solution of the differential equation $y_i' + \left(\lambda_i + \frac{\tau_i}{z}\right) y_i = 0$, while $v_i \in \mathbb{C}[[z^{-1}]]$ is invertible and is determined uniquely up to normalization.

5.2.2.2 Companion System, Prepared Form

Formal integrals have more natural foundations when differential equations of order one are considered. We therefore translate the ODE (5.4) into a one order ODE of dimension n by introducing $\mathbf{w} = \begin{pmatrix} w_1 \\ w_2 \\ \vdots \\ w_n \end{pmatrix} = \begin{pmatrix} w \\ w' \\ \vdots \\ w^{(n-1)} \end{pmatrix}$. We get the companion system

$$\partial \mathbf{w} + A\mathbf{w} = \mathbf{f}(z, \mathbf{w}), \tag{5.7}$$

with $A = \begin{pmatrix} 0 & -1 \cdots & & 0 \\ \vdots & \ddots & \ddots & \vdots \\ \vdots & & 0 & -1 \\ \frac{\alpha_n}{\alpha_0} + \frac{\beta_n}{z\alpha_0} & \cdots & \cdots & \frac{\alpha_1}{\alpha_0} + \frac{\beta_1}{z\alpha_0} \end{pmatrix}$ and $\mathbf{f}(z, \mathbf{w}) = \begin{pmatrix} 0 \\ \vdots \\ 0 \\ F(z, w_1)/\alpha_0 \end{pmatrix}$.

Since (5.4) has a unique formal solution $\widetilde{w}_0 \in \mathbb{C}[[z^{-1}]]$, $\operatorname{val} \widetilde{w}_0 \geq 2$, we may remark that (5.7) has a unique formal solution $\widetilde{\mathbf{w}}_0 \in \mathbb{C}^n[[z^{-1}]]$ as well, and in fact $\widetilde{\mathbf{w}}_0 \in z^{-2}(\mathbb{C}[[z^{-1}]])^n$.

Lemma 5.1. *There exists* $T_0(z) \in GL_n(\mathbb{C}\{z^{-1}\}[z])$ *so that the meromorphic gauge transform* $\mathbf{w} = T_0(z)\mathbf{v}$ *brings (5.7) into the* prepared form

$$\partial \mathbf{v} + B_0 \mathbf{v} = \mathbf{g}(z,\mathbf{v}), \quad B_0 = \begin{pmatrix} \lambda_1 + \frac{\tau_1}{z} & \cdots & 0 \\ \vdots & \ddots & \vdots \\ 0 & \cdots & \lambda_n + \frac{\tau_n}{z} \end{pmatrix}, \tag{5.8}$$

with $\mathbf{g}$ *a* $\mathbb{C}^n$*-valued function, holomorphic in a neighbourhood of* $(\infty, 0)$ *and* $\mathbf{g}(z,\mathbf{v}) = O(z^{-2}) + O(\|\mathbf{v}\|^2)$ *when* $z \to \infty$ *and* $\mathbf{v} \to 0$.
The prepared form (5.8) has a unique formal solution $\widetilde{\mathbf{v}}_0 \in (\mathbb{C}[[z^{-1}]])^n$ *and* $\widetilde{\mathbf{v}}_0 \in z^{-2}(\mathbb{C}[[z^{-1}]])^n$.

Proof. The proof is based on classical ideas for linear ODEs (see the first two volumes of this book [MS016, Lod016], see also [Bal000, Was65, Del70]). Looking at (5.6), we compare (5.7) with the linear equation

$$\partial \mathbf{u} + B_0 \mathbf{u} = 0, \quad B_0 = \begin{pmatrix} \lambda_1 + \frac{\tau_1}{z} & \cdots & 0 \\ \vdots & \ddots & \vdots \\ 0 & \cdots & \lambda_n + \frac{\tau_n}{z} \end{pmatrix} = \Lambda + \frac{1}{z} L, \tag{5.9}$$

whose general solution (holomorphic on $\widetilde{\mathbb{C}}$) is given in term of the fundamental matrix solution $z^{-L} e^{-z\Lambda}$,

$$\mathbf{u}(z) = z^{-L} e^{-z\Lambda} \mathbf{U} = \oplus_{i=1}^{n} z^{-\tau_i} e^{-z\lambda_i} \mathbf{U}, \qquad \mathbf{U} \in \mathbb{C}^n. \tag{5.10}$$

We remark that

$$\left(e^{-\lambda z} z^{-\tau}\right)^{(m)} = e^{-\lambda z} z^{-\tau} \sum_{j=0}^{m} \binom{m}{j} (-\lambda)^{m-j} \frac{(-\tau)_j}{z^j} \tag{5.11}$$

for $(\lambda, \tau) \in \mathbb{C}^2$ and $m \in \mathbb{N}$, where $(-\tau)_j = j! \binom{-\tau}{j}$ mimics the Pochhammer symbol:

$$(-\tau)_0 = 1 \text{ and } (-\tau)_j = (-1)^j \tau(\tau+1)\cdots(\tau+j-1) \text{ for } j \geq 1. \tag{5.12}$$

We thus set the meromorphic gauge transform $\mathbf{w} = T_0(z)\mathbf{v}$ with $T_0(z) \in GL_n(\mathbb{C}\{z^{-1}\}[z])$ of the form:

$$T_0(z) = \begin{pmatrix} 1 & \cdots & 1 \\ -\lambda_1 - \frac{\tau_1}{z} & \cdots & -\lambda_n - \frac{\tau_n}{z} \\ \vdots & & \vdots \\ \sum_{j=0}^{n-1} \binom{n-1}{j} (-\lambda_1)^{n-1-j} \frac{(-\tau_1)_j}{z^j} & \cdots & \sum_{j=0}^{n-1} \binom{n-1}{j} (-\lambda_n)^{n-1-j} \frac{(-\tau_n)_j}{z^j} \end{pmatrix}. \tag{5.13}$$

By its very definition, this gauge transform brings (5.7) into the differential equation:

$$\begin{aligned}\partial \mathbf{v} &= -\left[T_0^{-1}(\partial T_0) + T_0^{-1} A T_0\right] \mathbf{v} + T_0^{-1}\mathbf{f}(z, T_0\mathbf{v}) \\ &= -B_0\mathbf{v} + \mathbf{g}(z, \mathbf{v})\end{aligned} \tag{5.14}$$

where $\mathbf{g}$ has the properties described in the lemma. The fact that (5.8) has a unique formal solution $\widetilde{\mathbf{v}}_0 \in (\mathbb{C}[[z^{-1}]])^n$ is obvious. $\square$

Example 5.1. We have already seen that the companion system associated with (3.6) is (3.9). The gauge transform $\mathbf{w} = T_0(z)\mathbf{v}$, $T_0(z) = \begin{pmatrix} 1 & 1 \\ -1+\frac{3}{2z} & 1+\frac{3}{2z} \end{pmatrix}$, brings (3.9) into the prepared form:

$$\partial \mathbf{v} + \begin{pmatrix} 1-\frac{3}{2z} & 0 \\ 0 & -1-\frac{3}{2z} \end{pmatrix} \mathbf{v} = \frac{15}{8z^2} \begin{pmatrix} -1 & -1 \\ 1 & 1 \end{pmatrix} \mathbf{v} + \frac{1}{2} \begin{pmatrix} -F(z, v_1+v_2) \\ F(z, v_1+v_2) \end{pmatrix}. \tag{5.15}$$

Remark 5.1. Let us consider the action of the gauge transform $\mathbf{y} = T_0(z)\mathbf{u}$ on the differential equation $\partial \mathbf{u} + B_0\mathbf{u} = 0$. This differential equation is transformed into the system $\partial \mathbf{y} + A_0\mathbf{y} = 0$ with $A_0 = T_0 B_0 T_0^{-1} - (\partial T_0)T_0^{-1}$ of the form $A_0 = \begin{pmatrix} 0 & -1 & \cdots & 0 \\ \vdots & \ddots & \ddots & \vdots \\ \vdots & & 0 & -1 \\ p_n(z) & \cdots & \cdots & p_1(z) \end{pmatrix}$ with $p_n, \cdots, p_1 \in \mathbb{C}\{z^{-1}\}$, $p_n(z) = \frac{\alpha_n}{\alpha_0} + \frac{\beta_n}{z\alpha_0} + O(z^{-2})$, $\cdots$, $p_1(z) = \frac{\alpha_1}{\alpha_0} + \frac{\beta_1}{z\alpha_0} + O(z^{-2})$. The system $\partial \mathbf{y} + A_0\mathbf{y} = 0$ is the companion system for the one-dimensional homogeneous ODE of order n,

$$\partial^n y + p_1(z)\partial^{n-1} y + \cdots + p_n(z) y = 0, \tag{5.16}$$

whose general solution is $y(z) = \sum_{i=1}^n U_i e^{-\lambda_i z} z^{-\tau_i}$, $(U_1, \cdots, U_n) \in \mathbb{C}^n$.

5.2.2.3 Normal Form, Formal Reduction

We have previously reduced the companion system (5.7) to a prepared form through a meromorphic gauge transform. Under some conditions, one can go further, but through formal transformations, in the spirit of the Poincaré-Dulac theorem [Arn83] and the classification up to formal conjugation.

Proposition 5.1. *We consider the ODE (5.8) and we assume that the multipliers $\lambda_1, \cdots, \lambda_n$ are non-resonant. Then there exists a formal transformation $\mathbf{v} = \widetilde{T}(z, \mathbf{u})$,*

$$\widetilde{T}(z, \mathbf{u}) = \sum_{\mathbf{k} \in \mathbb{N}^n} \mathbf{u}^{\mathbf{k}} \widetilde{\mathbf{v}}_{\mathbf{k}}(z), \qquad \widetilde{\mathbf{v}}_{\mathbf{k}}(z) \in (\mathbb{C}[[z^{-1}]])^n, \tag{5.17}$$

which formally transforms (5.8) into the linear normal form *equation $\partial \mathbf{u} + B_0\mathbf{u} = 0$. In (5.17), $\widetilde{\mathbf{v}}_0$ stands for the unique formal solution of (5.8); for $j = 1, \ldots, n$, $\widetilde{\mathbf{v}}_{\mathbf{e}_j}$ is* *uniquely determined when one prescribes its constant term to be equal to* $\mathbf{e}_j$*; then the formal series* $\widetilde{\mathbf{v}}_{\mathbf{k}}$ *are unique for* $|\mathbf{k}| > 1$.

We will see in the sequel how this proposition can be shown. Here, we rather concentrate on its consequences.

One can refer to, e.g., [Kui003, Bra92] for a proof that extend to possibly nilpotent cases (but with no resonances), and to [Eca85] for a very general frame.

We know that the general solution of the normal form $\partial\mathbf{u} + B_0\mathbf{u} = 0$ is $\mathbf{u}(z) = \oplus_{i=1}^{n} z^{-\tau_i}\mathrm{e}^{-z\lambda_i}({}^t\mathbf{U})$, $\mathbf{U} = (U_1, \cdots, U_n) \in \mathbb{C}^n$. Through the action of the formal transformation $\mathbf{v} = \widetilde{T}(z, \mathbf{u})$, this provides the following general formal solution for the ODE (5.8):

$$\widetilde{\mathbf{v}}(z, \mathbf{U}) = \sum_{\mathbf{k}=(k_1,\cdots,k_n)\in\mathbb{N}^n} \prod_{i=1}^{n} (U_i z^{-\tau_i}\mathrm{e}^{-z\lambda_i})^{k_i} \widetilde{\mathbf{v}}_{\mathbf{k}}(z) = \sum_{\mathbf{k}\in\mathbb{N}^n} \mathbf{U}^{\mathbf{k}}\mathrm{e}^{-\boldsymbol{\lambda}.\mathbf{k}z} z^{-\boldsymbol{\tau}.\mathbf{k}} \widetilde{\mathbf{v}}_{\mathbf{k}}(z) \quad (5.18)$$

with $\boldsymbol{\lambda} = (\lambda_1, \cdots, \lambda_n) \in (\mathbb{C}^\star)^n$ and $\boldsymbol{\tau} = (\tau_1, \cdots, \tau_n) \in \mathbb{C}^n$.

Definition 5.3. The formal series (5.18) is called the *formal integral* of (5.8).

Of course, one can obtain the formal integral $\widetilde{\mathbf{w}}(z, \mathbf{U})$ of (5.7) as well, by the gauge transform $\widetilde{\mathbf{w}} = T_0(z)\widetilde{\mathbf{v}}$, with $T_0(z)$ given by (5.13). When finally returning to the n-th order ODE (5.4) of dimension 1 we started with, one gets its formal integral.

Definition 5.4. We assume that the multipliers are non-resonant. The *formal integral* $\widetilde{w}(z, \mathbf{U})$ of the ODE (5.4) is defined by:

$$\begin{aligned}\widetilde{w}(z, \mathbf{U}) &= \sum_{\mathbf{k}\in\mathbb{N}^n} \mathbf{U}^{\mathbf{k}}\mathrm{e}^{-\boldsymbol{\lambda}.\mathbf{k}z} z^{-\boldsymbol{\tau}.\mathbf{k}} \widetilde{w}_{\mathbf{k}}(z), \quad \widetilde{w}_{\mathbf{k}}(z) = \widetilde{\mathbf{v}}_{\mathbf{k}}(z).(1, \cdots, 1) \in \mathbb{C}[[z^{-1}]], \\ &= \widetilde{\Phi}(z, U_1\mathrm{e}^{-\lambda_1 z} z^{-\tau_1}, \cdots, U_n\mathrm{e}^{-\lambda_n z} z^{-\tau_n}) \end{aligned} \quad (5.19)$$

with $\widetilde{\Phi}(z, \mathbf{u}) = \sum_{\mathbf{k}\in\mathbb{N}^n} \mathbf{u}^{\mathbf{k}}\widetilde{w}_{\mathbf{k}}(z) \in \mathbb{C}[[z^{-1}, \mathbf{u}]]$. The formal transformation $w = \widetilde{\Phi}(z, \mathbf{u})$ formally transforms (5.4) into the normal form equation $\partial\mathbf{u} + B_0\mathbf{u} = 0$.

The formal integral (5.19), thus depending on the maximal n free parameters $\mathbf{U} = (U_1, \cdots, U_n) \in \mathbb{C}^n$, plays the role of the general formal solution for the ODE (5.4) of order n. Formal integrals can be defined as well for difference and differential-difference equations, see, e.g. [Eca85, Kui003]. This notion has been enlarged for nonlinear partial differential equations in [OSS003].

Remark 5.2. Although working at the formal level, one may wonder what is the chosen branch when we write $z^{-\boldsymbol{\tau}.\mathbf{k}}$. As a matter of fact, this is not relevant at this stage since moving from a determination to another one just translates into rescaling the free parameter $\mathbf{U}$.

Remark 5.3. Introducing $\mathbf{V}^{\mathbf{k}} = \mathbf{U}^{\mathbf{k}}\mathrm{e}^{-\boldsymbol{\lambda}.\mathbf{k}z} z^{-\boldsymbol{\tau}.\mathbf{k}}$, we remark the identity:

$$\partial_z(\mathbf{V}^{\mathbf{k}}\widetilde{w}_{\mathbf{k}}) = \left[\left(\partial_z - \sum_{i=1}^{n}(\lambda_i + \frac{\tau_i}{z})u_i\partial_{u_i}\right)(\mathbf{u}^{\mathbf{k}}\widetilde{w}_{\mathbf{k}})\right]|_{\mathbf{u}=\mathbf{V}}.$$

Looking at the equality

$$\widetilde{w}(z,\mathbf{U}) = \widetilde{\Phi}(z, U_1\mathrm{e}^{-\lambda_1 z}z^{-\tau_1}, \cdots, U_n\mathrm{e}^{-\lambda_n z}z^{-\tau_n}) \tag{5.20}$$

and since the formal integral (5.19) solves the differential equation (5.4), one deduces that $\widetilde{\Phi}$ satisfies:

$$P\Big(\partial_z - \sum_{i=1}^{n}(\lambda_i + \frac{\tau_i}{z})u_i\partial_{u_i}\Big)\widetilde{\Phi} + \frac{1}{z}Q\Big(\partial_z - \sum_{i=1}^{n}(\lambda_i + \frac{\tau_i}{z})u_i\partial_{u_i}\Big)\widetilde{\Phi} = F(z,\widetilde{\Phi}). \tag{5.21}$$

5.2.3 Formal Integral, General Considerations

Under convenient hypotheses, we have previously introduced the formal integral for the ODE (5.4), that is a n-parameters formal expansion of the form

$$w(z,\mathbf{U}) = \sum_{\mathbf{k}\in\mathbb{N}^n} \mathbf{U}^{\mathbf{k}}\mathrm{e}^{-\boldsymbol{\lambda}.\mathbf{k}z}z^{-\boldsymbol{\tau}.\mathbf{k}}w_{\mathbf{k}}(z), \qquad \boldsymbol{\lambda},\boldsymbol{\tau}\in\mathbb{C}^n, \tag{5.22}$$

Let us start with (5.22) and investigate the conditions to impose on the $w_{\mathbf{k}}$'s in order for (5.22) to be formally solution of (5.4).

We could start with (5.21) as well.

Using the identity (5.11) for $m\in\mathbb{N}$, one obtains from (5.22):

$$\begin{aligned} w^{(m)} &= \sum_{|\mathbf{k}|\geq 0}\mathbf{U}^{\mathbf{k}}\sum_{p=0}^{m}\binom{m}{p}(\mathrm{e}^{-\boldsymbol{\lambda}.\mathbf{k}z}z^{-\boldsymbol{\tau}.\mathbf{k}})^{(p)}w_{\mathbf{k}}^{(m-p)} \\ &= \sum_{|\mathbf{k}|\geq 0}\mathbf{U}^{\mathbf{k}}\mathrm{e}^{-\boldsymbol{\lambda}.\mathbf{k}z}z^{-\boldsymbol{\tau}.\mathbf{k}}T_{\mathbf{k},m+1}(w_{\mathbf{k}}) \end{aligned}$$

where $T_{\mathbf{0},m+1}(w_{\mathbf{0}}) = w_{\mathbf{0}}^{(m)}$ and, more generally for $\mathbf{k}\in\mathbb{N}^2$,

$$\begin{aligned} T_{\mathbf{k},m+1}(w_{\mathbf{k}}) &= \sum_{p=0}^{m}\binom{m}{p}\Big[\sum_{j=0}^{p}\binom{p}{j}(-\boldsymbol{\lambda}.\mathbf{k})^{p-j}\frac{(-\boldsymbol{\tau}.\mathbf{k})_j}{z^j}\Big]w_{\mathbf{k}}^{(m-p)} \\ &= \sum_{j=0}^{m}\binom{m}{j}\frac{(-\boldsymbol{\tau}.\mathbf{k})_j}{z^j}\Big[\sum_{q=0}^{m-j}\binom{m-j}{q}(-\boldsymbol{\lambda}.\mathbf{k})^{m-j-q}w_{\mathbf{k}}^{(q)}\Big], \end{aligned}$$

that is also

$$T_{\mathbf{k},m+1}(w_{\mathbf{k}}) = \sum_{j=0}^{m}\binom{m}{j}\frac{(-\boldsymbol{\tau}.\mathbf{k})_j}{z^j}\Big[(-\boldsymbol{\lambda}.\mathbf{k}+\partial)^{m-j}w_{\mathbf{k}}\Big]. \tag{5.23}$$

In what follows we will simply write $T_{\mathbf{k},m+1}$ instead of $T_{\mathbf{k},m+1}(w_{\mathbf{k}})$. We introduce the notation $\mathbf{V}^{\mathbf{k}} = \mathbf{U}^{\mathbf{k}} e^{-\boldsymbol{\lambda}.\mathbf{k}z} z^{-\boldsymbol{\tau}.\mathbf{k}}$ and we notice that for every $\mathbf{k}_1, \mathbf{k}_2 \in \mathbb{N}^n$, $\mathbf{V}^{\mathbf{k}_1}\mathbf{V}^{\mathbf{k}_2} = \mathbf{V}^{\mathbf{k}_1+\mathbf{k}_1}$. On the one hand,

$$P(\partial)w = \sum_{k=0}^{\infty} \mathbf{V}^{\mathbf{k}} \Big[\sum_{m=0}^{n} \alpha_{n-m} T_{\mathbf{k},m+1} \Big] = \sum_{|\mathbf{k}|\geq 0} \mathbf{V}^{\mathbf{k}} p_{\mathbf{k}}(\partial) w_{\mathbf{k}} \tag{5.24}$$

where for $|\mathbf{k}| \geq 0$,

$$\begin{aligned} p_{\mathbf{k}}(\partial) &= \sum_{m=0}^{n} \alpha_{n-m}(-\boldsymbol{\lambda}.\mathbf{k}+\partial)^m \\ &+ \sum_{m=1}^{n} \alpha_{n-m} \Big\{ \sum_{j=1}^{m} \binom{m}{j} \frac{(-\boldsymbol{\tau}.\mathbf{k})_j}{z^j} (-\boldsymbol{\lambda}.\mathbf{k}+\partial)^{m-j} \Big\}. \end{aligned}$$

In other words, for $|\mathbf{k}| \geq 0$,

$$p_{\mathbf{k}}(\partial) = P(-\boldsymbol{\lambda}.\mathbf{k}+\partial) + \sum_{j=1}^{n} \frac{1}{z^j} \binom{-\boldsymbol{\tau}.\mathbf{k}}{j} P^{(j)}(-\boldsymbol{\lambda}.\mathbf{k}+\partial). \tag{5.25}$$

Similarly

$$Q(\partial)w = \sum_{|\mathbf{k}|\geq 0} \mathbf{V}^{\mathbf{k}} q_{\mathbf{k}}(\partial) w_{\mathbf{k}} \tag{5.26}$$

with

$$q_{\mathbf{k}}(\partial) = Q(-\boldsymbol{\lambda}.\mathbf{k}+\partial) + \sum_{j=1}^{n-1} \frac{1}{z^j} \binom{-\boldsymbol{\tau}.\mathbf{k}}{j} Q^{(j)}(-\boldsymbol{\lambda}.\mathbf{k}+\partial). \tag{5.27}$$

On the other hand we consider the Taylor expansion of $F(z,w(z,\mathbf{U}))$ at $w_{\mathbf{0}}$, namely

$$F(z,w) = F(z,w_{\mathbf{0}}) + \sum_{\ell\geq 1} \frac{\left(\sum_{|\mathbf{k}|\geq 1} \mathbf{V}^{\mathbf{k}} w_{\mathbf{k}}\right)^{\ell}}{\ell!} \frac{\partial^{\ell} F(z,w_{\mathbf{0}})}{\partial w^{\ell}}. \tag{5.28}$$

We observe that for every $\ell \in \mathbb{N}^{\star}$,

$$\Big(\sum_{|\mathbf{k}|\geq 1} \mathbf{V}^{\mathbf{k}} w_{\mathbf{k}} \Big)^{\ell} = \sum_{|\mathbf{p}|\geq \ell} \mathbf{V}^{\mathbf{p}} \sum_{\substack{\mathbf{p}_1+\cdots+\mathbf{p}_{\ell}=\mathbf{p} \\ |\mathbf{p}_i|\geq 1,\, 1\leq i\leq \ell}} w_{\mathbf{p}_1} \cdots w_{\mathbf{p}_{\ell}}. \tag{5.29}$$

As a result, equation (5.28) reads

$$F(z,w) = F(z,w_{\mathbf{0}}) + \sum_{\substack{\ell\geq 1 \\ |\mathbf{p}|\geq \ell}} \mathbf{V}^{\mathbf{p}} \sum_{\substack{\mathbf{p}_1+\cdots+\mathbf{p}_{\ell}=\mathbf{p} \\ |\mathbf{p}_i|\geq 1,\, 1\leq i\leq \ell}} \frac{w_{\mathbf{p}_1} \cdots w_{\mathbf{p}_{\ell}}}{\ell!} \frac{\partial^{\ell} F(z,w_{\mathbf{0}})}{\partial w^{\ell}}. \tag{5.30}$$

Finally, plugging the formal expansion (5.22) into the differential equation (5.4), using the identities (5.24), (5.26), (5.30) and identifying the powers $\mathbf{V}^{\mathbf{k}}$, one gets the next lemma 5.2 which justifies the following definition.

Definition 5.5. For $\mathbf{k} \in \mathbb{N}^n$, we define

$$\begin{aligned} P_{\mathbf{k}}(\partial) &= P(-\boldsymbol{\lambda}.\mathbf{k}+\partial), \\ Q_{\mathbf{k}}(\partial) &= -\boldsymbol{\tau}.\mathbf{k}P'(-\boldsymbol{\lambda}.\mathbf{k}+\partial)+Q(-\boldsymbol{\lambda}.\mathbf{k}+\partial) \end{aligned} \tag{5.31}$$

$$R_{\mathbf{k}}(\partial) = \sum_{j=0}^{n-2} \frac{1}{z^j}\left[\binom{-\boldsymbol{\tau}.\mathbf{k}}{j+2} P^{(j+2)}(-\boldsymbol{\lambda}.\mathbf{k}+\partial) + \binom{-\boldsymbol{\tau}.\mathbf{k}}{j+1} Q^{(j+1)}(-\boldsymbol{\lambda}.\mathbf{k}+\partial)\right]. \tag{5.32}$$

For $\mathbf{k} \in \mathbb{N}^n$, we denote by $\mathfrak{D}_{\mathbf{k}} = \mathfrak{D}_{\mathbf{k}}(w_{\mathbf{0}})$ the linear differential operator

$$\mathfrak{D}_{\mathbf{k}} = P_{\mathbf{k}}(\partial) + \frac{1}{z}Q_{\mathbf{k}}(\partial) + \frac{1}{z^2}R_{\mathbf{k}} - \frac{\partial F(z,w_{\mathbf{0}})}{\partial w}$$

where $w_{\mathbf{0}}$ satisfies $P(\partial)w_{\mathbf{0}} + \frac{1}{z}Q(\partial)w_{\mathbf{0}} = F(z,w_{\mathbf{0}})$.
For $\mathbf{k} \in \mathbb{N}^n$, we denote by $\mathfrak{P}_{\mathbf{k}} = \mathfrak{P}_{\mathbf{k}}(w_{\mathbf{0}})$ the linear differential operator

$$\mathfrak{P}_{\mathbf{k}} = P(-\boldsymbol{\lambda}.\mathbf{k}+\partial) + \frac{1}{z}Q(-\boldsymbol{\lambda}.\mathbf{k}+\partial) - \frac{\partial F(z,w_{\mathbf{0}})}{\partial w}. \tag{5.33}$$

Lemma 5.2. *The n-parameters formal expansion*

$$w(z,\mathbf{U}) = \sum_{\mathbf{k}\in\mathbb{N}^n} \mathbf{U}^{\mathbf{k}} e^{-\boldsymbol{\lambda}.\mathbf{k}z} z^{-\boldsymbol{\tau}.\mathbf{k}} w_{\mathbf{k}}(z) \tag{5.34}$$

solves (5.4) if and only if :

$$P(\partial)w_{\mathbf{0}} + \frac{1}{z}Q(\partial)w_{\mathbf{0}} = F(z,w_{\mathbf{0}}), \tag{5.35}$$

$$\mathfrak{D}_{\mathbf{e}_i} w_{\mathbf{e}_i} = 0 \tag{5.36}$$

with $\mathbf{e}_i$ the i-th vector of the canonical base of $\mathbb{C}^n$, and for $|\mathbf{k}| \geq 2$,

$$\mathfrak{D}_{\mathbf{k}} w_{\mathbf{k}} = \sum_{\substack{\mathbf{k}_1+\cdots+\mathbf{k}_\ell=\mathbf{k} \\ |\mathbf{k}_i|\geq 1,\, \ell\geq 2}} \frac{w_{\mathbf{k}_1}\cdots w_{\mathbf{k}_\ell}}{\ell!} \frac{\partial^\ell F(z,w_{\mathbf{0}})}{\partial w^\ell}. \tag{5.37}$$

Remark 5.4. Notice that in lemma 5.2 we have neither supposed that $\boldsymbol{\lambda} = (\lambda_1,\cdots,\lambda_n)$ are the multipliers, nor that $\boldsymbol{\tau} = (\tau_1,\cdots,\tau_n)$ are such that $\tau_i = \dfrac{Q(-\lambda_i)}{P'(-\lambda_i)}$, $i = 1,\cdots,n$. However, these conditions will come in the next section.

Example 5.2. We consider equation (3.6) where $n = 2$, $P(\partial) = \partial^2 - 1$, $Q(\partial) = -3\partial$. Then, for every $\mathbf{k} \in \mathbb{N}^2$,

$$
\begin{aligned}
P_{\mathbf{k}}(\partial) &= \partial^2 - 2\boldsymbol{\lambda}.\mathbf{k}\partial + (\boldsymbol{\lambda}.\mathbf{k})^2 - 1, \\
Q_{\mathbf{k}}(\partial) &= (3 + 2\boldsymbol{\tau}.\mathbf{k})(-\partial + \boldsymbol{\lambda}.\mathbf{k}), \\
R_{\mathbf{k}}(\partial) &= \boldsymbol{\tau}.\mathbf{k}(\boldsymbol{\tau}.\mathbf{k} + 4).
\end{aligned} \tag{5.38}
$$

In particular, taking $\boldsymbol{\lambda} = (1, -1)$ (the zeros of $\zeta \mapsto P(-\zeta)$) and $\boldsymbol{\tau} = \left(-\frac{3}{2}, -\frac{3}{2}\right)$ (we take the values given by (5.5)), then writing $\mathbf{k} = (k_1, k_2)$:

$$
\begin{aligned}
P_{\mathbf{k}}(\partial) &= \partial^2 - 2(k_1 - k_2)\partial + (k_1 - k_2)^2 - 1, \\
Q_{\mathbf{k}}(\partial) &= 3(1 - k_1 - k_2)(-\partial + k_1 - k_2), \\
R_{\mathbf{k}}(\partial) &= \frac{9}{4}(k_1 + k_2)\left(k_1 + k_2 - \frac{8}{3}\right).
\end{aligned} \tag{5.39}
$$

We eventually mention some identities for later purposes, the proof of which being left as an exercise.

Lemma 5.3. *The operators $\mathfrak{P}_{\mathbf{k}}$ and $\mathfrak{D}_{\mathbf{k}}$ given by definition 5.5 satisfy the identities: for any $\mathbf{k}, \mathbf{k}_1, \mathbf{k}_2 \in \mathbb{N}^n$, $\mathrm{e}^{-\boldsymbol{\lambda}.\mathbf{k}_1 z}\mathfrak{P}_{\mathbf{k}_1}\mathrm{e}^{\boldsymbol{\lambda}.\mathbf{k}_1 z} = \mathrm{e}^{-\boldsymbol{\lambda}.\mathbf{k}_2 z}\mathfrak{P}_{\mathbf{k}_2}\mathrm{e}^{\boldsymbol{\lambda}.\mathbf{k}_2 z}$, $z^{-\boldsymbol{\tau}.\mathbf{k}}\mathfrak{D}_{\mathbf{k}} = \mathfrak{P}_{\mathbf{k}} z^{-\boldsymbol{\tau}.\mathbf{k}}$ and*

$$
(\mathrm{e}^{-\boldsymbol{\lambda}.\mathbf{k}_1 z} z^{-\boldsymbol{\tau}.\mathbf{k}_1})\mathfrak{D}_{\mathbf{k}_1}(\mathrm{e}^{-\boldsymbol{\lambda}.\mathbf{k}_1 z} z^{-\boldsymbol{\tau}.\mathbf{k}_1})^{-1} = (\mathrm{e}^{-\boldsymbol{\lambda}.\mathbf{k}_2 z} z^{-\boldsymbol{\tau}.\mathbf{k}_2})\mathfrak{D}_{\mathbf{k}_2}(\mathrm{e}^{-\boldsymbol{\lambda}.\mathbf{k}_2 z} z^{-\boldsymbol{\tau}.\mathbf{k}_2})^{-1}.
$$

Setting $W_{\mathbf{k}} = z^{-\boldsymbol{\tau}.\mathbf{k}} w_{\mathbf{k}}$ for $\mathbf{k} \in \mathbb{N}^n$ and the $w_{\mathbf{k}}$ given by lemma 5.2, one has $\mathfrak{P}_{\mathbf{e}_i} W_{\mathbf{e}_i} = 0$, $i = 1, 2$ while and for $|\mathbf{k}| \geq 2$,

$$
\mathfrak{P}_{\mathbf{k}} W_{\mathbf{k}} = \sum_{\substack{\mathbf{k}_1 + \cdots + \mathbf{k}_\ell = \mathbf{k} \\ |\mathbf{k}_i| \geq 1, \ell \geq 2}} \frac{W_{\mathbf{k}_1} \cdots W_{\mathbf{k}_\ell}}{\ell!} \frac{\partial^\ell F(z, w_{\mathbf{0}})}{\partial w^\ell}. \tag{5.40}
$$

5.3 First Painlevé Equation and Transseries Solutions

We partly describe in this section the contains of lemma 5.2 for the prepared form equation (3.6) associated with the first Painlevé equation. Thus $n = 2$, $P(\partial) = \partial^2 - 1$, $Q(\partial) = -3\partial$ and $F(z, w) = f_0(z) + f_1(z)w + f_2(z)w^2$. Also, we will for the moment specialise our study to only one-parameter formal expansions, that is we will assume that either $U_1 = 0$ or $U_2 = 0$ in (5.34). This study will be enough to get the truncated solutions. We will keep on our study of the formal integral associated with (3.6) in Sect. 5.4 where will we see the effects of resonances.

5.3.1 Transseries Solution - Statement

This section will be devoted to proving the following proposition.

Proposition 5.2. *We consider the prepared ODE (3.6). We set* $\boldsymbol{\lambda} = (\lambda_1, \lambda_2) = (1, -1)$ *where the* λ_i*'s are the multipliers, that is the roots of the polynomial* $\zeta \mapsto P(-\zeta)$. *We set* $\boldsymbol{\tau} = (\tau_1, \tau_2) = \left(-\frac{3}{2}, -\frac{3}{2}\right)$, *where* $\tau_i = \frac{Q(-\lambda_i)}{P'(-\lambda_i)}$, $i = 1, 2$.
Then for each $i = 1, 2$, *there exists a formal one-parameter solution of (3.6) in the graded algebra* $\bigoplus_{k \in \mathbb{N}} z^{-\tau_i k} \mathrm{e}^{-\lambda_i k z} \mathbb{C}[[z^{-1}]]$ *of the form:*

$$\widetilde{w}(z, U\mathbf{e}_i) = \sum_{k=0}^{\infty} U^k \mathrm{e}^{-\lambda_i k z} z^{-\tau_i k} \widetilde{w}_{k\mathbf{e}_i}(z), \qquad \widetilde{w}_{k\mathbf{e}_i} \in \mathbb{C}[[z^{-1}]]. \tag{5.41}$$

We have $\mathrm{val}\, \widetilde{w}_{k\mathbf{e}_i} = 2(k-1)$ *and the formal series (5.41) is unique once one fixes the normalization of* $\widetilde{w}_{\mathbf{e}_i}$ *to be* $\widetilde{w}_{\mathbf{e}_i}(z) = 1 + O(z^{-1})$. *Then* $\widetilde{w}_{k\mathbf{e}_i} \in \mathbb{R}[[z^{-1}]]$ *and* $\widetilde{w}_{k\mathbf{e}_i}(z) = \frac{k}{12^{k-1}} z^{-2(k-1)}(1 + O(z^{-1}))$ *for every* $k \geq 1$. *Furthermore changing the normalization of* $\widetilde{w}_{\mathbf{e}_i}$ *is equivalent to rescaling the parameter* $U \in \mathbb{C}$. *Eventually,* $\widetilde{w}_{k\mathbf{e}_1}(z) = \widetilde{w}_{k\mathbf{e}_2}(-z)$ *for every* $k \geq 0$.

Definition 5.6. The series (5.41) is called a formal *transseries*. The terms $\mathrm{e}^{-\lambda_i k z} z^{-\tau_i k}$ are (log-free) *transmonomials*. The formal series $\widetilde{w}_{k\mathbf{e}_i}$ are called the $k\mathbf{e}_i$*-th series* of the transseries. We set $\widetilde{W}_{k\mathbf{e}_i} = z^{-\tau_i k} \widetilde{w}_{k\mathbf{e}_i}$.

Remark 5.5. The term "transseries" is due to Ecalle [Eca92-1]. These are objects which are much used in resurgence theory, see, e.g. [DDP93, Sau95, KT96, KT011, HKT015] and references therein. More details on transseries can be founded in [Eca85, Eca92-1, Cos005, Cos009]. Transseries are also common objects in theoretical physics : these are the so-called "multi-instanton expansions", see e.g. [Zin89, JZ004-1, JZ004-2, JZ010, MSW009, GIKM012, ASV012, AS015, DU012, BDU013, DU014] and references therein.

In quantum mechanics or quantum field theory, an *instanton action* (the terminology of which is due to Gerard 't Hooft) is a classical solution of the equations of motion, with a finite and non-zero action. A well-known instanton effect in quantum mechanics is given by a particle in a double well potential. The tunneling effect provides a non-zero probability that the particle crosses the potential barrier. This gives rise to a tunneling amplitude proportionnal to the *instanton* $\mathrm{e}^{-S/\hbar}$ where S is the instanton action, $\hbar$ being the Planck constant or the coupling constant. For the bound states, this translates into the fact that they can be described at a formal level by a multi-instanton expansion, that is a transseries of the form $\sum_{k \geq 0} \widetilde{E}_k(\hbar) \mathrm{e}^{-kS/\hbar}$ where the perturbative fluctuations $\widetilde{E}_k(\hbar)$ are formal series in $\hbar$. The bound states are deduced from the multi-instanton expansion by (median) Borel-Laplace summation, see [Vor83, DDP97, DP97, DP98, DP99, DT000, Get011, Get013].

For later use, we mention a lemma which results from proposition 5.2 and lemma 5.3.

Lemma 5.4. *Under the conditions of proposition 5.2 and for any* $\mathbf{k} \in \mathbb{N}^2$*, the (so-called) general formal solution of the linear differential equation* $\mathfrak{P}_{\mathbf{k}}(\widetilde{w}_{\mathbf{0}})\widetilde{W} = 0$ *is* $\widetilde{W} = \mathrm{e}^{\boldsymbol{\lambda}.\mathbf{k}z}\big(C_1\mathrm{e}^{-\lambda_1 z}\widetilde{W}_{\mathbf{e}_1} + C_2\mathrm{e}^{-\lambda_2 z}\widetilde{W}_{\mathbf{e}_2}\big)$*,* $C_1, C_2 \in \mathbb{C}$*. For any* $\mathbf{k} \in \mathbb{N}^2$ *the (so-called) general formal solution of the linear differential equation* $\mathfrak{D}_{\mathbf{k}}(\widetilde{w}_{\mathbf{0}})\widetilde{w} = 0$ *is* $\widetilde{w}(z) = \mathrm{e}^{\boldsymbol{\lambda}.\mathbf{k}z} z^{\boldsymbol{\tau}.\mathbf{k}}\big(C_1\mathrm{e}^{-\lambda_1 z}\widetilde{W}_{\mathbf{e}_1} + C_2\mathrm{e}^{-\lambda_2 z}\widetilde{W}_{\mathbf{e}_2}\big)$*,* $C_1, C_2 \in \mathbb{C}$*.*

5.3.2 Transseries Solution - Proof

5.3.2.1 A Useful Lemma

We start with the following lemma which will be useful in the sequel.

Lemma 5.5. *We suppose* $n, N \in \mathbb{N}^\star$*. We consider the ordinary differential equation*

$$P(\partial)w + \frac{1}{z}R(\partial)w = \widetilde{f}(z),\ \widetilde{f}(z) = f_N z^{-N}(1 + O(z^{-1})) \in z^{-N}\mathbb{C}[[z^{-1}]],\ f_N \neq 0$$

with $P(\partial) = \sum_{m=0}^{n} \alpha_{n-m}\partial^m \in \mathbb{C}[\partial]$*,* $\alpha_n \neq 0$*,* $R(\partial) = \sum_{m=0}^{n-1} \gamma_{n-m}(z)\partial^m \in \mathbb{C}[[z^{-1}]][\partial]$*. This ODE has a unique solution* $\widetilde{w}$ *in* $\mathbb{C}[[z^{-1}]]$*, moreover* $\operatorname{val}\widetilde{w} = \operatorname{val}\widetilde{f}$ *and* $\widetilde{w}(z) = \frac{f_N}{P(0)} z^{-N}(1 + O(z^{-1}))$*.*

Proof. In the valuation ring $\mathbb{C}[[z^{-1}]]$ we consider the following map :

$$\begin{aligned} \mathsf{N} : \mathbb{C}[[z^{-1}]] &\to \mathbb{C}[[z^{-1}]] \\ w &\to \frac{1}{P(0)}\Big[\widetilde{f}(z) - \Big(P(\partial) - P(0)\Big)w - \frac{1}{z}R(\partial)w\Big]. \end{aligned}$$

(Remember that $P(0) = \alpha_n$ is nonzero). From the hypotheses made one easily observes that $\mathsf{N}(\mathbb{C}[[z^{-1}]]) \subset z^{-1}\mathbb{C}[[z^{-1}]]$ while, for every $p \in \mathbb{N}^\star$,

$$\text{if } u, v \in z^{-p}\mathbb{C}[[z^{-1}]],\ \text{then } \mathsf{N}(u) - \mathsf{N}(v) \in z^{-p-1}\mathbb{C}[[z^{-1}]].$$

This means that N is contractive in $\mathbb{C}[[z^{-1}]]$, thus the fixed point problem $w = \mathsf{N}(w)$ has a unique solution $\widetilde{w} = \lim_{p\to\infty} \mathsf{N}^p(0)$ in $\mathbb{C}[[z^{-1}]]$. Since $\mathsf{N}(0) = \widetilde{f}(z)/P(0)$ one gets $\widetilde{w}(z) = \frac{f_N}{P(0)} z^{-N}(1 + O(z^{-1}))$. □

5.3.2.2 Proof of Proposition 5.2

We precise as an introduction that the assertion $\widetilde{w}_{k\mathbf{e}_i} \in \mathbb{R}[[z^{-1}]]$ is just a consequence of the realness of equation (3.6). The relationships $\widetilde{w}_{(0,k)}(z) = \widetilde{w}_{(k,0)}(-z)$ for every $k \geq 0$, come from the property of equation (3.6) to be invariant under the change of variable $z \mapsto -z$ and to the chosen normalization of $\widetilde{w}_{\mathbf{e}_i}$, $i = 1, 2$.

The Return of the formal solution We remark that $w_{\mathbf{0}} = w_{(0,0)}$ needs to solve (5.35) which is nothing but the equation (3.6) one started with. In particular we know that this equation has a unique formal solution $\widetilde{w}_{\mathbf{0}} \in \mathbb{C}[[z^{-1}]]$ which has been investigated in the previous chapters.
In what follows, one will always replace $w_{\mathbf{0}}$ by this formal solution $\widetilde{w}_{\mathbf{0}}$. We mention the following obvious fact, essentially due to the property that val $\widetilde{w}_{\mathbf{0}} \geq 2$ and that for every $\ell = 0,1,2$, $\frac{\partial^\ell F(z,0)}{\partial w^\ell} \in z^{-2}\mathbb{C}\{z^{-1}\}$. (This is one place where it is interesting to work with a "well-prepared" equation, see what we have done in Sect. 3.1 to get (3.6) and exercise 3.1):

Lemma 5.6. *If $\widetilde{w}_{\mathbf{0}}(z) = \sum_{l\geq 2} a_{0,l} z^{-l} \in \mathbb{C}[[z^{-1}]]$ is the formal solution of (3.6), then for every $\ell = 0,1,2$, $\frac{\partial^\ell F(z,\widetilde{w}_{\mathbf{0}})}{\partial w^\ell} \in \mathbb{C}[[z^{-1}]]$ has valuation 2, and vanishes identically for every $\ell \geq 3$. Also, $\frac{\partial F(z,\widetilde{w}_{\mathbf{0}})}{\partial w} = -4z^{-2} + z^{-2}\widetilde{w}_{\mathbf{0}}$ is even and its coefficients are all real negative, and $\frac{\partial^2 F(z,\widetilde{w}_{\mathbf{0}})}{\partial w^2} = z^{-2}$.*

The cases $|k\mathbf{e}_i| = 1$ Formula (5.36) with $\mathbf{k} = \mathbf{e}_1$ provides

$$\mathfrak{D}_{\mathbf{e}_1} w_{\mathbf{e}_1} = 0 \tag{5.42}$$

where $\mathfrak{D}_{\mathbf{e}_1} = P_{\mathbf{e}_1}(\partial) + \dfrac{1}{z} Q_{\mathbf{e}_1}(\partial) + \dfrac{1}{z^2} R_{\mathbf{e}_1} - \dfrac{\partial F(z,\widetilde{w}_{\mathbf{0}})}{\partial w}$ with

$$\begin{aligned} P_{\mathbf{e}_1}(\partial) &= P(-\lambda_1 + \partial) = P(-\lambda_1) + P'(-\lambda_1)\partial + \frac{P''(-\lambda_1)}{2!}\partial^2 \\ Q_{\mathbf{e}_1}(\partial) &= -\tau_1 P'(-\lambda_1 + \partial) + Q(-\lambda_1 + \partial) \\ R_{\mathbf{e}_1} &= \tau_1(\tau_1 + 4) \end{aligned}$$

Assuming that $w_{\mathbf{e}_1} \in \mathbb{C}[[z^{-1}]]$, one observes that the right-hand side of (5.42) has valuation less or equal to (val $w_{\mathbf{e}_1}$) $- 2$, because of lemma 5.6. In order to get a non identically vanishing solution, one thus has to impose the condition $P(-\lambda_1) = 0$. Following our conventions, we set $\lambda_1 = 1$.
The same reasoning leads to impose furthermore that $-\tau_1 P'(-\lambda_1) + Q(-\lambda_1) = 0$, thus $\tau_1 = -\frac{3}{2}$. Therefore, $P_{\mathbf{e}_1}(\partial) = \partial^2 - 2\partial$, $Q_{\mathbf{e}_1}(\partial) = 0$, $R_{\mathbf{e}_1}(\partial) = -\frac{15}{4}$. Symmetrically for $\mathbf{k} = \mathbf{e}_2$, one gets $\lambda_2 = -1$, $\tau_2 = -\dfrac{3}{2}$ as a necessary condition and

$$\mathfrak{D}_{\mathbf{e}_2} w_{\mathbf{e}_2} = 0 \tag{5.43}$$

where $\mathfrak{D}_{\mathbf{e}_2} = P_{\mathbf{e}_2}(\partial) + \frac{1}{z} Q_{\mathbf{e}_2}(\partial) + \frac{1}{z^2} R_{\mathbf{e}_2} - \frac{\partial F(z,\widetilde{w}_{\mathbf{0}})}{\partial w}$ whereas $P_{\mathbf{e}_2}(\partial) = \partial^2 + 2\partial$, $Q_{\mathbf{e}_2}(\partial) = 0$, $R_{\mathbf{e}_2}(\partial) = -\frac{15}{4}$.

Lemma 5.7. *The linear homogeneous equations (5.42), (5.43) have both a one-parameter family of formal solutions $w_{\mathbf{e}_1} = U_1 \widetilde{w}_{\mathbf{e}_1}$ and $w_{\mathbf{e}_2} = U_2 \widetilde{w}_{\mathbf{e}_2}$ in $\mathbb{C}[[z^{-1}]]$, where $\widetilde{w}_{\mathbf{e}_1}$ and $\widetilde{w}_{\mathbf{e}_2}$ are uniquely determined by their given normalization $\widetilde{w}_{\mathbf{e}_i} = 1 + O(z^{-1})$. Moreover $\widetilde{w}_{\mathbf{e}_i} \in \mathbb{R}[[z^{-1}]]$ and $\widetilde{w}_{\mathbf{e}_2}(z) = \widetilde{w}_{\mathbf{e}_1}(-z)$. Furthemore, if*

$\widetilde{w}_{\mathbf{0}}(z) = \sum_{l\geq 0} a_{\mathbf{0},l} z^{-l}$ *and* $\widetilde{w}_{\mathbf{e}_1}(z) = \sum_{l\geq 0} a_{\mathbf{e}_1,l} z^{-l}$, *the following quadratic recursion relation is valid:*

$$\begin{cases} a_{\mathbf{e}_1,0} = 1, \\ a_{\mathbf{e}_1,l} = \dfrac{1}{8l}\left(-(2l-1)^2 a_{\mathbf{e}_1,l-1} + 4\sum_{p=0}^{l-1} a_{\mathbf{e}_1,p} a_{\mathbf{0},l-p-1}\right), \; l = 1,2,\cdots \end{cases} \tag{5.44}$$

Proof. We only examine (5.42). We look at this equation in the space of normalized formal series $\mathbb{C}[[z^{-1}]]$, namely

$$\begin{cases} (\partial - 2)\partial w_{\mathbf{e}_1} = \left(\dfrac{15}{4}\dfrac{1}{z^2} + \dfrac{\partial F(z,\widetilde{w}_{\mathbf{0}})}{\partial w}\right) w_{\mathbf{e}_1} \\ w_{\mathbf{e}_1} \in \mathbb{C}[[z^{-1}]], \quad w_{\mathbf{e}_1} = 1 + O(z^{-1}). \end{cases} \tag{5.45}$$

We remark that the restriction of the derivation operator ∂ to the maximal ideal $z^{-1}\mathbb{C}[[z^{-1}]]$ is a bijective operator between $z^{-1}\mathbb{C}[[z^{-1}]]$ and $z^{-2}\mathbb{C}[[z^{-1}]]$; we denote by ∂^{-1} the inverse operator,

$$z^{-1}\mathbb{C}[[z^{-1}]] \underset{\underset{\partial^{-1}}{\leftarrow}}{\overset{\overset{\partial}{\rightarrow}}{}} z^{-2}\mathbb{C}[[z^{-1}]].$$

We transform (5.45) into the equation $-2\partial w_{\mathbf{e}_1} = \left(-\partial^2 + \frac{15}{4}\frac{1}{z^2} + \frac{\partial F(z,\widetilde{w}_{\mathbf{0}})}{\partial w}\right) w_{\mathbf{e}_1}$ and we see that the right-hand side of this equation belongs to $z^{-2}\mathbb{C}[[z^{-1}]]$ once $w_{\mathbf{e}_1}$ belongs to $\mathbb{C}[[z^{-1}]]$, because of lemma 5.6 and to the choice of the coefficient τ_1. This means that the map

$$\begin{aligned} \mathsf{N} : \mathbb{C}[[z^{-1}]] &\to \mathbb{C}[[z^{-1}]] \\ w_{\mathbf{e}_1} &\to 1 - \frac{1}{2}\partial^{-1}\left(-\partial^2 + \frac{15}{4}\frac{1}{z^2} + \frac{\partial F(z,\widetilde{w}_{\mathbf{0}})}{\partial w}\right) w_{\mathbf{e}_1} \end{aligned}$$

is well defined and the problem (5.45) is equivalent to the fixed-point problem $w_{\mathbf{e}_1} = \mathsf{N}(w_{\mathbf{e}_1})$. One easily checks that the map N is contractive in $\mathbb{C}[[z^{-1}]]$ so that the fixed point problem $w_{\mathbf{e}_1} = \mathsf{N}(w_{\mathbf{e}_1})$ has a unique solution $\widetilde{w}_{\mathbf{e}_1}$ in $\mathbb{C}[[z^{-1}]]$.
From the fact that (5.42) is a homogeneous equation, one immediately concludes that $U_1 \widetilde{w}_{\mathbf{e}_1}$, $U_1 \in \mathbb{C}$, provides a one-parameter family of formal solutions.
The proof for the quadratic recursion relation (5.44) is left to the reader (see also [GIKM012, ASV012]). □

Remark 5.6. 1. The Newton polygon at infinity $\mathcal{N}_\infty(\mathfrak{D}_{\mathbf{e}_1})$ drawn on Fig. 5.2, has one horizontal side that corresponds to the operator -2∂. General nonsense in asymptotic theory (see the second volume of this book [Lod016], or [CL55, Koh99]) provides the existence of the formal (normalized) series solution $\widetilde{w}_{\mathbf{e}_1}$. The other (normalized) formal solution associated with the side of slope -1 is $e^{2z}\widetilde{w}_{\mathbf{e}_2}$ (see lemma 5.4) which, in our frame, is already incorporated in the other transseries solution.
2. From lemma 5.6 or (5.44), one easily shows that

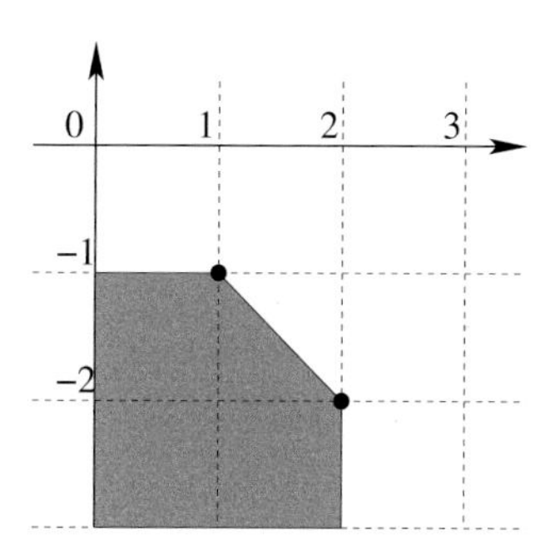

Fig. 5.2 The Newton polygon at infinity $\mathcal{N}_\infty(\mathfrak{D}_{\mathbf{e}_1})$ associated with the linear operator (5.45).

$$\widetilde{w}_{\mathbf{e}_1}(z) = 1 - \frac{1}{8}z^{-1} + \frac{9}{128}z^{-2} - \frac{341329}{1920000}z^{-3} + \cdots$$

is a real formal expansion, with coefficients that alternate in sign.

The cases $|k\mathbf{e}_i| \geq 2$

Lemma 5.8. *For any $\mathbf{k} = k\mathbf{e}_i$, $i = 1,2$ and $k \geq 2$, equation (5.37) has a unique formal solution $w_{k\mathbf{e}_i} = \widetilde{w}_{k\mathbf{e}_i}$ in $\mathbb{C}[[z^{-1}]]$. Moreover* $\operatorname{val} \widetilde{w}_{k\mathbf{e}_i} = 2(k-1)$.
Furthermore, when considering $U\widetilde{w}_{\mathbf{e}_i}$ instead of $\widetilde{w}_{\mathbf{e}_i}$ for the solution of (5.36), then the unique solution of (5.37) at rank $\mathbf{k} = k\mathbf{e}_i$, $k \geq 2$, is $U^k\widetilde{w}_{k\mathbf{e}_i}$. Also, $\widetilde{w}_{k\mathbf{e}_i} \in \mathbb{R}[[z^{-1}]]$, $\widetilde{w}_{k\mathbf{e}_i}(z) = \dfrac{k}{12^{k-1}}z^{-2(k-1)}(1+O(z^{-1}))$ and $\widetilde{w}_{(0,k)}(z) = \widetilde{w}_{(k,0)}(-z)$ for every $k \geq 2$.
Eventually, writing $\widetilde{w}_{k\mathbf{e}_1}(z) = \sum_{l \geq 0} a_{k\mathbf{e}_1,l} z^{-l}$, the coefficients are governed the following quadratic recursion relations, for every $k \geq 2$:

$$\begin{cases} a_{k\mathbf{e}_1,0} = a_{k\mathbf{e}_1,1} = 0, \\ (k^2-1)a_{k\mathbf{e}_1,l} = k(3k-2l-1)a_{k\mathbf{e}_1,l-1} - \frac{1}{4}(3k-2l)^2 a_{k\mathbf{e}_1,l-2} \\ +\sum_{p=0}^{l-2}\left(a_{k\mathbf{e}_1,p}a_{\mathbf{0},l-p-2} + \frac{1}{2}\sum_{\substack{k_1+k_2=k \\ k_1\geq 1, k_2 \geq 1}} a_{k_1\mathbf{e}_1,p}a_{k_2\mathbf{e}_1,l-p-2}\right), \; l = 2,3,\cdots \end{cases} \tag{5.46}$$

Proof. We only examine the case $\mathbf{k} = k\mathbf{e}_1$, $k \geq 2$.
The proof is done by induction on k. We first consider equation (5.37) for $k = 2$:

$$\mathfrak{D}_{2\mathbf{e}_1} w_{2\mathbf{e}_1} = \frac{\widetilde{w}_{\mathbf{e}_1}^2}{2!}\frac{\partial^2 F(z,\widetilde{w}_{\mathbf{0}})}{\partial w^2}, \tag{5.47}$$

with $\mathfrak{D}_{2\mathbf{e}_1} = P_{2\mathbf{e}_1}(\partial) + \frac{1}{z}Q_{2\mathbf{e}_1}(\partial) + \frac{1}{z^2}R_{2\mathbf{e}_1} - \frac{\partial F(z,\widetilde{w}_{\mathbf{0}})}{\partial w}$. We know that $P_{2\mathbf{e}_1}(0) = 3$ is nonzero since, by (5.31), $P_{2\mathbf{e}_1}(\partial) = P(-2\lambda_1 + \partial) = \partial^2 - 4\partial + 3$. Using lemma 5.6, one sees that lemma 5.5 can be applied to (5.47) and this provides a unique solution $\widetilde{w}_{2\mathbf{e}_1} \in \mathbb{C}[[z^{-1}]]$. Its valuation is 2 and explicit calculation gives:

$$\widetilde{w}_{2\mathbf{e}_1}(z) = \frac{1}{6}z^{-2} - \frac{11}{72}z^{-3} + \frac{53}{192}z^{-4} + \cdots, \qquad \widetilde{w}_{2\mathbf{e}_2}(z) = \widetilde{w}_{2\mathbf{e}_1}(-z).$$

One easily checks that replacing $\widetilde{w}_{\mathbf{e}_1}$ by $U\widetilde{w}_{\mathbf{e}_1}$ implies changing $\widetilde{w}_{2\mathbf{e}_1}$ into $U^2\widetilde{w}_{2\mathbf{e}_1}$. We now assume that the properties of lemma 5.8 are true for every $2 \leq k \leq K-1$. When considering equation (5.37) for K one gets :

$$\mathfrak{D}_{K\mathbf{e}_1} w_{K\mathbf{e}_1} = \sum_{\substack{k_1+k_2=K \\ k_1\geq 1, k_2\geq 1}} \frac{\widetilde{w}_{k_1\mathbf{e}_1}\widetilde{w}_{k_2\mathbf{e}_1}}{2!} \frac{\partial^2 F(z,\widetilde{w}_{\mathbf{0}})}{\partial w^2}, \tag{5.48}$$

with $\mathfrak{D}_{K\mathbf{e}_1} = P_{K\mathbf{e}_1}(\partial) + \frac{1}{z}Q_{K\mathbf{e}_1}(\partial) + \frac{1}{z^2}R_{K\mathbf{e}_1} - \frac{\partial F(z,\widetilde{w}_{\mathbf{0}})}{\partial w}$, $P_{K\mathbf{e}_1}(\partial) = \partial^2 - 2K\partial + (K^2-1)$. One deduces the conclusion of lemma 5.8 at the rank K by the arguments used previously. For the valuation, observe that val $\widetilde{w}_{k_1\mathbf{e}_1}\widetilde{w}_{k_2\mathbf{e}_1} \geq 2(k_1-1)+2(k_2-1)$ when $k_1+k_2=K$, thus val $\widetilde{w}_{k_1\mathbf{e}_1}\widetilde{w}_{k_2\mathbf{e}_1} \geq 2(K-2)$. As a matter of fact, for every $k \geq 2$, $\widetilde{w}_{(k,0)}(z) = b_k z^{-2(K-1)}(1+O(z^{-1}))$ with

$$\begin{cases} b_1 = 1, \\ b_k = \frac{1}{2(k-1)(k+1)} \sum_{p=1}^{k-1} b_p b_{k-p}, \quad k \geq 2, \end{cases}$$

which easily provides $b_k = \frac{k}{12^{k-1}}$ by induction. The reader will easily check that the recursive relations (5.46) are true. (See also [GIKM012, ASV012]). □

Remark 5.7. Here again, we are not interested in the whole formal fundamental solutions of equations (5.47), (5.48), which incorporate the general solution $(\mathrm{e}^{-\lambda_1 k z}z^{-\tau_1 k})^{-1}(C_1\mathrm{e}^{-\lambda_1 z}z^{-\tau_1}\widetilde{w}_{\mathbf{e}_1} + C_2\mathrm{e}^{-\lambda_2 z}z^{-\tau_2}\widetilde{w}_{\mathbf{e}_2})$ of the associated homogeneous linear ODEs $\mathfrak{D}_{(k,0)}w = 0$ (cf. lemma 5.4). Taking into account the term $(\cdots)\widetilde{w}_{\mathbf{e}_1}$ would imply a rescaling of U_1. The other term $(\cdots)\widetilde{w}_{\mathbf{e}_2}$ concerns the other transseries.

5.4 Formal Integral for the First Painlevé Equation

We made general considerations on formal integrals in Sect. 5.2. We started the study of the formal integral for the prepared equation (3.6) associated with the first Painlevé equation in Sect. 5.3 : this gave us the transseries described by proposition 5.2. When no resonances occur, one gets with quite similar arguments the formal integral. However, this is not that simple for the first Painlevé equation where we have to cope with resonances.

5.4.1 Notation and Preliminary Results

5.4.1.1 Notation

It will be useful for our purpose to introduce the following notation:

Definition 5.7. For any $n \in \mathbb{N}^\star$, we set $\mathbf{n} = n(1,1)$ and

$$\Xi_{n,0} = \{\mathbf{k} = (k_1, k_2) \in \mathbb{N}^2 \setminus \{\mathbf{0}\} \mid k_1 < n \text{ or } k_2 < n\} \cup \{\mathbf{n}\}.$$

We also set $\Xi_{0,0} = \{(0,0)\}$.

Example 5.3. $\Xi_{1,0} = (\mathbb{N}^\star \times \{0\}) \cup (\{0\} \times \mathbb{N}^\star) \cup \{(1,1)\}$,
$\Xi_{2,0} = (\mathbb{N}^\star \times \{0,1\}) \cup (\{0,1\} \times \mathbb{N}^\star) \cup \{(2,2)\}$.
Notice that for every $n \in \mathbb{N}$, $\Xi_{n+1,0} \setminus \Xi_{n,0} = \mathbf{n} + \Xi_{1,0}$.

5.4.1.2 Resonances : First Consequences

Equation (3.6) has the feature to have *positively resonant* multipliers $\lambda_1 = 1$, $\lambda_2 = -1$ because $\boldsymbol{\lambda}.\mathbf{n} = 0$, for every $n \in \mathbb{N}^\star$ (see definition 5.2). This brings semi-positively resonances, the cases of semi-positive resonances being all described by $\lambda_1 = \boldsymbol{\lambda}.(\mathbf{n} + \mathbf{e}_1)$ and $\lambda_2 = \boldsymbol{\lambda}.(\mathbf{n} + \mathbf{e}_2)$, for every $n \in \mathbb{N}^\star$.

We have already seen (proposition 5.2) that these properties have no consequence for the transseries but, as we shall see, this produces new phenomena when the formal integral is concerned, these being essentially consequences of the following fact, derived from lemma 5.3.

Lemma 5.9. *For every $n \in \mathbb{N}$, $\mathbf{k} \in \mathbb{N}^2$, the following identities are satisfied:*

$$\mathfrak{P}_{\mathbf{n}+\mathbf{k}} = \mathfrak{P}_{\mathbf{k}}, \qquad \mathfrak{D}_{\mathbf{n}+\mathbf{k}} = z^{\boldsymbol{\tau}.\mathbf{n}} \mathfrak{D}_{\mathbf{k}} z^{-\boldsymbol{\tau}.\mathbf{n}}, \qquad \boldsymbol{\tau}.\mathbf{n} = -3n.$$

5.4.1.3 Preliminary Lemmas

In a moment, we will have to deal with formal expansions of the type $\sum_{l=0}^{p} \log^l(z) \widetilde{f}_l(z)$, $p \in \mathbb{N}$, with the $\widetilde{f}_l$'s in $\mathbb{C}[[z^{-1}]]$.

Definition 5.8. We equip the graded algebra $\bigoplus_{l \in \mathbb{N}} \log^l(z) \mathbb{C}[[z^{-1}]]$ with the valuation val defined by: $\operatorname{val}\left(\sum_l \log^l(z) \widetilde{f}_l\right) = \min_l \{\operatorname{val} \widetilde{f}_l\}$.

Lemma 5.10. *We suppose $n, N \in \mathbb{N}^\star$ and $p \in \mathbb{N}$. We consider the ordinary differential equation*

$$P(\partial)w + \frac{1}{z} R(\partial)w = \widetilde{f}(z), \qquad \widetilde{f}(z) \in \bigoplus_{l=0}^{p} \log^l(z) \mathbb{C}[[z^{-1}]], \tag{5.49}$$

$$P(\partial) = \sum_{m=0}^{n} \alpha_{n-m} \partial^m \in \mathbb{C}[\partial],\ \alpha_n \neq 0, \qquad R(\partial) = \sum_{m=0}^{n-1} \gamma_{n-m}(z) \partial^m \in \mathbb{C}[[z^{-1}]][\partial]$$

Then (5.49) has a unique solution $\widetilde{w} \in \bigoplus_{l=0}^{p} \log^l(z)\mathbb{C}[[z^{-1}]]$ *and* $\mathrm{val}\,\widetilde{w} = \mathrm{val}\,\widetilde{f}$. *Moreover, if* $\widetilde{f} = \sum_{l=0}^{p} \log^l(z)\widetilde{f}_l$ *and* $\widetilde{w} = \sum_{l=0}^{p} \log^l(z)\widetilde{w}_l$, *then:*

1. $\widetilde{w}_p$ *solves the ODE:* $P(\partial)w + \frac{1}{z}R(\partial)w = \widetilde{f}_p$;
2. *if* $\mathrm{val}\,\widetilde{f}_p < \mathrm{val}\,\sum_{l=0}^{p-1} \log^l(z)\widetilde{f}_l$ *then* $\mathrm{val}\,\widetilde{w}_p < \mathrm{val}\,\sum_{l=0}^{p-1} \log^l(z)\widetilde{w}_l$.

Proof. One easily sees that the arguments used for the proof of lemma 5.5 can be extended, when observing that $\mathrm{val}\,\partial\left(\sum_l \log^l(z)\widetilde{f}_l\right) \leq \mathrm{val}\left(\sum_l \log^l(z)\widetilde{f}_l\right) + 1$. □

We have seen in lemma 5.7 that the operators $\mathfrak{D}_{\mathbf{e}_i}$, $i = 1,2$, have specific behaviours. This is the purpose of the following lemma.

Lemma 5.11. *We suppose* $p \in \mathbb{N}$ *and* $i \in \{1,2\}$. *We assume that* $\widetilde{f} = \sum_{l=0}^{p} \log^l(z)\widetilde{f}_l \in \bigoplus_{l=0}^{p} \log^l(z)\mathbb{C}[[z^{-1}]]$ *satisfies the conditions:*

1. $\mathrm{val}\,\widetilde{f}_p = 1$, $\widetilde{f}_p = f_{p1}z^{-1}(1 + 0(z^{-1}))$, $f_{p1} \neq 0$
2. $\mathrm{val}\left(\sum_{l=0}^{p-1} \log^l(z)\widetilde{f}_l\right) \geq 2$.

Then the equation $\mathfrak{D}_{\mathbf{e}_i} w = \widetilde{f}$ *has a unique solution* $\widetilde{w} = \sum_{l=0}^{p+1} \log^l(z)\widetilde{w}_l$ *in* $\bigoplus_{l=0}^{p+1} \log^l(z)\mathbb{C}[[z^{-1}]]$ *that satisfies the condition* $\mathrm{val}\left(\sum_{l=0}^{p} \log^l(z)\widetilde{w}_l\right) \geq 1$. *Moreover* $\widetilde{w}_{p+1} = \dfrac{f_{p1}}{(p+1)P'(-\lambda_i)}\widetilde{w}_{\mathbf{e}_i}$.
Otherwise, the general solution of the ODE $\mathfrak{D}_{\mathbf{e}_i} w = \widetilde{f}$ *in* $\bigoplus_{l=0}^{p+1} \log^l(z)\mathbb{C}[[z^{-1}]]$ *is of the form* $w = \widetilde{w} + U\widetilde{w}_{\mathbf{e}_i}$ *where* $U \in \mathbb{C}$.

Proof. We examine the case $i = 1$ only. The ODE $\mathfrak{D}_{\mathbf{e}_1} w = \widetilde{f}$ is equivalent to the equation :

$$P'(-\lambda_1)\partial w = \widetilde{f} + \left(-\partial^2 + \frac{15}{4}\frac{1}{z^2} + \frac{\partial F(z,\widetilde{w}_{\mathbf{0}})}{\partial w}\right) w, \quad P'(-\lambda_1) = -2.$$

By arguments already used in the proof of lemma 5.7, this problem amounts to looking for a formal solution which satisfies the fixed-point problem

$$w = U(z) + \frac{1}{P'(-\lambda_1)}\partial^{-1}\left(-\partial^2 + \frac{15}{4}\frac{1}{z^2} + \frac{\partial F(z,\widetilde{w}_{\mathbf{0}})}{\partial w}\right) w$$

where $U(z) = \partial^{-1}\left(\dfrac{\widetilde{f}}{P'(-\lambda_1)}\right) = \dfrac{f_{p1}}{(p+1)P'(-\lambda_1)}\log^{p+1}(z) + \sum_{l=0}^{p} \log^l(z)O(z^{-1})$.
Notice that we take the primitive with no constant term. This fixed-point problem has a unique formal solution under the form

$$\widetilde{w} = \frac{f_{p1}}{(p+1)P'(-\lambda_1)}\widetilde{w}_{\mathbf{e}_i}\log^{p+1}(z) + \sum_{l=0}^{p} \log^l(z)\widetilde{w}_l$$

and val $\left(\sum_{l=0}^{p}\log^l(z)\widetilde{w}_l\right)\geq 1$. Eventually one can add to this particular solution any solution of the homogeneous equation $\mathfrak{D}_{\mathbf{e}_i}w=0$, that is any term of the form $U\widetilde{w}_{\mathbf{e}_i}$ with $U\in\mathbb{C}$. □

5.4.2 Painlevé I: Formal Integral

We are now in position to detail the formal integral associated with the first Painlevé equation.

Theorem 5.1. *We consider the ODE (3.6). Let be* $\boldsymbol{\lambda}=(\lambda_1,\lambda_2)=(1,-1)$ *where the* λ_i*'s are the multipliers, and* $\boldsymbol{\tau}=(\tau_1,\tau_2)=\left(-\frac{3}{2},-\frac{3}{2}\right)$, $\tau_i=\frac{Q(-\lambda_i)}{P'(-\lambda_i)}$, $i=1,2$. *We set* $\mathbf{V}^{\mathbf{k}}=\mathbf{U}^{\mathbf{k}}\mathrm{e}^{-\boldsymbol{\lambda}.\mathbf{k}z}z^{-\boldsymbol{\tau}.\mathbf{k}}$ *for any* $\mathbf{k}\in\mathbb{N}^2$ *and any* $\mathbf{U}=(U_1,U_2)\in\mathbb{C}^2$. *We write* $\mathbf{n}=n(1,1)$ *for any* $n\in\mathbb{N}$.
There exists a two-parameter formal solution of (3.6), freely depending on $\mathbf{U}\in\mathbb{C}^2$, *of the form*

$$\widetilde{w}(z,\mathbf{U})=\widetilde{w}_{\mathbf{0}}(z)+\sum_{n=0}^{\infty}\sum_{\mathbf{k}\in\Xi_{n+1,0}\setminus\Xi_{n,0}}\mathbf{V}^{\mathbf{k}}\widetilde{w}_{\mathbf{k}}(z),\tag{5.50}$$

and uniquely determined by the following conditions:

1. $\widetilde{w}_{\mathbf{0}}\in\mathbb{C}[[z^{-1}]]$;
2. $\widetilde{w}_{\mathbf{k}}=\sum_{l=0}^{n}\log^l(z)\widetilde{w}_{\mathbf{k}}^{[l]}\in\bigoplus_{l=0}^{n}\log^l(z)\mathbb{C}[[z^{-1}]]$, *for every* $\mathbf{k}\in\Xi_{n+1,0}\setminus\Xi_{n,0}$, $n\in\mathbb{N}$;
3. *for* $i=1,2$, $\widetilde{w}_{\mathbf{e}_i}$ *satisfies* $\widetilde{w}_{\mathbf{e}_i}(z)=1+O(z^{-1})$;
4. *for every* $n\in\mathbb{N}^\star$ *and* $i=1,2$, $\widetilde{w}_{\mathbf{n}+\mathbf{e}_i}=\sum_{l=0}^{n}\log^l(z)\widetilde{w}_{\mathbf{n}+\mathbf{e}_i}^{[l]}$ *satisfies* val $\widetilde{w}_{\mathbf{n}+\mathbf{e}_1}^{[n]}<$ val $\left(\sum_{l=0}^{n-1}\log^l(z)\widetilde{w}_{\mathbf{n}+\mathbf{e}_i}^{[l]}\right)$.

Moreover, the following properties are satisfied:

5. *changing the normalization of* $\widetilde{w}_{\mathbf{e}_i}$, $i=1,2$, *is equivalent to rescaling the parameter* $\mathbf{U}\in\mathbb{C}^2$;
6. *for every* $n\in\mathbb{N}$ *and every* $\mathbf{k}\in\Xi_{n+1,0}\setminus\Xi_{n,0}$, $\widetilde{w}_{\mathbf{k}}\in\bigoplus_{l=0}^{n}\log^l(z)\mathbb{R}[[z^{-1}]]$. *Furthermore* $\widetilde{w}_{(k_1,k_2)}^{[l]}(z)=\widetilde{w}_{(k_2,k_1)}^{[l]}(-z)$ *for every* $l\in[0,n]$;
7. *for every* $n\in\mathbb{N}^\star$ *and every* $\mathbf{k}\in\Xi_{n+1,0}\setminus\Xi_{n,0}$,

$$\widetilde{w}_{\mathbf{k}}=\sum_{l=0}^{n}\frac{1}{l!}(\boldsymbol{\varkappa}.\mathbf{k})^l z^{\boldsymbol{\tau}.\mathbf{1}}\log^l(z)\widetilde{w}_{\mathbf{k}-\mathbf{1}}^{[0]}\tag{5.51}$$

where $\boldsymbol{\varkappa}=(\varkappa_1,\varkappa_2)=(\frac{5}{12},-\frac{5}{12})$ *is defined by:*

$$\varkappa_i = \frac{a^2}{P'(-\lambda_i)}\left(\frac{1}{P(0)} + \frac{1}{2!}\frac{1}{P(-2\lambda_i)}\right) = \frac{5}{12}\lambda_i, \quad i = 1,2, \tag{5.52}$$

whereas a is given by $\dfrac{\partial^2 F(z,0)}{\partial w^2} = az^{-2} + o(z^{-2})$. *As a consequence, for every* $n \in \mathbb{N}^\star$, $\widetilde{w}_{\mathbf{n}} \in \mathbb{R}[[z^{-1}]]$ *;*

8. *for every* $\mathbf{k} \in \mathbb{N}^2 \setminus \{\mathbf{0}\}$, $\operatorname{val} \widetilde{w}_{\mathbf{k}}^{[0]} = 2(|\mathbf{k}| - 1)$.

Proof. Once for all:

- the property 5. is easily derived by an argument of homogeneity;
- the realness and eveness in property 6. are just consequences of the realness of equation (3.6) and its property of being be invariant under the change of variable $z \mapsto -z$, and to the chosen normalizations.

In what follows, we investigate the terms under the form $\widetilde{w}_{\mathbf{k}}$ with $\mathbf{k} \in \Xi_{n+1,0} \setminus \Xi_{n,0}$ and $n \in \mathbb{N}$. We first look at what happens when $n = 0$ and $n = 1$, step by step so as to draw some conclusions, then we complete the proof by induction on n.

Case $n = 0$ and $\mathbf{k} = \mathbf{1}$ This is the first case where a resonance appears. However, this case yields no surprise. Indeed, equation (5.37) for $\mathbf{k} = \mathbf{1}$ reads

$$\begin{aligned} P_{\mathbf{1}}(\partial)w_{\mathbf{1}} + \frac{1}{z}Q_{\mathbf{1}}(\partial)w_{\mathbf{1}} &= \left(-\frac{1}{z^2}R_{\mathbf{1}} + \frac{\partial F(z,\widetilde{w}_{\mathbf{0}})}{\partial w}\right)w_{\mathbf{1}} \\ &\quad + \widetilde{w}_{\mathbf{e}_1}\widetilde{w}_{\mathbf{e}_2}\frac{\partial^2 F(z,\widetilde{w}_{\mathbf{0}})}{\partial w^2} \end{aligned} \tag{5.53}$$

with $P_{\mathbf{1}}(\partial) = P_{\mathbf{0}}(\partial) = \partial^2 - 1$. Therefore lemma 5.5 can be applied and one gets a unique solution $\widetilde{w}_{\mathbf{1}} \in \mathbb{C}[[z^{-1}]]$ with, moreover, $\operatorname{val} \widetilde{w}_{\mathbf{1}} = 2$ and $\widetilde{w}_{\mathbf{1}}(z) = \dfrac{a}{P(0)}z^{-2} + o(z^{-2})$ where $a = 1$ is given by: $\frac{\partial^2 F(z,0)}{\partial w^2} = az^{-2} + o(z^{-2})$.

Explicit calculation yields: $\widetilde{w}_{\mathbf{1}}(z) = -z^{-2} - \dfrac{9}{8}z^{-4} - \dfrac{902139}{80000}z^{-6} - \cdots$.

Cases $n = 1$ and $\mathbf{k} \in \Xi_{2,0} \setminus \Xi_{1,0}$

Cases $\mathbf{k} = \mathbf{1} + \mathbf{e}_i$, $i = 1,2$ These are the first cases of semi-positive resonances and are more serious.

Let us concentrate on the case $\mathbf{k} = \mathbf{1} + \mathbf{e}_1$ for which equation (5.37) is

$$\mathfrak{D}_{\mathbf{1}+\mathbf{e}_1} w_{\mathbf{1}+\mathbf{e}_1} = \left(\widetilde{w}_{\mathbf{1}}\widetilde{w}_{\mathbf{e}_1} + \widetilde{w}_{2\mathbf{e}_1}\widetilde{w}_{\mathbf{e}_2}\right)\frac{\partial^2 F(z,\widetilde{w}_{\mathbf{0}})}{\partial w^2},$$

that is also, from lemma 5.9 and proposition 5.2,

$$\mathfrak{D}_{\mathbf{e}_1}(z^3 w_{\mathbf{1}+\mathbf{e}_1}) = \widetilde{g}_{\mathbf{1}+\mathbf{e}_1}, \tag{5.54}$$

$$\begin{aligned}\widetilde{g}_{\mathbf{1}+\mathbf{e}_1} &= z^3\left(\widetilde{w}_{\mathbf{1}}\widetilde{w}_{\mathbf{e}_1} + \widetilde{w}_{2\mathbf{e}_1}\widetilde{w}_{\mathbf{e}_2}\right)\frac{\partial^2 F(z,\widetilde{w}_{\mathbf{0}})}{\partial w^2} \\ &= \left(\frac{1}{P(0)} + \frac{1}{2!}\frac{1}{P(-2\lambda_1)}\right)a^2 z^{-1} + O(z^{-2}) \\ &= -\frac{5}{6}z^{-1} + O(z^{-2}).\end{aligned}$$

The conditions of application of lemma 5.11 are fulfilled: equation (5.54) has a one-parameter family of formal solutions, depending on $U_{[1],1} \in \mathbb{C}$, of the form

$$\begin{aligned}&w_{\mathbf{1}+\mathbf{e}_1} = \widetilde{w}_{\mathbf{1}+\mathbf{e}_1} + U_{[1],1}z^{-3}\widetilde{w}_{\mathbf{e}_1}, \qquad \widetilde{w}_{\mathbf{1}+\mathbf{e}_1} = \widetilde{w}^{[1]}_{\mathbf{1}+\mathbf{e}_1}\log(z) + \widetilde{w}^{[0]}_{\mathbf{1}+\mathbf{e}_1}, \\ &\widetilde{w}^{[1]}_{\mathbf{1}+\mathbf{e}_1} = \varkappa_1 z^{-3}\widetilde{w}_{\mathbf{e}_1}, \qquad \operatorname{val}\widetilde{w}^{[0]}_{\mathbf{1}+\mathbf{e}_1} \geq 4. \\ &\varkappa_1 = \frac{a^2}{P'(-\lambda_1)}\left(\frac{1}{P(0)} + \frac{1}{2!}\frac{1}{P(-2\lambda_1)}\right) = \frac{5}{12}.\end{aligned} \tag{5.55}$$

Explicitly,

$$\widetilde{w}^{[0]}_{\mathbf{1}+\mathbf{e}_1}(z) = \frac{11}{72}z^{-4} - \frac{197}{576}z^{-5} + \frac{23903}{82944}z^{-6} - \cdots$$

Also remark that the property $\operatorname{val}\widetilde{w}^{[0]}_{\mathbf{1}+\mathbf{e}_1} \geq 4$ characterizes the particular solution $\widetilde{w}_{\mathbf{1}+\mathbf{e}_1}$ among the one-parameter family of solutions.

The case $\mathbf{k} = \mathbf{1} + \mathbf{e}_i$ is deduced from the above result from the invariance of (3.6) under the change of variable $z \mapsto -z$. One gets a one-parameter family of formal solutions, depending on $U_{[1],2} \in \mathbb{C}$, of the form

$$\begin{aligned}&w_{\mathbf{1}+\mathbf{e}_2} = \widetilde{w}_{\mathbf{1}+\mathbf{e}_2} + U_{[1],2}\widetilde{w}_{\mathbf{e}_2}, \qquad \widetilde{w}_{\mathbf{1}+\mathbf{e}_2} = \widetilde{w}^{[1]}_{\mathbf{1}+\mathbf{e}_2}\log(z) + \widetilde{w}^{[0]}_{\mathbf{1}+\mathbf{e}_2}, \\ &\widetilde{w}^{[1]}_{\mathbf{1}+\mathbf{e}_2}(z) = \widetilde{w}^{[1]}_{\mathbf{1}+\mathbf{e}_1}(-z) = \varkappa_2 z^{-3}\widetilde{w}_{\mathbf{e}_2}(z), \qquad \widetilde{w}^{[0]}_{\mathbf{1}+\mathbf{e}_2}(z) = \widetilde{w}^{[0]}_{\mathbf{1}+\mathbf{e}_1}(-z) \\ &\varkappa_2 = \frac{a^2}{P'(-\lambda_2)}\left(\frac{1}{P(0)} + \frac{1}{2!}\frac{1}{P(-2\lambda_2)}\right) = -\frac{5}{12}.\end{aligned} \tag{5.56}$$

In the sequel, we fix $U_{[1],1} = U_{[1],2} = 0$, that is we only consider the (well and uniquely defined) particular solutions $\widetilde{w}_{\mathbf{1}+\mathbf{e}_i}$, $i = 1,2$.

> We stress that adding terms of the form $U_{[1],1}\widetilde{w}_{\mathbf{e}_1}$ and $U_{[1],2}\widetilde{w}_{\mathbf{e}_2}$ has the effect to rescaling the parameter (U_1, U_2). In particular, changing the branch of the log has non consequence for the formal integral.

Cases $\mathbf{k} = \mathbf{1} + k\mathbf{e}_i$ One step further, we consider the case $\mathbf{k} = \mathbf{1} + 2\mathbf{e}_i$. We take $i = 1$ only for simplicity. From (5.37) and lemma 5.9, we get:

$$\mathfrak{D}_{2\mathbf{e}_1}(z^3 w_{\mathbf{1}+2\mathbf{e}_1}) = z^3\left(\widetilde{w}_{\mathbf{1}+\mathbf{e}_1}\widetilde{w}_{\mathbf{e}_1} + \widetilde{w}_{2\mathbf{e}_1}\widetilde{w}_{\mathbf{1}} + \widetilde{w}_{3\mathbf{e}_1}\widetilde{w}_{\mathbf{e}_2}\right)\frac{\partial^2 F(z,\widetilde{w}_{\mathbf{0}})}{\partial w^2}. \tag{5.57}$$

By proposition 5.2 and the above result, the right-hand side of equation (5.57) is a formal series expansion of the type $\widetilde{f} = \widetilde{f}^{[1]}\log(z) + \widetilde{f}^{[0]}$ with $\operatorname{val}\widetilde{f}^{[1]} = 2$ and val $\widetilde{f}^{[0]} = 3$. Applying lemma 5.10, we get for (5.57) a unique formal solution of the form $\widetilde{w}_{\mathbf{1}+2\mathbf{e}_1} = \widetilde{w}^{[1]}_{\mathbf{1}+2\mathbf{e}_1} \log(z) + \widetilde{w}^{[0]}_{\mathbf{1}+2\mathbf{e}_1} \in \bigoplus_{l=0}^{1} \log^l(z)\mathbb{C}[[z^{-1}]]$ with val $\widetilde{w}^{[1]}_{\mathbf{1}+2\mathbf{e}_1} = 5$ and val $\widetilde{w}^{[0]}_{\mathbf{1}+2\mathbf{e}_1} = 6$. Moreover, $\widetilde{w}^{[1]}_{\mathbf{1}+2\mathbf{e}_1}$ solves the ODE

$$\begin{aligned}\mathfrak{D}_{2\mathbf{e}_1}(z^3 w^{[1]}_{\mathbf{1}+2\mathbf{e}_1}) &= z^3 \widetilde{w}^{[1]}_{\mathbf{1}+\mathbf{e}_1} \widetilde{w}_{\mathbf{e}_1} \frac{\partial^2 F(z,\widetilde{w}_\mathbf{0})}{\partial w^2} \\ &= \varkappa_1 \widetilde{w}^2_{\mathbf{e}_1} \frac{\partial^2 F(z,\widetilde{w}_\mathbf{0})}{\partial w^2}.\end{aligned}$$

Comparing to (5.47), one concludes that

$$\begin{aligned}\widetilde{w}_{\mathbf{1}+2\mathbf{e}_1} &= \widetilde{w}^{[1]}_{\mathbf{1}+2\mathbf{e}_1} \log(z) + \widetilde{w}^{[0]}_{\mathbf{1}+2\mathbf{e}_1}, \\ \widetilde{w}^{[1]}_{\mathbf{1}+2\mathbf{e}_1} &= 2\varkappa_1 z^{-3} \widetilde{w}_{2\mathbf{e}_1}, \qquad \text{val } \widetilde{w}^{[0]}_{\mathbf{1}+2\mathbf{e}_1} = 6.\end{aligned}$$

We now reason by induction, assuming that for every $k \in [2, K-1]$ with $K \geq 3$, one has

$$\begin{aligned}\widetilde{w}_{\mathbf{1}+k\mathbf{e}_1} &= \widetilde{w}^{[1]}_{\mathbf{1}+k\mathbf{e}_1} \log(z) + \widetilde{w}^{[0]}_{\mathbf{1}+k\mathbf{e}_1}, \\ \widetilde{w}^{[1]}_{\mathbf{1}+k\mathbf{e}_1} &= k\varkappa_1 z^{-3} \widetilde{w}_{k\mathbf{e}_1}, \qquad \text{val } \widetilde{w}^{[0]}_{\mathbf{1}+k\mathbf{e}_1} = 2(k+1).\end{aligned}$$

Then, by (5.37) and lemma 5.9,

$$\begin{aligned}\mathfrak{D}_{K\mathbf{e}_1}(z^3 \widetilde{w}_{\mathbf{1}+K\mathbf{e}_1}) &= z^3 \sum_{\substack{\mathbf{k}_1+\mathbf{k}_2=\mathbf{1}+K\mathbf{e}_1 \\ |\mathbf{k}_1|\geq 1, |\mathbf{k}_2|\geq 1}} \frac{\widetilde{w}_{\mathbf{k}_1} \widetilde{w}_{\mathbf{k}_2}}{2} \frac{\partial^2 F(z,w_\mathbf{0})}{\partial w^2} \\ &= z^3 \sum_{\substack{k_1+k_2=K \\ k_1\geq 1, k_2\geq 1}} \widetilde{w}_{\mathbf{1}+k_1\mathbf{e}_1} \widetilde{w}_{k_2\mathbf{e}_1} \frac{\partial^2 F(z,w_\mathbf{0})}{\partial w^2} \\ &\quad + z^3 \left(\widetilde{w}_\mathbf{1} \widetilde{w}_{K\mathbf{e}_1} + \widetilde{w}_{(1+K)\mathbf{e}_1} \widetilde{w}_{\mathbf{e}_2}\right) \frac{\partial^2 F(z,w_\mathbf{0})}{\partial w^2}\end{aligned} \tag{5.58}$$

With the above reasoning, one gets a unique solution $\widetilde{w}_{\mathbf{1}+K\mathbf{e}_1} = \widetilde{w}^{[1]}_{\mathbf{1}+K\mathbf{e}_1} \log(z) + \widetilde{w}^{[0]}_{\mathbf{1}+K\mathbf{e}_1} \in \bigoplus_{l=0}^{1} \log^l(z)\mathbb{C}[[z^{-1}]]$ where $\widetilde{w}^{[1]}_{\mathbf{1}+K\mathbf{e}_1}$ solves the ODE

$$\begin{aligned}\mathfrak{D}_{K\mathbf{e}_1}(z^3 \widetilde{w}^{[1]}_{\mathbf{1}+K\mathbf{e}_1}) &= \varkappa_1 \sum_{\substack{k_1+k_2=K \\ k_1\geq 1, k_2\geq 1}} k_1 \widetilde{w}_{k_1\mathbf{e}_1} \widetilde{w}_{k_2\mathbf{e}_1} \frac{\partial^2 F(z,w_\mathbf{0})}{\partial w^2} \\ &= K\varkappa_1 \sum_{\substack{k_1+k_2=K \\ k_1\geq 1, k_2\geq 1}} \frac{\widetilde{w}_{k_1\mathbf{e}_1} \widetilde{w}_{k_2\mathbf{e}_1}}{2} \frac{\partial^2 F(z,w_\mathbf{0})}{\partial w^2}\end{aligned}$$

Comparing to (5.48), one concludes that

$$\begin{aligned}\widetilde{w}_{\mathbf{1}+K\mathbf{e}_1} &= \widetilde{w}^{[1]}_{\mathbf{1}+K\mathbf{e}_1} \log(z) + \widetilde{w}^{[0]}_{\mathbf{1}+K\mathbf{e}_1}, \\ \widetilde{w}^{[1]}_{\mathbf{1}+K\mathbf{e}_1} &= K\varkappa_1 z^{-3} \widetilde{w}_{K\mathbf{e}_1}, \qquad \text{val } \widetilde{w}^{[0]}_{\mathbf{1}+K\mathbf{e}_1} = 2(K+1).\end{aligned}$$

Case $\mathbf{k}=(2,2)$ What remains to do when $\mathbf{k}\in\Xi_{2,0}\setminus\Xi_{1,0}$ is to examine the case $\mathbf{k}=(2,2)$. By (5.37) and lemma 5.9,

$$\begin{aligned}&\mathfrak{D}_{\mathbf{1}}(z^3 w_2)=\\&z^3\left(\widetilde{w}_{\mathbf{1}+\mathbf{e}_1}\widetilde{w}_{\mathbf{e}_2}+\widetilde{w}_{\mathbf{1}+\mathbf{e}_2}\widetilde{w}_{\mathbf{e}_1}+\widetilde{w}_{2\mathbf{e}_1}\widetilde{w}_{2\mathbf{e}_2}+\tfrac{1}{2}\widetilde{w}_{\mathbf{1}}\widetilde{w}_{\mathbf{1}}\right)\frac{\partial^2 F(z,\widetilde{w}_{\mathbf{0}})}{\partial w^2}.\end{aligned}\tag{5.59}$$

We observe from (5.55) and (5.56) that

$$\widetilde{w}^{[1]}_{\mathbf{1}+\mathbf{e}_1}\widetilde{w}_{\mathbf{e}_2}+\widetilde{w}^{[1]}_{\mathbf{1}+\mathbf{e}_2}\widetilde{w}_{\mathbf{e}_1}=\varkappa_1 z^{-3}\widetilde{w}_{\mathbf{e}_1}\widetilde{w}_{\mathbf{e}_2}+\varkappa_2 z^{-3}\widetilde{w}_{\mathbf{e}_2}\widetilde{w}_{\mathbf{e}_1}=0.$$

Therefore the log-term disappears in the right-hand side of (5.59) as a consequence of the symmetries of the problem. Moreover

$$\mathrm{val}\,(\widetilde{w}^{[0]}_{\mathbf{1}+\mathbf{e}_1}\widetilde{w}_{\mathbf{e}_2}+\widetilde{w}^{[0]}_{\mathbf{1}+\mathbf{e}_2}\widetilde{w}_{\mathbf{e}_1}+\widetilde{w}_{2\mathbf{e}_1}\widetilde{w}_{2\mathbf{e}_2}+\frac{1}{2}\widetilde{w}_{\mathbf{1}}\widetilde{w}_{\mathbf{1}})\geq 4.$$

By lemma 5.10, we get $\widetilde{w}_2\in\mathbb{C}[[z^{-1}]]$ with val $\widetilde{w}_2=6$. Explicit calculation provides: $\widetilde{w}_2(z)=-\dfrac{5}{6}\dfrac{1}{z^6}-\dfrac{2177}{432}\dfrac{1}{z^8}-\dfrac{5288521}{54000}\dfrac{1}{z^{10}}+\cdots.$

Induction We assume that N is an integer ≥ 2 and we suppose that the properties announced in theorem 5.1 are true for any integer $n\in[0,N-1]$ and any $\mathbf{k}\in\Xi_{n+1,0}\setminus\Xi_{n,0}$.

We notice on the one hand that $\Xi_{N+1,0}\setminus\Xi_{N,0}=\mathbf{1}+\Xi_{N,0}\setminus\Xi_{N-1,0}$. On the other hand, for every $\mathbf{k}\in\Xi_{N,0}\setminus\Xi_{N-1,0}$,

$$\mathfrak{D}_{\mathbf{1}+\mathbf{k}}(\widetilde{w}_{\mathbf{1}+\mathbf{k}})=\sum_{\substack{\mathbf{k}_1+\mathbf{k}_2=\mathbf{1}+\mathbf{k}\\|\mathbf{k}_1|\geq 1,|\mathbf{k}_2|\geq 1}}\frac{\widetilde{w}_{\mathbf{k}_1}\widetilde{w}_{\mathbf{k}_2}}{2}\frac{\partial^2 F(z,w_{\mathbf{0}})}{\partial w^2}\tag{5.60}$$

We set $X=\log(z)$ and we consider X as an indeterminate. The right-hand side of (5.60) is of the form $\widetilde{f}=\sum\widetilde{f}^{[l]}X^l$ with

$$\begin{aligned}\partial_X\widetilde{f}&=\partial_X\sum_{\substack{\mathbf{k}_1+\mathbf{k}_2=\mathbf{1}+\mathbf{k}\\|\mathbf{k}_1|\geq 1,|\mathbf{k}_2|\geq 1}}\frac{\widetilde{w}_{\mathbf{k}_1}\widetilde{w}_{\mathbf{k}_2}}{2}\frac{\partial^2 F(z,w_{\mathbf{0}})}{\partial w^2}\\&=\sum_{\substack{\mathbf{k}_1+\mathbf{k}_2=\mathbf{1}+\mathbf{k}\\|\mathbf{k}_1|\geq 1,|\mathbf{k}_2|\geq 1}}\frac{(\partial_X\widetilde{w}_{\mathbf{k}_1})\widetilde{w}_{\mathbf{k}_2}+\widetilde{w}_{\mathbf{k}_1}(\partial_X\widetilde{w}_{\mathbf{k}_2})}{2}\frac{\partial^2 F(z,w_{\mathbf{0}})}{\partial w^2}.\end{aligned}$$

Using the induction hypothesis, when $\mathbf{1}+\mathbf{k}_1\in\Xi_{n+1,0}\setminus\Xi_{n,0}$, for any $n\in[0,N-1]$,

$$\partial_X\left(\sum_{l=1}^{n}\widetilde{w}^{[l]}_{\mathbf{1}+\mathbf{k}_1}X^l\right)=(\boldsymbol{\varkappa}.\mathbf{k}_1)z^{-3}\sum_{l=0}^{n-1}\widetilde{w}^{[l]}_{\mathbf{k}_1}X^l,$$

that is $\partial_X\widetilde{w}_{\mathbf{1}+\mathbf{k}_1}=(\boldsymbol{\varkappa}.\mathbf{k}_1)z^{-3}\widetilde{w}_{\mathbf{k}_1}$. Therefore:

$$\begin{aligned}\partial_X \widetilde{f} &= z^{-3} \sum_{\substack{\mathbf{k}_1+\mathbf{k}_2=\mathbf{k}\\ |\mathbf{k}_1|\geq 1, |\mathbf{k}_2|\geq 1}} (\boldsymbol{\varkappa}.\mathbf{k}_1)\widetilde{w}_{\mathbf{k}_1}\widetilde{w}_{\mathbf{k}_2}\frac{\partial^2 F(z,w_{\mathbf{0}})}{\partial w^2}\\ &= (\boldsymbol{\varkappa}.\mathbf{k})z^{-3} \sum_{\substack{\mathbf{k}_1+\mathbf{k}_2=\mathbf{k}\\ |\mathbf{k}_1|\geq 1, |\mathbf{k}_2|\geq 1}} \frac{\widetilde{w}_{\mathbf{k}_1}\widetilde{w}_{\mathbf{k}_2}}{2}\frac{\partial^2 F(z,w_{\mathbf{0}})}{\partial w^2}.\end{aligned}$$

Thus $\partial_X \widetilde{f} = (\boldsymbol{\varkappa}.\mathbf{k})z^{-3}\mathfrak{D}_{\mathbf{k}}(\widetilde{w}_{\mathbf{k}})$ and (5.60) provides:

$$\partial_X\left(\mathfrak{D}_{\mathbf{k}}(z^3\widetilde{w}_{\mathbf{1}+\mathbf{k}})\right) = (\boldsymbol{\varkappa}.\mathbf{k})\mathfrak{D}_{\mathbf{k}}(\widetilde{w}_{\mathbf{k}}).$$

Observing that $\partial_X\mathfrak{D}_{\mathbf{k}}\partial_X^{-1} = \mathfrak{D}_{\mathbf{k}}$, one easily gets $\widetilde{w}_{\mathbf{1}+\mathbf{k}}$ either from lemma 5.11 or lemma 5.10, with $\widetilde{w}_{\mathbf{1}+\mathbf{k}} = (\boldsymbol{\varkappa}.\mathbf{k})z^{-3}\partial_X^{-1}\widetilde{w}_{\mathbf{k}}$.
The property for $\widetilde{w}_{\mathbf{n}+\mathbf{1}}$ is easy and is left to the reader. This ends the proof of theorem 5.1. □

Definition 5.9. The two-parameter formal solution defined by theorem 5.1 is the *formal integral* of the prepared ODE (3.6) associated with the first Painlevé equation. The coefficients λ_i, τ_i and $\varkappa_i$, $i = 1,2$, are the *formal invariants*.
The formal series $\widetilde{w}_{\mathbf{k}}^{[0]}$ are called the **k**-*th series* of the formal integral. We set $\widetilde{W}_{\mathbf{k}}^{[0]} = z^{-\boldsymbol{\tau}.\mathbf{k}}\widetilde{w}_{\mathbf{k}}^{[0]}$ and $\widetilde{W}_{\mathbf{k}} = z^{-\boldsymbol{\tau}.\mathbf{k}}\widetilde{w}_{\mathbf{k}}$ for any $\mathbf{k}\in\mathbb{N}^2$.

Remark 5.8. Theorem 5.1 can be compared to [GIKM012] and specially to [ASV012], where the calculations made there translate into ours up to renormalization. We also mention obvious links between Theorem 5.1 and the instanton-type solutions of Kawai *et al* [KT96, AKT-96].

Definition 5.10. For any $\mathbf{k}\in\mathbb{N}^2$, one denotes by $\mathfrak{E}_{\mathbf{k}}$ and $\mathfrak{F}_{\mathbf{k}}$ the following operators:

$$\begin{aligned}\mathfrak{E}_{\mathbf{k}} &= \frac{\boldsymbol{\varkappa}.\mathbf{k}}{z^4}P'(\partial-\boldsymbol{\lambda}.\mathbf{k}) + \frac{\boldsymbol{\varkappa}.\mathbf{k}}{z^5}\left(Q'(\partial-\boldsymbol{\lambda}.\mathbf{k}) - \frac{\tau.(2\mathbf{k}-\mathbf{1})+1}{2!}P''(\partial-\boldsymbol{\lambda}.\mathbf{k})\right)\\ &= 2\frac{\boldsymbol{\varkappa}.\mathbf{k}}{z^4}(\partial-\boldsymbol{\lambda}.\mathbf{k}) - \frac{\boldsymbol{\varkappa}.\mathbf{k}}{z^5}\left(\tau.(2\mathbf{k}-\mathbf{1})+4\right),\\ \mathfrak{F}_{\mathbf{k}} &= \frac{1}{2!}\frac{(\boldsymbol{\varkappa}.\mathbf{k})^2}{z^8}P''(\partial-\boldsymbol{\lambda}.\mathbf{k}) = \frac{(\boldsymbol{\varkappa}.\mathbf{k})^2}{z^8}.\end{aligned}$$

We need hardly mention the analogue of lemma 5.9.

Lemma 5.12. *For every* $n\in\mathbb{N}$, $\mathbf{k}\in\mathbb{N}^2$,

$$\mathfrak{E}_{\mathbf{n}+\mathbf{k}} = z^{\boldsymbol{\tau}.\mathbf{n}}\mathfrak{E}_{\mathbf{k}}z^{-\boldsymbol{\tau}.\mathbf{n}}, \qquad \mathfrak{F}_{\mathbf{n}+\mathbf{k}} = z^{\boldsymbol{\tau}.\mathbf{n}}\mathfrak{F}_{\mathbf{k}}z^{-\boldsymbol{\tau}.\mathbf{n}}.$$

We finally give a corollary stemming from theorem 5.1.

Corollary 5.1. *The formal integral (5.50) associated with the prepared ODE (3.6) can be written under the form:*

$$\widetilde{w}(z,\mathbf{U}) = \sum_{\mathbf{k}\in\mathbb{N}^2} \mathbf{V}^{\mathbf{k}} \widetilde{w}_{\mathbf{k}}^{[0]}, \qquad \mathbf{V}^{\mathbf{k}} = \mathbf{U}^{\mathbf{k}} \mathrm{e}^{-(\boldsymbol{\lambda}.\mathbf{k})z+(\boldsymbol{\varkappa}.\mathbf{k})\mathbf{U}^{\mathbf{1}}\log(z)} z^{-\boldsymbol{\tau}.\mathbf{k}}. \tag{5.61}$$

Equivalently, $\widetilde{w}(z,\mathbf{U}) = \widetilde{\Phi}(z, U_1 \mathrm{e}^{-\lambda_1 z - (\tau_1 - \varkappa_1 \mathbf{U}^{\mathbf{1}})\log(z)}, U_2 \mathrm{e}^{-\lambda_2 z - (\tau_2 - \varkappa_2 \mathbf{U}^{\mathbf{1}})\log(z)})$ *where* $\widetilde{\Phi}(z,\mathbf{u}) = \sum_{\mathbf{k}\in\mathbb{N}^2} \mathbf{u}^{\mathbf{k}} \widetilde{w}_{\mathbf{k}}^{[0]}(z) \in \mathbb{C}[[z^{-1},\mathbf{u}]]$ *is solution of the equation:*

$$P\Big(\partial_z - \sum_{i=1}^{2}(\lambda_i + \frac{\tau_i - \varkappa_i \mathbf{u}^{\mathbf{1}}}{z}) u_i \partial_{u_i}\Big)\widetilde{\Phi} + \frac{1}{z} Q\Big(\partial_z - \sum_{i=1}^{2}(\lambda_i + \frac{\tau_i - \varkappa_i \mathbf{u}^{\mathbf{1}}}{z}) u_i \partial_{u_i}\Big)\widetilde{\Phi} = F(z,\widetilde{\Phi}). \tag{5.62}$$

The formal series $\widetilde{w}_{\mathbf{k}}^{[0]} \in z^{-2|\mathbf{k}|+2}\mathbb{R}[[z^{-1}]]$ *satisfy:*

- *for any* $\mathbf{k} \in \Xi_{1,0} \setminus \Xi_{0,0}$, $\mathfrak{D}_{\mathbf{k}} \widetilde{w}_{\mathbf{k}}^{[0]} = \sum_{\substack{\mathbf{k}_1+\mathbf{k}_2=\mathbf{k} \\ |\mathbf{k}_i|\geq 1}} \frac{w_{\mathbf{k}_1}^{[0]} w_{\mathbf{k}_2}^{[0]}}{2!} \frac{\partial^2 F(z,\widetilde{w}_{\mathbf{0}})}{\partial w^2}$;
- *for any* $\mathbf{k} \in \Xi_{2,0} \setminus \Xi_{1,0}$, $\mathfrak{D}_{\mathbf{k}} \widetilde{w}_{\mathbf{k}}^{[0]} + \mathfrak{E}_{\mathbf{k}} \widetilde{w}_{\mathbf{k}-\mathbf{1}}^{[0]} = \sum_{\substack{\mathbf{k}_1+\mathbf{k}_2=\mathbf{k} \\ |\mathbf{k}_i|\geq 1}} \frac{w_{\mathbf{k}_1}^{[0]} w_{\mathbf{k}_2}^{[0]}}{2!} \frac{\partial^2 F(z,\widetilde{w}_{\mathbf{0}})}{\partial w^2}$;
- *otherwise,* $\mathfrak{D}_{\mathbf{k}} \widetilde{w}_{\mathbf{k}}^{[0]} + \mathfrak{E}_{\mathbf{k}} \widetilde{w}_{\mathbf{k}-\mathbf{1}}^{[0]} + \mathfrak{F}_{\mathbf{k}} \widetilde{w}_{\mathbf{k}-\mathbf{2}}^{[0]} = \sum_{\substack{\mathbf{k}_1+\mathbf{k}_2=\mathbf{k} \\ |\mathbf{k}_i|\geq 1}} \frac{w_{\mathbf{k}_1}^{[0]} w_{\mathbf{k}_2}^{[0]}}{2!} \frac{\partial^2 F(z,\widetilde{w}_{\mathbf{0}})}{\partial w^2}$;

Proof. Let us examine (5.50) more closely. The formal integral can be written as follows:

$$\widetilde{w}(z,\mathbf{U}) = \sum_{n=0}^{\infty} \mathbf{V}^{\mathbf{n}} \widetilde{w}_{\mathbf{n}}(z) + \sum_{i=1,2}\sum_{k=1}^{\infty}\sum_{n=0}^{\infty} \mathbf{V}^{\mathbf{n}+k\mathbf{e}_i} \widetilde{w}_{\mathbf{n}+k\mathbf{e}_i}(z), \tag{5.63}$$

that is we consider the sums along the direction given by the vector $(1,1)$ that determines the resonance. We set $\mathbf{T}^{\mathbf{k}} = \mathbf{U}^{\mathbf{k}} \mathrm{e}^{-(\boldsymbol{\lambda}.\mathbf{k})z+(\boldsymbol{\varkappa}.\mathbf{k})\mathbf{U}^{\mathbf{1}}\log(z)} z^{-\boldsymbol{\tau}.\mathbf{k}}$.
For the first sum we know that each $\widetilde{w}_{\mathbf{n}}(z)$ belongs to $\mathbb{C}[[z^{-1}]]$ and $\sum_{n=0}^{\infty} \mathbf{V}^{\mathbf{n}} \widetilde{w}_{\mathbf{n}} = \sum_{n=0}^{\infty} \mathbf{T}^{\mathbf{n}} \widetilde{w}_{\mathbf{n}}$ because $\boldsymbol{\varkappa}.\mathbf{n} = 0$.
We now look at the other sums and we use the relations given by (5.51). We get for $i = 1,2$,

$$\begin{aligned}
\sum_{k=1}^{\infty}\sum_{n=0}^{\infty} \mathbf{V}^{\mathbf{n}+k\mathbf{e}_i} \widetilde{w}_{\mathbf{n}+k\mathbf{e}_i} &= \sum_{k=1}^{\infty} \mathbf{V}^{k\mathbf{e}_i} \sum_{n=0}^{\infty} \mathbf{V}^{\mathbf{n}} \sum_{l=0}^{n} \frac{1}{l!}\left(\varkappa_i k z^{-3}\log(z)\right)^l \widetilde{w}_{\mathbf{n}-\mathbf{l}+k\mathbf{e}_i}^{[0]} \\
&= \sum_{n=0}^{\infty} \mathbf{V}^{\mathbf{n}} \sum_{k=1}^{\infty} \mathbf{V}^{k\mathbf{e}_i} \mathrm{e}^{\left(\varkappa_i k \mathbf{U}^{\mathbf{1}}\log(z)\right)} \widetilde{w}_{\mathbf{n}+k\mathbf{e}_i}^{[0]}. \\
&= \sum_{n=0}^{\infty}\sum_{k=1}^{\infty} \mathbf{T}^{\mathbf{n}+k\mathbf{e}_i} \widetilde{w}_{\mathbf{n}+k\mathbf{e}_i}^{[0]}.
\end{aligned}$$

The equation (5.62) is obtained by the arguments developed in remark 5.3. The reader will check that equation (5.62) is equivalent to the given hierarchy of equations. □

Let us write $u_1(z) = U_1 \mathrm{e}^{-\lambda_1 z - (\tau_1 - \varkappa_1 \mathbf{U}^{\mathbf{1}})\log(z)}$, $u_2(z) = U_2 \mathrm{e}^{-\lambda_2 z - (\tau_2 - \varkappa_2 \mathbf{U}^{\mathbf{1}})\log(z)}$ and observe that ${}^t(u_1, u_2)$ provides the general analytic solution for a non linear differential equation that only depends on the formal invariants:

$$\partial \begin{pmatrix} u_1 \\ u_2 \end{pmatrix} + \begin{pmatrix} \lambda_1 + \frac{\tau_1}{z} & 0 \\ 0 & \lambda_2 + \frac{\tau_2}{z} \end{pmatrix} \begin{pmatrix} u_1 \\ u_2 \end{pmatrix} = \begin{pmatrix} \frac{\varkappa_1}{z^4} u_1 u_2 & 0 \\ 0 & \frac{\varkappa_2}{z^4} u_1 u_2 \end{pmatrix} \begin{pmatrix} u_1 \\ u_2 \end{pmatrix}. \tag{5.64}$$

This means that corollary 5.1 can be written in term of formal classification and of (canonical) normal form:

Corollary 5.2. *There exists a formal transformation* $w = \widetilde{\Phi}(z, \mathbf{u})$ *of the form*

$$\widetilde{\Phi}(z, \mathbf{u}) = \sum_{\mathbf{k} \in \mathbb{N}^2} \mathbf{u}^{\mathbf{k}} \widetilde{w}_{\mathbf{k}}^{[0]}(z), \qquad \widetilde{w}_{\mathbf{k}}^{[0]} \in \mathbb{C}[[z^{-1}]], \tag{5.65}$$

that formally transforms the prepared ODE (3.6) into the normal form *equation:*

$$\partial \mathbf{u} + B_0(z)\mathbf{u} = B_1(z, \mathbf{u})\mathbf{u} \tag{5.66}$$

$$B_0 = \begin{pmatrix} \lambda_1 + \frac{\tau_1}{z} & 0 \\ 0 & \lambda_2 + \frac{\tau_2}{z} \end{pmatrix}, \quad B_1(z, \mathbf{u}) = \frac{\mathbf{u}^{\mathbf{1}}}{z^4} \begin{pmatrix} \varkappa_1 & 0 \\ 0 & \varkappa_2 \end{pmatrix}, \quad \mathbf{u}^{\mathbf{1}} = u_1 u_2.$$

5.5 Comments

Analogues of proposition 5.1 can be stated for differential equations, *resp.* difference equations, of order 1 and dimension n, with one level and no resonance, given in prepared form :

$$\partial \mathbf{v} + B_0(z)\mathbf{v} = \mathbf{g}(z, \mathbf{v}) \tag{5.67}$$

with $B_0(z) = \bigoplus_j (\lambda_j I_{n_j} + z^{-1} M_j)$, $\sum_j n_j = n$, *resp.*

$$\mathbf{v}(z+1) = B_0(z)\mathbf{v}(z) + \mathbf{g}(z, \mathbf{v}) \tag{5.68}$$

with $B_0(z) = \bigoplus_j \mathrm{e}^{-\lambda_j z}(1 + z^{-1})^{M_j}$. In each case, there exists a formal transformation of the type $\mathbf{v} = \widetilde{T}(z, \mathbf{u})$, $\widetilde{T}(z, \mathbf{u}) = \sum_{\mathbf{k} \in \mathbb{N}^n} \mathbf{u}^{\mathbf{k}} \widetilde{\mathbf{v}}_{\mathbf{k}}(z)$, $\widetilde{\mathbf{v}}_{\mathbf{k}}(z) \in \mathbb{C}^n[[z^{-1}]]$ that brings the equation to the linear normal form $\partial \mathbf{u} + B_0(z)\mathbf{u} = 0$, *resp.* $\mathbf{u}(z+1) = B_0(z)\mathbf{u}(z)$.

To be correct, the upshot for difference equations is more subtle.

This property is still valid for differential equations with more than one level, see [Kui003, Bra92, Cos009] and references therein. In particular, the whole set of formal invariants is already given by the linear part (in Jordan form) of the equation.

When resonances occur and as we saw with the first Painlevé equation, the normal form equation is nonlinear and incorporates new formal invariants. This is essentially a consequence of the Poincaré-Dulac theorem [Arn83]; for instance in (5.66), one recognizes the effect of the positively resonance of order 3 with the resonance monomials $u_1^2 u_2$ and $u_1 u_2^2$. The classification is detailed in [Eca85], see also [Eca92-2] where the notion of (so-called) moulds and arborification are used (a good introduction of which is [Sau009]).

Acknowledgements I am indebted to my student Julie Belpaume for helping me to working out this chapter. I thank Jean Ecalle for interesting discussions on phenomena induced by resonances.

References

AKT-96. T. Aoki, T. Kawai, Y. Takei, *WKB analysis of Painlevé transcendents with a large parameter. II. Multiple-scale analysis of Painlevé transcendents.* Structure of solutions of differential equations (Katata/Kyoto, 1995), 1-49, World Sci. Publ., River Edge, NJ, 1996.

AS015. I. Aniceto, R. Schiappa, *Nonperturbative ambiguities and the reality of resurgent transseries.* Comm. Math. Phys. **335** (2015), no. 1, 183-245.

ASV012. I. Aniceto, R. Schiappa, M. Vonk, *The resurgence of instantons in string theory.* Commun. Number Theory Phys. **6** (2012), no. 2, 339-496.

Arn83. V. I. Arnol'd, *Geometrical methods in the theory of ordinary differential equations.* Grundlehren der Mathematischen Wissenschaften, 250. Springer-Verlag, New York-Berlin, 1983.

Bal000. W. Balser, *Formal power series and linear systems of meromorphic ordinary differential equations.* Universitext. Springer-Verlag, New York, 2000.

BDU013. G. Başar, G. Dunne, M. Ünsal, *Resurgence theory, ghost-instantons, and analytic continuation of path integrals.* J. High Energy Phys. 2013, no. 10, 041, front matter+34 pp.

Bra92. B.L.J. Braaksma, *Multisummability of formal power series solutions of nonlinear meromorphic differential equations.* Ann. Inst. Fourier (Grenoble) **42** (1992), no. 3, 517-540.

CL55. E.A. Coddington, N. Levinson, *Theory of ordinary differential equations.* McGraw-Hill Book Company 1955.

Cos005. O. Costin, *Topological construction of transseries and introduction to generalized Borel summability.* Analyzable functions and applications, 137-175, Contemp. Math., **373**, Amer. Math. Soc., Providence, RI, 2005.

Cos009. O. Costin, *Asymptotics and Borel summability*, Chapman & Hall/CRC Monographs and Surveys in Pure and Applied Mathematics, 141. CRC Press, Boca Raton, FL, 2009.

DDP93. E. Delabaere, H. Dillinger, F. Pham, *Résurgence de Voros et périodes des courbes hyperelliptiques*, Annales de l'Institut Fourier **43** (1993), no. 1, 163-199.

DDP97. E. Delabaere, H. Dillinger, F. Pham, *Exact semi-classical expansions for one dimensional quantum oscillators*, Journal Math. Phys. **38** (1997), 12, 6126-6184.

DP97. E. Delabaere, F. Pham, *Unfolding the quartic oscillator*, Ann. Physics **261** (1997), no. 2, 180-218.

DP98. E. Delabaere, F. Pham, *Eigenvalues of complex Hamiltonians with PT-symmetry*, Phys. Lett. A **250** (1998), no. 1-3, 25-32.

DP99. E. Delabaere, F. Pham, *Resurgent methods in semi-classical asymptotics*, Ann. Inst. Henri Poincaré, Sect. A **71** (1999), no 1, 1-94.

DT000. E. Delabaere, D.T. Trinh, *Spectral analysis of the complex cubic oscillator,* J.Phys. A: Math. Gen. **33** (2000), no. 48, 8771-8796.

Del70. P. Deligne, *Equations différentielles à points singuliers réguliers.* Lecture Notes in Mathematics, Vol. 163. Springer-Verlag, Berlin-New York, 1970.

DU012. G. Dunne, M. Ünsal, *Resurgence and trans-series in quantum field theory: the* $\mathbb{CP}^{N-1}$ *model.* J. High Energy Phys. 2012, no. 11, 170, front matter + 84 pp.

DU014. G. Dunne, M. Ünsal, *Uniform WKB, multi-instantons, and resurgent trans-series.* Phys. Rev. D **89**, 105009 (2014)

Eca85. J. Ecalle, *Les fonctions résurgentes. Tome III : l'équation du pont et la classification analytique des objets locaux.* Publ. Math. d'Orsay, Université Paris-Sud, 1985.05 (1985).

Eca92-1. J. Écalle, *Fonctions analysables et preuve constructive de la conjecture de Dulac.* Actualités mathématiques, Hermann, Paris (1992).

Eca92-2. J. Écalle, *Singularités non abordables par la géométrie.* Ann. Inst. Fourier (Grenoble) **42** (1992), no. 1-2, 73-164.

Get011. A. Getmanenko, *Resurgent analysis of the Witten Laplacian in one dimension.* Funkcial. Ekvac. **54** (2011), no. 3, 383-438.

Get013. A. Getmanenko, *Resurgent analysis of the Witten Laplacian in one dimension-II.* Funkcial. Ekvac. **56** (2013), no. 1, 121-176.

GIKM012. S. Garoufalidis, A. Its, A. Kapaev, M. Mariño, *Asymptotics of the instantons of Painlevé I.* Int. Math. Res. Not. IMRN, 2012, no. 3, 561-606.

HKT015. N. Honda, T. Kawai, Y. Takei, *Virtual turning points.* Springer Briefs in Mathematical Physics, 4. Springer, Tokyo, 2015.

JZ004-1. U. Jentschura, J. Zinn-Justin, *Instantons in quantum mechanics and resurgent expansions.* Phys. Lett. B **596** (2004), no. 1-2, 138-144.

JZ004-2. U. Jentschura, J. Zinn-Justin, *Multi-instantons and exact results. II. Specific cases, higher-order effects, and numerical calculations.* Ann. Physics **313** (2004), no. 2, 269-325.

JZ010. U. Jentschura, J. Zinn-Justin, *Multi-instantons and exact results. III. Unification of even and odd anharmonic oscillators.* Ann. Physics **325** (2010), no. 5, 1135-1172.

KT96. T. Kawai, Y. Takei, *WKB analysis of Painlevé transcendents with a large parameter. I.* Adv. Math. **118** (1996), no. 1, 1-33.

KT011. T. Kawai, Y. Takei, *WKB analysis of higher order Painlevé equations with a large parameter. II. Structure theorem for instanton-type solutions of* $(PJ)_m$ $(J = I, 34, II-2 \text{ or } IV)$ *near a simple P-turning point of the first kind.* Publ. Res. Inst. Math. Sci. **47** (2011), no. 1, 153-219.

Koh99. M. Kohno, *Global analysis in linear differential equations.* Mathematics and its Applications, 471. Kluwer Academic Publishers, Dordrecht, 1999.

Kui003. G.R. Kuik, *Transseries in Difference and Differential Equations.* PhD thesis, Rijksuniversiteit Groningen (2003).

Lod016. M. Loday-Richaud, *Divergent Series, Summability and Resurgence II. Simple and Multiple Summability.* Lecture Notes in Mathematics, **2154**. Springer, Heidelberg, 2016.

MS016. C. Mitschi, D. Sauzin, *Divergent Series, Summability and Resurgence I. Monodromy and Resurgence.* Lecture Notes in Mathematics, **2153**. Springer, Heidelberg, 2016.

MSW009. M. Mariño, R. Schiappa, M. Weiss, *Multi-instantons and multicuts.* J. Math. Phys. **50** (2009), no. 5, 052301, 31 pp.

OSS003. C. Olivé, D. Sauzin, T. Seara, *Resurgence in a Hamilton-Jacobi equation.* Proceedings of the International Conference in Honor of Frédéric Pham (Nice, 2002). Ann. Inst. Fourier (Grenoble) **53** (2003), no. 4, 1185-1235.

Sau95. D. Sauzin, *Résurgence paramétrique et exponentielle petitesse de l'écart des séparatrices du pendule rapidement forcé.* Ann. Inst. Fourier (Grenoble) **45** (1995), no. 2, 453-511.

Sau009. D. Sauzin, *Mould expansions for the saddle-node and resurgence monomials.* In Renormalization and Galois theories, 83-163, IRMA Lect. Math. Theor. Phys., 15, Eur. Math. Soc., Zürich, 2009.

Vor83. A. Voros, *The return of the quartic oscillator: the complex WKB method.* Ann. Inst. H. Poincaré Sect. A (N.S.) **39** (1983), no. 3, 211-338.

Was65. W. Wasow, *Asymptotic expansions for ODE.* Reprint of the 1965 edition. Robert E. Krieger Publishing Co., Huntington, N.Y., 1976.

Zin89. J. Zinn-Justin, *Quantum field theory and critical phenomena.* Oxford Univ. Press (1989).

Chapter 6
Truncated Solutions For The First Painlevé Equation

Abstract In the previous chapters, we studied the unique formal solution of the first Painlevé equation then we introduced its formal integral. In this chapter, we show that formal series components of the formal integral are 1-Gevrey and their minors have analytic properties quite similar to those for the minor of the formal series solution we started with (Sect. 6.1). We then make a focus on the transseries solution and we show their Borel-Laplace summability (Sect. 6.2). This provides the truncated solutions by Borel-Laplace summation (Sect. 6.4).

6.1 Borel-Laplace Summability of the k-th Series and Beyond

We described with theorem 5.1 and its corollary 5.1 the formal integral $\widetilde{w}(z,\mathbf{U}) = \sum_{\mathbf{k}\in\mathbb{N}^2} \mathbf{V}^{\mathbf{k}} \widetilde{w}_{\mathbf{k}}^{[0]}$ associated with the first Painlevé equation. Our goal in this section is mainly to show the following assertion.

Theorem 6.1. *For every $\mathbf{k} \in \mathbb{N}^2$, the* $\mathbf{k}$*-th series $\widetilde{w}_{\mathbf{k}}^{[0]}$ is 1-Gevrey, its minor $\widehat{w}_{\mathbf{k}}^{[0]}$ defines a holomorphic function on $\overset{\bullet}{\mathscr{R}}{}^{(0)}$ with at most exponential growth of order 1 at infinity. Moreover, $\widehat{w}_{\mathbf{k}}^{[0]}$ can be analytically continued to the Riemann surface $\mathscr{R}^{(1)}$, with at most exponential growth of order 1 at infinity on $\mathscr{R}^{(1)}$.*

We already know by theorem 3.3 and theorem 4.2 that $\widehat{w}_{\mathbf{0}} = \widehat{w}_{\mathbf{0}}^{[0]}$ enjoyes the above properties. Our task comes down to studying the other $\mathbf{k}$-th series. This is what we do in what follows and we start with some preliminaries.

6.1.1 Preliminary Results

In what follows we use a notation introduced in definition 5.5.

E. Delabaere, *Divergent Series, Summability and Resurgence III*,
Lecture Notes in Mathematics 2155, DOI 10.1007/978-3-319-29000-3_6

Lemma 6.1. *We set $P(\partial) = \partial^2 - 1$ and for every $\mathbf{k} \in \mathbb{N}^2$, $P_{\mathbf{k}}(\partial) = P(-\boldsymbol{\lambda}.\mathbf{k}+\partial)$ with $\boldsymbol{\lambda} = (\lambda_1, \lambda_2) = (1,-1)$. For $i = 1,2$, we denote by $\tilde{P}_{\mathbf{e}_i}(\partial)$ the operator defined by $P_{\mathbf{e}_i}(\partial) = \tilde{P}_{\mathbf{e}_i}(\partial)\partial$ so that $\tilde{P}_{\mathbf{e}_i}(-\lambda_i) \neq 0$. Then, for any $\rho \in]0,1[$, there exists $M_{\rho,(0)} > 0$ such that, for every $\zeta \in \mathbb{C} \setminus \bigcup_{m \in \mathbb{Z}^\star} \overline{D}(m, m\rho)$:*

1. *for $i = 1,2$, $\left|\dfrac{1}{\tilde{P}_{\mathbf{e}_i}(-\zeta)}\right| \leq M_{\rho,(0)}$;*
2. *for every $\mathbf{k} \in \Xi_{1,0}$ with $|\mathbf{k}| \geq 2$, for $m = 0,1$, $\left|\dfrac{(\zeta+\boldsymbol{\lambda}.\mathbf{k})^m}{P_{\mathbf{k}}(-\zeta)}\right| \leq \dfrac{M_{\rho,(0)}}{|\mathbf{k}|-1}$ and, for $\mathbf{k} \neq (1,1)$, $\left|\dfrac{1}{P_{\mathbf{k}}(-\zeta)}\right| \leq \dfrac{M^2_{\rho,(0)}}{|\mathbf{k}|^2-1}$.*

Moreover one can choose $M_{\rho,(0)} = \frac{1}{\rho}$.

Proof. We only examine the case $\mathbf{k} \in \Xi_{1,0} \setminus \{(1,1)\}$ with $|\mathbf{k}| > 1$. With no loss of generality, we can assume that $\mathbf{k} = (k,0)$ with $k \geq 2$. Thus $P_{\mathbf{k}}(-\zeta) = (\zeta+k-1)(\zeta+k+1)$, $\zeta + \boldsymbol{\lambda}.\mathbf{k} = \zeta + k$ and we notice that $|\zeta+k-1| \geq (k-1)\rho$ and $|\zeta+k+1| \geq (k+1)\rho$ for $\zeta \in \mathbb{C} \setminus \bigcup_{m\in\mathbb{Z}^\star} \overline{D}(m,m\rho)$. Therefore, $\frac{1}{|P_{\mathbf{k}}(-\zeta)|} \leq \frac{1}{(k^2-1)\rho^2}$ for $\zeta \in \mathbb{C}\setminus\bigcup_{m\in\mathbb{Z}^\star} \overline{D}(m,m\rho)$. Now either $\Re(\zeta+k) \geq 0$, thus $|\zeta+k+1| \geq \max\{1,|\zeta+k|\}$ and therefore $\frac{\max\{1,|\zeta+\boldsymbol{\lambda}.\mathbf{k}|\}}{|P_{\mathbf{k}}(-\zeta)|} \leq \frac{1}{(k-1)\rho}$; or else $\Re(\zeta+k) \leq 0$, which implies $|\zeta+k-1| \geq \max\{1,|\zeta+k|\}$ and finally $\frac{\max\{1,|\zeta+\boldsymbol{\lambda}.\mathbf{k}|\}}{|P_{\mathbf{k}}(-\zeta)|} \leq \frac{1}{(k+1)\rho}$. □

Lemma 6.2. *We follow the conditions of lemma 6.1. We set $Q(\partial) = -3\partial$, while $Q_{\mathbf{k}}(\partial)$, $R_{\mathbf{k}}(\partial)$ are given by (5.31), (5.32) with $\tau = \left(-\frac{3}{2}, -\frac{3}{2}\right)$. Then, for every $\mathbf{k} \in \Xi_{1,0} \setminus \{(1,1)\}$ with $|\mathbf{k}| > 1$, for every $\zeta \in \overset{\bullet}{\mathscr{R}}{}^{(0)}_{\rho}$,*

$$\frac{|Q_{\mathbf{k}}|(|\zeta|)}{|P_{\mathbf{k}}(-\zeta)|} \leq 3M_{\rho,(0)}, \qquad \frac{|R_{\mathbf{k}}|(|\zeta|)}{|P_{\mathbf{k}}(-\zeta)|} \leq \frac{9}{4}M^2_{\rho,(0)}.$$

Proof. We notice that lemma 6.1 can be applied for $\zeta \in \overset{\bullet}{\mathscr{R}}{}^{(0)}_{\rho}$.
We have $|Q_{\mathbf{k}}|(\xi) = 3(|\mathbf{k}|-1)\left|\xi + \boldsymbol{\lambda}.\mathbf{k}\right|$ (see (5.39)), Therefore, by lemma 6.1, $\frac{|Q_{\mathbf{k}}|(|\zeta|)}{|P_{\mathbf{k}}(-\zeta)|} \leq 3M_{\rho,(0)}$. In the same way, one easily sees that $|R_{\mathbf{k}}(\partial)| \leq \frac{9}{4}|\mathbf{k}|(|\mathbf{k}|-1)$ (cf. (5.39)), thus the result by lemma 6.1. □

We eventually introduce the following notation that complements definition 3.10.

Definition 6.1. Assume that $G(\zeta,\mathbf{w}) = \sum_{|\mathbf{l}|\geq 0} c_{\mathbf{l}}(\zeta)\mathbf{w}^{\mathbf{l}}$ is an analytic function on the open polydisc $\Delta_{\mathbf{r}} = \prod_{i=0}^{n} D(0,r_i)$. One defines the function $|G|$, analytic on $\Delta_{\mathbf{r}}$, by $|G|(\xi,\mathbf{w}) = \sum_{l\geq 0} |c_l|(\xi)\mathbf{w}^{\mathbf{l}}$.

6.1.2 The $\mathbf{e}_i$-th Series

We start our proof of theorem 6.1 by paying special attention to $\widetilde{w}_{\mathbf{e}_i} = \widetilde{w}_{\mathbf{e}_i}^{[0]}$.

Lemma 6.3. *The $\mathbf{e}_i$-st series $\widetilde{w}_{\mathbf{e}_i}$ is 1-Gevrey. Its formal Borel transform reads $\mathscr{B}(\widetilde{w}_{\mathbf{e}_i}) = \delta + \widehat{w}_{\mathbf{e}_i}$ and $\widehat{w}_{\mathbf{e}_i}$ is holomorphic on $\overset{\bullet}{\mathscr{R}}^{(0)}$ with at most exponential growth of order 1 at infinity. More precisely, for every $\rho \in]0,1[$, there exist $A > 0$ and $\tau > 0$ such that for every $\zeta \in \overset{\bullet}{\mathscr{R}}_{\rho}^{(0)}$, $|\widehat{w}_{\mathbf{e}_i}(\zeta)| \leq A\mathrm{e}^{\tau|\zeta|}$. In the above upper bounds one can choose $A = \tau = \frac{5.81}{\rho}$. Moreover, $\widehat{w}_{\mathbf{e}_i}$ can be analytically continued to the Riemann surface $\mathscr{R}^{(1)}$, with at most exponential growth of order 1 at infinity on $\mathscr{R}^{(1)}$.*

Proof. It is enough to study $\widetilde{w}_{\mathbf{e}_1}$ since $\widetilde{w}_{\mathbf{e}_2}(z) = \widetilde{w}_{\mathbf{e}_1}(-z)$. We know that $\widetilde{w}_{\mathbf{e}_1}$ solves (5.45), namely:

$$\partial \tilde{P}_{\mathbf{e}_1}(\partial)\widetilde{w}_{\mathbf{e}_1} = \left(\frac{15}{4}\frac{1}{z^2} + \frac{\partial F(z,\widetilde{w}_{\mathbf{0}})}{\partial w}\right)\widetilde{w}_{\mathbf{e}_1}, \qquad \tilde{P}_{\mathbf{e}_i} = \partial - 2. \tag{6.1}$$

The formal Borel transform of $\widetilde{w}_{\mathbf{e}_1}$ reads $\mathscr{B}(\widetilde{w}_{\mathbf{e}_1}) = \delta + \widehat{w}_{\mathbf{e}_1}$ where the minor $\widehat{w}_{\mathbf{e}_1}(\zeta) \in \mathbb{C}[[\zeta]]$ satisfies the following convolution equation, deduced from (6.1):

$$\widehat{\partial}\tilde{P}_{\mathbf{e}_1}(\widehat{\partial})\widehat{w}_{\mathbf{e}_1} = \left(\frac{15}{4}\zeta + \frac{\partial \widehat{F}(\zeta,\widehat{w}_{\mathbf{0}})}{\partial w}\right) * (\delta + \widehat{w}_{\mathbf{e}_1}). \tag{6.2}$$

In this equation, we use the notation:

$$\frac{\partial \widehat{F}(\zeta,\widehat{w}_{\mathbf{0}})}{\partial w} = \widehat{f}_1(\zeta) + 2\widehat{f}_2 * \widehat{w}_{\mathbf{0}}(\zeta) = -4\zeta + \zeta * \widehat{w}_{\mathbf{0}}(\zeta). \tag{6.3}$$

Equation (6.2) can be thought of as a linear differential equation with a *regular singular* point at 0.

Instead of (6.2), consider the convolution equation $\widehat{\partial}\tilde{P}_{\mathbf{e}_1}(\widehat{\partial})\widehat{w} = \left(a_1\zeta + a_2\frac{\zeta^2}{2!}\right) * (\delta + \widehat{w})$. Set $\widehat{g} = \widehat{\partial}\tilde{P}_{\mathbf{e}_1}(\widehat{\partial})\widehat{w} = \zeta(\zeta+2)\widehat{w}$. For $\zeta \neq 0$, one gets $\widehat{g} = \left(a_1\zeta + a_2\frac{\zeta^2}{2!}\right) * \left(\delta + \frac{\widehat{g}}{\zeta(\zeta+2)}\right)$. This implies by differentiation that $\widehat{g}^{(4)} = a_1\left(\frac{\widehat{g}}{\zeta(\zeta+2)}\right)^{(2)} + a_2\left(\frac{\widehat{g}}{\zeta(\zeta+2)}\right)^{(1)}$ where $\widehat{g}^{(i)} = \frac{\mathrm{d}^i\widehat{g}}{\mathrm{d}\zeta^i}$. The last ODE has a regular singular point at 0. One can apply the same trick to (6.2) but for the fact of getting an infinite order differential operator.

Equation (6.2) can be analyzed with the tools developed in Sect. 3.3.2. We introduce $\widehat{G}(\zeta) = \frac{15}{4}\zeta + \frac{\partial \widehat{F}(\zeta,\widehat{w}_{\mathbf{0}})}{\partial w} = -\frac{\zeta}{4} + \zeta * \widehat{w}_{\mathbf{0}}(\zeta)$ and we remark that $\widehat{G}$ belongs to the maximal ideal $\mathscr{M}\mathscr{O}(\overset{\bullet}{\mathscr{R}}_{\rho}^{(0)})$ of $\mathscr{O}(\overset{\bullet}{\mathscr{R}}_{\rho}^{(0)})$ for any $\rho \in]0,1[$, thus $\widehat{\partial}^{-1}\widehat{G} \in \mathscr{O}(\overset{\bullet}{\mathscr{R}}_{\rho}^{(0)})$ is well-defined. We set $\widehat{w}_{\mathbf{e}_1} = \tilde{P}_{\mathbf{e}_1}^{-1}(\widehat{\partial})\widehat{\partial}^{-1}\widehat{G} + \widehat{v}_{\mathbf{e}_1}$ and (6.2) becomes

$$\widehat{\partial}\tilde{P}_{\mathbf{e}_1}(\widehat{\partial})\widehat{v}_{\mathbf{e}_1} = \widehat{G} * \left(\tilde{P}_{\mathbf{e}_1}^{-1}(\widehat{\partial})\widehat{\partial}^{-1}\widehat{G}\right) + \widehat{G} * \widehat{v}_{\mathbf{e}_1}. \tag{6.4}$$

Observe that $\widehat{G} * \left(\tilde{P}_{\mathbf{e}_1}^{-1}(\widehat{\partial})\widehat{\partial}^{-1}\widehat{G}\right)$ belongs to $\mathscr{M}\mathscr{O}(\overset{\bullet}{\mathscr{R}}_\rho^{(0)})$. Let $R > 0$ be any real positive number, U_R be the star-shaped domain $U_R = D(0,R) \cap \overset{\bullet}{\mathscr{R}}_\rho^{(0)}$ and we set $B_r = \{\widehat{v} \in \mathscr{O}(\overline{U_R}), \|\widehat{v}\|_\nu \leq r\}$, for $r > 0$ and $\nu > 0$. By proposition 3.9 and lemma 6.1, when $\nu \to \infty$,

$$\|\tilde{P}_{\mathbf{e}_1}^{-1}(\widehat{\partial})\widehat{\partial}^{-1}\left(\widehat{G} * \left(\tilde{P}_{\mathbf{e}_1}^{-1}(\widehat{\partial})\widehat{\partial}^{-1}\widehat{G}\right)\right)\|_\nu \to 0.$$

Explicitly

$$\begin{aligned}\|\tilde{P}_{\mathbf{e}_1}^{-1}(\widehat{\partial})\widehat{\partial}^{-1}\left(\widehat{G} * \left(\tilde{P}_{\mathbf{e}_1}^{-1}(\widehat{\partial})\widehat{\partial}^{-1}\widehat{G}\right)\right)\|_\nu &\leq \frac{M_{\rho,(0)}}{R}\|\widehat{\partial}^{-1}\left(\widehat{G} * \left(\tilde{P}_{\mathbf{e}_1}^{-1}(\widehat{\partial})\widehat{\partial}^{-1}\widehat{G}\right)\right)\|_\nu \\ &\leq \frac{M_{\rho,(0)}}{\nu R^2}\|\widehat{\partial}^{-1}\widehat{G}\|_0\|\tilde{P}_{\mathbf{e}_1}^{-1}(\widehat{\partial})\widehat{\partial}^{-1}\widehat{G}\|_\nu.\end{aligned}$$

Also, $\|\tilde{P}_{\mathbf{e}_1}^{-1}(\widehat{\partial})\widehat{\partial}^{-1}\left(\widehat{G} * \widehat{v}_{\mathbf{e}_1}\right)\|_\nu \leq \frac{M_{\rho,(0)}}{\nu R^2}\|\widehat{\partial}^{-1}\widehat{G}\|_0\|\widehat{v}_{\mathbf{e}_1}\|_\nu$, Equation (6.4) thus translates into a fixed point problem $\widehat{v}_{\mathbf{e}_1} = \mathsf{N}(\widehat{v}_{\mathbf{e}_1})$ where $\mathsf{N} : B_r \to B_r$ is a contractive mapping for ν large enough. This ensures the existence and uniquess of $\widehat{w}_{\mathbf{e}_1} \in \mathscr{O}(\overset{\bullet}{\mathscr{R}}^{(0)})$. The same reasoning can be applied for showing that $\widehat{w}_{\mathbf{e}_1}$ can be analytically continued to $\mathscr{R}^{(1)}$, in application of proposition 4.5 and theorem 4.2.

To get upper bounds, we notice by (6.3) and lemma 3.3 that for every $\zeta \in \overset{\bullet}{\mathscr{R}}_\rho^{(0)}$, $\left|\widehat{\partial}^{-1}\widehat{G}(\zeta)\right| \leq \frac{1}{4} + 1 * \widehat{W}_{\mathbf{0}}(|\zeta|)$ where $\widehat{W}_{\mathbf{0}}(\xi) = A\mathrm{e}^{\tau\xi}$ stands for the majorant function of $\widehat{w}_{\mathbf{0}}$ given by theorem 3.3 and corollary 3.1, thus with $A = 4.22$ and $\tau = \frac{4.22}{\rho}$. Viewing the Grönwall-like lemma 3.9, one sees that for every $\zeta \in \overset{\bullet}{\mathscr{R}}_\rho^{(0)}$, $|\widehat{w}_{\mathbf{e}_1}(\zeta)| \leq \widehat{W}_{\mathbf{e}_1}(|\zeta|)$ where $\widehat{W}_{\mathbf{e}_1}$ solves the convolution equation:

$$\frac{1}{M_{\rho,(0)}}\widehat{W}_{\mathbf{e}_1} = \left(\frac{1}{4} + 1 * \widehat{W}_{\mathbf{0}}\right) * (\delta + \widehat{W}_{\mathbf{e}_1}). \tag{6.5}$$

This means that $\widehat{W}_{\mathbf{e}_1}$ has an analytic Laplace transform under the form[1]:

$$\widetilde{W}_{\mathbf{e}_1}(z) = \sum_{n\geq 1}\frac{1}{\rho^n}\left(\frac{1}{4z} + \frac{1}{z}\frac{A}{z-\tau}\right)^n, \quad A = 4.22,\ \tau = \frac{4.22}{\rho}.$$

When assuming $|z| \geq \dfrac{5.81}{\rho}$, for instance, one gets $\left|\dfrac{1}{\rho}\left(\dfrac{1}{4z} + \dfrac{1}{z}\dfrac{A}{z-\tau}\right)\right| \leq 0.5$ (since $\rho < 1$), thus $|\widetilde{W}_{\mathbf{e}_1}(z)| \leq 1$. Therefore by lemma 3.5, for any $0 < \rho < 1$, for every $\zeta \in \overset{\bullet}{\mathscr{R}}_\rho^{(0)}$, $|\widehat{w}_{\mathbf{e}_1}(\zeta)| \leq \frac{5.81}{\rho}\mathrm{e}^{\frac{5.81}{\rho}|\xi|}$. One shows in the same way that $\widehat{w}_{\mathbf{e}_1}$ has at most

[1] We recall that $\mathscr{B}\left(\dfrac{A}{z-\tau}\right) = A\mathrm{e}^{\tau\xi}$.

exponential growth of order 1 at infinity on $\mathscr{R}^{(1)}$, using lemma 4.11 and theorem 4.2.
□

6.1.3 The $k\mathbf{e}_i$-th Series

We now turn to the $k\mathbf{e}_i$-th series, that is the terms $\widetilde{w}_{k\mathbf{e}_i} = \widetilde{w}^{[0]}_{k\mathbf{e}_i}$ of the transseries, for $k \geq 2$.

Lemma 6.4. *For every integer $k \geq 2$, the k-th series $\widetilde{w}_{k\mathbf{e}_i} \in z^{-2(k-1)}\mathbb{C}[[z^{-1}]]$ is 1-Gevrey, its minor $\widehat{w}_{k\mathbf{e}_i}$ defines a holomorphic function on $\overset{\bullet}{\mathscr{R}}{}^{(0)}$ with at most exponential growth of order 1 at infinity. Moreover, $\widehat{w}_{k\mathbf{e}_i}$ can be analytically continued to the Riemann surface $\mathscr{R}^{(1)}$, with at most exponential growth of order 1 at infinity on $\mathscr{R}^{(1)}$.*

Proof. Once again from the invariance of the equation (3.6) under the symmetry $z \mapsto -z$, there is no loss of generality in studying only the $k\mathbf{e}_i$-th series $\widehat{w}_{k\mathbf{e}_1}$.
We know that $\widehat{w}_{\mathbf{0}}, \widehat{w}_{\mathbf{e}_1}$ are holomorphic on $\overset{\bullet}{\mathscr{R}}{}^{(0)}$ and can be analytically continued to $\mathscr{R}^{(1)}$. Moreover, for every $\zeta \in \overset{\bullet}{\mathscr{R}}{}^{(0)}_{\rho}$, $|\widehat{w}_{\mathbf{0}}(\zeta)| \leq \widehat{W}_{\mathbf{0}}(\xi)$, $|\widehat{w}_{\mathbf{e}_1}(\zeta)| \leq \widehat{W}_{\mathbf{e}_1}(\xi)$, $\xi = |\zeta|$ and for every $\zeta \in \overline{\mathscr{R}}{}^{(1)}_{\rho}$, $|\widehat{w}_{\mathbf{0}}(\zeta)| \leq \widehat{W}_{\mathbf{0}}(\xi)$, $|\widehat{w}_{\mathbf{e}_1}(\zeta)| \leq \widehat{W}_{\mathbf{e}_1}(\xi)$, $\xi = \mathrm{leng}_{\rho}(\zeta)$, where $\widehat{W}_{\mathbf{0}}$ and $\widehat{W}_{\mathbf{e}_1}$ are entire functions, real positive and non-decreasing on $\mathbb{R}^+$, with at most exponential growth of order 1 at infinity.
We know from lemma 5.8 and (5.48) that for every $k \geq 2$,

$$\widetilde{w}_{k\mathbf{e}_1}(z) = \sum_{l \geq 0} a_{k\mathbf{e}_1,l} z^{-l} \in z^{-2(k-1)}\mathbb{C}[[z^{-1}]]$$

solves the differential equation

$$\mathfrak{D}_{k\mathbf{e}_1}\widetilde{w}_{k\mathbf{e}_1} = \sum_{\substack{k_1+k_2=k\\ k_1\geq 1, k_2\geq 1}} \frac{\widetilde{w}_{k_1\mathbf{e}_1}\widetilde{w}_{k_2\mathbf{e}_1}}{2!}\frac{\partial^2 F(z,\widetilde{w}_{\mathbf{0}})}{\partial w^2}. \tag{6.6}$$

We deduce that the formal Borel transform $\mathscr{B}(\widetilde{w}_{k\mathbf{e}_1}) = a_{k\mathbf{e}_1,0}\delta + \widehat{w}_{k\mathbf{e}_1}$ has its minor which satisfies the identity[2]:

$$\mathfrak{D}_{k\mathbf{e}_1}\widehat{w}_{k\mathbf{e}_1} = \sum_{\substack{k_1+k_2=k\\ k_1\geq 1, k_2\geq 1}} \frac{(a_{k_1\mathbf{e}_1,0}\delta + \widehat{w}_{k_1\mathbf{e}_1}) * (a_{k_2\mathbf{e}_1,0}\delta + \widehat{w}_{k_2\mathbf{e}_1})}{2!} * \frac{\partial^2 \widehat{F}(\zeta,\widehat{w}_{\mathbf{0}})}{\partial w^2} \tag{6.7}$$

where $\dfrac{\partial^2 \widehat{F}(\zeta,\widehat{w}_{\mathbf{0}})}{\partial w^2} = 2\widehat{f}_2(\zeta) = \zeta$, whereas

[2] Remember that $a_{k\mathbf{e}_1,0} = 0$ as a rule, apart from the case $k = 1$ where $a_{\mathbf{e}_1,0} = 1$.

$$\mathfrak{D}_{k\mathbf{e}_1}\widehat{w}_{k\mathbf{e}_1} = P_{k\mathbf{e}_1}(\widehat{\partial})\widehat{w}_{k\mathbf{e}_1} + 1 * Q_{k\mathbf{e}_1}(\widehat{\partial})\widehat{w}_{k\mathbf{e}_1} + \left(\zeta R_{k\mathbf{e}_1} - \frac{\partial \widehat{F}(\zeta,\widehat{w}_{\mathbf{0}})}{\partial w}\right) * \widehat{w}_{k\mathbf{e}_1} \tag{6.8}$$

with $\dfrac{\partial \widehat{F}(\zeta,\widehat{w}_{\mathbf{0}})}{\partial w}$ given by (6.3).
These equations (6.7) can be seen as linear differential equations with a *regular* point at 0. They are all of the type

$$p(\widehat{\partial})\widehat{w} + 1 * [q(\widehat{\partial})\widehat{w}] = \zeta * [r(\widehat{\partial})\widehat{w}] + \sum_{n=0}^{N} \widehat{f}_n * \widehat{w}^{*n} \tag{6.9}$$

investigated in Sect. 3.3.2 and Sect. 4.5. We use the methods introduced there and make a proof by induction on k, considering the operators N_k defined as follows:

$$\mathsf{N}_k\widehat{v} = P^{-1}_{(k,0)}(\widehat{\partial})\left[-1 * \left[Q_{(k,0)}(\widehat{\partial})\widehat{v}\right] + \left(-\zeta R_{(k,0)} + \frac{\partial \widehat{F}(\zeta,\widehat{w}_{\mathbf{0}})}{\partial w}\right) * \widehat{v}\right.$$
$$\left. + \sum_{\substack{k_1+k_2=K \\ k_1\geq 1, k_2\geq 1}} \frac{(a_{k_1\mathbf{e}_1,0}\delta + \widehat{w}_{k_1\mathbf{e}_1}) * (a_{k_2\mathbf{e}_1,0}\delta + \widehat{w}_{k_2\mathbf{e}_1})}{2!} * \frac{\partial^2 \widehat{F}(\zeta,\widehat{w}_{\mathbf{0}})}{\partial w^2}\right].$$

Case $k=2$ Let $R>0$ be a real positive number, $\rho \in]0,1[$ and U_R be the star-shaped domain $U_R = D(0,R) \cap \overset{\bullet}{\mathscr{R}}{}^{(0)}_{\rho}$. We set $B_r = \{\widehat{v} \in \mathscr{O}(\overline{U_R}), \|\widehat{v}\|_\nu \leq r\}$ for $r>0$ and $\nu > 0$, and we look at the mapping $\mathsf{N}_2 : \widehat{v} \in B_r \mapsto \mathsf{N}_2\widehat{v}$. We know that $\widehat{w}_{(1,0)} \in \mathscr{O}(\overset{\bullet}{\mathscr{R}}{}^{(0)})$ while $\dfrac{\partial \widehat{F}(\zeta,\widehat{w}_{\mathbf{0}})}{\partial w}$ and $\dfrac{\partial^2 \widehat{F}(\zeta,\widehat{w}_{\mathbf{0}})}{\partial w^2}$ belong to $\mathscr{M}\mathscr{O}(\overset{\bullet}{\mathscr{R}}{}^{(0)}_{\rho})$. Using lemma 6.1 and arguments already used in Sect. 3.3.2.3, one easily shows that N_2 is a contractive map. Thus equation (6.7), $k=2$ has a unique solution in B_r. This shows, by uniqueness, that $\widehat{w}_{2\mathbf{e}_1}$ can be continued holomorphically on $\overset{\bullet}{\mathscr{R}}{}^{(0)}$.
When replacing U_R by the open set of L-points $\mathscr{U} = \mathscr{U}_{\rho,L} \subset \mathscr{R}^{(1)}$ and arguing like what have been done for the proof of theorem 4.2, one shows that $\widehat{w}_{2\mathbf{e}_1}$ can be holomorphically continued to the Riemann surface $\mathscr{R}^{(1)}$.

To get upper bounds, we notice that for every $\zeta \in \overset{\bullet}{\mathscr{R}}{}^{(0)}_{\rho}$, $\left|\frac{\partial \widehat{F}(\zeta,\widehat{w}_{\mathbf{0}})}{\partial w}\right| \leq \left|\frac{\partial \widehat{F}}{\partial w}\right|(\xi,\widehat{W}_{\mathbf{0}})$ and $\left|\frac{\partial^2 \widehat{F}(\zeta,\widehat{w}_{\mathbf{0}})}{\partial w^2}\right| \leq \left|\frac{\partial^2 \widehat{F}}{\partial w^2}\right|(\xi,\widehat{W}_{\mathbf{0}})$ with $\xi = |\zeta|$, $\left|\frac{\partial^2 \widehat{F}}{\partial w^2}\right|(\xi,\widehat{W}_{\mathbf{0}}) = 2|f_2|(\xi) = \xi$ and $\left|\frac{\partial \widehat{F}}{\partial w}\right|(\xi,\widehat{W}_{\mathbf{0}}) = |\widehat{f}_1|(\xi) + 2|\widehat{f}_2| * \widehat{W}_{\mathbf{0}}(\xi) = 4\xi + \xi * \widehat{W}_{\mathbf{0}}(\xi)$. Using lemma 6.2 and the Grönwall lemma 3.9, we sees that for every $\zeta \in \overset{\bullet}{\mathscr{R}}{}^{(0)}_{\rho}$, $|\widehat{w}_{2\mathbf{e}_1}(\zeta)| \leq \widehat{W}_{2\mathbf{e}_1}(\xi)$ where $\widehat{W}_{2\mathbf{e}_1}$ is the entire function, real positive on $\mathbb{R}^+$, with at most exponential growth of order 1 at infinity, satisfying the linear equation:

$$\frac{1}{M_{\rho,(0)}}\widehat{W}_{2\mathbf{e}_1} = \left(3+\frac{9}{4}M_{\rho,(0)}\xi + \left|\frac{\partial \widehat{F}}{\partial w}\right|(\xi,\widehat{W}_{\mathbf{0}})\right) * \widehat{W}_{2\mathbf{e}_1} + \frac{(\delta+\widehat{W}_{\mathbf{e}_1})^{*2}}{2!} * \left|\frac{\partial^2 \widehat{F}}{\partial w^2}\right|(\xi,\widehat{W}_{\mathbf{0}}). \tag{6.10}$$

When working on $\mathscr{R}^{(1)}$, one rather argues with the Grönwall lemma 4.11, thus getting $|\widehat{w}_{2\mathbf{e}_1}(\zeta)| \leq \widehat{W}_{2\mathbf{e}_1}(\xi)$ for every $\zeta \in \overline{\mathscr{R}}_\rho^{(1)}$. In these estimates, $\xi = \mathrm{leng}_\rho(\zeta)$, and $\widehat{W}_{2\mathbf{e}_1}$ is the entire function, real positive and non-decreasing on $\mathbb{R}^+$, with at most exponential growth of order 1 at infinity, satisfying the linear equation:

$$\frac{1}{M_{\rho,(1)}}\widehat{W}_{2\mathbf{e}_1} = \left(3+\frac{9}{4}M_{\rho,(1)}\xi + \left|\frac{\partial \widehat{F}}{\partial w}\right|(\xi,\widehat{W}_{\mathbf{0}})\right) * \widehat{W}_{2\mathbf{e}_1} + \frac{(\delta+\widehat{W}_{\mathbf{e}_1})^{*2}}{2!} * \left|\frac{\partial^2 \widehat{F}}{\partial w^2}\right|(\xi,\widehat{W}_{\mathbf{0}}). \tag{6.11}$$

Induction Let $K \geq 3$ be an integer greater than 3. We assume that for every integer $k \in [0,K[$, $\widehat{w}_{k\mathbf{e}_1}$ is holomorphic on $\overset{\bullet}{\mathscr{R}}^{(0)}$ and can be analytically continued to $\mathscr{R}^{(1)}$. Furthermore,

$$\text{for every } \zeta \in \overset{\bullet}{\mathscr{R}}_\rho^{(0)},\ |\widehat{w}_{k\mathbf{e}_1}(\zeta)| \leq \widehat{W}_{k\mathbf{e}_1}(\xi), \qquad \xi = |\zeta|,$$

$$\text{for every } \zeta \in \overline{\mathscr{R}}_\rho^{(1)},\ |\widehat{w}_{k\mathbf{e}_1}(\zeta)| \leq \widehat{W}_{k\mathbf{e}_1}(\xi), \qquad \xi = \mathrm{leng}_\rho(\zeta),$$

where, in each case, $\widehat{W}_{k\mathbf{e}_1}$ is an entire function, real positive and non-decreasing on $\mathbb{R}^+$, with at most exponential growth of order 1 at infinity.

One easily shows that the mapping $\mathsf{N}_K : \widehat{v} \in B_r \mapsto \mathsf{N}_K\widehat{v}$ is a contractive, either working in $(\mathscr{O}(\overline{U_R}), \|.\|_\nu)$ or in $(\mathscr{O}(\overline{\mathscr{U}_{\rho,L}}), \|.\|_\nu)$. Thus, by uniqueness, $\widehat{w}_{K\mathbf{e}_1}$ is holomorphic on $\overset{\bullet}{\mathscr{R}}^{(0)}$ and can by analytically continued to $\mathscr{R}^{(1)}$.

We get upper bounds, either in $\overset{\bullet}{\mathscr{R}}_\rho^{(0)}$ with the Grönwall lemma 3.9, or in $\overline{\mathscr{R}}_\rho^{(1)}$ with the Grönwall lemma 4.11. We get that for every $\zeta \in \overset{\bullet}{\mathscr{R}}_\rho^{(0)}$ $|\widehat{w}_{K\mathbf{e}_1}(\zeta)| \leq \widehat{W}_{K\mathbf{e}_1}(\xi)$ with $\xi = |\zeta|$, where $\widehat{W}_{K\mathbf{e}_1}$ is the entire function, real positive on $\mathbb{R}^+$, with at most exponential growth of order 1 at infinity, satisfying the linear equation:

$$\frac{1}{M_{\rho,(0)}}\widehat{W}_{K\mathbf{e}_1} = \left(3+\frac{9}{4}M_{\rho,(0)}\xi + \left|\frac{\partial \widehat{F}}{\partial w}\right|(\xi,\widehat{W}_{\mathbf{0}})\right) * \widehat{W}_{K\mathbf{e}_1} + \sum_{\substack{k_1+k_2=K \\ k_1\geq 1, k_2\geq 1}} \frac{(a_{k_1\mathbf{e}_1,0}\delta+\widehat{W}_{k_1\mathbf{e}_1}) * (a_{k_2\mathbf{e}_1,0}\delta+\widehat{W}_{k_2\mathbf{e}_1})}{2!} * \left|\frac{\partial^2 \widehat{F}}{\partial w^2}\right|(\xi,\widehat{W}_{\mathbf{0}}). \tag{6.12}$$

Also, for every $\zeta \in \overline{\mathscr{R}}_\rho^{(1)}$, $|\widehat{w}_{K\mathbf{e}_1}(\zeta)| \leq \widehat{W}_{K\mathbf{e}_1}(\xi)$ where $\xi = \text{leng}_\rho(\zeta)$, with $\widehat{W}_{K\mathbf{e}_1}$ an entire function, real positive and nondecreasing on $\mathbb{R}^+$, with at most exponential growth of order 1 at infinity, satisfying the linear equation:

$$\frac{1}{M_{\rho,(1)}}\widehat{W}_{K\mathbf{e}_1} = \left(3+\frac{9}{4}M_{\rho,(1)}\xi + \left|\frac{\partial \widehat{F}}{\partial w}\right|(\xi,\widehat{W}_{\mathbf{0}})\right) * \widehat{W}_{K\mathbf{e}_1}$$
$$+ \sum_{\substack{k_1+k_2=K \\ k_1\geq 1, k_2\geq 1}} \frac{(a_{k_1\mathbf{e}_1,0}\delta + \widehat{W}_{k_1\mathbf{e}_1}) * (a_{k_2\mathbf{e}_1,0}\delta + \widehat{W}_{k_2\mathbf{e}_1})}{2!} * \left|\frac{\partial^2 \widehat{F}}{\partial w^2}\right|(\xi,\widehat{W}_{\mathbf{0}}).$$

This ends the proof of lemma 6.4. □

6.1.4 The Other **k***-th Series*

Looking at (5.53), one easily see that the above methods can be applied to study the minor $\widehat{w}_{\mathbf{1}} = \widehat{w}_{\mathbf{1}}^{[0]}$ of the $(1,1)$-series $\widetilde{w}_{\mathbf{1}}$. Thus, theorem 6.1 is shown for $\mathbf{k} = \mathbf{0}$ any $\mathbf{k} \in \Xi_{n+1,0} \setminus \Xi_{n,0}$ and with $n = 1$. The rest of the proof is made by induction on n, using the hierarchy of equations given in corollary 5.1 and the reasoning made above. This part holds no surprise and is left to the reader. This ends the proof of theorem 6.1.

6.2 Borel-Laplace Summability of the Transseries

We now restrict ourself to the transseries solution of the ODE (3.6), having in view of analyzing their Borel-Laplace summability. From the invariance of the equation (3.6) under the symmetry $z \mapsto -z$, it is enough to only focus on the transseries (5.41) associated with the multiplier $\lambda_1 = 1$, namely:

$$\widetilde{w}(z, U\mathbf{e}_1) = \sum_{k=0}^{\infty} V^k \widetilde{w}_{k\mathbf{e}_1}(z), \qquad V^k = U^k \mathrm{e}^{-\lambda_1 k z} z^{-\tau_1 k}. \tag{6.13}$$

6.2.1 A Useful Supplement

We complete lemma 6.4 with the following result.

Lemma 6.5. *For every $\rho \in]0,1[$, there exist $A = A(\rho) > 0$, $\tau = \tau(\rho) > 0$ and a sequence $\left(\widehat{W}_{k\mathbf{e}_1}\right)_{k\geq 2}$ of entire functions, real positive on $\mathbb{R}^+$, with the following properties:*

- *for every integer* $k \geq 2$, $\widehat{W}_{k\mathbf{e}_1}(\xi) \in \xi^{2k-3}\mathbb{C}\{\xi\}$;
- *for every* $\xi \in \mathbb{C}$, $|\widehat{W}_{k\mathbf{e}_1}(\xi)| \leq \left(\frac{3\sqrt{\rho}}{2}\right)^k A\mathrm{e}^{\tau|\xi|}$, *and for every integer* $m \in [1, 2k-3]$, $|\widehat{W}_{k\mathbf{e}_1}(\xi)| \leq \left(\frac{3\sqrt{\rho}}{2}\right)^k A^{m+1}\left(\frac{\zeta^{m-1}}{(m-1)!} * \mathrm{e}^{\tau\zeta}\right)(|\xi|)$.
- *for every* $\zeta \in \overset{\bullet}{\mathscr{R}}{}^{(0)}_\rho$, $|\widehat{w}_{k\mathbf{e}_1}(\zeta)| \leq \widehat{W}_{k\mathbf{e}_1}(\xi)$ *with* $\xi = |\zeta|$.

Moreover one can choose $A = \tau = \frac{27}{4\rho}$ *in the above estimates.*

Proof. We know by theorem 3.3, lemma 6.3 and lemma 6.4 that, for every integer $k \in \mathbb{N}$, $\widehat{w}_{k\mathbf{e}_1}$ is holomorphic on $\overset{\bullet}{\mathscr{R}}{}^{(0)}$. Also, for every $\rho \in]0,1[$, for every $\zeta \in \overset{\bullet}{\mathscr{R}}{}^{(0)}_\rho$, $|\widehat{w}_{k\mathbf{e}_1}(\zeta)| \leq \widehat{W}_{k\mathbf{e}_1}(\xi)$ with $\xi = |\zeta|$ where $\widehat{W}_{\mathbf{0}}(\xi) = A_{\mathbf{0}}\mathrm{e}^{\tau_{\mathbf{0}}\xi}$ and $\widehat{W}_{\mathbf{e}_1}(\xi) = A_{\mathbf{e}_1}\mathrm{e}^{\tau_{\mathbf{e}_1}\xi}$ are convenient majorant functions while, for any integer $k \geq 2$, $\widehat{W}_{k\mathbf{e}_1}$ solves the convolution equation (6.12). One first shows that for any integer $k \geq 2$, $\widehat{W}_{k\mathbf{e}_1}(\xi)$ belongs to $\xi^{2k-3}\mathbb{C}\{\xi\}$ and we reason by induction: using the fact that $\left|\frac{\partial^2 \widehat{F}}{\partial w^2}\right|(\xi, \widehat{W}_{\mathbf{0}}) = O(\xi)$, one sees that $(\delta + \widehat{W}_{\mathbf{e}_1})^{*2} * \left|\frac{\partial^2 \widehat{F}}{\partial w^2}\right|(\xi, \widehat{W}_{\mathbf{0}}) = O(\xi)$, thus $\widehat{W}_{2\mathbf{e}_1}(\zeta) = O(\zeta)$; then, by an induction hypothesis, we check that integer $k \geq 3$ of the form $k = k_1 + k_2$ with $k_1, k_2 \in \mathbb{N}^\star$, $(a_{k_1\mathbf{e}_1,0}\delta + \widehat{W}_{k_1\mathbf{e}_1}) * (a_{k_2\mathbf{e}_1,0}\delta + \widehat{W}_{k_2\mathbf{e}_1}) = O(\xi^{2k-5})$ (we recall that $a_{k\mathbf{e}_1,0} = 0$ apart from $a_{\mathbf{e}_1,0} = 1$), thus $\widehat{W}_{k\mathbf{e}_1}(\zeta) = O(\xi^{2k-3})$ by (6.12).

We then introduce the generating function $\widehat{W}(\xi, V) = \sum_{k=2}^{\infty} V^k \widehat{W}_{k\mathbf{e}_1}(\xi)$ and we deduce from (6.12) that $\widehat{W}$ satisfies the identity:

$$\frac{1}{M_{\rho,(0)}}\widehat{W} = \left(3 + \frac{9}{4}M_{\rho,(0)}\xi + \left|\frac{\partial \widehat{F}}{\partial w}\right|(\xi, \widehat{W}_{\mathbf{0}})\right) * \widehat{W}$$
$$+ \sum_{k=2}^{\infty} V^k \sum_{\substack{k_1+k_2=k \\ k_1\geq 1, k_2 \geq 1}} \frac{(a_{k_1\mathbf{e}_1,0}\delta + \widehat{W}_{k_1\mathbf{e}_1}) * (a_{k_2\mathbf{e}_1,0}\delta + \widehat{W}_{k_2\mathbf{e}_1})}{2!} * \left|\frac{\partial^2 \widehat{F}}{\partial w^2}\right|(\xi, \widehat{W}_{\mathbf{0}}).$$

This can be written also as follows (remember: $a_{k\mathbf{e}_1,0} = 0$ apart from $a_{\mathbf{e}_1,0} = 1$):

$$\frac{1}{M_{\rho,(0)}}\widehat{W} =$$
$$\left(3 + \frac{9}{4}M_{\rho,(0)}\xi + \left|\frac{\partial \widehat{F}}{\partial w}\right|(\xi, \widehat{W}_{\mathbf{0}})\right) * \widehat{W} + \frac{\left(V(\delta + \widehat{W}_{\mathbf{e}_1}) + \widehat{W}\right)^{*2}}{2!} * \left|\frac{\partial^2 \widehat{F}}{\partial w^2}\right|(\xi, \widehat{W}_{\mathbf{0}}).$$

Explicitly, one can choose $M_{\rho,(0)} = \frac{1}{\rho}$ (lemma 6.1), $\widehat{W}_{\mathbf{0}}(\xi) = 4.22\mathrm{e}^{\frac{4.22}{\rho}\xi}$ (theorem 3.3), $\widehat{W}_{\mathbf{e}_1}(\xi) = \frac{5.81}{\rho}\mathrm{e}^{\frac{5.81}{\rho}\xi}$ (lemma 6.3), and we recall that $\left|\frac{\partial^2 \widehat{F}}{\partial w^2}\right|(\xi, \widehat{W}_{\mathbf{0}}) = \xi$ while $\left|\frac{\partial \widehat{F}}{\partial w}\right|(\xi, \widehat{W}_{\mathbf{0}}) = 4\xi + \xi * \widehat{W}_{\mathbf{0}}(\xi)$. Therefore, $\widehat{W}$ solves the convolution equation:

$$\rho\widehat{W} = \left(3+\left(4+\frac{9}{4\rho}\right)\xi + 4.22\xi * \mathrm{e}^{\frac{4.22}{\rho}\xi}\right) * \widehat{W} + \frac{\xi}{2!} * \left(V(\delta + \frac{4.63}{\rho}\mathrm{e}^{\frac{4.63}{\rho}\xi}) + \widehat{W}\right)^{*2}.$$

The generating function $\widehat{W}(\xi,V)$ is thus the Borel transform of $\widetilde{W}(\zeta,V)$, solution of the algebraic equation

$$\begin{aligned}\rho\widetilde{W} = &\left(\frac{3}{z}+\left(4+\frac{9}{4\rho}\right)\frac{1}{z^2}+\frac{4.22}{z^2}\frac{1}{z-\frac{4.22}{\rho}}\right)\widetilde{W} \\ &+\frac{1}{2z^2}\left[V\left(1+\frac{5.81}{\rho}\frac{1}{z-\frac{5.81}{\rho}}\right)+\widetilde{W}\right]^2\end{aligned} \tag{6.14}$$

with $\widetilde{W}(z,V) \simeq \frac{1}{2\rho}\left[\frac{V}{z}\left(1+\frac{5.81}{\rho}\frac{1}{z-\frac{5.81}{\rho}}\right)\right]^2$ when $V \to 0$ with $|z|$ large enough. We view (6.14) as a fixed point problem $W = \mathsf{N}(W)$. We set $U = D(\infty, \frac{4\rho}{27}) \times D(0, \frac{2}{3\sqrt{\rho}})$, we equip the space $\mathscr{O}(\overline{U})$ with the maximum norm and we consider the closed ball $B_1 = \{W \in \mathscr{O}(\overline{U}),\ \|W\| \le 1\}$ of the Banach algebra $\left(\mathscr{O}(\overline{U}), \|\ \|\right)$. One easily shows that $\mathsf{N} : B_1 \to B_1$ is a contractive map (remember that $\rho < 1$), hence the fixed-point problem $W = \mathsf{N}(W)$ has a unique solution $\widetilde{W} = \widetilde{W}(z,V)$ in B_1. Its Taylor expansion with respect to V at 0 reads $\widetilde{W}(z,V) = \sum_{k=2}^{\infty} V^k \widetilde{W}_{k\mathbf{e}_1}(z)$, where $(\widetilde{W}_{k\mathbf{e}_1})_{k\ge 2}$ is a sequence of holomorphic functions on the disc $D(\infty, \frac{4\rho}{27})$ and, by the Cauchy inequalities, for every integer $k \ge 2$, $\sup_{|z|>\frac{27}{4\rho}} |\widetilde{W}_{k\mathbf{e}_1}(z)| \le \left(\frac{3\sqrt{\rho}}{2}\right)^k$. Moreover, since $\widehat{W}_{k\mathbf{e}_1}(\xi) = O(\xi^{2k-3})$, $\widetilde{W}_{k\mathbf{e}_1}(z) = O(z^{-2(k-1)})$. We end the proof with lemma 3.5: $\widehat{W}_{k\mathbf{e}_1}$ is an entire function, for every $\xi \in \mathbb{C}$, $|\widehat{W}_{k\mathbf{e}_1}(\xi)| \le \left(\frac{3\sqrt{\rho}}{2}\right)^k \frac{27}{4\rho}\mathrm{e}^{\frac{27}{4\rho}|\xi|}$ and for every positive integer $1 \le m \le 2k-3$,

$$|\widehat{W}_{k\mathbf{e}_1}(\xi)| \le \left(\frac{3\sqrt{\rho}}{2}\right)^k \left(\frac{27}{4\rho}\right)^{m+1} \left(\frac{\zeta^{m-1}}{(m-1)!} * \mathrm{e}^{\frac{27}{4\rho}\zeta}\right)(|\xi|).$$

This ends the proof. □

6.2.2 Borel-Laplace Summability of the Transseries

Before keeping on, we lay down a definition, see also the first volume of this book [MS016].

Definition 6.2. Let $(\widetilde{g}_k)_{k\ge 0}$ be a sequence of formal series $\widetilde{g}_k(z) \in \mathbb{C}[[z^{-1}]]$. One says that the transseries $\widetilde{g}(z,V) = \sum_{k=0}^{\infty} V^k \widetilde{g}_k(z)$ is Borel-Laplace summable in a direction $\theta \in \mathbb{S}_1$ if each $\widetilde{g}_k$ is Borel-Laplace summable in that direction and if the series of functions $\sum_{k=0}^{\infty} V^k \mathscr{S}^\theta \widetilde{g}_k(z)$ converges uniformaly on any compact subset of a domain of the form $\overset{\bullet}{\Pi}{}^\theta_\tau \times \mathscr{V}$. In that case, one denotes by $\mathscr{S}^\theta \widetilde{g}(z,V) \in \mathscr{O}(\overset{\bullet}{\Pi}{}^\theta_\tau \times \mathscr{V})$ its sum, called the *Borel-Laplace sum of the transseries*.

In the sequel, we have in mind to analyze the Borel-Laplace summability of the transseries given by proposition 5.2. This means analyzing the Borel-Laplace summability of the transseries $\sum_{k=0}^{\infty} V^k \widetilde{w}_{k\mathbf{e}_1}(z)$, *resp.* $\sum_{k=0}^{\infty} V^k \widetilde{w}_{k\mathbf{e}_2}(z)$, then substituting $V = U\mathrm{e}^{-z} z^{3/2}$, *resp.* $V = U\mathrm{e}^{z} z^{3/2}$, in the Borel-Laplace sum. Notice however that the mapping $z \mapsto \mathrm{e}^{\pm z} z^{3/2}$ is ill-defined on $\mathbb{C}$ but should be considered on the Riemann surface of the square root or on its universal covering $\widetilde{\mathbb{C}}$. This justifies the use of domains of the form $\Pi^\theta_\tau \in \widetilde{\mathbb{C}}$, $\theta \in \widetilde{\mathbb{S}}^1$ (see definition 3.19) in what follows.

Definition 6.3. Let $g : \widetilde{\mathbb{C}} \to \mathbb{C}$ and $\kappa : \mathbb{R} \to \mathbb{R}^{+\star}$ be two continuous functions, $\theta \in \widetilde{\mathbb{S}}^1$ and $\tau \in \mathbb{R}$. We set $\mho^\theta(g,\tau,\kappa) = \bigcup_{c>\tau} \{z \in \Pi^\theta_c,\ |g(z)| < \kappa(c)\} \subset \widetilde{\mathbb{C}}$. Let $I \subset \widetilde{\mathbb{S}}^1$ be an open arc, $\gamma : I \to \mathbb{R}$ a locally bounded function and $\mathscr{K} : I \to \mathscr{C}^0(\mathbb{R}, \mathbb{R}^{+\star})$ a continuous function. We denote by $\mathscr{V}(I,g,\gamma,\mathscr{K})$ the domain of $\widetilde{\mathbb{C}}$ defined as follows:

$$\mathscr{V}(I,g,\gamma,\mathscr{K}) = \bigcup_{\theta \in I} \mho^\theta(g,\gamma(\theta),\mathscr{K}(\theta)) \subset \widetilde{\mathbb{C}}.$$

Theorem 6.2. *The transseries solutions of the prepared equation (3.6) associated with the first Painlevé equation,*

$$\widetilde{w}(z,U\mathbf{e}_i) = \sum_{k=0}^{\infty} [V_i(z,U)]^k \widetilde{w}_{k\mathbf{e}_i}(z), \quad V_i(z,U) = U\mathrm{e}^{-\lambda_i z} z^{-\tau_i}, \quad i = 1,2, \tag{6.15}$$

are Borel-Laplace summable and their Borel-Laplace sums are holomorphic solutions of (3.6). More precisely, for any $R > 0$, for any open arc $I_j =]j\pi,(j+1)\pi[\subset \widetilde{\mathbb{S}}^1$, $j \in \mathbb{Z}$, the sum

$$\mathscr{S}^{I_j} \widetilde{w}(z,U\mathbf{e}_i) := \sum_{k=0}^{\infty} [V_i(z,U)]^k \mathscr{S}^{\overset{\bullet}{I}_j} \widetilde{w}_{k\mathbf{e}_i}(\overset{\bullet}{z}), \tag{6.16}$$

with $\overset{\bullet}{I}_j = \pi(I_j) \subset \mathbb{S}_1$ and $\overset{\bullet}{z} = \pi(z) \in \mathbb{C}^\star$, converges to a function of (z,U) holomorphic on $\mathscr{V}(I_j, V_i(R), \tau, \mathscr{K}) \times D(0,R)$ where one can choose $\tau(\theta) = \frac{27}{4|\sin(\theta)|}$ and $\mathscr{K}(\theta) : c \in \mathbb{R} \mapsto \frac{2c^2}{3\tau(\theta)^2\sqrt{\sin(\theta)}}$. Moreover, the sum $\mathscr{S}^{I_j}\widetilde{w}$ is solution of equation (3.6).

Proof. This theorem is a consequence of theorem 3.3, lemma 6.3, lemma 6.4 and lemma 6.5. Let us precise the reasoning for $i = 1$ and the open arc $I_0 =]0,\pi[\subset \widetilde{\mathbb{S}}^1$. We know from lemmas 6.4 and 6.5 (applied with $m = 2k-3$) that for any $\delta \in]0, \frac{\pi}{2}[$ and any integer $k \geq 2$, for every $\zeta \in \overset{\bullet}{\Delta}{}^\infty_0(]\delta, \pi - \delta[)$,

$$|\widehat{w}_{k\mathbf{e}_1}(\zeta)| \leq \left(\frac{3\sqrt{\sin(\delta)}}{2}\right)^k A_\delta^{2k-2} \left(\frac{\xi^{2k-4}}{(2k-4)!} * \mathrm{e}^{\tau_\delta \xi}\right)(\xi), \quad \xi = |\zeta|, \tag{6.17}$$

with $A_\delta = \tau_\delta = \dfrac{27}{4\sin(\delta)}$. We now fix a direction $\theta \in I_0$ and for $k \geq 2$, we consider the Borel-Laplace sum

$$\mathscr{S}^{\dot{\theta}}\, \widetilde{w}_{k\mathbf{e}_1}(\dot{z}) = \int_0^{\infty e^{i\dot{\theta}}} e^{-\dot{z}\zeta}\, \widehat{w}_{k\mathbf{e}_1}(\zeta)\, d\zeta = \int_0^{+\infty} e^{-\dot{z}\xi e^{i\dot{\theta}}}\, \widehat{w}_{k\mathbf{e}_1}(\xi e^{i\dot{\theta}})\, e^{i\dot{\theta}} d\xi.$$

For any $c > \tau_\theta$ and any $z \in \overline{\Pi}_c^\theta$, $|e^{-\dot{z}\xi e^{i\dot{\theta}}}| \leq e^{-c\xi}$, for $\xi \geq 0$. Therefore, for $z \in \overline{\Pi}_c^\theta$ and $\xi \geq 0$,

$$\left| e^{-\dot{z}\xi e^{i\dot{\theta}}}\, \widehat{w}_{k\mathbf{e}_1}(\xi e^{i\dot{\theta}})\, e^{i\dot{\theta}} \right| \leq \left(\frac{3\sqrt{\sin(\theta)}}{2} \right)^k A_\theta^{2k-2} e^{-c\xi} \left(\frac{\xi^{2k-4}}{(2k-4)!} * e^{\tau_\theta \xi} \right)(\xi).$$

The function $\mathscr{S}^\theta\, \widetilde{w}_{k\mathbf{e}_1}(z) := \mathscr{S}^{\dot{\theta}}\, \widetilde{w}_{k\mathbf{e}_1}(\dot{z})$ is thus holomorphic on Π_c^θ and, for every $z \in \overline{\Pi}_c^\theta$,

$$|\mathscr{S}^\theta\, \widetilde{w}_{k\mathbf{e}_1}(z)| \leq \left(\frac{3\sqrt{\sin(\theta)}}{2} \right)^k \left(\frac{A_\theta}{c} \right)^{2k-2} \frac{c}{c - \tau_\theta}.$$

We turn to the series of function $\sum_{k\geq 2} \left(U e^{-z} z^{3/2} \right)^k \mathscr{S}^\theta\, \widetilde{w}_{k\mathbf{e}_1}(z)$. From what precedes, for any $R > 0$, for any $c' > c > \tau_\theta$, for every $(z, U) \in \Pi_{c'}^\theta \times D(0, R)$, the series is normally convergent when $|Re^{-z} z^{3/2}| \leq \frac{2c^2}{3A_\theta^2 \sqrt{\sin(\theta)}}$. We end with theorem 3.3 and lemma 6.3: for any direction $\theta \in I_0$, for any $c > \tau_\theta$, the series of functions $\sum_{k\geq 0} \left(U e^{-z} z^{3/2} \right)^k \mathscr{S}^\theta\, \widetilde{w}_{k\mathbf{e}_1}(z)$ defines a holomorphic function on the do-

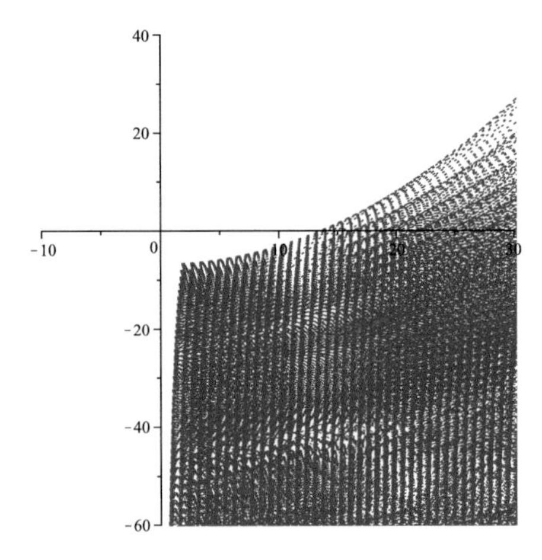

Fig. 6.1 The (shaded) domain $\mathscr{V}(I_0, V_1(0.5), \tau, \mathscr{K})$ on projection, for $\tau(\theta) = \frac{27}{4|\sin(\theta)|}$, $(\mathscr{K}(\theta))(c) = \frac{2c^2}{3\tau(\theta)^2\sqrt{\sin(\theta)}}$ and $V_1(z, U) = U e^{-z} z^{3/2}$.

main $\mho^\theta \times D(0,R)$ with $\mho^\theta = \bigcup_{c>\tau_\theta} \{z \in \Pi_c^\theta,\ |Re^{-z}z^{3/2}| < \frac{2c^2}{3A_\theta^2\sqrt{\sin(\theta)}}\}$. Making θ varying on I_0, these functions glue together to provide a holomorphic function $\mathscr{S}^{I_0}\widetilde{w}(z,U\mathbf{e}_1)$ on the domain $\mathscr{V}(I_0,V_1(R),\tau,\mathscr{K})\times D(0,R)$ with $\tau(\theta) = \frac{27}{4|\sin(\theta)|}$ and $\mathscr{K}(\theta) : c \in \mathbb{R} \mapsto \frac{2c^2}{3\tau(\theta)^2\sqrt{\sin(\theta)}}$ (since $A_\theta = \tau_\theta$), see Fig. 6.1. Finally, we encourage the reader to show that $\mathscr{S}^{I_j}\widetilde{w}$ solves the ODE (3.6). □

Remark 6.1. The theorem 6.2 can be shown by other means, see the comments in Sect. 6.5.

6.2.3 Remarks

In what follows we set $w_{tru,j,i}(z,U) = \mathscr{S}^{I_j}\widetilde{w}(z,U\mathbf{e}_i)$.

1. We know by proposition 5.2 that $\widetilde{w}_{k\mathbf{e}_2}(z) = \widetilde{w}_{k\mathbf{e}_1}(-z)$ for every $k \geq 0$. One deduces that for any $j \in \mathbb{Z}$, for any $\theta \in I_j$, for every $z \in \Pi^{\pi-\theta}_{\tau(\pi-\theta)}$, $ze^{i\pi} \in \Pi^{-\theta}_{\tau(-\theta)}$ and $\mathscr{S}^{\pi-\theta}\,\widetilde{w}_{k\mathbf{e}_2}(z) = \mathscr{S}^{-\theta}\,\widetilde{w}_{k\mathbf{e}_1}(ze^{i\pi})$. Therefore, for any $\theta \in I_j$, for every $z \in \Pi^{\pi-\theta}_{\tau(\pi-\theta)}$, $\mathscr{S}^{\pi-\theta}\widetilde{w}(z,U\mathbf{e}_2) = \mathscr{S}^{-\theta}\widetilde{w}(ze^{i\pi},Ue^{i\pi/2}\mathbf{e}_1)$ and, as a consequence, for any $j \in \mathbb{Z}$:

$$\begin{aligned}
&\text{for every } z \in \mathscr{V}(I_j,V_2(U),\tau,\mathscr{K}),\ w_{tru,j,2}(z,U) = w_{tru,j-1,1}(ze^{i\pi},Ue^{i\pi/2}),\\
&\text{for every } z \in \mathscr{V}(I_j,V_1(U),\tau,\mathscr{K}),\ w_{tru,j,1}(z,U) = w_{tru,j-1,2}(ze^{i\pi},Ue^{i\pi/2}).
\end{aligned} \tag{6.18}$$

2. Here we adopt the convention : for $z = re^{i\alpha} \in \widetilde{\mathbb{C}}$, we set $\overline{z} = r^{-i\alpha} \in \widetilde{\mathbb{C}}$.
We know by proposition 5.2 that $\widetilde{w}_{k\mathbf{e}_i}(z) \in \mathbb{R}[[z^{-1}]]$ for any $k \in \mathbb{N}$, $i = 1,2$. Thus, for any $j \in \mathbb{Z}$ and any $\theta \in I_j$, for $z \in \Pi^\theta_{\tau(\theta)}$, $\overline{\mathscr{S}^\theta\,\widetilde{w}_{k\mathbf{e}_i}(z)} = \mathscr{S}^{-\theta}\,\widetilde{w}_{k\mathbf{e}_i}(\overline{z})$. Therefore, for any $j \in \mathbb{Z}$, for every $z \in \mathscr{V}(I_j,V_i(U),\tau,\mathscr{K})$,

$$\overline{w_{tru,j,i}}(z,U) = w_{tru,(-j-1),i}(\overline{z},\overline{U})$$

and with (6.18) we deduce that, for every $z \in \mathscr{V}(I_j,V_1(U),\tau,\mathscr{K})$ and $z \in \mathscr{V}(I_j,V_2(U),\tau,\mathscr{K})$ respectively,

$$\begin{aligned}
\overline{w_{tru,j,1}}(z,U) &= w_{tru,j,2}\left(\overline{z}e^{-(2j+1)i\pi},\overline{U}e^{-(j+1/2)i\pi}\right)\\
\overline{w_{tru,j,2}}(z,U) &= w_{tru,j,1}\left(\overline{z}e^{-(2j+1)i\pi},\overline{U}e^{-(j+1/2)i\pi}\right).
\end{aligned} \tag{6.19}$$

6.2.4 Considerations on the Domain

Viewing (6.18) and (6.19), it will be enough for our purpose to consider the domain $\mathscr{V}(I_0,V_1,\tau,\mathscr{K})$ with $I_0 =]0,\pi[$, $\big(V_1(U)\big)(z) = Ue^{-z}z^{3/2}$ with $|U| > 0$,

$\tau(\theta) = \frac{27}{4|\sin(\theta)|}$, $\big(\mathscr{K}(\theta)\big)(c) = \frac{2c^2}{3\tau(\theta)^2\sqrt{\sin(\theta)}}$. We would like to describe the boundary of this domain. As a matter of fact, we will restrict ourself to describing its subdomain $\mho^\theta\big(V_1(U), \tau(\theta), \mathscr{K}(\theta)\big)$ with $\theta = \pi/2$. Considered by projection on $\mathbb{C}$, this domain reads: $z = x + \mathrm{i}y$, $(x,y) \in \mathbb{R}^2$, belongs to $\mho^{\frac{\pi}{2}}\big(V_1, \tau(\frac{\pi}{2}), \mathscr{K}(\frac{\pi}{2})\big)$ if and only if there exists $\lambda > 1$ so that

$$\begin{cases} y < -\dfrac{27}{4}\lambda \\ |U|\mathrm{e}^{-x}(x^2+y^2)^{3/4} < \dfrac{2}{3}\lambda^2. \end{cases}$$

(We take $c = \frac{27}{4}\lambda > \tau(\pi/2)$). We now fix $y = -\frac{27}{4}\lambda$ with $\lambda > 1$ and we remark that $z = x + \mathrm{i}y$ belongs to $\mho^{\frac{\pi}{2}}\big(V_1(U), \tau(\frac{\pi}{2}), \mathscr{K}(\frac{\pi}{2})\big)$ iff $x > X$ with X such that

$$|U|\mathrm{e}^{-X}(X^2+y^2)^{3/4} = \frac{2}{3}\left(\frac{4}{27}y\right)^2. \tag{6.20}$$

Indeed, just see that the real mapping $x \mapsto \mathrm{e}^{-x}(x^2+y^2)^p$ is decreasing when $|y| \geq p$, and use an argument of continuity. With the implicit function theorem, these arguments show the existence of a unique solution $X : y \in]-\infty, -\frac{3}{4}[\mapsto X(y)$ of (6.20), of class $\mathscr{C}^\infty$ and increasing with y, which can be described as follows. The above equality is equivalent to writing

$$\left(1 + \frac{X^2}{y^2}\right)^3 = \alpha y^2 \mathrm{e}^{4X}, \qquad \alpha = \left(\frac{32}{2187|U|}\right)^4. \tag{6.21}$$

and we can remark that $X(-\alpha^{-1/2}) = 0$ if $-\alpha^{-1/2} < -\frac{3}{4}$. When assuming $y^2 \gg X^2$, we get $X = -\dfrac{\ln(\alpha y^2)}{4} + \varepsilon$, $\varepsilon = o(1)$ as a first approximation. Plugging this in (6.21), one gets

$$X = -\frac{\ln(\alpha y^2)}{4} + 3\frac{\ln^2(\alpha y^2)}{4^2 y^2} + o(y^{-2})$$

and one can keep on this way to get an asymptotic expansion at any order of the solution[3]. To put it in a nutshell:

Corollary 6.1. *In theorem 6.2, the sum* $w_{tru,0,1}(z,U) = \mathscr{S}^{l_0}\widetilde{w}(z, U\mathbf{e}_1)$ *defines, for any* $U \in \mathbb{C}^\star$, *a holomorphic function with respect to* z *on a domain which contains, by projection on* $\mathbb{C}$, *a subdomain of the form* $\left\{z = x + \mathrm{i}y,\ y < -\dfrac{27}{4},\ x > X(y)\right\}$ *where* X *is an increasing* $\mathscr{C}^\infty$ *function on* $]-\infty, -\frac{3}{4}[$, *whose asymptotics when* $y \to -\infty$ *is given by:*

[3] One can also describe the solution in term of the Lambert function, the compositional inverse of the function $x\mathrm{e}^x$.

$$X(y) = -\frac{\ln(\alpha y^2)}{4} + 3\frac{\ln^2(\alpha y^2)}{4^2 y^2} + o(y^{-2}), \qquad \alpha = \left(\frac{32}{2187|U|}\right)^4 \tag{6.22}$$

and so that $X(-\alpha^{-1/2}) = 0$ *if* $-\alpha^{-1/2} < -\frac{3}{4}$.

6.3 Summability of the Formal Integral

We saw with corollary 5.2 that the formal integral can be interpreted as a formal transformation $w = \widetilde{\Phi}(z, \mathbf{u})$,

$$\widetilde{\Phi}(z, \mathbf{u}) = \sum_{\mathbf{k} \in \mathbb{N}^2} \mathbf{u}^{\mathbf{k}} \widetilde{w}_{\mathbf{k}}^{[0]}(z), \tag{6.23}$$

that formally transforms the prepared ODE (3.6) into the normal form equation (5.66). It is then natural to wonder whether this formal transformation gives rise to an analytic transformations $\Phi_\theta(z, \mathbf{u})$ by Borel-Laplace summation,

$$\Phi_\theta(z, \mathbf{u}) = \mathscr{S}^\theta \widetilde{\Phi}(z, \mathbf{u}) = \sum_{\mathbf{k} \in \mathbb{N}^2} \mathbf{u}^{\mathbf{k}} \mathscr{S}^\theta \widetilde{w}_{\mathbf{k}}^{[0]}(z),$$

with a definition of the sum similar to that of definition 6.2. One could give a positive answer to this question, for the price of some further effort.

One has to extend lemma 6.5 to the whole $\mathbf{k}$-th series $\widetilde{w}_{\mathbf{k}}^{[0]}$. It is worth for this matter to complete the Banach spaces detailed by proposition 3.9 by other "focusing algebras" for which we refer to [Cos009], in particular those based on L^1_ν-norms.

This does not mean that the formal integral is Borel-Laplace summable : this is wrong, due to the effect of the exponentials. Only the restrictions of the formal integral to convenient submanifolds is 1-summable, which means here just considering one of the two transseries. However, the sums of the two transseries share no common domain of convergence and *a fortiori* the formal integral cannot be summed by Borel-Laplace summation.

We do not pursue toward this direction and we conclude this chapter with the truncated solutions.

6.4 Truncated Solutions for the First Painlevé Equation

We know from theorem 6.2 that the sum $w_{tru,j,i}(z,U) = \mathscr{S}^{I_j} \widetilde{w}(z, U\mathbf{e}_i)$, $j \in \mathbb{Z}$ and $i = 1, 2$, is a holomorphic solution of (3.6), for z on a domain of the form $\mathscr{V}(I_j, V_i(U), \tau, \mathscr{K})$. From its very definition and from corollary 6.1, the domain $\mathscr{V}(I_j, V_i(U), \tau, \mathscr{K})$ contains a sectorial neighbourhood of infinity with aperture $\breve{I}_{j,i}$ where (see Fig. 6.1):

- when $i=1$, $\breve{I}_{j,1}=]-\frac{1}{2}\pi,+\frac{1}{2}\pi[-j\pi$ for j even, $\breve{I}_{j,1}=]-\frac{3}{2}\pi,-\frac{1}{2}\pi[-j\pi$ for j odd;
- when $i=2$, $\breve{I}_{j,2}=]-\frac{1}{2}\pi,+\frac{1}{2}\pi[-j\pi$ for j odd, $\breve{I}_{j,2}=]-\frac{3}{2}\pi,-\frac{1}{2}\pi[-j\pi$ for j even.

To go back to the the first Painlevé equation (2.1), we use the transformation $\mathscr{T}$ of definition 3.20.

Definition 6.4. The conformal mapping $\mathscr{T}$ sends the domain $\mathscr{V}(I,g,\gamma,\mathscr{K})$ onto the domain $\mathscr{T}\left(\mathscr{V}(I,g,\gamma,\mathscr{K})\right)$ and we set

$$\mathscr{S}(I,g,\gamma,\mathscr{K})=\mathscr{T}\left(\mathscr{V}(I,g,\gamma,\mathscr{K})\right),\quad \overset{\bullet}{\mathscr{S}}(I,g,\gamma,\mathscr{K})=\pi\Big(\mathscr{S}(I,g,\gamma,\mathscr{K})\Big). \tag{6.24}$$

The domain $\mathscr{S}(I_j,V_i(U),\tau,\mathscr{K})$ contains a sectorial neighbourhood of infinity with aperture $K_{j,i}$ (see Fig. 6.2):

- when $i=1$, $K_{j,1}=]-\frac{7}{5}\pi,-\frac{3}{5}\pi[-\frac{4}{5}j\pi$ for j even, $K_{j,1}=]-\frac{11}{5}\pi,-\frac{7}{5}\pi[-\frac{4}{5}j\pi$ for j odd;
- when $i=2$, $K_{j,2}=]-\frac{7}{5}\pi,-\frac{3}{5}\pi[-\frac{4}{5}j\pi$ for j odd, $K_{j,2}=]-\frac{11}{5}\pi,-\frac{7}{5}\pi[-\frac{4}{5}j\pi$ for j even.

In any case, the domains $\mathscr{S}(I_j,V_i(U),\tau,\mathscr{K})$ are in connection: for every $j\in\mathbb{Z}$,

$$\mathscr{S}(I_{j+1},V_2(U),\tau,\mathscr{K})=\mathrm{e}^{-4\mathrm{i}\pi/5}\mathscr{S}(I_j,V_1(U),\tau,\mathscr{K}).$$

From (3.4), (2.6), (2.7), the transformation

$$\begin{aligned}&z\in\mathscr{V}(I_j,V_i(U),\tau,\mathscr{K})\leftrightarrow x\in\mathscr{S}(I_j,V_i(U),\tau,\mathscr{K})\\ &\qquad w_{tru,j,i}(z,U)\leftrightarrow u_{tru,j,i}(x,U)=\frac{\mathrm{e}^{i\frac{\pi}{2}}x^{\frac{1}{2}}}{\sqrt{6}}\left(1-\frac{4}{25\left(\mathscr{T}^{-1}(x)\right)^2}+\frac{w_{tri,j,i}\left(\mathscr{T}^{-1}(x),U\right)}{\left(\mathscr{T}^{-1}(x)\right)^2}\right)\end{aligned}$$

provides the solutions $u_{tru,j,i}(x,U)$ for the first Painlevé equation. These are the *truncated solutions.*

The property (6.18) translates into the following relationships between truncated solutions: for any $j\in\mathbb{Z}$, for every $x\in\mathscr{S}(I_j,V_1(U),\tau,\mathscr{K})$, *resp.* $x\in\mathscr{S}(I_j,V_2(U),\tau,\mathscr{K})$,

$$\begin{aligned}u_{tru,j,1}(x,U)&=\mathrm{e}^{2\mathrm{i}\pi/5}u_{tru,j+1,2}(x\mathrm{e}^{-4\mathrm{i}\pi/5},U\mathrm{e}^{-\mathrm{i}\pi/2})\\ u_{tru,j,2}(x,U)&=\mathrm{e}^{2\mathrm{i}\pi/5}u_{tru,j+1,1}(x\mathrm{e}^{-4\mathrm{i}\pi/5},U\mathrm{e}^{-\mathrm{i}\pi/2})\end{aligned} \tag{6.25}$$

These are the symmetries discussed in Sect. 2.5. In the same way from (6.19), for any $j\in\mathbb{Z}$, for every $x\in\mathscr{S}S(I_j,V_1(U),\tau,\mathscr{K})$, respectively $x\in\mathscr{S}(I_j,V_2(U),\tau,\mathscr{K})$,

$$\begin{aligned}\overline{u_{tru,j,1}}(x,U)&=\mathrm{e}^{\frac{2}{5}(2j+1)\mathrm{i}\pi}u_{tru,j,2}(\overline{x}\mathrm{e}^{-\frac{2}{5}(4j+7)\mathrm{i}\pi},\overline{U}\mathrm{e}^{-(j+1/2)\mathrm{i}\pi}),\\ \overline{u_{tru,j,2}}(x,U)&=\mathrm{e}^{\frac{2}{5}(2j+1)\mathrm{i}\pi}u_{tru,j,1}(\overline{x}\mathrm{e}^{-\frac{2}{5}(4j+7)\mathrm{i}\pi},\overline{U}\mathrm{e}^{-(j+1/2)\mathrm{i}\pi}).\end{aligned} \tag{6.26}$$

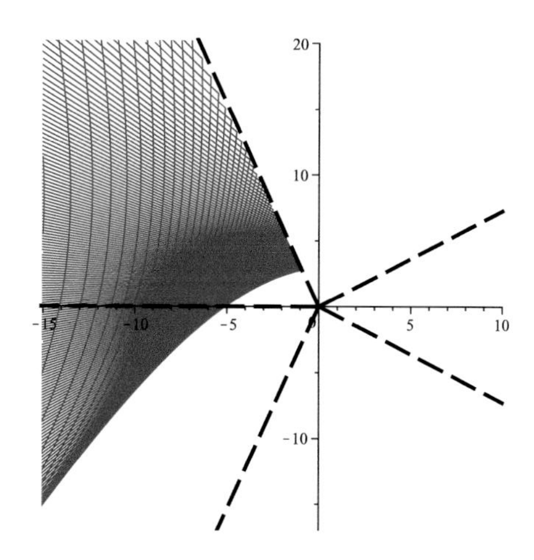

Fig. 6.2 The (shaded) domain $\overset{\bullet}{\mathscr{S}}(I_0, V_1(U), \tau, \mathscr{K})$ for $\tau(\theta) = \frac{27}{4|\sin(\theta)|}$, $(\mathscr{K}(\theta))(c) = \frac{2c^2}{3\tau(\theta)^2\sqrt{\sin(\theta)}}$ and $V_1(z,U) = U\mathrm{e}^{-z}z^{3/2}$.

6.5 Comments

We mentioned in Sect. 5.5 the existence of formal transforms of the type $\mathbf{v} = \widetilde{T}(z,\mathbf{u})$, $\widetilde{T}(z,\mathbf{u}) = \sum_{\mathbf{k}\in\mathbb{N}^n} \mathbf{u}^{\mathbf{k}}\widetilde{\mathbf{v}}_{\mathbf{k}}(z)$, $\widetilde{\mathbf{v}}_{\mathbf{k}}(z) \in \mathbb{C}^n[[z^{-1}]]$ that brings differential and difference systems to their linear normal form, under some convenient hypotheses. For differential equations of type (5.67), the series $\widetilde{\mathbf{v}}_{\mathbf{k}}$ are in general not 1-summable but multisummable [Lod016]. The first results in that direction, concerning the multisummability of the formal series solutions, have been obtained by Braaksma [Bra92] then by Ramis & Sibuya [RS94]. A resurgent approach for 1-level differential equations is undertaken by Costin in [Cos98], with the proof of the 1-summability of the formal integral on restriction to convenient submanifolds. These results have been generalized to differential and difference equations, see e.g. [Bra001, Kui003, CC001, Cos99] and references therein, at least for the cases where no resonance occurs. The question of the (multi)summability of the above formal transforms may be delicate, even for 1-level differential systems or ODEs, when *quasi-resonance* occurs, giving rise to *small divisors*.

If $\boldsymbol{\lambda} = (\lambda_1, \cdots, \lambda_n)$ stands for the multipliers and in absence of resonance, it may happen that $\boldsymbol{\lambda}.\mathbf{k}$ comes close to one multiplier, for some $\mathbf{k} \in \mathbb{N}^n$. Thus, the construction of the formal integral gives rise to division by small factors. One has "quasi-resonance" when there exists an increasing sequence $(\mathbf{k}_j \in \mathbb{N}^n)$ such that $\lim_{j\to\infty} \boldsymbol{\lambda}.\mathbf{k}_j = 0$ fast enough, a condition that translates into diophantine relations on the sequence.

More details on this subject can be found in [Eca85].

We finally mention a general upshot, that of the formation of singularities near the anti-Stokes rays. Considering the Borel-Laplace sum of a transseries stemming from (resurgent) 1-level differential or difference equations, it is possible, as shown in [CC001] (see also [Cos009]) to analyze its behavior on the boundary of its domain of convergence, by a suitable use of a multi-scale analysis. This is detailed in [Cos99] for the first Painlevé equation.

Acknowledgements I warmly thank my student Julie Belpaume who helped me to working out this chapter.

References

Bra92. B.L.J. Braaksma, *Multisummability of formal power series solutions of nonlinear meromorphic differential equations.* Ann. Inst. Fourier (Grenoble) **42** (1992), no. 3, 517-540.

Bra001. B.L.J. Braaksma, *Transseries for a class of nonlinear difference equations. In memory of W. A. Harris, Jr.* J. Differ. Equations Appl. **7** (2001), no. 5, 717-750.

CNP93-1. B. Candelpergher, C. Nosmas, F. Pham, *Approche de la résurgence*, Actualités mathématiques, Hermann, Paris (1993).

Cos98. O. Costin, *On Borel summation and Stokes phenomena for rank-1 nonlinear systems of ordinary differential equations.* Duke Math. J. **93** (1998), no. 2, 289-344.

Cos99. O. Costin, *Correlation between pole location and asymptotic behavior for Painlevé I solutions.* Comm. Pure Appl. Math. **52** (1999), no. 4, 461-478.

Cos009. O. Costin, *Asymptotics and Borel summability*, Chapman & Hall/CRC Monographs and Surveys in Pure and Applied Mathematics, 141. CRC Press, Boca Raton, FL, 2009.

CC001. O. Costin, R. Costin, *On the formation of singularities of solutions of nonlinear differential systems in antistokes directions.* Invent. Math. **145** (2001), no. 3, 425-485.

Eca85. J. Ecalle, *Les fonctions résurgentes. Tome III : l'équation du pont et la classification analytique des objets locaux.* Publ. Math. d'Orsay, Université Paris-Sud, 1985.05 (1985).

Kui003. G.R. Kuik, *Transseries in Difference and Differential Equations.* PhD thesis, Rijksuniversiteit Groningen (2003).

Lod016. M. Loday-Richaud, *Divergent Series, Summability and Resurgence II. Simple and Multiple Summability.* Lecture Notes in Mathematics, **2154**. Springer, Heidelberg, 2016.

MS016. C. Mitschi, D. Sauzin, *Divergent Series, Summability and Resurgence I. Monodromy and Resurgence.* Lecture Notes in Mathematics, **2153**. Springer, Heidelberg, 2016.

RS94. J.-P. Ramis, Y Sibuya, *A new proof of multisummability of formal solutions of nonlinear meromorphic differential equations.* Ann. Inst. Fourier (Grenoble) **44** (1994), no. 3, 811-848.

Chapter 7
Supplements To Resurgence Theory

Abstract This chapter is devoted to some general nonsense in resurgence theory which will be useful to study furthermore the first Painlevé equation from the resurgence viewpoint. We define sectorial germs of holomorphic functions (Sect. 7.2) and we introduce the sheaf of microfunctions (Sect. 7.3). This provides an approach to the notion of singularities which is the purpose of Sect. 7.4. We define the formal Laplace transform for microfunctions and for singularities and conversely, the formal Borel transform acting on asymptotic classes (Sect. 7.5). The main properties of the Laplace transform needed in this course are developed to Sect. 7.6. We then introduce some spaces of resurgent functions and define the alien operators (Sect. 7.7 to 7.9).

7.1 Introduction

In this introduction, we assume that the reader has a previous acquaintance with 1-summability theory, much discussed in the second volume of this book [Lod016] to which we refer.

At its very root, one can rely the Borel-Laplace summation scheme to the simple formula

$$\frac{1}{z^n} = \mathscr{L}^\theta\left(\frac{\zeta^{n-1}}{\Gamma(n)}\right) = \int_0^{\infty e^{i\theta}} e^{-z\zeta}\frac{\zeta^{n-1}}{\Gamma(n)}d\zeta, \qquad n\in\mathbb{N}^\star,\;\; z\in\dot{\Pi}_0^\theta.$$

Let $\widehat{\varphi}\in\mathscr{O}(D(0,R))$ be a holomorphic function and $\sum_{n\geq 1} a_n\frac{\zeta^{n-1}}{\Gamma(n)}$ be its Taylor series at the origin. We choose an open arc $I =]-\alpha+\theta,\theta+\alpha[$, $0<\alpha\leq\pi/2$, bisected by the direction θ, and we set $I^\star =]-\alpha-\theta,-\theta+\alpha[\subseteq\check{\theta}$. For some $r\geq 0$, we set $\dot{\mathfrak{s}}^\infty = \dot{\mathfrak{s}}_r^\infty(I^\star)$. For any cut-off $\kappa\in]0,R[$, the truncated Laplace integral $\varphi_\kappa(z) = \int_0^{\kappa e^{i\theta}} e^{-z\zeta}\widehat{\varphi}(\zeta)d\zeta$ provides an element of $\overline{\mathscr{A}}_1(\dot{\mathfrak{s}}^\infty)$ whose 1-Gevrey

E. Delabaere, *Divergent Series, Summability and Resurgence III*,
Lecture Notes in Mathematics 2155, DOI 10.1007/978-3-319-29000-3_7

asymptotics $T_{1,\overset{\bullet}{\mathcal{S}}^\infty} \varphi_\kappa(z)$ in $\overset{\bullet}{\mathcal{S}}^\infty$ is given by the 1-Gevrey series $\sum_{n\geq 1} \frac{a_n}{z^n} \in \mathbb{C}[[z^{-1}]]_1$. This is essentially the Borel-Ritt theorem for 1-Gevrey asymptotics. For two cut-off points $\kappa_1, \kappa_2 \in]0,R[$, the difference $\varphi_{\kappa_1} - \varphi_{\kappa_2}$ belongs to $\overline{\mathscr{A}}^{\leq -1}(\overset{\bullet}{\mathcal{S}}^\infty)$, the differential ideal of $\overline{\mathscr{A}}_1(\overset{\bullet}{\mathcal{S}}^\infty)$ made of 1-exponentially flat functions on $\overset{\bullet}{\mathcal{S}}^\infty$.
One gets this way a morphism $\mathscr{L}(I) : \widehat{\varphi} \in \mathscr{O}_0 \mapsto \mathrm{cl}(\varphi_\kappa) \in \mathscr{A}_1(I^\star)/\mathscr{A}^{\leq -1}(I^\star)$, where here $\mathscr{O}_0$ stands for the constant sheaf (of convolution algebras) over $\mathbb{S}^1$. By (obvious) compatibility with the restriction maps, one obtains[1] a morphism of sheaves of differential algebras, $\mathscr{L} : \mathscr{O}_0 \to \mathscr{A}_1/\mathscr{A}^{\leq -1}$, where the quotient sheaf $\mathscr{A}_1/\mathscr{A}^{\leq -1}$ over $\mathbb{S}^1$ is known to be isomorphic to the constant sheaf $\mathbb{C}[[z^{-1}]]_1$ (Borel-Ritt theorem 3.4, see [Lod016, Mal95]). The formal Laplace transfom $\mathscr{L}$ is an isomorphism, the inverse morphism being the formal Borel transform $\mathscr{B} : \mathbb{C}[[z^{-1}]]_1 \to \mathscr{O}_0$ (seen as a morphism of sheaves).

One can extend the theory by considering Laplace integrals defined along Hankel contours. For instance, standard formulae provide

$$\Gamma(\sigma) = \frac{1}{1-\mathrm{e}^{-2i\pi\sigma}} \int_{\gamma_{[-2\pi,0],\varepsilon}} \mathrm{e}^{-\zeta} \zeta^{\sigma-1} d\zeta, \qquad \sigma \in \mathbb{C}\setminus\mathbb{N}, \tag{7.1}$$

where the integration contour $\gamma_{[-2\pi,0],\varepsilon}$ is the (endless) Hankel contour drawn on Fig. 7.1, while $\zeta^{\sigma-1} = \mathrm{e}^{(\sigma-1)\log\zeta}$ and $\log\zeta$ is the branch of the logarithm so that $\arg(\log\zeta) \in]-2\pi, 0[$. Performing a change of variable, one gets the identity

$$\frac{1}{z^\sigma} = \mathscr{L}^0 \overset{\vee}{I}_\sigma(z) = \int_{\gamma_{[-2\pi,0],\varepsilon}} \mathrm{e}^{-z\zeta} \overset{\vee}{I}_\sigma(\zeta) d\zeta, \quad z \in \overset{\bullet}{\Pi}{}^0_0, \tag{7.2}$$

with $z^\sigma = \mathrm{e}^{\sigma \log z}$ where this time $\log z$ is the branch of the logarithm so that $\arg(\log z) \in]-\pi, \pi[$, while

$$\overset{\vee}{I}_\sigma(\zeta) = \begin{cases} \dfrac{\zeta^{\sigma-1}\log(\zeta)}{2i\pi\Gamma(\sigma)} & \text{for } \sigma - 1 \in \mathbb{N} \\[2ex] \dfrac{\zeta^{\sigma-1}}{(1-\mathrm{e}^{-2i\pi\sigma})\Gamma(\sigma)} & \text{for } \sigma - 1 \in \mathbb{C}\setminus\mathbb{N}. \end{cases}$$

The form of $\overset{\vee}{I}_\sigma$ that we give for $\sigma - 1 \in \mathbb{C}\setminus\mathbb{N}$ is well-defined when $-\sigma \notin \mathbb{N}$. It can be analytically continued to the case $-\sigma \in \mathbb{N}$ by the reflection formula.

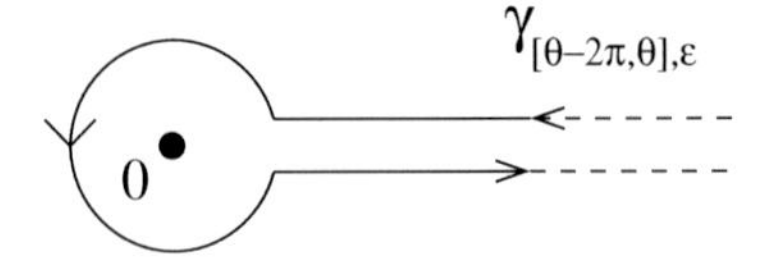

Fig. 7.1 The Hankel contour $\gamma_{[\theta-2\pi,\theta],\varepsilon}$ for $\theta = 0$.

[1] Modulo the quite innocent complex conjugation $I \to I^\star$.

This example, already detailed in the first volume of this book [MS016], provides another one which will be used later on : for any $m \in \mathbb{N}$, any $\sigma \in \mathbb{C} \setminus \mathbb{N}^\star$, for $z \in \overset{\bullet}{\Pi}{}^0_0$, $(-1)^m z^{-\sigma} (\log z)^m = \mathscr{L}^0 \overset{\vee}{J}_{\sigma,m}$, $\overset{\vee}{J}_{\sigma,m} = \left(\frac{\partial}{\partial \sigma}\right)^m \overset{\vee}{I}_\sigma$ with the above convention for the $\log z$. Remark however that $\mathscr{L}^0 \overset{\vee}{I}_\sigma = \mathscr{L}^0 \big(\overset{\vee}{I}_\sigma + \mathrm{hol} \big)$ when hol is any holomorphic function on a half-strip containing the origin, with at most exponential growth of order 1 at infinity. This justifies the introduction of the spaces of microfunctions and singularities that we do in the next sections.

This chapter can be seen as a sequel of the resurgence theory developed in the first volume of this book [MS016]. For most of the materials presented here, we mainly refer to [Eca81-1, Eca92, Eca93-1, CNP93-1, MS016], see also [DP99, Sau006, OSS003]. Another approach to resurgence theory is provided in [SS96].

7.2 Sectorial Germs

7.2.1 Sectors

We precise our notation for sectors on $(\widetilde{\mathbb{C}}, \pi)$, the Riemann surface of the logarithm (Compare with definition 3.3).

Definition 7.1. Let $I \subset \widetilde{\mathbb{S}}^1$ be an open arc. For $0 \leq r < R \leq \infty$, we denote by $\mathcal{S}_r^R(I)$ the simply connected domain of $\widetilde{\mathbb{C}}$ of the form $\mathcal{S}_r^R(I) = \{\zeta = \xi e^{i\theta} \mid \theta \in I, \xi \in]r, R[\}$. One denotes by $\bar{\mathcal{S}}_r^R(I)$ the closure of $\mathcal{S}_r^R(I)$ in $\widetilde{\mathbb{C}}$ We use abridged notation $\mathcal{S}_0(I)$, $\bar{\mathcal{S}}_0(I)$, $\mathcal{S}^\infty(I)$ and $\bar{\mathcal{S}}^\infty(I)$ for sectors, when R or r is unspecified.
For any continuous function $\mathrm{R} : \widetilde{\mathbb{S}}^1 \to]0, +\infty[$, we denote by $\mathcal{S}_0^{\mathrm{R}}$ the simply connected domain defined by $\mathcal{S}_0^{\mathrm{R}} = \{\zeta = re^{i\theta} \mid \theta \in \widetilde{\mathbb{S}}^1, 0 < r < \mathrm{R}(\theta)\} \subset \widetilde{\mathbb{C}}$. We simply write $\mathcal{S}_0$ for such a domain, when there is no need to specify the function R.

7.2.2 Sectorial Germs

Definition 7.2. Let $I \subset \mathbb{S}^1$ be an open arc. One says that two functions $\varphi_1 \in \mathscr{O}(\overset{\bullet}{\mathcal{S}}{}_0^{R_1}(I))$, $\varphi_2 \in \mathscr{O}(\overset{\bullet}{\mathcal{S}}{}_0^{R_2}(I))$ define the same *sectorial germ* $\overset{\vee}{\varphi}$ *of direction* I *at* 0, when φ_1 and φ_2 coincide on a same domain of type $\overset{\bullet}{\mathcal{S}}_0(I)$. We denote by $\overline{\mathscr{O}}^0(I) = \varinjlim_{R \to 0} \mathscr{O}(\overset{\bullet}{\mathcal{S}}{}_0^R(I))$ the space of germs of direction I at 0, and by $\mathscr{O}^0$ the sheaf over $\mathbb{S}^1$ associated with the presheaf $\overline{\mathscr{O}}^0$.

As a rule in this paper for the (pre)sheafs one encounters, the restriction maps are the usual restrictions of functions. We warn the reader that the presheaf $\overline{\mathscr{O}}^0$ is not a sheaf over $\mathbb{S}^1$ (see for instance a counter example given in the second volume of this book [Lod016]) :

for an open arc I, a section $\overset{\vee}{\varphi}\in \mathscr{O}^0(I)=\Gamma(I,\mathscr{O}^0)$ is a collection of holomorphic functions $\varphi_i \in \mathscr{O}(\overset{\bullet}{\Delta}{}_0^{R_i}(I_i))$ that glue together on their intersection domains, the set $\{I_i\}$ being an open covering of I

Example 7.1. We denote by $\mathbb{C}\{\zeta,\zeta^{-1}\}$ the space of Laurent series $\sum_{n\in\mathbb{Z}} a_n\zeta^n$ which converge on a punctured disc $D(0,R)^\star$. This space can also be seen as a constant sheaf over $\mathbb{S}^1$ and the space $\mathscr{O}^0(\mathbb{S}^1)$ of global sections of $\mathscr{O}^0$ on $\mathbb{S}^1$ coincides with $\mathbb{C}\{\zeta,\zeta^{-1}\}$.
For $n \in \mathbb{N}^\star$ and a given direction $\theta_0 \in \mathbb{S}^1$, let us consider the sectorial germ $\overset{\vee}{\varphi}_{\theta_0}(\zeta) = \frac{\zeta^{n-1}\log(\zeta)}{2i\pi\Gamma(n)} \in \mathscr{O}^0_{\theta_0}$, for any given determination of the log. Here $\mathscr{O}^0_{\theta_0}$ denotes the stalk at θ_0 of the sheaf $\mathscr{O}^0$. When making θ varying from θ_0 on $I=]-\pi+\theta_0,\theta_0+\pi[\subset \mathbb{S}^1$, the sectorial germs $\overset{\vee}{\varphi}_\theta\in \mathscr{O}^0_\theta$ glue together and defined a section $\overset{\vee}{\varphi}\in \Gamma(I,\mathscr{O}^0)$ which cannot be prolonged to a global section.

This last example illustrates the need for defining sectorial germs for functions defined on sectors of $\widetilde{\mathbb{C}}$. The covering map $\pi : \widetilde{\mathbb{S}}^1 \to \mathbb{S}^1$ allows to consider the sheaf $\pi^\star\mathscr{O}^0$ over $\widetilde{\mathbb{S}}^1$, that is the inverse image by π of the sheaf $\mathscr{O}^0$ (see [CNP93-1, God73]). For J an open arc of $\widetilde{\mathbb{S}}^1$, an element $\overset{\vee}{\varphi}$ of $\pi^\star\mathscr{O}^0(J)$ appears as an element of the *space $\Gamma(J,\mathscr{O}^0)$ of multivalued sections* of $\mathscr{O}^0$ on J, that is $\overset{\vee}{\varphi}= s(J)$ where s is a continuous map such that the following diagram commutes:

$$\begin{array}{ccc} & & \mathscr{O}^0 = \bigsqcup_{\theta\in\mathbb{S}^1} \mathscr{O}^0_\theta \\ & {}^{s}\nearrow & \downarrow \\ \widetilde{\mathbb{S}}^1 & \underset{\pi}{\longrightarrow} & \mathbb{S}^1 \end{array}$$

. We say that in another way in the following definition:

Definition 7.3. Let $J \in \widetilde{\mathbb{S}}^1$ be an open arc. One says that two functions $\varphi_1 \in \mathscr{O}(\Delta_0^{R_1}(J))$, $\varphi_2 \in \mathscr{O}(\Delta_0^{R_2}(J))$ define the same sectorial germ $\overset{\vee}{\varphi}$ of direction J at 0 when φ_1 and φ_2 coincide on a same domain of type $\Delta_0(J)$. We denote by $\Gamma(J,\mathscr{O}^0)$ the space of multivalued sections of germs of direction J.

Remark 7.1. For any $\omega \in \mathbb{C}$ and by translation, one can of course define $\mathscr{O}^\omega$, the sheaf over $\mathbb{S}^1$ of sectorial germs at ω, associated with the presheaf $\overline{\mathscr{O}}^\omega$.

7.3 Microfunctions

We introduce the sheaf of microfunctions $\mathscr{C}_\omega$ at $\omega \in \mathbb{C}$, in the spirit of [CNP93-1] to whom we refer. Since $\mathscr{C}_\omega$ is deduced from $\mathscr{C} = \mathscr{C}_0$ by translation, we make the focus on the case $\omega = 0$.

7.3.1 Microfunctions - Definitions

We complete definition 3.13.

Definition 7.4. Let θ be a direction and $I =]\alpha,\beta[$ be an open arc (of $\mathbb{S}^1$ or $\widetilde{\mathbb{S}}^1$). we set:

1. $\theta^* = -\theta$ and $I^\star =]-\beta,-\alpha[$ the complex conjugate arc;
2. $\breve{\theta} =]-\frac{\pi}{2}-\theta, -\theta+\frac{\pi}{2}[$ and $\breve{I} = \bigcup_{\theta\in I}\breve{\theta}$;
3. $\check{\theta} =]\theta-3\pi/2, \theta-\pi/2[$ the *copolar* of θ;
4. $\check{I} =]\alpha-3\pi/2, \beta-\pi/2[= \bigcup_{\theta\in I}\check{\theta}$ the *copolar* of I;
5. when $|I| > \pi$, $\widehat{I} =]\alpha+\pi/2, \beta-\pi/2[$; when $|I| < \pi$, $\widehat{I} =]\beta-\pi/2, \alpha+\pi/2[$. When $|I| = \pi$, we set $\widehat{I} = \{\beta-\pi/2\}$.

We would like to define "microfunctions of codirection I at 0". For any open arc $I \subset \mathbb{S}^1$ of length $\leq \pi$, we notice that its copolar $\check{I}$ is of length $\leq \pi$, thus can be seens as an arc of $\mathbb{S}^1$. For such an arc, we set $\check{\overline{\mathscr{O}}}^0(I) = \overline{\mathscr{O}}^0(\check{I})$.
We now remark that for two arcs $I_2 \subseteq I_1$ of lengths $\leq \pi$, one has $\check{I}_2 \subseteq \check{I}_1$. The restriction map $\rho_{\check{I}_2,\check{I}_1} : \overline{\mathscr{O}}^0(\check{I}_1) \to \overline{\mathscr{O}}^0(\check{I}_2)$ gives rise to a restriction map $\check{\rho}_{I_2,I_1} = \rho_{\check{I}_2,\check{I}_1}$ from $\check{\overline{\mathscr{O}}}^0(I_1)$ into $\check{\overline{\mathscr{O}}}^0(I_2)$. This justifies the following definition.

Definition 7.5. Let $I \subset \mathbb{S}^1$ be any open arc of length $\leq \pi$.
One sets $\check{\overline{\mathscr{O}}}^0(I) = \overline{\mathscr{O}}^0(\check{I})$ and $\check{\overline{\mathscr{O}}}^0(I)$ is called the *space of germs of codirection I at* 0.
We denote by $\check{\mathscr{O}}^0$ the corresponding sheaf over $\mathbb{S}^1$.
Viewing $\mathscr{O}_0$ as a constant sheaf over $\mathbb{S}^1$, we set $\mathscr{C} = \check{\mathscr{O}}^0/\mathscr{O}_0$. This quotient sheaf over $\mathbb{S}^1$ is the sheaf of *microfunctions* at 0 and $\mathscr{C}(I) = \Gamma(I,\mathscr{C})$ is the space of sections of *microfunctions of codirection I* at 0.

The sheaf of microfunctions $\mathscr{C}$ makes allusion to Sato's microlocal analysis, see, e.g. [Pha005, KKK86, Mor93]. We mention that microfunctions depending on parameters can be also defined, see for instance [DP99] for a resurgent context.

We mention that $\mathscr{C}(I) = \check{\mathscr{O}}^0(I)/\mathscr{O}_0$, that is the quotient sheaf coincide with the pre-quotient sheaf, because $\mathscr{O}_0$ is a constant sheaf.

In what follows, we transpose with some abuse the notation for singularities introduced in the first volume of this book [MS016] to that for microfunctions.

Definition 7.6. Let $I \subset \mathbb{S}^1$ be any open arc of length $\leq \pi$. We denote by $\overset{\triangledown}{\varphi} = \mathrm{sing}_0^I \overset{\vee}{\varphi} \in \mathscr{C}(I)$ the microfunction of codirection I at 0 defined by the sectorial germ $\overset{\vee}{\varphi} \in \check{\overline{\mathscr{O}}}^0(I)$ of codirection I.

When I is an arc of length $> \pi$, then $\check{I}$ is of length larger than 2π and should be seen as an arc of $\widetilde{\mathbb{S}}^1$. In that case, a microfunction $\overset{\triangledown}{\varphi}$ of $\mathscr{C}(I)$ is represented by an element $\overset{\vee}{\varphi}$ of $\Gamma(\check{I}, \mathscr{O}^0)$.

For any arc $I \subset \mathbb{S}^1$ of length $> \pi$, one can define the variation map var :

$$\mathrm{var} : \overset{\triangledown}{\varphi} \in \mathscr{C}(I) \mapsto \widehat{\varphi} \in \Gamma(\widehat{I}, \mathscr{O}^0), \quad \widehat{\varphi}(\zeta) = \overset{\vee}{\varphi}(\zeta) - \overset{\vee}{\varphi}(\zeta e^{-2i\pi}).$$

Example 7.2. 1. For any $n \in \mathbb{N}$, the sectorial germ $\overset{\vee}{I}_{-n}(\zeta) = \dfrac{(-1)^n}{2i\pi}\dfrac{n!}{\zeta^{n+1}}$ can be seen as a global section of the sheaf $\mathscr{O}^0$. The associated microfunction is equally denoted by $\overset{\triangledown}{I}_{-n}$, $\delta^{(n)}$ or by $\operatorname{sing}_0 \overset{\vee}{I}_{-n}$.
Notice that for any holomorphic germ $\widehat{\varphi} \in \mathscr{O}_0$, the sectorial germ $\widehat{\varphi}\, \overset{\vee}{I}_0$ defines a microfunction $\operatorname{sing}_0(\widehat{\varphi}\, \overset{\vee}{I}_0)$ equal to $\widehat{\varphi}(0)\delta^{(0)} = \widehat{\varphi}(0)\delta$.
2. More generally, the constant sheaf $\mathbb{C}\{\zeta,\zeta^{-1}\}$ over $\mathbb{S}^1$ can be seen as a subsheaf of $\mathscr{C}$ (of vector spaces). Any microfunction $\overset{\triangledown}{\psi}$ of $\mathbb{C}\{\zeta,\zeta^{-1}\}$ can be written as a sum $\sum_{n\geq 0} a_n \overset{\triangledown}{I}_{-n} = \sum_{n\geq 0} a_n \delta^{(n)}$, where the Laurent series $\overset{\vee}{\psi}(\zeta) = \sum_{n\geq 0} a_n \frac{(-1)^n}{2i\pi}\frac{n!}{\zeta^{n+1}}$ converges for $|\zeta| > 0$.
3. We assume that $\widehat{\varphi} \in \mathscr{O}_0$ is a germ of holomorphic function. For any given direction $\theta_0 \in \mathbb{S}^1$, we consider the microfunction $\overset{\triangledown}{\phi}_{\theta_0} = \operatorname{sing}_0^{\theta_0}\left(\widehat{\varphi}\frac{\log}{2i\pi}\right) \in \mathscr{C}_{\theta_0}$ (where $\mathscr{C}_{\theta_0}$ is the stalk at θ_0 of the sheaf $\mathscr{C}$), represented by the sectorial germ $\overset{\vee}{\phi}_{\theta_0} = \widehat{\varphi}\frac{\log}{2i\pi} \in \check{\mathscr{O}}^0_{\theta_0}$, for any given determination of the log (remark that $\overset{\triangledown}{\phi}_{\theta_0}$ does not depend on the chosen determination). Making θ varying from θ_0 up to $\theta_0 + 2\pi$ on $\mathbb{S}^1$, the microfunctions $\overset{\triangledown}{\phi}_{\theta} = \operatorname{sing}_0^{\theta}\left(\widehat{\varphi}\frac{\log}{2i\pi}\right) \in \mathscr{C}_\theta$ glue together and $\overset{\triangledown}{\phi}_{\theta_0} = \overset{\triangledown}{\phi}_{\theta_0+2\pi}$. This provides a global section $\overset{\triangledown}{\phi} = \operatorname{sing}_0\left(\widehat{\varphi}\frac{\log}{2i\pi}\right) \in \Gamma(\mathbb{S}^1,\mathscr{C})$ which does not depend of the chosen determination of the log one started with.
It can be shown (through the variation map) that the space of global sections $\Gamma(\mathbb{S}^1,\mathscr{C})$ of the sheaf of microfunctions, is composed of microfunctions of the form $\overset{\triangledown}{\psi} + \operatorname{sing}_0\left(\widehat{\varphi}\frac{\log}{2i\pi}\right)$, with $\overset{\triangledown}{\psi} \in \mathbb{C}\{\zeta,\zeta^{-1}\}$ and $\widehat{\varphi} \in \mathscr{O}_0$, see [CNP93-1].
4. We suppose $\sigma - 1 \in \mathbb{C}\setminus\mathbb{N}$ and let $\theta \in \mathbb{S}^1$ be a direction. The microfunction $\overset{\triangledown}{\phi}_{\theta} = \operatorname{sing}_0^{\theta}\left(\overset{\vee}{I}_\sigma\right)$, represented by the sectorial germ $\overset{\vee}{I}_\sigma(\zeta) = \dfrac{\zeta^{\sigma-1}}{(1-\mathrm{e}^{-2i\pi\sigma})\Gamma(\sigma)}$, is well-defined once the determination of the log has been chosen. Let us now fix the arc $I =]0,2\pi[$, consider the arc $\check{I} =]-3\pi/2, 3\pi/2[$ as an arc of $\widetilde{\mathbb{S}}^1$ and $\overset{\vee}{I}_\sigma \in \Gamma(\check{I},\mathscr{O}^0)$ as a (uniquely well-defined) multivalued section of $\mathscr{O}^0$ on $\check{I}$. One can apply to its associated microfunction $\overset{\triangledown}{I}_\sigma \in \mathscr{C}(I)$ the variation map and $\operatorname{var}(\overset{\triangledown}{I}_\sigma) = \widehat{I}_\sigma \in \Gamma(\widehat{I},\mathscr{O}^0)$, $\widehat{I} =]\pi/2, 3\pi/2[$, is given by $\widehat{I}_\sigma(\zeta) = \dfrac{\zeta^{\sigma-1}}{\Gamma(\sigma)}$.

7.3.2 Microfunctions and Convolution Product

This subsection is devoted to convolution products of microfunctions. We start with some geometrical preliminaries.

7.3.2.1 Geometrical Preliminaries

Definition 7.7. Let $\varepsilon > 0$ be a real psoitive number and $I \subset \mathbb{S}^1$ be an open sector of length $< \pi$. We set $S_\varepsilon(\widehat{I}) = \bigcup_{\eta \in \overset{\bullet}{\mathcal{S}}{}_0^\infty(\widehat{I})} D(\eta, \varepsilon)$, the "$\varepsilon$-neighbourhood" in $\mathbb{C}$ of the sector $\overset{\bullet}{\mathcal{S}}{}_0^\infty(\widehat{I})$. When the open arc I is of length $= \pi$, then $\widehat{I} = \{\theta\}$ and we set $S_\varepsilon(\widehat{I}) = \bigcup_{s \in \mathbb{R}^+} D(se^{i\theta}, \varepsilon)$. We set $\overset{\bullet}{\mathfrak{S}}_\varepsilon(I) = \mathbb{C} \setminus \overline{S}_\varepsilon(\widehat{I})$ and we denote by $-\partial \overset{\bullet}{\mathfrak{S}}_\varepsilon(I) = \partial S_\varepsilon(\widehat{I})$ the oriented boundary. We denote by $\Gamma_{I,\varepsilon,\eta_1,\eta_2}$ the curve that follows the oriented boundary $-\partial \overset{\bullet}{\mathfrak{S}}_\varepsilon(I)$ from η_1 to η_2. We denote by $\Gamma_{I,\varepsilon}$ the endless curve that follows the oriented boundary $-\partial \overset{\bullet}{\mathfrak{S}}_\varepsilon(I)$.

Lemma 7.1. *Let $\zeta - \overline{S}_\varepsilon(\widehat{I})$ be the convex domain deduced from $\overline{S}_\varepsilon(\widehat{I})$ by the point reflection centered on $\zeta/2 \in \mathbb{C}$. If $\mathrm{dist}(\zeta, S_\varepsilon(\widehat{I})) \geq 2\varepsilon$, then $\zeta - \overline{S}_\varepsilon(\widehat{I}) \subset \overset{\bullet}{\mathfrak{S}}_\varepsilon(I)$. In particular, for every $\zeta \in \overset{\bullet}{\mathfrak{S}}_{2\varepsilon}(I)$, for every $\eta \in (-\partial \overset{\bullet}{\mathfrak{S}}_\varepsilon(I))$, one has $\zeta - \eta \in \overset{\bullet}{\mathfrak{S}}_\varepsilon(I)$.*

Proof. We only consider the case where $I \subset \mathbb{S}^1$ is an open arc of length $< \pi$. We pick an open sector $\overset{\bullet}{\mathcal{S}}{}_0^\infty(\widehat{I})$ and $\zeta \in \mathbb{C} \setminus \overset{\bullet}{\mathcal{S}}{}_0^\infty(\widehat{I})$. Then $\zeta/2 \in \mathbb{C} \setminus \overset{\bullet}{\mathcal{S}}{}_0^\infty(\widehat{I})$ as well. We denote by $\zeta - \overset{\bullet}{\mathcal{S}}{}_0^\infty(\widehat{I})$ the convex domain deduced from $\overset{\bullet}{\mathcal{S}}{}_0^\infty(\widehat{I})$ by the point reflection centered on $\zeta/2 \in \mathbb{C}$. One sees that for every $\xi \in \zeta - \overset{\bullet}{\mathcal{S}}{}_0^\infty(\widehat{I})$, for every $\eta \in \overset{\bullet}{\mathcal{S}}{}_0^\infty(\widehat{I})$, $\mathrm{dist}(\zeta, \overset{\bullet}{\mathcal{S}}{}_0^\infty(\widehat{I})) \leq \mathrm{dist}(\xi, \eta)$ (dist is the euclidean distance). Indeed, by the projection theorem for convex sets, there exist a unique point η_0 on the closure of $\overset{\bullet}{\mathcal{S}}{}_0^\infty(\widehat{I})$ so that $\mathrm{dist}(\zeta, \eta_0) = \mathrm{dist}\big(\zeta, \overset{\bullet}{\mathcal{S}}{}_0^\infty(\widehat{I})\big)$, see Fig. 7.2. One easily shows that the perpendicular

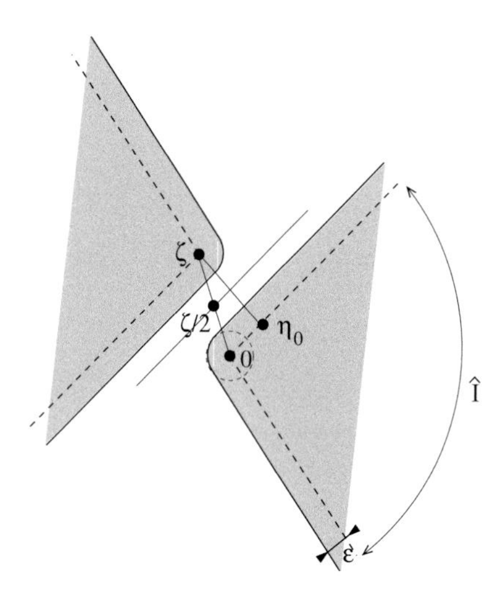

Fig. 7.2 The domain $S_\varepsilon(\widehat{I})$ (left-hand side shaded domain), the domain $\zeta - \overline{S}_\varepsilon(\widehat{I})$ (right-hand side shaded domain.

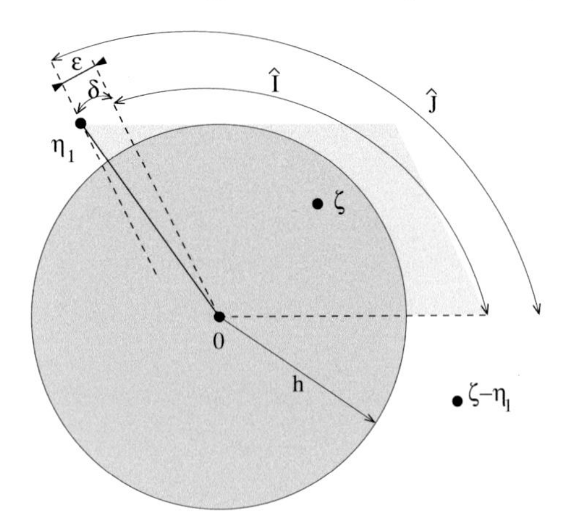

Fig. 7.3 Picture associated with the proof of lemma 7.2.

bisector of the segment $[\zeta,\eta_0]$ separates the two convex sets $\overset{\bullet}{\mathcal{S}}{}_0^{\infty}(\widehat{I})$ and $\zeta-\overset{\bullet}{\mathcal{S}}{}_0^{\infty}(\widehat{I})$. Therefore, if $\mathrm{dist}(\zeta,S_\varepsilon(\widehat{I}))\geq 2\varepsilon$, then $\zeta-\overline{S}_\varepsilon(\widehat{I})\subset\overset{\bullet}{\mathfrak{S}}_\varepsilon(I)$. □

Lemma 7.2. *Let* $I=]\alpha,\beta[\subset\mathbb{S}^1$ *be an open sector of length* $\leq\pi$ *and* $\varepsilon>0$. *We consider* $\eta_1\in(-\partial\overset{\bullet}{\mathfrak{S}}_\varepsilon(I))$ *and we set* $r=|\eta_1|$. *We suppose that* $(\varepsilon/r)<1$ *and we set* $\delta=\arcsin(\varepsilon/r)\in]0,\pi/2[$.

1. *if* $\widehat{J}=]\beta-\pi/2,\alpha+\pi/2+\delta[$ *is an open sector of length* $<\pi$, *we set* $h=r\sin(\widehat{J})$. *Then, for any* $\zeta\in D(0,h)$, $\zeta-\eta_1\in\overset{\bullet}{\mathcal{S}}{}_0^{\infty}(\check{I})$.
2. *if* $\widehat{J}=]\beta-\pi/2,\alpha+\pi/2+\delta[$ *is an open sector of length* $\leq\pi/2$, *then, for any* $\zeta\in D(0,r)$, $\zeta-\eta_1\in\overset{\bullet}{\mathcal{S}}{}_0^{\infty}(\check{I})$.

Proof. Left as an easy exercise. Just look at Fig. 7.3. □

7.3.2.2 Convolution Product of Microfunctions

We pick two microfunctions $\overset{\triangledown}{\varphi}$ and $\overset{\triangledown}{\psi}$ of codirection I, where I is an open arc of length $<\pi$. For any strict subarc $I_1\Subset I$, these microfunctions can be represented by functions $\overset{\vee}{\varphi}$ and $\overset{\vee}{\psi}$ belonging to $\mathscr{O}\big(\overset{\bullet}{\bar{\mathcal{S}}}{}_0^{R+r}(\check{I}_1)\big)$ with $R>r>0$ small enough.
In what follows, we choose $\varepsilon\in]0,\frac{r}{2}\sin(\pi-|\widehat{I}|)[$. We remark that both $\overset{\bullet}{\mathfrak{S}}_{2\varepsilon}(I)\cap D(0,r)$ and $\overset{\bullet}{\mathfrak{S}}_\varepsilon(I_1)\cap D(0,R)$ are non empty domains and $\overset{\bullet}{\mathfrak{S}}_\varepsilon(I_1)\cap D(0,R)\subset\overset{\bullet}{\bar{\mathcal{S}}}{}_0^{R+r}(\check{I}_1)$.
We consider a path $\Gamma=\Gamma_{I_1,\varepsilon,\eta_1,\eta_2}$ that follows the oriented boundary $-\partial\overset{\bullet}{\mathfrak{S}}_\varepsilon(I_1)$ from η_1 to η_2 with $r<|\eta_1|<R$, $r<|\eta_2|<R$, drawn on Fig. 7.4.

For any $\eta\in\Gamma_{I_1,\varepsilon,\eta_1,\eta_2}$ and any $\zeta\in\overset{\bullet}{\mathfrak{S}}_{2\varepsilon}(I)\cap D(0,r)$, $|\zeta-\eta|<R+r$ and we know by lemma 7.1 that $\zeta-\eta\in\overset{\bullet}{\mathfrak{S}}_\varepsilon(I)$. Therefore, the function

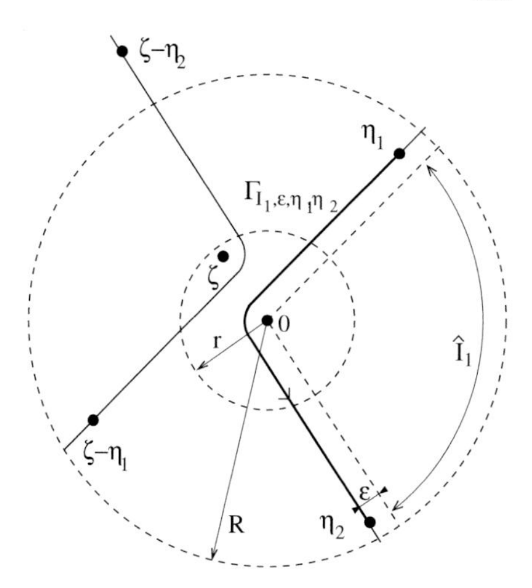

Fig. 7.4 The path of integration $\Gamma_{I_1,\varepsilon,\eta_1,\eta_2}$.

$$\overset{\vee}{\chi}_{I_1,\varepsilon,\eta_1,\eta_2}(\zeta) = \int_{\Gamma_{I_1,\varepsilon,\eta_1,\eta_2}} \overset{\vee}{\varphi}(\eta)\,\overset{\vee}{\psi}(\zeta-\eta)d\eta \tag{7.3}$$

is well-defined for all $\zeta \in \dot{\mathfrak{S}}_{2\varepsilon}(I) \cap D(0,r)$ and is holomorphic on this domain (which is non empty since $2\varepsilon < r$).

Notice that $\overset{\vee}{\chi}_{I_1,\varepsilon,\eta_1,\eta_2}$ can be analytically continued to $\dot{\mathfrak{S}}_{2\varepsilon}(I) \cup D(0,r)$ when $\overset{\vee}{\psi}$ is holomorphic on $D(0,R+r)$, because $|\zeta-\eta| < R+r$ for η on the integration contour and $\zeta \in D(0,r)$. Thus, by linearity, adding to $\overset{\vee}{\psi}$ an element of $\mathscr{O}\big(D(0,R+r)\big)$ results in the addition of an element of $\mathscr{O}\big(D(0,r)\big)$ for $\overset{\vee}{\chi}_{I_1,\varepsilon,\eta_1,\eta_2}$. Similarly when $\overset{\vee}{\varphi}$ is holomorphic on $D(0,R+r)$, then $\overset{\vee}{\chi}_{I_1,\varepsilon,\eta_1,\eta_2}$ can be analytically continued to $\dot{\mathfrak{S}}_{2\varepsilon}(I) \cup D(0,r)$: through an homotopy in $D(0,R)$, just deform the contour $\Gamma_{I_1,\varepsilon,\eta_1,\eta_2}$ into an arc Γ' running from η_1 to η_2 in $\{\eta = s\mathrm{e}^{\mathrm{i}\theta} \,|\, s \in]r,R[,\ \theta \in \widehat{I}\} \subset \overline{S}_\varepsilon(\widehat{I})$; by Cauchy, the two functions $\int_{\Gamma_{I_1,\varepsilon,\eta_1,\eta_2}} \overset{\vee}{\varphi}(\eta)\,\overset{\vee}{\psi}(\zeta-\eta)d\eta$ and $\int_{\Gamma'} \overset{\vee}{\varphi}(\eta)\,\overset{\vee}{\psi}(\zeta-\eta)d\eta$ coincide for $\zeta \in \dot{\mathfrak{S}}_{2\varepsilon}(I) \cap D(0,r)$, while the second integral is holomorphic on $D(0,r)$.

Replacing η_1, η_2 by η_1', η_2' on $-\partial\dot{\mathfrak{S}}_\varepsilon(I_1)$, with $r < |\eta_1'| < R$, $r < |\eta_2'| < R$, results in modifying $\overset{\vee}{\chi}_{I_1,\varepsilon,\eta_1,\eta_2}$ by an element of $\mathscr{O}\big(D(0,h)\big)$ for $h > 0$ small enough: the difference

$$\overset{\vee}{\chi}_{I_1,\varepsilon,\eta_1,\eta_2}(\zeta) - \overset{\vee}{\chi}_{I_1,\varepsilon,\eta_1',\eta_2'} = \left(\int_{\eta_1}^{\eta_1'} + \int_{\eta_2'}^{\eta_2}\right) \overset{\vee}{\varphi}(\eta)\,\overset{\vee}{\psi}(\zeta-\eta)d\eta \tag{7.4}$$

can be analytically continued from $\dot{\mathfrak{S}}_{2\varepsilon}(I) \cap D(0,r)$ to $D(0,h)$. Indeed, using the condition on ε and by lemma 7.2, we see that for η on the two segment contours and for $\zeta \in D(0,h)$ with $0 < h \leq r\sin(\widehat{I})$, $\zeta - \eta$ remains in $\overset{\bullet}{\mathcal{S}}_0^\infty(\check{I}_1) \cap D(0,R+r)$ where $\overset{\vee}{\psi}$ is holomorphic.

Finally replacing ε by a another $\varepsilon' \in]0, \frac{r}{2}\sin(\pi - |\widehat{I}|)[$ yields the same conclusion : for ζ on the intersection domain $\overset{\bullet}{\mathfrak{S}}_{2\varepsilon}(I) \cap \overset{\bullet}{\mathfrak{S}}_{2\varepsilon'}(I) \cap D(0,r)$, one can compare the two functions $\overset{\vee}{\chi}_{I_1,\varepsilon,\eta_1,\eta_2}$ and $\overset{\vee}{\chi}_{I_1,\varepsilon',\eta_1',\eta_2'}$. By Cauchy, the difference reads like (7.4) with the same conclusion.

In particular, we can let $\varepsilon \to 0$ in the above construction: the family of functions $\overset{\vee}{\chi}_{I_1,\varepsilon,\eta_1,\eta_2}$ glue together modulo the elements of $\mathcal{O}_0$, thus providing a microfunction of codirection I_1. Making the arcs $I_1 \subset I$ recovering I, one sees that these microfunctions glue together to give a microfunction of codirection I.

Definition 7.8. Let be I an open arc I of length $< \pi$. We consider two microfunctions of codirection I, $\overset{\triangledown}{\varphi}$ and $\overset{\triangledown}{\psi}$, represented by the sectorial germ of codirection I, $\overset{\vee}{\varphi}$ and $\overset{\vee}{\psi}$ respectively. For a covering of I by open arcs $I_1 \subset I$, the family of functions

$$\overset{\vee}{\varphi} *_\Gamma \overset{\vee}{\psi}(\zeta) = \int_{\Gamma_{I_1,\varepsilon,\eta_1,\eta_2}} \overset{\vee}{\varphi}(\eta) \overset{\vee}{\psi}(\zeta - \eta) d\eta \tag{7.5}$$

with $\Gamma = \Gamma_{I_1,\varepsilon,\eta_1,\eta_2}$, glue together modulo $\mathcal{O}_0$ and provide a microfunction of codirection I denoted by $\overset{\triangledown}{\varphi} * \overset{\triangledown}{\psi}$. It is called the *convolution product of* $\overset{\triangledown}{\varphi}$ *and* $\overset{\triangledown}{\psi}$.

Proposition 7.1. *The sheaf of microfunctions* $\mathscr{C}$ *is a sheaf of* $\mathbb{C}$*-differential convolution algebras, for the derivation* $\overset{\triangledown}{\partial}$: $\mathrm{sing}_0^I(\overset{\vee}{\psi}) \mapsto \mathrm{sing}_0^I(-\zeta \overset{\vee}{\psi})$. *These algebras are commutative, associative and with unit* $\delta = \mathrm{sing}_0\left(\frac{1}{2i\pi}\frac{1}{\zeta}\right)$.

Proof. In what follows we use the previous notation : $\overset{\triangledown}{\varphi}$ and $\overset{\triangledown}{\psi}$ are two microfunctions of codirection I, an open arc of length $< \pi$. One pick a subarc $I_1 \Subset I$ and the microfunctions can be represented by functions $\overset{\vee}{\varphi}$ and $\overset{\vee}{\psi}$ belonging to $\mathcal{O}\big(\overset{\bullet}{\bar{\mathcal{S}}}_0^{R+r}(\check{I}_1)\big)$ with $R > r > 0$ small enough.

We consider the microfunction $\overset{\triangledown}{\psi}_0 = \delta \in \mathscr{C}(\mathbb{S}^1)$ that we represent by $\overset{\vee}{\psi}_0(\zeta) = \widehat{\varphi}_0(\zeta)\overset{\vee}{I}_0(\zeta) = \frac{\widehat{\varphi}_0(\zeta)}{2i\pi\zeta}$ with $\widehat{\varphi}_0 \in \mathcal{O}\big(D(0,R+r)\big)$ and subject to the condition $\widehat{\varphi}_0(0) = 1$. Thus $\overset{\vee}{\varphi} *_\Gamma \overset{\vee}{\psi}_0$ reads:

$$\overset{\vee}{\varphi} *_\Gamma \overset{\vee}{\psi}_0(\zeta) = \frac{1}{2i\pi}\int_{\Gamma_{I_1,\varepsilon,\eta_1,\eta_2}} \overset{\vee}{\varphi}(\eta)\frac{\widehat{\varphi}_0(\zeta - \eta)}{\zeta - \eta} d\eta.$$

By Cauchy and the residue formula, one easily gets that for all $\zeta \in \overset{\bullet}{\bar{\mathcal{S}}}_0^{R+r}(\check{I}_1) \cap D(0,r)$, $\overset{\vee}{\varphi} *_\Gamma \overset{\vee}{\psi}_0 = \overset{\vee}{\varphi} + \mathrm{hol}$, where hol can be analytically continued to $D(0,r)$. This implies that $\overset{\triangledown}{\varphi} * \delta = \overset{\triangledown}{\varphi}$.

We then consider the integral:

$$\overset{\vee}{\varphi} *_{\Gamma\times\Gamma'} \overset{\vee}{\psi}(\zeta) = \frac{1}{2i\pi}\int_{\Gamma\times\Gamma'} \frac{\widehat{\varphi}_0(\zeta-(\xi_1+\xi_2))}{\zeta-(\xi_1+\xi_2)} \overset{\vee}{\varphi}(\xi_1)\overset{\vee}{\psi}(\xi_2)d\xi_1 d\xi_2, \quad (7.6)$$
$$\widehat{\varphi}_0 \in \mathscr{O}\big(D(0,R+r)\big), \quad \widehat{\varphi}_0(0)=1,$$

where $\Gamma = \Gamma_{I_1,\varepsilon,\eta_1,\eta_2}$, $\Gamma' = \Gamma_{I_1,\varepsilon',\eta_1',\eta_2'}$. We remark that for any $(\xi_1,\xi_2)\in\Gamma\times\Gamma'$ one has $(\xi_1+\xi_2)\in \overline{S}_{\varepsilon+\varepsilon'}(\widehat{I}_1)\cap D(0,2R)$. Thus $\overset{\vee}{\varphi} *_{\Gamma\times\Gamma'} \overset{\vee}{\psi}$ defines a holomorphic function on the simply connected domain $\overset{\bullet}{\mathfrak{S}}_{\varepsilon+\varepsilon'}(I_1)$: just apply the Lebesgue dominated convergence theorem for ζ on any connected compact subset of $\overset{\bullet}{\mathfrak{S}}_{\varepsilon+\varepsilon'}(I_1)$. This also allows to use the Fubini theorem:

$$\overset{\vee}{\varphi} *_{\Gamma\times\Gamma'} \overset{\vee}{\psi}(\zeta) = \int_{\Gamma}\left(\frac{1}{2i\pi}\int_{\Gamma'} \frac{\widehat{\varphi}_0(\zeta-(\xi_1+\xi_2))}{\zeta-(\xi_1+\xi_2)}\overset{\vee}{\psi}(\xi_2)d\xi_2\right)\overset{\vee}{\varphi}(\xi_1)d\xi_1$$
$$= \int_{\Gamma'}\left(\frac{1}{2i\pi}\int_{\Gamma} \frac{\widehat{\varphi}_0(\zeta-(\xi_1+\xi_2))}{\zeta-(\xi_1+\xi_2)}\overset{\vee}{\varphi}(\xi_1)d\xi_1\right)\overset{\vee}{\psi}(\xi_2)d\xi_2.$$

From the previous considerations, we recognize $\overset{\vee}{\varphi} *_{\Gamma\times\Gamma'} \overset{\vee}{\psi} = \overset{\vee}{\varphi} *_{\Gamma} \overset{\vee}{\psi}$ +hol for the first equality, $\overset{\vee}{\varphi} *_{\Gamma\times\Gamma'} \overset{\vee}{\psi} = \overset{\vee}{\psi} *_{\Gamma'} \overset{\vee}{\varphi}$ +hol for the second equality, where hol is a holomorphic function that can be analytically continued to a neighbourhood of 0. As a consequence,

$$\overset{\triangledown}{\varphi} * \overset{\triangledown}{\psi} = \overset{\triangledown}{\psi} * \overset{\triangledown}{\varphi},$$

that is the convolution product of microfunctions is commutative. One easily shows in the same way that the convolution product of microfunctions is associative. The fact that $\overset{\triangledown}{\partial}$ is a derivation is obvious. □

We have previously seen two kind of integral representations, $\overset{\vee}{\varphi} *_{\Gamma} \overset{\vee}{\psi}$ (equation (7.5)) and $\overset{\vee}{\varphi} *_{\Gamma\times\Gamma'} \overset{\vee}{\psi}$ (equation (7.6)) for the convolution product $\overset{\triangledown}{\varphi} * \overset{\triangledown}{\psi}$ of two microfunctions. Other representations can be obtained under convenient hypotheses as exemplified by the next proposition.

Proposition 7.2. *Let $\overset{\triangledown}{\psi}$ be a microfunction of codirection I, an open arc of length $<\pi$, represented by the sectorial germ $\overset{\vee}{\psi}$ of codirection I. Let be $\overset{\triangledown}{\varphi}\in\Gamma(\mathbb{S}^1,\mathscr{C})$ a microfunction of the form* $\text{sing}_0\left(\widehat{\varphi}\frac{\log}{2i\pi}\right)$ *with $\widehat{\varphi}\in\mathscr{O}_0$. Then, the microfunction $\overset{\triangledown}{\varphi} * \overset{\triangledown}{\psi}$ of codirection I can be represented modulo $\mathscr{O}_0$ by a family of functions of the form*

$$\int_0^{\eta_1}\widehat{\varphi}(\eta)\overset{\vee}{\psi}(\zeta-\eta)d\eta \quad \textit{and} \quad \int_0^{\eta_2}\widehat{\varphi}(\eta)\overset{\vee}{\psi}(\zeta-\eta)d\eta \qquad (7.7)$$

with η_1, η_2 as for definition 7.8.

The proof is left as an exercise. (See [Sau006]). Starting with the integral representation (7.5), the idea is to decompose the path $\Gamma_{I_1,\varepsilon,\eta_1,\eta_2}$ as on Fig. 7.5 and to use the integrability of the log at the origin.

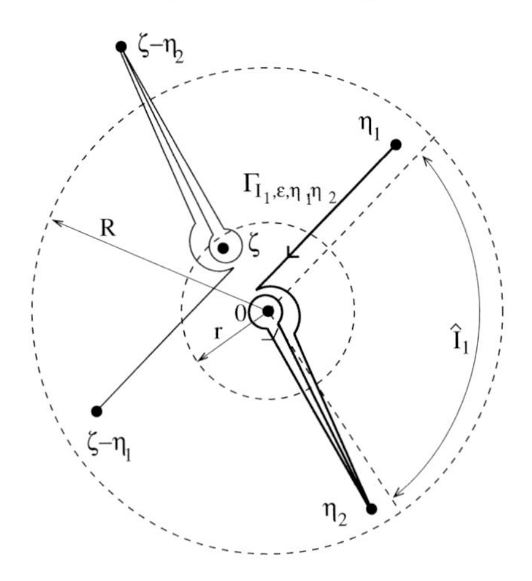

Fig. 7.5 Decomposition of the path $\Gamma_{I_1,\varepsilon,\eta_1,\eta_2}$.

7.4 Space of Singularities

The reader will recognize in what follows classical notions and notation in resurgence theory already encountered in the first volume of this book [MS016], see also [Eca92, Eca93-1, Sau006, OSS003].

7.4.1 Singularities

Definition 7.9. Let $\theta \in \mathbb{R}$ be a direction and $\alpha > 0$. We denote by $\mathrm{ANA}_{\theta,\alpha}$ the space of sections $\Gamma(\check{J}, \mathscr{O}^0)$ where $\check{J} =]\theta - \alpha - 2\pi, \theta + \alpha[\subset \widetilde{\mathbb{S}}^1$, and by $\mathrm{ANA} = \Gamma(\widetilde{\mathbb{S}}^1, \mathscr{O}^0)$ the space of global sections.

Thus, ANA is the space of sectorial germs at 0 that are represented by functions $\overset{\vee}{\varphi}$ holomorphic on a simply connected domain of the form $\boldsymbol{\Delta}_0$.

Definition 7.10. One sets $\mathrm{SING}_{\theta,\alpha} = \mathrm{ANA}_{\theta,\alpha}/\mathscr{O}_0$ and $\mathrm{SING} = \mathrm{ANA}/\mathscr{O}_0$. The elements of these quotient spaces are called *singularities* at 0. One denotes by sing_0 the canonical projection,

$$\mathrm{sing}_0 : \begin{cases} \mathrm{ANA} \to \mathrm{SING} \\ \overset{\vee}{\varphi} \quad\;\; \mapsto \overset{\triangledown}{\varphi} \end{cases}, \qquad \mathrm{sing}_0 : \begin{cases} \mathrm{ANA}_{\theta,\alpha} \to \mathrm{SING}_{\theta,\alpha} \\ \overset{\vee}{\varphi} \qquad\quad \mapsto \overset{\triangledown}{\varphi} \end{cases}.$$

If $\mathrm{sing}_0(\overset{\vee}{\varphi}) = \overset{\triangledown}{\varphi}$, then $\overset{\vee}{\varphi}$ is called a *major* of the singularity $\overset{\triangledown}{\varphi}$.

In particular, with this notation:

Proposition 7.3. *The space of singularities* $\mathrm{SING}_{\theta,\alpha}$ *can be identified with the space* $\Gamma(J, \mathscr{C})$ *of multivalued sections of* $\mathscr{C}$ *by* π*, with* $J =]-\frac{\pi}{2} - \alpha + \theta, \theta + \alpha + \frac{\pi}{2}[$.

Definition 7.11. One defines the spaces SING_{ω}, *resp.* $\mathrm{SING}_{\omega,\theta,\alpha}$ of singularities at $\omega \in \mathbb{C}$, by translation from SING, *resp.* $\mathrm{SING}_{\theta,\alpha}$.

It is of course enough to study the spaces of singularities at 0 and this is what we do in what follows.

Notice that $\mathrm{SING}_{\theta,\alpha}$ and SING are naturally $\mathscr{O}_0$-modules.

Definition 7.12. Let $f \in \mathscr{O}_0$ be a germ of holomorphic functions and let $\overset{\triangledown}{\varphi} = \mathrm{sing}_0 \overset{\vee}{\varphi}$ be a singularity in SING, *resp.* $\mathrm{SING}_{\theta,\alpha}$. One defines the product $f \overset{\triangledown}{\varphi}$ in SING, *resp.* $\mathrm{SING}_{\theta,\alpha}$, by $f \overset{\triangledown}{\varphi} = \mathrm{sing}_0(f \overset{\vee}{\varphi})$.

Definition 7.13. The so-called *variation map* is defined by:

$$\mathrm{var} : \begin{cases} \mathrm{SING} & \to \mathrm{ANA} \\ \overset{\triangledown}{\varphi} = \mathrm{sing}_0(\overset{\vee}{\varphi}) \mapsto \widehat{\varphi}, & \widehat{\varphi}(\zeta) = \overset{\vee}{\varphi}(\zeta) - \overset{\vee}{\varphi}(\zeta e^{-2i\pi}), \end{cases}$$

and $\widehat{\varphi} = \mathrm{var}(\overset{\triangledown}{\varphi})$ is called the *minor* of the singularity $\overset{\triangledown}{\varphi}$.

The variation map var operates similarly on every element $\overset{\triangledown}{\varphi} \in \mathrm{SING}_{\theta,\alpha}$, with $\widehat{\varphi} = \mathrm{var}(\overset{\triangledown}{\varphi})$ in $\Gamma(\widehat{J}, \mathscr{O}^0)$, where $\widehat{J} =]\theta - \alpha, \theta + \alpha[\subset \widetilde{\mathbb{S}}^1$.

A minor is said to be *regular* when it belongs to $\mathscr{O}_0$.

We illustrate the notion of singularities by the following examples. (The reader will recognize sectorial germs used in the introduction of this chapter).

Definition 7.14. The singularities $\overset{\triangledown}{I}_{\sigma}, \overset{\triangledown}{J}_{\sigma,m} \in \mathrm{SING}$, $\sigma \in \mathbb{C}$, $m \in \mathbb{N}$ are defined as follows.

- For $\sigma \in \mathbb{C} \setminus \mathbb{N}^{\star}$, $\overset{\triangledown}{I}_{\sigma} = \mathrm{sing}_0(\overset{\vee}{I}_{\sigma})$ where $\overset{\vee}{I}_{\sigma}(\zeta) = \frac{\zeta^{\sigma-1}}{(1-e^{-2i\pi\sigma})\Gamma(\sigma)}$.

  In particular, $\overset{\triangledown}{I}_{-n} = \delta^{(n)} = \mathrm{sing}_0\left(\dfrac{(-1)^n}{2i\pi}\dfrac{n!}{\zeta^{n+1}}\right)$, $n \in \mathbb{N}$.
- For $n \in \mathbb{N}^{\star}$, $\overset{\triangledown}{I}_n = \mathrm{sing}_0(\overset{\vee}{I}_n)$ with $\overset{\vee}{I}_n(\zeta) = \frac{\zeta^{n-1}\log(\zeta)}{2i\pi\Gamma(n)}$.
- For $m \in \mathbb{N}$ and $\sigma \in \mathbb{C}$, $\overset{\triangledown}{J}_{\sigma,m} = \left(\frac{\partial}{\partial\sigma}\right)^m \overset{\triangledown}{I}_{\sigma}$.

It is useful to define the following subspaces of "integrable singularities", $\mathrm{SING}^{\mathrm{int}} \subset \mathrm{SING}$ and $\mathrm{SING}^{\mathrm{int}}_{\theta,\alpha} \subset \mathrm{SING}_{\theta,\alpha}$.

Definition 7.15. An *integrable minor* is a germ $\widehat{\varphi} \in \mathrm{ANA}$ holomorphic in the domain $\mathcal{S}_0 \subset \widetilde{\mathbb{C}}$ which has a primitive $\widehat{\phi}$ such that $\widehat{\phi} \to 0$ uniformaly in any proper subsector $\bar{\mathcal{S}}'_0 \Subset \mathcal{S}_0$. The space of integrable minors is denoted by $\mathrm{ANA}^{\mathrm{int}}$.

An *integrable singularity* is a singularity $\overset{\triangledown}{\varphi} \in \mathrm{SING}$ which admits a major $\overset{\vee}{\varphi}$ holomorphic in the domain $\mathcal{S}_0 \subset \widetilde{\mathbb{C}}$ such that $\lim_{\zeta \to 0} \zeta \overset{\vee}{\varphi}(\zeta) = 0$ uniformaly in any proper subsector $\bar{\mathcal{S}}'_0 \Subset \mathcal{S}_0$. One denotes by $\mathrm{SING}^{\mathrm{int}}$ the space of integrable singularities.

There is a natural injection $\mathcal{O}_0 \hookrightarrow \mathrm{ANA}^{\mathrm{int}}$ from the space of germs of holomorphic functions to the space $\mathrm{ANA}^{\mathrm{int}}$ of integrable minors. The space $\mathrm{ANA}^{\mathrm{int}}$ can be equipped with a convolution product, by extending the usual law convolution on $\mathcal{O}_0$.

It is not hard to show that integrable singularities satisfy the following property:

Proposition 7.4. *By restriction, the variation map* var *induces a linear isomorphism* $\mathrm{SING}^{\mathrm{int}} \to \mathrm{ANA}^{\mathrm{int}}$. *The inverse map is denoted by* $^\flat : \widehat{\varphi} \in \mathrm{ANA}^{\mathrm{int}} \mapsto {}^\flat\widehat{\varphi} \in \mathrm{SING}^{\mathrm{int}}$.

This allows to transports the convolution law from $\mathrm{ANA}^{\mathrm{int}}$ to $\mathrm{SING}^{\mathrm{int}}$ by the variation map.

Definition 7.16. The convolution product of $\widehat{\varphi}_1, \widehat{\varphi}_2 \in \mathrm{ANA}^{\mathrm{int}}$ is defined by $\widehat{\varphi}_1 * \widehat{\varphi}_2(\zeta) = \int_0^\zeta \widehat{\varphi}_1(\eta)\widehat{\varphi}_1(\zeta-\eta)d\eta$. The convolution of two integrable singularities $\overset{\triangledown}{\varphi}_1 = {}^\flat\widehat{\varphi}_1, \overset{\triangledown}{\varphi}_2 = {}^\flat\widehat{\varphi}_2 \in \mathrm{SING}^{\mathrm{int}}$ is given by : $\overset{\triangledown}{\varphi}_1 * \overset{\triangledown}{\varphi}_2 = {}^\flat\big(\widehat{\varphi}_1 * \widehat{\varphi}_2\big)$.

Quite similarly:

Definition 7.17. A minor $\widehat{\varphi}$ holomorphic on the domain $\mathcal{S}_0(\widehat{I}) \subset \widetilde{\mathbb{C}}$ is said to be *integrable* if $\widehat{\varphi}$ has a primitive $\widehat{\phi}$ such that $\widehat{\phi} \to 0$ uniformaly in any proper subsector $\bar{\mathcal{S}}_0' \Subset \mathcal{S}_0(\widehat{I})$. One denotes by $\mathrm{ANA}^{\mathrm{int}}_{\theta,\alpha}$ the space of these integrable minors.
An *integrable singularity* is a singularity $\overset{\triangledown}{\varphi} \in \mathrm{SING}_{\theta,\alpha}$ which has a major $\overset{\vee}{\varphi}$ holomorphic in the domain $\mathcal{S}_0(\check{I}) \subset \widetilde{\mathbb{C}}$ and such that $\lim_{\zeta\to 0} \zeta \overset{\vee}{\varphi}(\zeta) = 0$ uniformaly in any proper subsector $\bar{\mathcal{S}}_0' \Subset \mathcal{S}_0(\check{I})$. One denotes $\mathrm{SING}^{\mathrm{int}}_{\theta,\alpha}$ the space of these integrable singularities.

Proposition 7.5. *By restriction, the variation map* var *induces a linear isomorphism* $\mathrm{SING}^{\mathrm{int}}_{\theta,\alpha} \to \mathrm{ANA}^{\mathrm{int}}_{\theta,\alpha}$. *The inverse map is denoted by* $^\flat : \widehat{\varphi} \in \mathrm{ANA}^{\mathrm{int}}_{\theta,\alpha} \mapsto {}^\flat\widehat{\varphi} \in \mathrm{SING}^{\mathrm{int}}_{\theta,\alpha}$.

We end with further definitions.

Definition 7.18. Any singularity $\overset{\triangledown}{\varphi}$ of the form $\overset{\triangledown}{\varphi} = a\delta + {}^\flat\widehat{\varphi}$ with $\widehat{\varphi} \in \mathcal{O}_0$ is said to be *simple*. The space of simple singularities is denoted by $\mathrm{SING}^{\mathrm{simp}}$.
The space $\mathrm{SING}^{\mathrm{s.ram}}$ of *simply ramified* singularities is the vector space spanned by $\mathrm{SING}^{\mathrm{simp}}$ and the set of singularities $\{\overset{\triangledown}{I}_{-n}, n \in \mathbb{N}\}$.

7.4.2 Convolution Product of Singularities

The resurgence theory asserts that the space of singularities SING can be equipped with a convolution product [Eca81-1, Eca85, MS016], see also [CNP93-1, Ou012]. Since $\mathrm{SING}_{\theta,\alpha}$ can be identified with the space $\Gamma(J,\mathscr{C})$ of multivalued sections of $\mathscr{C}$ by π, with $J =]-\frac{\pi}{2}-\alpha+\theta, \theta+\alpha+\frac{\pi}{2}[$, the convolution product for microfunctions (proposition 7.1) allows to transport this product to $\mathrm{SING}_{\theta,\alpha}$: for any two singularities $\overset{\triangledown}{\varphi}, \overset{\triangledown}{\psi} \in \mathrm{SING}_{\theta,\alpha}$ and any strict subarc $I \Subset J$ of length $< \pi$, one can find two majors $\overset{\vee}{\varphi}, \overset{\vee}{\psi} \in \mathrm{ANA}_{\theta,\alpha}$ that can be represented by holomorphic functions on a sector $\mathcal{S}_0(\check{I})$. By projection on $\mathbb{C}$, one can think of $\overset{\vee}{\varphi}, \overset{\vee}{\psi}$ as belonging to $\mathcal{O}(\overset{\bullet}{\mathcal{S}}_0(\check{I}))$, that is sectorial germs of codirection I. By restriction, $\overset{\triangledown}{\varphi}, \overset{\triangledown}{\psi}$ are seen as microfunctions of codirection I, whose convolution product $\overset{\triangledown}{\varphi} * \overset{\triangledown}{\psi} \in \Gamma(I, \mathscr{C})$ can be represented either by

$$\overset{\vee}{\varphi} *_\Gamma \overset{\vee}{\psi}(\zeta) = \int_\Gamma \overset{\vee}{\varphi}(\eta)\, \overset{\vee}{\psi}(\zeta-\eta)\,d\eta \tag{7.8}$$

or by

$$\overset{\vee}{\varphi} *_{\Gamma\times\Gamma} \overset{\vee}{\psi}(\zeta) = \frac{1}{2i\pi}\int_{\Gamma\times\Gamma} \frac{f(\zeta-(\xi_1+\xi_2))}{\zeta-(\xi_1+\xi_2)}\, \overset{\vee}{\varphi}(\xi_1)\, \overset{\vee}{\psi}(\xi_2)\,d\xi_1 d\xi_2, \tag{7.9}$$

with $f \in \mathcal{O}_0$ and $f(0) = 1$ (cf. (7.5) and (7.6)), where $\Gamma = \Gamma_{I,\varepsilon,\eta_1,\eta_2}$ is as in definition 7.7. When considering a covering of J by such arcs I, these sections glue together to give the convolution product $\overset{\triangledown}{\varphi} * \overset{\triangledown}{\psi}$ as a multivalued section of $\mathscr{C}$ over J.

Proposition 7.6. *The space* SING *can be equipped with a convolution product denoted by* $*$ *that makes it a commutative convolution algebra, with unit* $\delta = \mathrm{sing}_0\left(\dfrac{1}{2i\pi\zeta}\right) = \overset{\triangledown}{I}_0$. *Moreover:*

1. *the linear operator,* $\overset{\triangledown}{\partial}: \overset{\triangledown}{\varphi} = \mathrm{sing}_0(\overset{\vee}{\varphi}) \in \mathrm{SING} \mapsto \overset{\triangledown}{\partial}\overset{\triangledown}{\varphi} = \mathrm{sing}_0(-\zeta\, \overset{\vee}{\varphi}) \in \mathrm{SING}$, *is a derivation.*
2. *if* $\overset{\triangledown}{\varphi}$ *and* $\overset{\triangledown}{\psi}$ *belong to* $\mathrm{SING}^{\mathrm{int}}$, *then* $\overset{\triangledown}{\varphi} * \overset{\triangledown}{\psi}$ *belongs to* $\mathrm{SING}^{\mathrm{int}}$ *and* ${}^\flat\widehat{\varphi} * {}^\flat\widehat{\varphi} = {}^\flat(\widehat{\varphi} * \widehat{\varphi})$. *In particular, the space of simple singularities* $\mathrm{SING}^{\mathrm{simp}}$ *is a convolution subalgebra.*

Theses properties remain true when one considers $\mathrm{SING}_{\theta,\alpha}$ *instead of* SING.

Proof. We have already shown that $\mathrm{SING}_{\theta,\alpha}$ (thus SING) is a commutative convolution algebra for the convolution product with unit δ. The equality ${}^\flat\widehat{\varphi} * {}^\flat\widehat{\varphi} = {}^\flat(\widehat{\varphi} * \widehat{\varphi})$ for integrable singularities, emerges from considerations on integrals and is left as an exercise. (Start with proposition 7.2. See [Sau006]). □

7.5 Formal Laplace Transform, Formal Borel Transform

7.5.1 Formal Laplace Transform for Microfunctions

We start with the following definition.

Definition 7.19. Let $I \subset \mathbb{S}^1$ be an open arc and $r \geq 0$ be a nonegative real number. we denote by:

1. $\overline{\mathscr{A}}^{\leq 0}(\dot{\mathcal{S}}_r^\infty(I))$ the $\mathbb{C}$-differential algebra of holomorphic functions φ on $\dot{\mathcal{S}}_r^\infty(I)$ that satisfy the property : for any proper subdomain $\bar{\dot{\mathcal{S}}}^\infty \Subset \dot{\mathcal{S}}_r^\infty(I)$, for any $\varepsilon > 0$, there exists $C > 0$ so that for all $z \in \bar{\dot{\mathcal{S}}}^\infty$, $|\varphi(z)| \leq C e^{\varepsilon |z|}$;
2. $\overline{\mathscr{A}}^{\leq 0}(I) = \varinjlim_{r \to \infty} \overline{\mathscr{A}}^{\leq 0}(\dot{\mathcal{S}}_r^\infty(I))$. This defines a presheaf $\overline{\mathscr{A}}^{\leq 0}$;
3. $\mathscr{A}^{\leq 0}$ the sheaf over $\mathbb{S}^1$ associated with the presheaf $\overline{\mathscr{A}}^{\leq 0}$.

Remark 7.2. The fact that $\mathscr{A}^{\leq 0}$ is indeed a sheaf of differential algebras is an exercise left to the reader. (We stress that the derivation considered is the usual one for holomorphic functions).
The sheaf $\mathscr{A}^{\leq 0}$ should not be confused with the sheaf $\mathscr{A}^{<0}$ of flat germs at infinity (definition 3.17). As a matter of fact, $\overline{\mathscr{A}}^{<0}(I) \subset \overline{\mathscr{A}}(I) \subset \overline{\mathscr{A}}^{\leq 0}(I)$ where $\overline{\mathscr{A}}$ stands for the presheaf of asymptotic functions (see definition 3.17 and [Lod016, Mal91, Mal95]).
We mention that our definition of $\mathscr{A}^{\leq 0}$ differs from that of Malgrange in [Mal91] where $\mathscr{A}^{\leq 0}$ is defined as the sheaf of sectorial germs that admit an asymptotics belonging to the formal Nilsson class, that is of the form $\sum \widetilde{w}(z) \frac{\log^m(z)}{z^\sigma}$, $\sigma \in \mathbb{C}$, $m \in \mathbb{N}$, $\widetilde{w} \in \mathbb{C}[[z^{-1}]]$. Our sheaf $\mathscr{A}^{\leq 0}$ contains this sheaf as a subsheaf. However, the constructions in the sequel resemble in much aspects to that of Malgrange [Mal91].

The following Lemma is left to the reader as an exercise. This will allow us in a moment to properly define the quotient sheaf $\mathscr{A}^{\leq 0}/\mathscr{A}^{\leq -1}$ over $\mathbb{S}^1$.

Lemma 7.3. *The space $\overline{\mathscr{A}}^{\leq -1}(\dot{\mathcal{S}}^\infty)$, resp. $\overline{\mathscr{A}}^{\leq -1}(I)$, of 1-exponentially flat functions on $\dot{\mathcal{S}}^\infty$, resp. of 1-exponentially flat germs at infinity over I, is a differential ideal of $\overline{\mathscr{A}}^{\leq 0}(\dot{\mathcal{S}}^\infty(I))$ –resp. of $\overline{\mathscr{A}}^{\leq 0}$.*

Definition 7.20. Let θ be any direction (of $\mathbb{S}^1$ or $\widetilde{\mathbb{S}}^1$). We denote by R_θ the ray $]0, e^{i\theta}\infty[$. For $\kappa > \varepsilon \geq 0$, we set $R_{\theta,\varepsilon} =]\varepsilon e^{i\theta}, e^{i\theta}\infty[$ and $R_{\theta,\varepsilon;\kappa} =]\varepsilon e^{i\theta}, \kappa e^{i\theta}[$.
For any closed arc $\bar{J} = [\theta_1, \theta_2]$, we denote by $\gamma_{\bar{J},\varepsilon}$, *resp.* $\gamma_{\bar{J},\varepsilon;\kappa}$, the Hankel contour, *resp.* truncated Hankel contour, which consists in following:

1. $R_{\theta_1,\varepsilon}$, *resp.* $R_{\theta_1,\varepsilon;\kappa}$, backward,
2. then the circular arc $\delta_{\bar{J},\varepsilon} = \{\varepsilon e^{i\theta} \,|\, \theta \in \bar{J}\}$ oriented in the anti-clockwise way,
3. finally $R_{\theta_2,\varepsilon}$, *resp.* $R_{\theta_2,\varepsilon;\kappa}$, forward.

Let us pick an open arc I of $\mathbb{S}^1$ of length $\leq \pi$, and a microfunction $\overset{\triangledown}{\varphi} \in \mathscr{C}(I)$ of codirection I, represented by the germ $\overset{\vee}{\varphi} \in \check{\mathscr{O}}^0(I)$. For any open arc $I_1 =]\alpha_1, \beta_1[$ with $\bar{I}_1 \Subset I$, one can find $R > 0$ so that the restriction of $\overset{\vee}{\varphi}$ to $\check{I}_1 =]\alpha_1 - 3\pi/2, \beta_1 - \pi/2[\subset \mathbb{S}^1$ is represented by a function (still denoted by $\overset{\vee}{\varphi}$) holomorphic in the sector $\mathcal{S}_0^R(\check{I}_1)$. We consider another open arc $I_2 =]\alpha_2, \beta_2[$, $\bar{I}_2 \subset I_1$, so that $\check{I}_1 \setminus \check{\bar{I}}_2$ has two connected

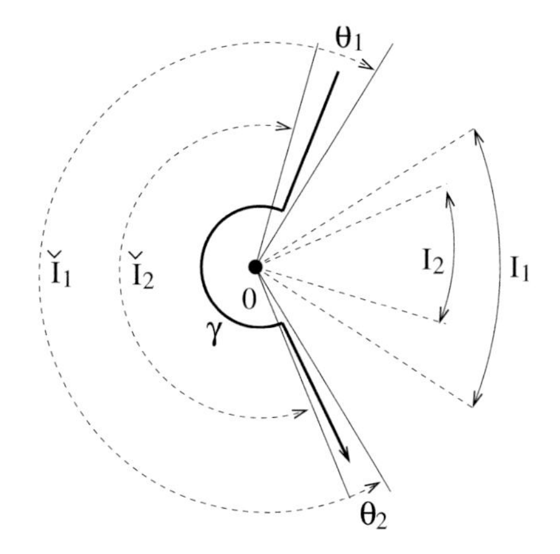

Fig. 7.6 Formal Laplace transform. The open arcs I_1, I_2, $\check{I}_1$, $\check{\check{I}}_2$, and the path $\gamma = \gamma_{[\theta_1,\theta_2],\varepsilon;\kappa}$.

components. We choose one arbitrary direction in each component, $\theta_1 \in]\alpha_1 - 3\pi/2, \alpha_2 - 3\pi/2[$, $\theta_2 \in]\beta_2 - \pi/2, \beta_1 - \pi/2[$. For $R > \kappa > \varepsilon > 0$, we consider the truncated Laplace integral $\varphi_{\theta_1,\theta_2,\kappa}(z) = \int_{\gamma_{[\theta_1,\theta_2],\varepsilon;\kappa}} \mathrm{e}^{-z\zeta} \overset{\vee}{\varphi}(\zeta)d\zeta$, see Fig. 7.6.

The function $\varphi_{\theta_1,\theta_2,\kappa}$ satisfies the following properties:

- $\varphi_{\theta_1,\theta_2,\kappa}$ is an entire function, since one integrates on a (relatively) compact path of the domain of holomorphy of $\overset{\vee}{\varphi}$.
- for $\varepsilon > 0$ chosen as small as we want, we set $M = \sup_{\bar{\mathbf{\Delta}}_\varepsilon^\kappa(]\theta_1,\theta_2[)} |\overset{\vee}{\varphi}|$. then:
  - for all $z \in \mathbb{C}$, $\left| \int_{\delta_{[\theta_1,\theta_2],\varepsilon}} \mathrm{e}^{-z\zeta} \overset{\vee}{\varphi}(\zeta)d\zeta \right| \leq \varepsilon |\check{I}_1| M \mathrm{e}^{\varepsilon|z|}$ where $|\check{I}_1| = \beta_1 - \alpha_1 + \pi$;
  - for any $r > 0$, for every $z \in \overset{\bullet}{\Pi}{}_r^{\theta_1}$, $\left| \int_{R_{\theta_1,\varepsilon;\kappa}} \mathrm{e}^{-z\zeta} \overset{\vee}{\varphi}(\zeta)d\zeta \right| \leq \kappa M \mathrm{e}^{-\varepsilon r}$. Similarly,

    for every $z \in \overset{\bullet}{\Pi}{}_r^{\theta_2}$, $\left| \int_{R_{\theta_2,\varepsilon;\kappa}} \mathrm{e}^{-z\zeta} \overset{\vee}{\varphi}(\zeta)d\zeta \right| \leq \kappa M \mathrm{e}^{-\varepsilon r}$.
  - the domain $\overset{\bullet}{\Pi}{}_r^{\theta_1}$ contains any closed sector of the form $\overset{\bullet}{\bar{\mathbf{\Delta}}}{}_{r'}^{\infty}(J_1)$ with J_1 an open arc so that $\bar{J}_1 \subset]-\frac{\pi}{2} - \theta_1, -\theta_1 + \frac{\pi}{2}[$ and $r' > 0$ large enough. Since $\beta_2 - \frac{\pi}{2} < \theta_1 < \alpha_2 + \frac{\pi}{2}$, one deduces that $\overset{\bullet}{\Pi}{}_r^{\theta_1}$ contains any closed sector of the form $\overset{\bullet}{\bar{\mathbf{\Delta}}}{}_{r'}^{\infty}(I_2^\star)$ with $r' > 0$ large enough. Similarly, $\overset{\bullet}{\Pi}{}_r^{\theta_2}$ contains any closed sector of the form $\overset{\bullet}{\bar{\mathbf{\Delta}}}{}_{r'}^{\infty}(I_2^\star)$ with $r' > 0$ large enough.

  From this analysis, since $\varepsilon > 0$ can be chosen arbitrarily small, we retain that $\varphi_{\theta_1,\theta_2,\kappa}$ belongs to the space $\overline{\mathscr{A}}^{\leq 0}(\overset{\bullet}{\mathbf{\Delta}}{}_r^{\infty}(I_2^\star))$, $r > 0$ large enough.
- Furthermore, looking at the above analysis and by Cauchy, we may observe that for two cut-off points $\kappa, \kappa' \in]\varepsilon, R[$, for two directions $\theta_1' \in]\alpha_1 - 3\pi/2, \alpha_2 - 3\pi/2[$, $\theta_2' \in]\beta_2 - \pi/2, \beta_1 - \pi/2[$ the difference $\varphi_{\theta_1,\theta_2,\kappa} - \varphi_{\theta_1',\theta_2',\kappa'}$ belongs to $\overline{\mathscr{A}}^{\leq -1}(\overset{\bullet}{\mathbf{\Delta}}{}_r^{\infty}(I_2^\star))$

with $r > 0$ large enough. We finally remark that adding to $\overset{\vee}{\varphi}$ a function holomorphic on $D(0,R)$ only affects $\varphi_{\theta_1,\theta_2,\kappa}(z)$ by the addition of an element of $\overline{\mathscr{A}}^{\leq -1}(\overset{\bullet}{\mathcal{S}}{}_r^{\infty}(I_2^\star))$, $r > 0$ large enough.

In this way, one obtains a morphism, $\mathscr{L}(I,I_2) : \overset{\triangledown}{\varphi} \in \mathscr{C}(I) \mapsto \overset{\vartriangle}{\varphi} \in \overline{\mathscr{A}}^{\leq 0}(I_2^\star)/\overline{\mathscr{A}}^{\leq -1}(I_2^\star)$, $\overset{\vartriangle}{\varphi} = \mathrm{cl}(\varphi_{\theta_1,\theta_2,\kappa})$, which is obviosuly compatible with the restriction maps. This allows to move up to stalks, $\mathscr{L}_\alpha : \mathscr{C}_\alpha \to \left(\mathscr{A}^{\leq 0}/\mathscr{A}^{\leq -1}\right)_{\alpha^*}$ and finally[2] to a morphism of sheaves $\mathscr{L} : \mathscr{C} \to \mathscr{A}^{\leq 0}/\mathscr{A}^{\leq -1}$.

Definition 7.21. One calls *formal Laplace transform* for microfunctions at 0, the morphism of sheaves $\mathscr{L} : \mathscr{C} \to \mathscr{A}^{\leq 0}/\mathscr{A}^{\leq -1}$. The quotient sheaf $\mathscr{A}^{\leq 0}/\mathscr{A}^{\leq -1}$ over $\mathbb{S}^1$ is called the *sheaf of asymptotic classes*. An asymptotic class is usually denoted by $\overset{\vartriangle}{\varphi}$.

The term "sheaf of asymptotic classes" is borrowed from [CNP93-1] where the sheaf $\mathscr{A}^{\leq 0}$ is denoted by $\mathscr{E}^0$, and the sheaf $\mathscr{A}^{\leq -1}$ is denoted by $\mathscr{E}^-$. The notation $\overset{\vartriangle}{\varphi}$ is own.

Example 7.3. For $(\sigma,m) \in \mathbb{C} \times \mathbb{N}$ and $I =]-\pi/2, \pi/2[\in \mathbb{S}^1$, we consider the microfunction $\overset{\triangledown}{J}_{\sigma,m} = \mathrm{sing}_0^I \left(\overset{\vee}{J}_{\sigma,m}\right) \in \mathscr{C}(I)$ represented by the sectorial germ $\overset{\vee}{J}_{\sigma,m} = \left(\frac{\partial}{\partial \sigma}\right) \overset{\vee}{I}_\sigma \in \check{\mathscr{O}}^0(I)$ and the branch of the log such that $\arg(\log \zeta) \in \check{I} =]-2\pi, 0[$. By standard formulae recalled in Sect. 7.1, one readily gets that its formal Laplace transform $\overset{\vartriangle}{J}_{\sigma,m} = \mathscr{L}(I)\, \overset{\triangledown}{J}_{\sigma,m}$ is an asymptotic class that can be represented by the (sectorial germ at infinity of) holomorphic function(s) $(-1)^m \frac{\log^m(z)}{z^\sigma} \in \overline{\mathscr{A}}^{\leq 0}(I^\star)$, $I^\star =]-\pi/2, \pi/2[$ with the determination of the log so that $\arg(\log z) \in I^\star$.

The following proposition is a straight consequence of the very construction of the formal Laplace transform.

Proposition 7.7. *The formal Laplace transform $\mathscr{L} : \mathscr{C} \to \mathscr{A}^{\leq 0}/\mathscr{A}^{\leq -1}$ satisfies the identity : $\mathscr{L} \circ \overset{\triangledown}{\partial} = \partial \circ \mathscr{L}$.*

7.5.2 Formal Borel transform for asymptotic classes

Let $I^\star \subset \mathbb{S}^1$ be an open arc with length $\leq \pi$ and $\varphi \in \mathscr{A}^{\leq 0}(I^\star)$ be a sectorial germ at infinity. For any open arc $I_1^\star \Subset I^\star$, one can find $r > 0$ so that the restriction of φ to $I_1^\star$ is (represented by) a holomorphic function (still denoted by φ) on the domain $\overset{\bullet}{\mathcal{S}}{}_r^{\infty}(I_1^\star)$. We set $\overset{\vee}{\varphi}_{z_1,\alpha}(\zeta) = -\frac{1}{2i\pi}\int_{R_{\alpha,z_1}} e^{z\zeta}\varphi(z)dz$ for any $z_1 \in \overset{\bullet}{\mathcal{S}}{}_r^{\infty}(I_1^\star)$ and any direction $\alpha \in I_1^\star$, see Fig. 7.7. We can make the following observations about this Laplace integral $\overset{\vee}{\varphi}_{z_1,\alpha}$:

[2] Modulo complex conjugation

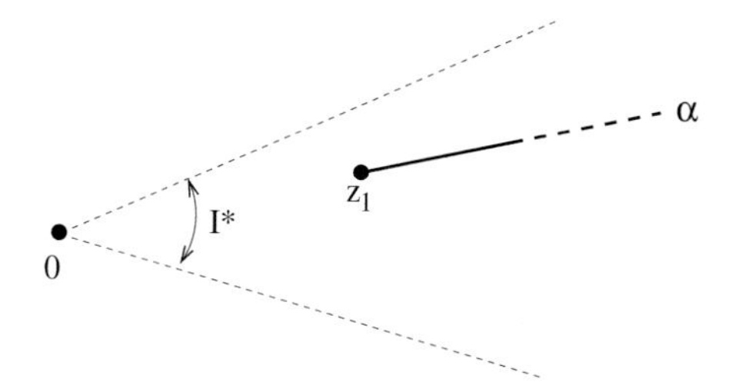

Fig. 7.7 Formal Borel transform. The open arcs $I^\star$, and the path R_{α,z_1}.

- since φ belongs to $\overline{\mathscr{A}}^{\le 0}(\overset{\bullet}{\boldsymbol{\Delta}}_r^\infty(I_1^\star))$, we know that for any proper subsector $\overset{\bullet}{\bar{\boldsymbol{\Delta}}}_{r_1}^\infty(J^\star) \Subset \overset{\bullet}{\boldsymbol{\Delta}}_r^\infty(I_1^\star)$ and any $\varepsilon > 0$, there exists $C > 0$ so that, for all $z \in \overset{\bullet}{\bar{\boldsymbol{\Delta}}}_{r_1}^\infty(J^\star)$, $|\varphi(z)| \le C e^{\varepsilon|z|}$. Therefore $\overset{\vee}{\varphi}_{z_1,\alpha}$ belongs to $\mathscr{O}(\overset{\bullet}{\Pi}_\varepsilon^{\alpha+\pi})$ when $z_1 \in \overset{\bullet}{\bar{\boldsymbol{\Delta}}}_{r_1}^\infty(J^\star)$ and $\alpha \in \bar{J}^\star$. Making α varying in $J^\star$ and since $\varepsilon > 0$ can be chosen arbitrarily small, these functions glue together by Cauchy, and provide a holomorphic function $\overset{\vee}{\varphi}_{z_1,J^\star}$ on $\overset{\bullet}{\mathscr{D}}(J^\star, 0) = \overset{\bullet}{\boldsymbol{\Delta}}_0^\infty(\check{J})$. Notice that for two points $z_1, z_2 \in \overset{\bullet}{\bar{\boldsymbol{\Delta}}}_{r_1}^\infty(J^\star)$, the difference $\overset{\vee}{\varphi}_{z_2,J^\star} - \overset{\vee}{\varphi}_{z_1,J^\star}$ defines an entire function (with at most exponential growth of order 1 at infinity). Therefore, localising near the origin, we get a sectorial germ $\overset{\vee}{\varphi}_{z_1,I^\star} \in \mathscr{O}(\check{I}) = \check{\mathscr{O}}(I)$, defined modulo the elements of $\mathscr{O}_0$, that is a microfunction of codirection I;
- when φ belongs to $\overline{\mathscr{A}}^{\le -1}(I^\star)$, one easily sees from the above analysis that $\overset{\vee}{\varphi}_{z_1,I^\star}$ is holomorphic on a domain containing a full neighbourhood of the origin, thus by localisation, an element of $\mathscr{O}_0$.

To conclude, we have defined a morphism (of $\mathbb{C}$-differential algebras), $\mathscr{B}(I^\star) : \overset{\triangle}{\varphi} \in \mathscr{A}^{\le 0}(I^\star)/\mathscr{A}^{\le -1}(I^\star) \mapsto \overset{\triangledown}{\varphi} = \mathrm{cl}\,(\overset{\vee}{\varphi}_{z_1,I^\star}) \in \mathscr{C}(I)$ whose compatibility with the restriction maps is easy to check.

Definition 7.22. The morphism of sheaves $\mathscr{B} : \mathscr{A}^{\le 0}/\mathscr{A}^{\le -1} \to \mathscr{C}$ is called the *formal Borel transform*

The formal Laplace transform for microfunctions and the formal Borel transform for asymptotic classes are isomorphisms of sheaves, as shown in [CNP93-1] to whom we refer:

Proposition 7.8. *The morphisms $\mathscr{L} : \mathscr{C} \to \mathscr{A}^{\le 0}/\mathscr{A}^{\le -1}$ and $\mathscr{B} : \mathscr{A}^{\le 0}/\mathscr{A}^{\le -1} \to \mathscr{C}$ are isomorphisms of sheaves and* $\mathscr{L} \circ \mathscr{B} = \mathrm{Id}$, $\mathscr{B} \circ \mathscr{L} = \mathrm{Id}$.

Remark 7.3. The morphism of sheaves $\widehat{\varphi} \in \mathscr{O}_0 \mapsto \overset{\triangledown}{\varphi} = \mathrm{sing}_0^I\left(\widehat{\varphi}\frac{\log}{2i\pi}\right) \in \mathscr{C}(I)$ is injective as already mentioned. The following commutative diagram makes a link between the formal Laplace transform for regular minor, *resp.* formal Borel transform for 1-Gevrey formal series, and the formal Laplace transform for microfunctions, *resp.* formal Borel transform for asymptotic classes:

$$\begin{array}{ccc} \mathscr{O}_0 & \hookrightarrow & \mathscr{C} \\ \mathscr{L} \downarrow\uparrow \mathscr{B} & & \mathscr{B} \uparrow\downarrow \mathscr{L} \\ \mathscr{A}_1/\mathscr{A}^{\le -1} & \hookrightarrow & \mathscr{A}^{\le 0}/\mathscr{A}^{\le -1} \end{array}.$$

7.5.3 *Formal Laplace Transform for Singularities and Back to Convolution Product*

In the sequel, we translate to singularities what have obtained so far for microfunctions.

7.5.3.1 Formal Laplace Transform for Singularities

We start with two definitions.

Definition 7.23. Let $\theta \in \widetilde{\mathbb{S}}^1$ be a direction and $\alpha > 0$. We denote by $\mathrm{ASYMP}_{\theta,\alpha}$ the space of asymptotic classes defined as multivalued sections of $\mathscr{A}^{\leq 0}/\mathscr{A}^{\leq -1}$ on $J^\star =]-\pi/2-\alpha-\theta, -\theta+\alpha+\pi/2[$. We denote by ASYMP the space of asymptotic classes given by global sections of $\mathscr{A}^{\leq 0}/\mathscr{A}^{\leq -1}$ on $\widetilde{\mathbb{S}}^1$.

Definition 7.24. Let $\sigma \in \mathbb{C}$ be a complex number and $m \in \mathbb{N}$. We denote by $\overset{\vartriangle}{I}_\sigma \in$ ASYMP the asymptotic class represented by $1/z^\sigma$. We denote by $\overset{\vartriangle}{J}_{\sigma,m} \in$ ASYMP the asymptotic class represented by $(-1)^m \frac{\log^m(z)}{z^\sigma}$. We often simply write $1/z^\sigma$ instead of $\overset{\vartriangle}{I}_\sigma$ and similarly for $\overset{\vartriangle}{J}_{\sigma,m}$.

We have already said that the space of singularities $\mathrm{SING}_{\theta,\alpha}$ can be identified with the space $\Gamma(J,\mathscr{C})$ of multivalued sections of $\mathscr{C}$ by π, with $J =]-\frac{\pi}{2}-\alpha+\theta, \theta+\alpha+\frac{\pi}{2}[$. The formal Laplace transform for microfunctions thus extends to singularities, by inverse image:

$$\begin{array}{ccccc} & & \bigsqcup_{\dot\beta\in\mathbb{S}^1}\mathscr{C}_{\dot\beta} & \overset{\mathscr{L}}{\to} & \bigsqcup_{\dot\beta^\star\in\mathbb{S}^1}\left(\mathscr{A}^{\leq 0}/\mathscr{A}^{\leq -1}\right)_{\dot\beta^\star} \\ & s\nearrow & \downarrow & & \downarrow \\ \widetilde{\mathbb{S}}^1 \supset J \ni \beta & \underset{\pi}{\longrightarrow} & \mathbb{S}^1 \ni \dot\beta & \underset{\star}{\to} & \mathbb{S}^1 \ni \dot\beta^\star \end{array}$$

When returning to the very construction of the formal Laplace transform (Sect. 7.5.1), one sees that for any singularity $\overset{\triangledown}{\varphi} \in \mathrm{SING}_{\theta,\alpha}$, for any direction $\beta \in \widehat{J} =]-\alpha+\theta, \theta+\alpha[$, setting $\breve{\beta}^\star =]-\frac{\pi}{2}+\beta, \beta+\frac{\pi}{2}[$, the formal Laplace transform $\mathscr{L}(\breve{\beta}^\star)\,\overset{\triangledown}{\varphi}$ is given as the class $\overset{\vartriangle}{\varphi} = \mathrm{cl}(\varphi_{\beta-2\pi,\beta,\kappa}) \in \mathscr{A}^{\leq 0}(\breve{\beta})/\mathscr{A}^{\leq -1}(\breve{\beta})$ where $\breve{\beta} =]-\frac{\pi}{2}-\beta, -\beta+\frac{\pi}{2}[$ and $\varphi_{\beta-2\pi,\beta,\kappa}(z) = \displaystyle\int_{\gamma_{[\beta-2\pi,\beta],\varepsilon;\kappa}} \mathrm{e}^{-z\zeta}\,\overset{\vee}{\varphi}(\zeta)d\zeta$, with $\overset{\vee}{\varphi}$ any major of $\overset{\triangledown}{\varphi}$. This introduces the following definition. (Notice that $\breve{\widehat{J}} = J^\star$).

Definition 7.25. The morphism $\mathscr{L}^\beta = \mathscr{L}(\breve{\beta}^\star) : \mathrm{SING}_{\theta,\alpha} \to \mathscr{A}^{\leq 0}(\breve{\beta})/\mathscr{A}^{\leq -1}(\breve{\beta})$ is called the *formal Laplace transform* in the direction $\beta \in \widehat{J} =]-\alpha+\theta, \theta+\alpha[$.
For any singularity $\overset{\triangledown}{\varphi} \in \mathrm{SING}_{\theta,\alpha}$, one denotes by $\mathscr{L}^{\widehat{J}}\,\overset{\triangledown}{\varphi} \in \mathrm{ASYMP}_{\theta,\alpha}$ the asymptotic class given by the collection $\left(\mathscr{L}^\beta\,\overset{\triangledown}{\varphi}\right)_{\beta\in\widehat{J}}$.

Example 7.4. We continue the example 7.3 but for the fact that we now consider $\overset{\triangledown}{J}_{\sigma,m}$ as a singularity in $\mathrm{SING}_{0,\pi}$. The formal Laplace transform $\mathscr{L}^{]-\pi,\pi[}\,\overset{\triangledown}{J}_{\sigma,m}$ is the asymptotic class $\overset{\triangle}{J}_{\sigma,m}\in \mathrm{ASYMP}_{0,\pi}$ seen by restriction as an element of $\Gamma(]-3\pi/2,3\pi/2[,\mathscr{A}^{\leq 0}/\mathscr{A}^{\leq -1})$.

We linger for a moment at the cases of singularities of the form $\overset{\triangledown}{\varphi}= {}^\flat\widehat{\varphi}\in \mathrm{SING}^{\mathrm{int}}_{\theta,\alpha}$. For any direction $\beta\in]-\alpha+\theta,\theta+\alpha[$, the formal Laplace transform $\overset{\triangle}{\varphi}=\mathscr{L}^{\beta}\,\overset{\triangledown}{\varphi}$, $\overset{\triangle}{\varphi}\in\mathscr{A}^{\leq 0}(\breve{\beta})/\mathscr{A}^{\leq -1}(\breve{\beta})$, can be represented by the function

$$\varphi_{\beta-2\pi,\beta,\kappa}(z)=\int_{\gamma_{[\beta-2\pi,\beta],\varepsilon;\kappa}}\mathrm{e}^{-z\zeta}\,\overset{\vee}{\varphi}(\zeta)d\zeta=\int_{R_{\beta,0;\kappa}}\mathrm{e}^{-z\zeta}\,\widehat{\varphi}(\zeta)d\zeta, \tag{7.10}$$

and we thus recover the "usual" formal Laplace transform (see Sect. 7.1). In particular, we recall that we have extended the convolution law to $\mathrm{SING}^{\mathrm{int}}_{\theta,\alpha}$ by the variation map: for $\overset{\triangledown}{\varphi}_1= {}^\flat\widehat{\varphi}_1,\overset{\triangledown}{\varphi}_2= {}^\flat\widehat{\varphi}_2\in\mathrm{SING}^{\mathrm{int}}_{\theta,\alpha}$, $\overset{\triangledown}{\varphi}_1*\overset{\triangledown}{\varphi}_2= {}^\flat\big(\widehat{\varphi}_1*\widehat{\varphi}_2\big)$. The above remark (7.10) shows that $\mathscr{L}^{\beta}(\overset{\triangledown}{\varphi}_1*\overset{\triangledown}{\varphi}_2)=(\mathscr{L}^{\beta}\,\overset{\triangledown}{\varphi}_1)(\mathscr{L}^{\beta}\,\overset{\triangledown}{\varphi}_2)$, by the properties of the "usual" formal Laplace transform.

We now assume that $\overset{\triangledown}{\varphi}$ is a simple singularity, $\overset{\triangledown}{\varphi}=a\delta+{}^\flat\widehat{\varphi}\in\mathrm{SING}^{\mathrm{simp}}$ with $\widehat{\varphi}\in\mathscr{O}_0$. For any open arc $\widehat{J}=]-\alpha+\theta,\theta+\alpha[$, the formal Laplace transform $\overset{\triangle}{\varphi}=\mathscr{L}^{\widehat{J}}(a\delta+\overset{\triangledown}{\varphi})$ is an asymptotic class which belongs to $\Gamma(J^\star,\mathscr{A}_1/\mathscr{A}^{\leq -1})$. This again comes from (an analogue of) the identity (7.10) and classical arguments recalled in the introduction of this chapter.

Definition 7.26. One denotes by $\mathrm{ASYMP}^{\mathrm{simp}}$ the subspace of asymptotic classes obtained by injection of the global sections $\Gamma(\widetilde{\mathbb{S}}^1,\mathscr{A}_1/\mathscr{A}^{\leq -1})$ into ASYMP.

Proposition 7.9. *The restriction of the formal Laplace transform $\mathscr{L}$ to* $\mathrm{SING}^{\mathrm{simp}}$ *has* $\mathrm{ASYMP}^{\mathrm{simp}}$ *for its range.*

Remark 7.4. Consider a formal series $\widetilde{\varphi}\in\mathbb{C}[[z^{-1}]]$ and an open arc of the form $J^\star=]-\pi/2-\alpha-\theta,-\theta+\alpha+\pi/2[\subset\widetilde{\mathbb{S}}^1$. By the Borel-Ritt theorem, there are infinitely many $\varphi\in\overline{\mathscr{A}}(J^\star)$ whose Poincaré asymptotics $T(J^\star)\varphi$ is given by $\widetilde{\varphi}$ on $J^\star$. These various φ differ by flat germs, that is elements of $\overline{\mathscr{A}}^{<0}(J^\star)$. Therefore as a rule, these germs φ represent different asymptotic classes $\overset{\triangle}{\varphi}\in\mathrm{ASYMP}_{\theta,\alpha}$.
Now suppose that $\widetilde{\varphi}$ is 1-Gevrey and choose a (good) covering (I_i) of $J^\star$ where each I_i is an open arc of length less than π. By the Borel-Ritt theorem for 1-Gevrey asymptotics and for each subscript i, there exists $\varphi_i\in\overline{\mathscr{A}}_1(I_i)$ whose 1-Gevrey asymptotics $T_1(I_i)\varphi_i$ is φ. Moreover, each φ_i is uniquely defined this way up to 1-exponentially flat germs, that is up to elements of $\overline{\mathscr{A}}^{\leq -1}(I_i)$. One thus gets a uniquely defined section $\overset{\triangle}{\varphi}\in\Gamma(J^\star,\mathscr{A}_1/\mathscr{A}^{\leq -1})$ that can be thought of as an asymptotic class. One can characterize another way this asymptotic class $\overset{\triangle}{\varphi}\in\mathrm{ASYMP}^{\mathrm{simp}}$

by settling $\overset{\triangle}{\varphi}= \mathscr{L}(a\delta+ \overset{\triangledown}{\varphi})$ where $\overset{\triangledown}{\varphi}=^{\flat}\widehat{\varphi}$ with $\widehat{\varphi}$ the minor of $\widetilde{\varphi}$ while a is its constant term.

Definition 7.27. The mapping $^{\natural} : \widetilde{\varphi} \in \mathbb{C}[[z^{-1}]]_1 \mapsto \overset{\triangle}{\varphi}=^{\natural}\widetilde{\varphi} \in \mathrm{ASYMP}^{\mathrm{simp}}$ is defined by $\overset{\triangle}{\varphi}= \mathscr{L}(a\delta+ \overset{\triangledown}{\varphi})$ where $\overset{\triangledown}{\varphi}=^{\flat}\widehat{\varphi}$, whereas $\widehat{\varphi}$ stands for the minor of $\widetilde{\varphi}$ and a its constant term.

Obviously, the mapping $^{\natural}$ is an isomorphism, the inverse map being the 1-Gevrey Taylor map. This allows to merge $^{\natural}\widetilde{\varphi}$ with $\widetilde{\varphi}$ in practice.

7.5.3.2 Back to Convolution Product

We have said without proof that $\mathscr{L}$ and $\mathscr{B}$ are morphisms of sheaves of algebras. It is thus certainly worthy to prove the following proposition.

Proposition 7.10. *For any two singularities* $\overset{\triangledown}{\varphi}_1, \overset{\triangledown}{\varphi}_2 \in \mathrm{SING}_{\theta,\alpha}$ *and any direction* $\beta \in]-\alpha+\theta, \theta+\alpha[$, *the following properties hold:*
$(\mathscr{L}^{\beta}\, \overset{\triangledown}{\varphi}_1)(\mathscr{L}^{\beta}\, \overset{\triangledown}{\varphi}_1) = \mathscr{L}^{\beta}(\overset{\triangledown}{\varphi}_1 * \overset{\triangledown}{\varphi}_2)$ *and* $\mathscr{L}^{\beta}(\overset{\triangledown\triangledown}{\partial\varphi_1}) = \partial \mathscr{L}^{\beta}\, \overset{\triangledown}{\varphi}_1$.

Proof. (Adapted from [CNP93-1]). Let $\overset{\triangledown}{\varphi}_1, \overset{\triangledown}{\varphi}_2 \in \mathrm{SING}_{\theta,\alpha}$ be two singularities with majors $\overset{\vee}{\varphi}_1, \overset{\vee}{\varphi}_2$. We pick a direction $\beta \in \widehat{J} =]-\alpha+\theta, \theta+\alpha[$ and we consider the formal Laplace transforms $\overset{\triangle}{\varphi}_1= \mathscr{L}^{\beta}\, \overset{\triangledown}{\varphi}_1$ and $\overset{\triangle}{\varphi}_2= \mathscr{L}^{\beta}\, \overset{\triangledown}{\varphi}_2$. These are elements of $\mathscr{A}^{\leq 0}(\breve{\beta})/\mathscr{A}^{\leq -1}(\breve{\beta})$ which can be represented respectively by

$$\varphi_1(z) = \int_{\gamma_1} \mathrm{e}^{-z\zeta}\, \overset{\vee}{\varphi}_1(\zeta)d\zeta \in \overline{\mathscr{A}}^{\leq 0}(\overset{\bullet}{\Delta}{}^{\infty}_r(\breve{\beta})),\ \varphi_2(z) = \int_{\gamma_2} \mathrm{e}^{-z\zeta}\, \overset{\vee}{\varphi}_2(\zeta)d\zeta \in \overline{\mathscr{A}}^{\leq 0}(\overset{\bullet}{\Delta}{}^{\infty}_r(\breve{\beta})),$$

with $\gamma_1 = \gamma_{[\beta-2\pi,\beta],\varepsilon_1;\kappa_1}$, $\gamma_2 = \gamma_{[\beta-2\pi,\beta],\varepsilon_2;\kappa_2}$ and some $r>0$ large enough. The product $\overset{\triangle}{\varphi}_1\overset{\triangle}{\varphi}_2 \in \mathscr{A}^{\leq 0}(\breve{\beta})/\mathscr{A}^{\leq -1}(\breve{\beta})$ is thus represented by

$$\varphi_1\varphi_2(z) = \int_{\gamma_1\times\gamma_2} \mathrm{e}^{-z(\zeta_1+\zeta_2)}\, \overset{\vee}{\varphi}_1(\zeta_1)\, \overset{\vee}{\varphi}_2(\zeta_2)d\zeta_1 d\zeta_2 \in \overline{\mathscr{A}}^{\leq 0}(\overset{\bullet}{\Delta}{}^{\infty}_r(\breve{\beta})).$$

Let us look at the formal Borel transform $\mathscr{B}(\breve{\beta})(\overset{\triangle}{\varphi}_1\overset{\triangle}{\varphi}_2) \in \mathscr{C}(\breve{\beta}^{\star})$. This Borel transform can be represented by the integral $\overset{\vee}{(\varphi_1\varphi_2)}_{z_1,\alpha_1}(\zeta) = -\dfrac{1}{2i\pi}\displaystyle\int_{R_{\alpha_1,z_1}} \mathrm{e}^{z\zeta}\varphi_1\varphi_2(z)dz$ with $z_1 \in \overset{\bullet}{\Delta}{}^{\infty}_{r_1}(\overset{\star}{\breve{\beta}})$, $r_1>r$, and for any direction $\alpha_1 \in \breve{\beta}^{\star}$. The function $\overset{\vee}{(\varphi_1\varphi_2)}_{z_1,\alpha_1}(\zeta)$ is holomorphic on $\overset{\bullet}{\Pi}{}_0^{\alpha_1+\pi}$ (go back to the construction of the formal Borel transform, Sect. 7.5.2). Taking $\zeta \in \overset{\bullet}{\Pi}{}_{2\varepsilon}^{\alpha_1+\pi}$ with $\varepsilon > \varepsilon_1+\varepsilon_2$, we can apply Fubini.

Remark that $\zeta_1 + \zeta_2$ (or rather $\overset{\bullet}{\zeta}_1 + \overset{\bullet}{\zeta}_2$) remains in the bounded strip $\{\zeta \in \mathbb{C} \mid \mathrm{dist}(\zeta, \mathrm{e}^{i\beta}[0,\kappa]) \leq \varepsilon_1 + \varepsilon_2\}$, for $(\zeta_1, \zeta_2) \in \gamma_1 \times \gamma_2$. Thus $\zeta - (\zeta_1 + \zeta_2)$ remains in the domain $\overset{\bullet}{\Pi}{}_{\varepsilon}^{\alpha_1+\pi}$ for $\zeta \in \overset{\bullet}{\Pi}{}_{2\varepsilon}^{\alpha_1+\pi}$ and this ensures the integrability conditions.

This way, we get:

$$\begin{aligned}
\overset{\vee}{(\varphi_1 \varphi_2)}_{z_1,\alpha_1}(\zeta) &= -\frac{1}{2i\pi}\int_{R_{\alpha_1,z_1}} \mathrm{e}^{z\zeta}\left(\int_{\gamma_1\times\gamma_2} \mathrm{e}^{-z(\zeta_1+\zeta_2)}\overset{\vee}{\varphi}_1(\zeta_1)\overset{\vee}{\varphi}_2(\zeta_2)d\zeta_1 d\zeta_2\right)dz \\
&= \int_{\gamma_1\times\gamma_2} \frac{\mathrm{e}^{z_1(\zeta-\zeta_1-\zeta_2)}}{2i\pi(\zeta-\zeta_1-\zeta_2)}\overset{\vee}{\varphi}_1(\zeta_1)\overset{\vee}{\varphi}_2(\zeta_2)d\zeta_1 d\zeta_2 \\
&= \int_{\gamma_1}\left(\int_{\gamma_2} \frac{\mathrm{e}^{z_1(\zeta-\zeta_1-\zeta_2)}}{2i\pi(\zeta-\zeta_1-\zeta_2)}\overset{\vee}{\varphi}_2(\zeta_2)d\zeta_2\right)\overset{\vee}{\varphi}_1(\zeta_1)d\zeta_1
\end{aligned}$$

Returning to the very construction of the convolution product for singularities, we see that $\overset{\vee}{(\varphi_1 \varphi_2)}_{z_1,\alpha_1}$ is nothing but a major of the singularity $\mathrm{sing}_0\left(\frac{\mathrm{e}^{z_1\zeta}}{2i\pi\zeta}\right) * \overset{\triangledown}{\varphi}_1 * \overset{\triangledown}{\varphi}_2$. But $\mathrm{sing}_0\left(\frac{\mathrm{e}^{z_1\zeta}}{2i\pi\zeta}\right) = \delta$ and therefore $\mathrm{sing}_0\left(\overset{\vee}{(\varphi_1 \varphi_2)}_{z_1,\alpha_1}\right) = \overset{\triangledown}{\varphi}_1 * \overset{\triangledown}{\varphi}_2$. From Proposition 7.8, we know that $\mathscr{B} \circ \mathscr{L} = \mathrm{Id}$ (when considering $\mathscr{B}$ and $\mathscr{L}$ as morphisms of sheaves), thus the conclusion. The last statement as been already seen. □

Example 7.5. We know by theorem 3.3 that the formal series $\widetilde{w}_{(0,0)}$ solution of the prepared ODE (3.6) associated with the first Painlevé equation, is 1-Gevrey. Its minor $\widehat{w}_{(0,0)} = \mathscr{B}\widetilde{w}_{(0,0)}$ is thus a germ of holomorphic functions at the origin and we set $\overset{\triangledown}{w}_{(0,0)} = {}^{\flat}\widehat{w}_{(0,0)} \in \mathrm{SING}^{\mathrm{simp}}$. We now consider the singularity $\overset{\triangledown}{I}_\sigma * \overset{\triangledown}{w}_{(0,0)} \in \mathrm{SING}$, for any $\sigma \in \mathbb{C}$. By proposition 7.10, for an arbitrary direction $\beta \in \widetilde{\mathbb{S}}^1$, the formal Laplace transform $\mathscr{L}^\beta(\overset{\triangledown}{I}_\sigma * \overset{\triangledown}{w}_{(0,0)}) \in \mathscr{A}^{\leq 0}(\breve{\beta})/\mathscr{A}^{\leq -1}(\breve{\beta})$ is the asymptotic class of direction $\breve{\beta}$ which reads also as:

$$\mathscr{L}^\beta(\overset{\triangledown}{I}_\sigma * \overset{\triangledown}{w}_{(0,0)}) = \mathscr{L}^\beta(\overset{\triangledown}{I}_\sigma)\mathscr{L}^\beta(\overset{\triangledown}{w}_{(0,0)}).$$

On the one hand, $\mathscr{L}^\beta \overset{\triangledown}{I}_\sigma$ is the asymptotic class $\overset{\vartriangle}{I}_\sigma \in \Gamma(\breve{\beta}, \mathscr{A}^{\leq 0}/\mathscr{A}^{\leq -1})$. On the other hand, $\mathscr{L}^\beta \overset{\triangledown}{w}_{(0,0)} = {}^{\natural}\widetilde{w}_{(0,0)}$. Therefore, $\mathscr{L}^\beta(\overset{\triangledown}{I}_\sigma * \overset{\triangledown}{w}_{(0,0)}) = \overset{\vartriangle}{I}_\sigma {}^{\natural}\widetilde{w}_{(0,0)}$ that can be identified with $\frac{1}{z^\sigma}\widetilde{w}_{(0,0)}$ with the branch of z^σ determined by the condition $\arg z \in \breve{\beta}$.

Example 7.6. We now use the notation of Sect. 3.4.2.2 but for the fact that we consider arcs on $\widetilde{\mathbb{S}}^1$. We write $\widehat{I_0} =]0,\pi[$ and $I_0^\star =]-3\pi/2, \pi/2[\subset \widetilde{\mathbb{S}}^1$ and in what follows with think of the Laplace-Borel sum $w_{tri,0} = \mathscr{S}^{\widehat{I_0}}\widetilde{w}_{(0,0)}$ as (representing) a multivalued section of $\mathscr{A}_1$ on $I_0^\star$. Similarly, we set $\widehat{I_1} =]\pi, 2\pi[$ and $I_1^\star =]-5\pi/2, -\pi/2[\subset \widetilde{\mathbb{S}}^1$ and think of $w_{tri,1} = \mathscr{S}^{\widehat{I_1}}\widetilde{w}_{(0,0)}$ as an element of $\Gamma(I_1^\star, \mathscr{A}_1)$. Notice that $I_0^\star \cap I_1^\star =]-3\pi/2, -\pi/2[$ on $\widetilde{\mathbb{S}}^1$. Since both $w_{tri,0}$ and $w_{tri,1}$ are asymptotic to the

1-Gevrey series $\widetilde{w}_{(0,0)}$, we know that the difference $w_{tri,0} - w_{tri,1}$ is a multivalued section of $\mathscr{A}^{\leq -1}$ on $I_0^\star \cap I_1^\star$. Therefore, for any $\sigma \in \mathbb{C}$, $\frac{1}{z^\sigma} w_{tri,0}$ and $\frac{1}{z^\sigma} w_{tri,1}$ glue together to give a multivalued section of $\mathscr{A}^{\leq 0}/\mathscr{A}^{\leq -1}$ on $I_0^\star \cup I_1^\star$, that can be identified with the asymptotic class $\overset{\triangle}{I}_\sigma \natural \widetilde{w}_{(0,0)} \in \mathrm{ASYMP}_{\pi,\pi}$. The formal Borel transform $\mathscr{B}(I_0^\star)\left(\overset{\triangle}{I}_\sigma \natural \widetilde{w}_{(0,0)}\right)$ is the multivalued section of $\mathscr{C}$ on $I_0 =]-\pi/2, 3\pi/2[$ which can be thought of as a singularity in $\mathrm{SING}_{\pi/2,\pi/2}$, and is given by $\mathscr{B}(I_0^\star)\left(\overset{\triangle}{I}_\sigma \natural \widetilde{w}_{(0,0)}\right) = \overset{\triangledown}{I}_\sigma * \overset{\triangledown}{w}_{(0,0)}$. Similarly, the formal Borel transform $\mathscr{B}(I_1^\star)\left(\overset{\triangle}{I}_\sigma \natural \widetilde{w}_{(0,0)}\right)$ is the multivalued section of $\mathscr{C}$ on $I_1 =]\pi/2, 5\pi/2[$ which provides a singularity in $\mathrm{SING}_{3\pi/2,\pi/2}$, of the form $\mathscr{B}(I_1^\star)\left(\overset{\triangle}{I}_\sigma \natural \widetilde{w}_{(0,0)}\right) = \overset{\triangledown}{I}_\sigma * \overset{\triangledown}{w}_{(0,0)}$. These two singularities glue together as the element $\overset{\triangledown}{I}_\sigma * \overset{\triangledown}{w}_{(0,0)}$ of $\mathrm{SING}_{\pi,\pi}$.

7.5.3.3 Formal Laplace Transform for Singularities at ω

The spaces SING_ω, *resp.* $\mathrm{SING}_{\omega,\theta,\alpha}$ of singularities at $\omega \in \mathbb{C}$ are the translated of SING, *resp.* $\mathrm{SING}_{\theta,\alpha}$. (See definition 7.11). By its very construction, the formal Laplace transform brings the translation into the multilplication by an exponential.

Definition 7.28. The formal Laplace transform $\mathscr{L}$ sends SING_ω, *resp.* $\mathrm{SING}_{\omega,\theta,\alpha}$, onto the space denoted by $\mathrm{e}^{-\omega z}\mathrm{ASYMP}$, *resp.* $\mathrm{e}^{-\omega z}\mathrm{ASYMP}_{\theta,\alpha}$, made of *asymptotic classes with support based at* ω.

We mention the following result that can be thought of as an analogue of the Watson's lemma [Lod016].

Lemma 7.4. *For any* $\omega \in \mathbb{C}^\star$, *the sum of the two* $\mathbb{C}$*-vector spaces* $\mathrm{ASYMP}_{\theta,\alpha}$ *and* $\mathrm{e}^{-\omega z}\mathrm{ASYMP}_{\theta,\alpha}$ *is direct.*

Proof. We consider an asymptotic class $\overset{\triangle}{\varphi} \in \mathrm{ASYMP}_{\theta,\alpha}$. By definition, one can find a (good) open covering (J_j) of $J^\star =]-\pi/2-\alpha-\theta, -\theta+\alpha+\pi/2[$ and a "0-cochain" $\left(\varphi_j \in \mathscr{A}^{\leq 0}(J_j)\right)_j$ with associated "1-coboundary" $\left(\varphi_{j+1} - \varphi_j \in \mathscr{A}^{\leq -1}(J_{j+1} \cap J_j)\right)_j$ that represents $\overset{\triangle}{\varphi}$. Now assume that $\overset{\triangle}{\varphi}$ also belongs to $\mathrm{e}^{-\omega z}\mathrm{ASYMP}_{\theta,\alpha}$. Considering a refinement of (J_j) if necessary, one deduces that $\varphi_j \in \mathscr{A}^{\leq -1}(J_j)$ for at least one j, since $J^\star$ is an arc of length $> \pi$. This implies that the formal Borel transform $\overset{\triangledown}{\varphi} \in \mathrm{SING}_{\theta,\alpha}$ has a major $\overset{\vee}{\varphi}$ which can be analytically continued to 0, thus $\overset{\triangledown}{\varphi} = 0$ and as a consequence $\overset{\triangle}{\varphi} = 0$. $\square$

7.6 Laplace Transforms

We develop here only matters convenient for this course. For more general nonsense on Laplace transforms in the framework of resurgent analysis, see [CNP93-1, Cos009, Eca81-1, Eca85, Mal91].

7.6.1 Laplace Transforms

Definition 7.29. Let $I \subset \mathbb{S}^1$ be an open arc and $r \geq 0$. We denote by:

1. $\mathscr{E}^{\leq 1}(I)$ the $\mathbb{C}$-differential algebra of holomorphic functions φ on $\overset{\bullet}{\mathcal{S}}{}_0^\infty(I)$ with 1-exponential growth at infinity on the direction I : for any proper subsector $\overset{\bullet}{\bar{\mathcal{S}}}{}^\infty \Subset \overset{\bullet}{\mathcal{S}}{}_0^\infty(I)$, there exist $C > 0$ and $\tau > 0$ so that, for all $z \in \overset{\bullet}{\bar{\mathcal{S}}}{}^\infty$, $|\varphi(z)| \leq C\mathrm{e}^{\tau|z|}$;
2. when I is of length $\leq \pi$, $\check{\mathscr{E}}^{\leq 1}(I) = \mathscr{E}^{\leq 1}(\check{I})$ is the space of holomorphic functions φ on $\overset{\bullet}{\mathcal{S}}{}_0^\infty(\check{I})$, with 1-exponential growth at infinity on the codirection I.
3. $\mathscr{E}^{\leq 1}$, *resp.* $\check{\mathscr{E}}^{\leq 1}$, the sheaf over $\mathbb{S}^1$ corresponding to the family $\left(\mathscr{E}^{\leq 1}(I)\right)$, *resp.* $\check{\mathscr{E}}^{\leq 1}(I)$;
4. $\mathscr{O}(\mathbb{C})^{\leq 1}$ the space of entire functions with 1-exponential growth at infinity on every direction.

Pick an open arc $I \subset \mathbb{S}^1$ of length $\leq \pi$, and a function $\overset{\vee}{\varphi} \in \check{\mathscr{E}}^{\leq 1}(I)$. Thus $\overset{\vee}{\varphi}$ is holomorphic on $\overset{\bullet}{\mathcal{S}}{}_0^\infty(\check{I})$ and for any open arc I_1 so that $\bar{I}_1 \subset I$, for any $\varepsilon > 0$, there exist $C > 0$ and $\tau > 0$ so that, for all $\zeta \in \overset{\bullet}{\bar{\mathcal{S}}}{}_\varepsilon^\infty(\check{I}_1)$, $|\overset{\vee}{\varphi}(\zeta)| \leq C\mathrm{e}^{\tau|\zeta|}$. We consider the following Laplace integral,

$$\varphi_{I_1}(z) = \int_{\gamma_{[\theta_1,\theta_2],\varepsilon}} \mathrm{e}^{-z\zeta} \overset{\vee}{\varphi}(\zeta)d\zeta = \left(-\int_{R_{\theta_1,\varepsilon}} + \int_{\delta_{[\theta_1,\theta_2],\varepsilon}} + \int_{R_{\theta_2,\varepsilon}}\right) \overset{\vee}{\varphi}(\zeta)d\zeta,$$

where $\check{I}_1 =]\theta_1, \theta_2[$ (for the contour of integration, see definition 7.20). This Laplace integral can be decomposed as follows:

- by classical arguments, the integral $\displaystyle\int_{R_{\theta_1,\varepsilon}} \mathrm{e}^{-z\zeta} \overset{\vee}{\varphi}(\zeta)d\zeta$ defines a holomorphic function on $\overset{\bullet}{\Pi}{}_\tau^{\theta_1}$ and we observe that for any $r > \tau$, for every $z \in \overset{\bullet}{\overline{\Pi}}{}_r^{\theta_1}$,

$$\left|\int_{R_{\theta_1,\varepsilon}} \mathrm{e}^{-z\zeta} \overset{\vee}{\varphi}(\zeta)d\zeta\right| \leq \int_\varepsilon^\infty \mathrm{e}^{-sr} C\mathrm{e}^{\tau s} ds \leq \frac{C}{r-\tau} \mathrm{e}^{-\varepsilon(r-\tau)}.$$

In the same way, the integral $\int_{R_{\theta_2,\varepsilon}} \mathrm{e}^{-z\zeta} \overset{\vee}{\varphi}(\zeta)d\zeta$ defines a holomorphic function on $\dot{\Pi}_\tau^{\theta_2}$ and for any $r > \tau$, for every $z \in \overline{\dot{\Pi}}_r^{\theta_2}$, $\left|\int_{R_{\theta_2,\varepsilon}} \mathrm{e}^{-z\zeta} \overset{\vee}{\varphi}(\zeta)d\zeta\right| \leq \frac{C}{r-\tau}\mathrm{e}^{-\varepsilon(r-\tau)}$;

- the integral $\int_{\delta_{[\theta_1,\theta_2],\varepsilon}} \mathrm{e}^{-z\zeta} \overset{\vee}{\varphi}(\zeta)d\zeta$ defines an entire function and $\left|\int_{\delta_{[\theta_1,\theta_2],\varepsilon}} \mathrm{e}^{-z\zeta} \overset{\vee}{\varphi}(\zeta)d\zeta\right| \leq C|\check{I}_1|\varepsilon \mathrm{e}^{\tau\varepsilon}\mathrm{e}^{\varepsilon|z|}$.
- by arguments already encounter (see Sect. 7.5.1), both $\dot{\Pi}_\tau^{\theta_1}$ and $\dot{\Pi}_\tau^{\theta_2}$ contains any proper subsector $\dot{\bar{\Delta}}^\infty$ of $\dot{\Delta}_r^\infty(I_1^\star)$, once $r > 0$ is chosen large enough.

Therefore, φ_{I_1} belongs to the space $\overline{\mathscr{A}}^{\leq 0}(\dot{\Delta}_{r_1}^\infty(I_1^\star))$ for $r_1 > 0$ large enough, because $\varepsilon > 0$ can be chosen arbitrarily small.
It is easy to see that adding to $\overset{\vee}{\varphi}$ any element of $\mathscr{O}(\mathbb{C})^{\leq 1}$, does not affect the function φ_{I_1} (just deform the contour of integration, by Cauchy).
The family of functions $(\varphi_{I_1})_{I_1\subset I}$ obtained this way glue together analytically, by Cauchy.
The above construction gives a morphism, $\mathscr{L}(I) : \check{\mathscr{E}}^{\leq 1}(I)/\mathscr{O}(\mathbb{C})^{\leq 1} \to \mathscr{A}^{\leq 0}(I^\star)$, compatible with the restriction maps, which provides a morphism of sheaves[3].

Definition 7.30. The morphism of sheaves $\mathscr{L} : \check{\mathscr{E}}^{\leq 1}/\mathscr{O}(\mathbb{C})^{\leq 1} \to \mathscr{A}^{\leq 0}$ is called the *strict Laplace transform*[4].

We return to the construction we did to get the formal Borel transform, Sect. 7.5.2. We pick an open arc $I^\star \subset \mathbb{S}^1$ of length $\leq \pi$ and $\varphi \in \mathscr{A}^{\leq 0}(I^\star)$. For $z_1 \in \dot{\Delta}_r^\infty(I^\star)$, $r > 0$ large enough, for any direction $\alpha \in I^\star$, we set $\overset{\vee}{\varphi}_{z_1,\alpha}(\zeta) = -\frac{1}{2i\pi}\int_{R_{\alpha,z_1}} \mathrm{e}^{z\zeta}\varphi(z)dz$.

We have seen that, making α varying, one gets an element of $\check{\mathscr{E}}^{\leq 1}(I)$, while $\overset{\vee}{\varphi}_{z_1,\alpha}$ depends on z_1 only modulo an element of $\mathscr{O}(\mathbb{C})^{\leq 1}$. We thus get a morphism of sheaves $\mathscr{B} : \mathscr{A}^{\leq 0} \to \check{\mathscr{E}}^{\leq 1}/\mathscr{O}(\mathbb{C})^{\leq 1}$ which has the following property (we refer to [CNP93-1] for the proof):

Proposition 7.11. *The morphisms of sheaves* $\mathscr{L} : \check{\mathscr{E}}^{\leq 1}/\mathscr{O}(\mathbb{C})^{\leq 1} \to \mathscr{A}^{\leq 0}$ *and* $\mathscr{B} : \mathscr{A}^{\leq 0} \to \check{\mathscr{E}}^{\leq 1}/\mathscr{O}(\mathbb{C})^{\leq 1}$ *are isomorphisms of sheaves of* $\mathbb{C}$*-differential algebras, and* $\mathscr{L} \circ \mathscr{B} = \mathrm{Id}$, $\mathscr{B} \circ \mathscr{L} = \mathrm{Id}$.

---

[3] As usual, modulo complex conjugation

[4] We abide a notation of [CNP93-1], although the construction therein slightly differs from ours.

7.6.2 Singularities and Laplace Transform

7.6.2.1 Summable Singularities

We recall that $\mathrm{SING}_{\theta,\alpha}$ can be identified with the space $\Gamma(J,\mathscr{C})$ of multivalued sections of $\mathscr{C}$ over $J =]-\pi/2-\alpha+\theta, \theta+\alpha+\pi/2[\subset \widetilde{\mathbb{S}}^1$. In particular, any singularity $\overset{\triangledown}{\varphi} \in \mathrm{SING}_{\theta,\alpha}$ can be represented by a major $\overset{\vee}{\varphi} \in \mathrm{ANA}_{\theta,\alpha} = \Gamma(\check{J},\mathscr{O}^0)$, with $\check{J} =]\theta-\alpha-2\pi, \theta+\alpha[\subset \widetilde{\mathbb{S}}^1$.

Definition 7.31. An element $\overset{\vee}{\varphi} \in \mathrm{ANA}_{\theta,\alpha} = \Gamma(\check{J},\mathscr{O}^0)$ is said *summable* in the direction $\beta \in \widehat{J} =]-\alpha+\theta, \theta+\alpha[$ if there exists a neighbourhood $\widehat{J}_1 \subset \widehat{J}$ of β so that the two restrictions $\overset{\vee}{\varphi}_1 \in \Gamma(\widehat{J}_1,\mathscr{O}^0)$ and $\overset{\vee}{\varphi}_2 \in \Gamma(\widehat{J}_2,\mathscr{O}^0)$ of $\overset{\vee}{\varphi}$ over $\widehat{J}_1$ and $\widehat{J}_2 = -2\pi+\widehat{J}_1$ respectively, can be represented by elements of $\Gamma(\widehat{J}_1,\mathscr{E}^{\leq 1})$ and $\Gamma(\widehat{J}_2,\mathscr{E}^{\leq 1})$ respectively. A singularity $\overset{\triangledown}{\varphi} \in \mathrm{SING}_{\theta,\alpha}$ is *summable* in the direction $\widehat{J}$ if for any $\beta \in \widehat{J}$, the singularity $\overset{\triangledown}{\varphi}$ has a major $\overset{\vee}{\varphi} \in \mathrm{ANA}_{\theta,\alpha}$ which summable in the direction β. We denote by $\mathrm{SING}^{\mathrm{sum}}_{\theta,\alpha}$ the space of singularities $\overset{\triangledown}{\varphi} \in \mathrm{SING}_{\theta,\alpha}$ which are summable in the direction $\widehat{J}$.

7.6.2.2 Laplace Transforms of Summable Singularities

We consider a singularity $\overset{\triangledown}{\varphi} \in \mathrm{SING}^{\mathrm{sum}}_{\theta,\alpha}$ and a direction $\beta \in \widehat{J} =]-\alpha+\theta, \theta+\alpha[$. Let $\overset{\vee}{\varphi}$ be a major of $\overset{\triangledown}{\varphi}$ which is summable in the direction β and set $\widehat{\varphi} = \mathrm{var}\, \overset{\triangledown}{\varphi}$. Using the notation of definition 7.31, we consider the following Laplace integral where $\varepsilon > 0$ is chosen small enough:

$$\begin{aligned}
\varphi_\beta(z) &= \int_{\gamma_{[\beta-2\pi,\beta],\varepsilon}} \mathrm{e}^{-z\zeta} \overset{\vee}{\varphi}(\zeta)d\zeta \qquad (7.11)\\
&= \int_{\delta_{[\beta-2\pi,\beta],\varepsilon}} \mathrm{e}^{-z\zeta} \overset{\vee}{\varphi}(\zeta)d\zeta - \int_{R_{\beta-2\pi,\varepsilon}} \mathrm{e}^{-z\zeta} \overset{\vee}{\varphi}_2(\zeta)d\zeta + \int_{R_{\beta,\varepsilon}} \mathrm{e}^{-z\zeta} \overset{\vee}{\varphi}_1(\zeta)d\zeta\\
&= \int_{\delta_{[\beta-2\pi,\beta],\varepsilon}} \mathrm{e}^{-z\zeta} \overset{\vee}{\varphi}(\zeta)d\zeta + \int_{R_{\beta,\varepsilon}} \mathrm{e}^{-z\zeta} \widehat{\varphi}(\zeta)d\zeta.
\end{aligned}$$

From the arguments used in Sect. 7.6.1, we see that φ_β defines an element of $\overline{\mathscr{A}}^{\leq 0}(\breve{\beta})$. Moreover, if $\overset{\vee}{\psi}$ is another major of $\overset{\triangledown}{\varphi}$ which is summable in the direction β (for instance $\overset{\vee}{\varphi} - \overset{\vee}{\psi} \in \mathscr{O}(\mathbb{C})^{\leq 1}$), then its Laplace integral ψ_β coincide with φ_β as elements of $\overline{\mathscr{A}}^{\leq 0}(\breve{\beta})$. Thus φ_β is independent of the chosen summable major and only depends on $\overset{\triangledown}{\varphi} \in \mathrm{SING}^{\mathrm{sum}}_{\theta,\alpha}$. This allows us to write $\varphi_\beta = \mathscr{L}^\beta \overset{\triangledown}{\varphi}$.

Making β varying in $\widehat{J}$, the functions $\mathscr{L}^\beta \overset{\triangledown}{\varphi}$ obviously glue together analytically (by Cauchy and using the independence of $\mathscr{L}^\beta \overset{\triangledown}{\varphi}$ with respect to the chosen summable major), to give and element $\mathscr{L}^{\widehat{J}} \overset{\triangledown}{\varphi}$ of $\Gamma(J^\star, \mathscr{A}^{\leq 0})$.

Definition 7.32. The morphism $\mathscr{L}^\beta : \mathrm{SING}^{\mathrm{sum}}_{\theta,\alpha} \to \overline{\mathscr{A}}^{\leq 0}(\breve{\beta})$ is called the *Laplace transform* in the direction $\beta \in \widehat{J} =]-\alpha+\theta, \theta+\alpha[$.
The morphism $\mathscr{L}^{\widehat{J}} : \mathrm{SING}^{\mathrm{sum}}_{\theta,\alpha} \to \Gamma(J^\star, \mathscr{A}^{\leq 0})$ is called the Laplace transform in the direction $\widehat{J} =]-\alpha+\theta, \theta+\alpha[$.

We recover with the following proposition the examples given in the introduction of the chapter, see also [MS016].

Proposition 7.12. *The singularities* $\overset{\triangledown}{I}_\sigma$ *and* $\overset{\triangledown}{J}_{\sigma,m}$ *belong to* $\mathrm{SING}^{\mathrm{sum}}_{\theta,\alpha}$ *for any direction* θ *and any* $\alpha > 0$. *Moreover, for any direction* $\beta \in \widetilde{\mathbb{S}}^1$,

$$\mathscr{L}^\beta \overset{\triangledown}{I}_\sigma (z) = \frac{1}{z^\sigma}, \qquad \mathscr{L}^\beta \overset{\triangledown}{J}_{\sigma,m} (z) = (-1)^m \frac{\log^m(z)}{z^\sigma}, \qquad z \in \Pi_0^\beta \subset \widetilde{\mathbb{C}}.$$

This has the following consequences:

Proposition 7.13. *For all* $\sigma_1, \sigma_2 \in \mathbb{C}$, *for all* $m_1, m_2 \in \mathbb{N}$ $\overset{\triangledown}{I}_{\sigma_1} * \overset{\triangledown}{I}_{\sigma_2} = \overset{\triangledown}{I}_{\sigma_1+\sigma_2}$ *and* $\overset{\triangledown}{J}_{\sigma_1,m_1} * \overset{\triangledown}{J}_{\sigma_2,m_2} = \overset{\triangledown}{J}_{\sigma_1+\sigma_2,m_1+m_2}$.

Proof. From proposition 7.12, we deduce that $\mathscr{L} \overset{\triangledown}{I}_{\sigma_1} = \dfrac{1}{z^{\sigma_1}}$ and $\mathscr{L} \overset{\triangledown}{I}_{\sigma_2} = \dfrac{1}{z^{\sigma_2}}$. Thus by proposition 7.10, $\mathscr{L} \overset{\triangledown}{I}_{\sigma_1} * \overset{\triangledown}{I}_{\sigma_2} = \dfrac{1}{z^{\sigma_1+\sigma_2}}$ and one concludes by formal Borel transform. Same proof for the other equality. □

In definition 7.32, we meant morphisms of vector spaces. As a matter of fact, these are morphisms of $\mathbb{C}$-differential algebras. This is the matter of the following proposition.

Proposition 7.14. *The space* $\mathrm{SING}^{\mathrm{sum}}_{\theta,\alpha}$ *is a commutative and associative algebra with unit* δ. *The Laplace transform* $\mathscr{L}^\beta : \mathrm{SING}^{\mathrm{sum}}_{\theta,\alpha} \to \overline{\mathscr{A}}^{\leq 0}(\breve{\beta})$ *is compatible with the convolution of singularities:* $\mathscr{L}^\beta \overset{\triangledown}{\varphi} * \overset{\triangledown}{\psi} = \left(\mathscr{L}^\beta \overset{\triangledown}{\varphi}\right)\left(\mathscr{L}^\beta \overset{\triangledown}{\psi}\right)$. *Moreover,* $\mathscr{L}^\beta (\overset{\triangledown}{\partial} \overset{\triangledown}{\varphi}) = \partial \mathscr{L}^\beta \overset{\triangledown}{\varphi}$.

Proof. We go back to the very definition of the convolution product of microfunctions and singularities. For $\overset{\triangledown}{\varphi}, \overset{\triangledown}{\psi} \in \mathrm{SING}_{\theta,\alpha}$, for any $\beta \in \widehat{J} =]-\alpha+\theta, \theta+\alpha[$, the convolution product $\overset{\triangledown}{\varphi} * \overset{\triangledown}{\psi}$ can be represented, for $\zeta \in \overset{\bullet}{\mathfrak{S}}_{2\varepsilon}(]\beta - 2\pi, \beta[)$ with $\varepsilon > 0$ as small as we want, by

$$\overset{\vee}{\varphi} *_{\Gamma\times\Gamma} \overset{\vee}{\psi}(\zeta) = \frac{1}{2i\pi}\int_{\Gamma\times\Gamma} \frac{e^{\nu(\zeta-(\xi_1+\xi_2))}}{\zeta-(\xi_1+\xi_2)} \overset{\vee}{\varphi}(\xi_1)\overset{\vee}{\psi}(\xi_2)d\xi_1 d\xi_2, \tag{7.12}$$

(see 7.9), where $\Gamma = \Gamma_{\beta,\varepsilon,\eta_1,\eta_2}$ is as in definition 7.7 and where $\overset{\vee}{\varphi}, \overset{\vee}{\psi}$ are thought of as belonging to $\mathscr{O}(\overset{\bullet}{\mathbb{A}}_0(]\beta-2\pi,\beta[))$. In (7.12), $\nu \in \mathbb{C}$ is a free parameter which can be chosen at our convenience.
We now assume that $\overset{\triangledown}{\varphi}, \overset{\triangledown}{\psi} \in \mathrm{SING}^{\mathrm{sum}}_{\theta,\alpha}$ and that $\overset{\vee}{\varphi}, \overset{\vee}{\psi}$ are summable majors in the direction β. In that case, choosing $\nu = |\nu|e^{-i\beta}$ with $|\nu|$ large enough to ensure the integrability, one can rather consider the convolution product $\overset{\triangledown}{\varphi} * \overset{\triangledown}{\psi}$ as represented by (7.12), but this time with an endless path $\Gamma = \Gamma_{\beta,\varepsilon}$ (see definition 7.7). This construction gives a major of $\overset{\triangledown}{\varphi}, \overset{\triangledown}{\psi}$ which is summable in the direction β. Moreover, the arguments used in the proof of the proposition 7.10 show that $\mathscr{L}^{\beta}\, \overset{\triangledown}{\varphi} * \overset{\triangledown}{\psi} = \big(\mathscr{L}^{\beta}\, \overset{\triangledown}{\varphi}\big)\big(\mathscr{L}^{\beta}\, \overset{\triangledown}{\psi}\big)$. □

Example 7.7. We consider the formal Borel transform $\widehat{w}_{(0,0)} = \mathscr{B}\widetilde{w}_{(0,0)}$ where $\widetilde{w}_{(0,0)}$ is the formal series solution of the prepared ODE (3.6) associated with the first Painlevé equation. We know by theorem 3.3 that $\widehat{w}_{(0,0)}$ can be analytically continued to the star-shaped domain $\overset{\bullet}{\mathscr{R}}{}^{(0)}$ with at most exponential growth of order 1 at infinity along non-horizontal directions. We set $\overset{\triangledown}{w}_{(0,0)} = {}^{\flat}\widehat{w}_{(0,0)} \in \mathrm{SING}^{\mathrm{int}}$. Then $\overset{\triangledown}{w}_{(0,0)} \in \mathrm{SING}^{\mathrm{sum}}_{\pi/2,\pi/2}$ (or $\overset{\triangledown}{w}_{(0,0)} \in \mathrm{SING}^{\mathrm{sum}}_{-\pi/2,\pi/2}$) : just consider the major $\overset{\vee}{w}_{(0,0)}(\zeta) = \widehat{w}_{(0,0)}(\zeta)\frac{\log(\zeta)}{2i\pi}$. The Laplace transform $\mathscr{L}^{]0,\pi[}\, \overset{\triangledown}{w}_{(0,0)}$ is well-defined and gives a section of $\mathscr{A}^{\leq 0}$ on $]-3\pi/2,\pi/2[$. As a matter of fact,

$$\mathscr{L}^{]0,\pi[}\, \overset{\triangledown}{w}_{(0,0)} = \mathscr{L}^{]0,\pi[}\widehat{w}_{(0,0)} = \mathscr{S}^{]0,\pi[}\widetilde{w}_{(0,0)}$$

and $\mathscr{L}^{]0,\pi[}\, \overset{\triangledown}{w}_{(0,0)}$ can be thought of as belonging to the space of sections $\Gamma(]-3\pi/2,\pi/2[,\mathscr{A}_1)$. We now consider the singularity $\overset{\triangledown}{I}_{\sigma} * \overset{\triangledown}{w}_{(0,0)}$, for any $\sigma \in \mathbb{C}$. Using propositions 7.12 and 7.14, this singularity belongs (for instance) to $\mathrm{SING}^{\mathrm{sum}}_{\pi/2,\pi/2}$ and $\mathscr{L}^{]0,\pi[}\, \overset{\triangledown}{I}_{\sigma} * \overset{\triangledown}{w}_{(0,0)} = \big(\mathscr{L}^{]0,\pi[}\, \overset{\triangledown}{I}_{\sigma}\big)\big(\mathscr{L}^{]0,\pi[}\, \overset{\triangledown}{w}_{(0,0)}\big) = \frac{1}{z^{\sigma}}\mathscr{S}^{]0,\pi[}\widetilde{w}_{(0,0)}$, this time viewed as a multivalued section $\mathscr{A}^{\leq 0}$ on $]-3\pi/2,\pi/2[\subset \widetilde{\mathbb{S}}^1$.

7.7 Spaces of Resurgent Functions

7.7.1 Preliminaries

We refer the reader to [CNP93-1] (Pré I.3, lemme 3.0) for the proof of the following key-lemma, the idea of which being due to Ecalle.

Lemma 7.5. *Let $R_0 > 0$ be a real positive number and $\Gamma \subset \mathbb{C}$ be an embedded curve, transverse to the circles $|\zeta| = R$ for all $R \geq R_0$. Let Φ be a holomorphic function on a neighbourhood of Γ. Then, for any continuous function $m : \mathbb{R}^+ \to \mathbb{R}^+$ so that $\inf\{m([0,\xi])\} > 0$ for all $\xi > 0$, there exists $\Psi \in \mathscr{O}(\mathbb{C})$ such that, for all $\zeta \in \Gamma$, $|\Phi(\zeta) + \Psi(\zeta)| \leq m(|\zeta|)$.*

In what follows, we use the notation introduced in definition 7.7. We also recall that $\widetilde{\mathbb{C}\setminus\mathbb{Z}}$ stands for the universal covering of $\mathbb{C}\setminus\mathbb{Z}$. One may also think of $\widetilde{\mathbb{C}\setminus\mathbb{Z}}$ as the universal covering of $\widetilde{\mathbb{C}} \setminus \bigcup_{\theta=\pi k, k\in\mathbb{Z}} \{m e^{i\theta} \mid m \in \mathbb{N}^\star\}$.

Lemma 7.6. *Let $\overset{\nabla}{\varphi} \in$ SING be a singularity which can be determined by a major analytically continuable to $\widetilde{\mathbb{C}\setminus\mathbb{Z}}$. Then, for any direction θ and any $\varepsilon > 0$ small enough, the singularity $\overset{\nabla}{\varphi}$ has a major $\overset{\vee}{\varphi}$ with the following properties:*

1. *the restriction of $\overset{\vee}{\varphi}$ as a sectorial germ of codirection $I =]-\pi/2+\theta, \theta+\pi/2[$, can be represented by a function Φ holomorphic on the cut plane $\mathbb{C}\setminus [0, e^{i\theta}\infty[= \overset{\bullet}{\mathcal{S}}{}_0^\infty(\check{I})$, $\check{I} =]-2\pi+\theta, \theta[$;*
2. *Φ is bounded on $\overset{\bullet}{\mathfrak{S}}_{\varepsilon'}(I)$, for every $\varepsilon' > \varepsilon$.*
3. *Φ can be analytically continued to $\widetilde{\mathbb{C}\setminus\mathbb{Z}}$.*

Proof. Let $\overset{\vee}{\varphi}_1$ be a major of $\overset{\nabla}{\varphi}$ which can be analytically continued to $\widetilde{\mathbb{C}\setminus\mathbb{Z}}$. This major can be represented by a function Φ_1 holomorphic on $\overset{\bullet}{\mathcal{S}}{}_0^R(\check{I}) \cup S_{2\varepsilon}(\widehat{I}) \setminus [0, e^{i\theta}\infty[$, for $R > 0$ and $\varepsilon > 0$ small enough, and Φ_1 can be analytically continued to $\widetilde{\mathbb{C}\setminus\mathbb{Z}}$. The boundary $\Gamma_{I,\varepsilon} = -\partial \overset{\bullet}{\mathfrak{S}}_\varepsilon(I)$ can be seen as an embedded curve $H_0 : \mathbb{R} \to \mathbb{C}$ that fulfills the condition of lemma 7.5 : one can find a function $\Psi_1 \in \mathscr{O}(\mathbb{C})$ so that $\Phi_2 = \Phi_1 + \Psi_1$ satisfies $|\Phi_2(\eta)| \leq \exp(-|\eta|)$ for all $\eta \in \Gamma_{I,\varepsilon}$. One can also assume that $|H_0'(s)|$ is bounded and these conditions ensure the integrability for the integral $\Phi(\zeta) = \dfrac{1}{2i\pi}\displaystyle\int_{H_0} \frac{\Phi_2(\eta)}{\zeta-\eta} d\eta$ which thus, defines a holomorphic function on $\overset{\bullet}{\mathfrak{S}}_\varepsilon(I)$. Moreover, one easily sees by Cauchy that $\Phi = \Phi_2 + \Psi_2$ where $\Psi_2 \in \mathscr{O}_0$. One observes that $|\zeta - \eta| \geq \varepsilon' - \varepsilon$ for $(\zeta, \eta) \in \overset{\bullet}{\mathfrak{S}}_{\varepsilon'}(I) \times \Gamma_{I,\varepsilon}$, with $\varepsilon' > \varepsilon$. Thus Φ is bounded on $\overset{\bullet}{\mathfrak{S}}_{\varepsilon'}(I)$. Notice that Φ_2 inherits from Φ_1 the property of being analytically continuable to $\widetilde{\mathbb{C}\setminus\mathbb{Z}}$. Thus one can analytically continue Φ to $\widetilde{\mathbb{C}\setminus\mathbb{Z}}$ by Cauchy, by deformation of the contour by isotopies[5] $H : (s,t) \in \mathbb{R} \times [0,1] \mapsto H(s,t) = H_t(s) \in \mathbb{C}\setminus\mathbb{Z}$ that are equal to the identity in a neighbourhood of infinity, Fig. 7.8.
Finally, from the fact that $\Phi = \Phi_1 + \Psi$ with $\Psi_1 + \Psi_2 \in \mathscr{O}_0$, we see that Φ defines a sectorial germ $\overset{\vee}{\varphi}$ of codirection $I =]-\pi/2+\theta, \theta+\pi/2[$ whose associated microfunction coincides with the restriction of $\overset{\nabla}{\varphi}$ to the codirection I. □

---

[5] That is H is a homotopy and for each $t \in [0,1]$, H_t is an embedding. Remember that we see $\Gamma_{I,\varepsilon}$ as an embedded curve $H_0 : \mathbb{R} \to \mathbb{C}$.

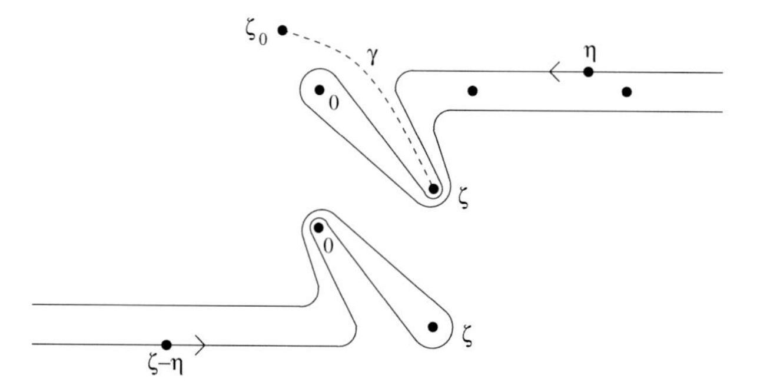

Fig. 7.8 Deformation of the contour $\Gamma_{I,\varepsilon}$ by an isotopy equal to the identity in a neighbourhood of infinity, for $\theta = 0$.

Lemma 7.7. *Let* $\overset{\triangledown}{\varphi} \in$ SING *be a singularity which can be determined by a major analytically continuable to* $\widetilde{\mathbb{C}\setminus\mathbb{Z}}$. *Then, for any direction* θ *and for any* $\varepsilon > 0$ *small enough, the singularity* $\overset{\triangledown}{\varphi}$ *has a major* $\overset{\vee}{\varphi}$ *with the following properties:*

1. *the restriction of* $\overset{\vee}{\varphi}$ *as a sectorial germ of codirection* $I =]-\pi/2+\theta, \theta+\pi/2[$, *can be represented by a function* Φ *holomorphic on the cut plane* $\mathbb{C}\setminus[0, \mathrm{e}^{\mathrm{i}\theta}\infty[= \overset{\bullet}{\mathcal{S}}{}_0^{\infty}(\check{I})$, $\check{I} =]-2\pi+\theta, \theta[$;
2. $|\Phi(\eta)| \leq \exp(-|\eta|)$ *for all* $\eta \in \Gamma_{I,\varepsilon}$, *where* $\Gamma_{I,\varepsilon} = -\partial\overset{\bullet}{\mathfrak{S}}_{\varepsilon}(I) \subset \overset{\bullet}{\mathcal{S}}{}_0^{\infty}(\check{I})$;
3. Φ *can be analytically continued to* $\widetilde{\mathbb{C}\setminus\mathbb{Z}}$.

Proof. Just consider first the function Φ_1 given by lemma 7.6, then use lemma 7.5 to define Φ from Φ_1. □

The above lemmas 7.6 and 7.7 motivate the introduction of new Riemann surfaces that will be used in a moment.

Definition 7.33. Let $\theta \in \mathbb{S}^1$ be a direction. We set $\overset{\bullet}{\mathscr{R}}{}^{\theta,(0)} = \mathbb{C}\setminus[0, \mathrm{e}^{\mathrm{i}\theta}\infty[$. Let ζ_0 be a complex number in $\overset{\bullet}{\mathscr{R}}{}^{\theta,(0)} \setminus \mathbb{Z}$. We denote by $\mathfrak{A}_{\theta,\zeta_0}$ (*resp.* $\mathfrak{B}_{\zeta_0}$) the set of paths in $\overset{\bullet}{\mathscr{R}}{}^{\theta,(0)}$ (*resp.* $\mathbb{C}\setminus\mathbb{Z}$) originating from ζ_0, endowed with the equivalence relation $\sim_{\overset{\bullet}{\mathscr{R}}{}^{\theta,(0)}}$ (*resp.* $\sim_{\mathbb{C}\setminus\mathbb{Z}}$) of homotopy of paths with fixed extremities.

We set $\mathfrak{R}_{\theta,\zeta_0} = \mathfrak{A}_{\theta,\zeta_0} \cup \mathfrak{B}_{\zeta_0}$ and we denote by $\overset{(\theta,\zeta_0)}{\sim}$ the relation on $\mathfrak{R}_{\theta,\zeta_0}$ defined as follows. For any two $\gamma_1, \gamma_2 \in \mathfrak{R}_{\theta,\zeta_0}$, $\gamma_1 \overset{(\theta,\zeta_0)}{\sim} \gamma_2$ when one of the following conditions is satisfied: either $\gamma_1 \sim_{\overset{\bullet}{\mathscr{R}}{}^{\theta,(0)}} \gamma_2$ or $\gamma_1 \sim_{\mathbb{C}\setminus\mathbb{Z}} \gamma_2$; or else there exists $\gamma_3 \in \mathfrak{A}_{\theta,\zeta_0} \cap \mathfrak{B}_{\zeta_0}$ such that $\begin{cases} \gamma_1 \sim_{\overset{\bullet}{\mathscr{R}}{}^{\theta,(0)}} \gamma_3 \\ \gamma_2 \sim_{\mathbb{C}\setminus\mathbb{Z}} \gamma_3 \end{cases}$ or $\begin{cases} \gamma_1 \sim_{\mathbb{C}\setminus\mathbb{Z}} \gamma_3 \\ \gamma_2 \sim_{\overset{\bullet}{\mathscr{R}}{}^{\theta,(0)}} \gamma_3 \end{cases}$.

Let γ be an element of $\mathfrak{R}_{\theta,\zeta_0}$. We denote by $\mathrm{cl}_{\theta,\zeta_0}(\gamma)$ its equivalence class for the relation $\overset{(\theta,\zeta_0}{\sim}$. We finally set:

$$\mathscr{R}^{\theta}_{\mathbb{Z},\zeta_0} = \{\mathrm{cl}_{\theta,\zeta_0}(\gamma) \mid \gamma \in \mathfrak{R}_{\theta,\zeta_0}\} \text{ and } \mathfrak{p}_{\theta,\zeta_0} : \mathrm{cl}_{\theta,\zeta_0}(\gamma) \mapsto \underline{\underline{\gamma}}(1) \in \overset{\bullet}{\mathscr{R}}^{\theta,(0)}. \quad (7.13)$$

Proposition 7.15. *The space $\mathscr{R}^{\theta}_{\mathbb{Z},\zeta_0}$ can be equipped with a separated topology which makes $(\mathscr{R}^{\theta}_{\mathbb{Z},\zeta_0}, \mathfrak{p}_{\theta,\zeta_0})$ an étalé space. The space $\mathscr{R}^{\theta}_{\mathbb{Z},\zeta_0}$ is arc-connected and simply connected, thus defines a Riemann surface by pulling back by $\mathfrak{p}_{\theta,\zeta_0}$ the complex structure of $\mathbb{C}$. Moreover, for two points $\zeta_0, \zeta_1 \in \overset{\bullet}{\mathscr{R}}^{\theta,(0)} \setminus \mathbb{Z}$, the two Riemann surfaces $\mathscr{R}^{\theta}_{\mathbb{Z},\zeta_0}$ and $\mathscr{R}^{\theta}_{\mathbb{Z},\zeta_0}$ are isomorphic.*

The proof of proposition 7.15 is left as an exercise. (Just copy what have been done in Sect. 4.2.2). We complete the above proposition with a definition.

Definition 7.34. The class of isomorphisms of the Riemann surfaces $(\mathscr{R}^{\theta}_{\mathbb{Z},\zeta_0}, \mathfrak{p}_{\theta,\zeta_0})$ is denoted by $(\mathscr{R}^{\theta}_{\mathbb{Z}}, \mathfrak{p}_{\theta})$. We often use abridged notation $(\mathscr{R}^{\theta}, \mathfrak{p})$. We call *principal sheet* the unique domain $\mathscr{R}^{\theta,(0)} \subset \mathscr{R}^{\theta}$ so that the resctriction $\mathfrak{p}|_{\mathscr{R}^{\theta,(0)}}$ realizes a homeomorphism between $\mathscr{R}^{\theta,(0)}$ and the simply connected domain $\overset{\bullet}{\mathscr{R}}^{\theta,(0)}$.

7.7.2 *Resurgent Functions*

Various spaces of so-called resurgent functions can be defined and used according to the context. We start with the notion of resurgent singularities.

7.7.2.1 Resurgent Singularities, Resurgent Asymptotic Classes

Definition 7.35. A singularity $\overset{\triangledown}{\varphi} \in \mathrm{SING}$ is said to be $\mathbb{Z}$*-resurgent* when it can be determined by a major $\overset{\vee}{\varphi} \in \mathrm{ANA}$ which can be analytically continued to $\widetilde{\mathbb{C} \setminus \mathbb{Z}}$. We denote by $\mathrm{RES}_{\mathbb{Z}}$ or simply RES the space of $\mathbb{Z}$-resurgent singularities.

A $\mathbb{Z}$-resurgent singularity is often simply called a $\mathbb{Z}$-resurgent function. Throughout this course we will usually write "resurgent singularity" in place of $\mathbb{Z}$-resurgent singularity.

Remark 7.5. It is important to keep in mind that the minor $\widehat{\varphi}$ of any resurgent singularity $\overset{\triangledown}{\varphi} \in \mathrm{RES}$, can be analytically continued to $\widetilde{\mathbb{C} \setminus \mathbb{Z}}$, since the minor $\widehat{\varphi}$ does not depend on the chosen major.

Definition 7.36. One says that $\overset{\triangledown}{\varphi} \in \mathrm{RES}$ is a *resurgent constant* when $\overset{\triangledown}{\varphi}$ has a major which can be analytically continued to $\widetilde{\mathbb{C}}$. The space of resurgent constants is denoted by CONS.

Definition 7.37. An asymptotic class $\overset{\triangle}{\varphi} \in \mathrm{ASYMP}$ is called a $\mathbb{Z}$*-resurgent asymptotic class, resp.* a *resurgent constant*, when its formal Borel transform $\overset{\triangledown}{\varphi}$ is a $\mathbb{Z}$-resurgent singularity, *resp.* a resurgent constant. We denote by $\widetilde{\mathrm{RES}}_{\mathbb{Z}}$ or simply $\widetilde{\mathrm{RES}}$

the space made of $\mathbb{Z}$-resurgent asymptotic classes. We denote by $\widetilde{\text{CONS}}$ the subspace of resurgent constants.

A $\mathbb{Z}$-resurgent asymptotic class is often simply called a $\mathbb{Z}$-resurgent function or even a resurgent function.

Example 7.8. The singularities $\overset{\triangledown}{I}_\sigma$ and $\overset{\triangledown}{J}_{\sigma,m}$ are resurgent constants, as well as their associated asymptotic classes $\overset{\vartriangle}{I}_\sigma$ and $\overset{\vartriangle}{J}_{\sigma,m}$.

7.7.2.2 Resurgent Functions, Resurgent Series

We recall the following simple definition, for objects much discussed in [MS016].

Definition 7.38. The $\mathbb{C}$-differential commutative and associative convolution algebra $\mathbb{C}\delta \oplus \widehat{\mathscr{R}}_\mathbb{Z}$ with unit δ, is called a space of $\mathbb{Z}$-resurgent functions. We denote by $\overset{\triangledown}{\mathscr{R}}_\mathbb{Z} \subset$ RES the $\mathbb{C}$-differential commutative and associative convolution algebra made of resurgent singularities of the form $\overset{\triangledown}{\varphi} = a\delta + {}^\flat\widehat{\varphi}$ with $\widehat{\varphi} \in \widehat{\mathscr{R}}_\mathbb{Z}$.

Since $\mathbb{C}\delta \oplus \widehat{\mathscr{R}}_\mathbb{Z}$ is a convolution algebra, the identity ${}^\flat\widehat{\varphi} * {}^\flat\widehat{\varphi} = {}^\flat (\widehat{\varphi} * \widehat{\varphi})$ (proposition 7.6) implies that $\overset{\triangledown}{\mathscr{R}}_\mathbb{Z}$ is indeed a convolution algebra. One usually uses abridged notation $\overset{\triangledown}{\mathscr{R}}$ in this course.

Definition 7.39. A series expansion $\widetilde{\varphi} \in \mathbb{C}[[z^{-1}]]$ is a $\mathbb{Z}$-resurgent series when its formal Borel transform $\mathscr{B}\widetilde{\varphi}$ is a $\mathbb{Z}$-resurgent function or, equivalently, when the asymptotic class ${}^\natural\widetilde{\varphi}$ belongs to $\widetilde{\text{RES}}_\mathbb{Z}$. We denote by $\widetilde{\mathscr{R}}_\mathbb{Z}$ the $\mathbb{C}$-differential commutative and associative algebra made of $\mathbb{Z}$-resurgent series.

Throughout this course we usually simply write "resurgent functions" or "resurgent series" instead of $\mathbb{Z}$-resurgent functions or $\mathbb{Z}$-resurgent series, since there is no risk of misunderstanding.

7.7.2.3 Resurgent Singularities and Convolution

Theorem 7.1. *The space* RES *is a $\mathbb{C}$-differential commutative and associative convolution algebra with unit δ, and* CONS $\subset$ RES *is a subalgebra. Therefore, the space* $\widetilde{\text{RES}}$ *is a $\mathbb{C}$-differential commutative and associative algebra and* $\widetilde{\text{CONS}} \subset \widetilde{\text{RES}}$ *is a subalgebra.*

Proof. (Adapted from [Eca85, CNP93-1]. The reader should look before at the reasoning made for the proof of proposition 4.6).
It is enough to only show that RES is a convolution space. We take two singularities $\overset{\triangledown}{\varphi}, \overset{\triangledown}{\psi} \in$ RES, we choose a direction θ and we suppose $0 < \varepsilon \ll 1$.
By lemma 7.7, *resp.* lemma 7.6, $\overset{\triangledown}{\varphi}$, *resp.* $\overset{\triangledown}{\psi}$, has a major such that its restriction as a sectorial germ of codirection $I =]-\pi/2+\theta, \theta+\pi/2[$, can be represented by a

function $\overset{\vee}{\varphi}$, *resp.* $\overset{\vee}{\psi}$, holomorphic on $\overset{\bullet}{\mathscr{R}}^{\theta,(0)}$, that can be analytically continued to the Riemann surface $(\mathscr{R}^{\theta},\mathfrak{p})$ and moreover, satisfies the condition:

1. $|\overset{\vee}{\varphi}(\eta)| \leq \exp(-|\eta|)$ for all $\eta \in \Gamma_{I,\varepsilon}$, where $\Gamma_{I,\varepsilon} = -\partial\overset{\bullet}{\mathfrak{S}}_{\varepsilon}(I) \subset \overset{\bullet}{\mathscr{R}}^{\theta,(0)}$;
2. $\overset{\vee}{\psi}$ is bounded on $\overset{\bullet}{\mathfrak{S}}_{\varepsilon}(I)$.

We know by lemma 7.1 that $\zeta - \Gamma_{I,\varepsilon} \subset \overset{\bullet}{\mathfrak{S}}_{\varepsilon}(I)$ for every $\zeta \in \overset{\bullet}{\mathfrak{S}}_{2\varepsilon}(I)$. We also think of $\Gamma_{I,\varepsilon}$ as an embedded curve $H_0 : \mathbb{R} \to \mathbb{C}$ with $|H_0'(s)|$ bounded. Therefore, the above properties and the dominated Lebesgue theorem, ensure that the integral

$$\chi(\zeta) = \overset{\vee}{\varphi} *_{H_0} \overset{\vee}{\psi}(\zeta) = \int_{H_0} \overset{\vee}{\varphi}(\eta)\overset{\vee}{\psi}(\zeta-\eta)d\eta \tag{7.14}$$

defines a holomorphic function on $\overset{\bullet}{\mathfrak{S}}_{2\varepsilon}(I) \subset \overset{\bullet}{\mathscr{R}}^{\theta,(0)}$ which by (7.8), represents the convolution product $\overset{\triangledown}{\varphi} * \overset{\triangledown}{\psi}$. We want to show that χ can be analytically continued onto the Riemann surface $(\mathscr{R}^{\theta},\mathfrak{p})$ (thus to $\widetilde{\mathbb{C}\setminus\mathbb{Z}}$ as well). We choose a point $\zeta_0 \in \overset{\bullet}{\mathfrak{S}}_{2\varepsilon}(I)$ so that $\{\zeta_0 - H_0\} \cap \mathbb{Z} = \emptyset$, and we view χ as a germ of holomorphic functions at ζ_0: for $\xi \in \mathbb{C}$ close to 0, $\chi(\zeta_0+\xi) = \int_{H_0} \overset{\vee}{\varphi}(\eta)\overset{\vee}{\psi}(\xi+\zeta_0-\eta)d\eta$. We take a smooth path $\gamma : [0,1] \to \mathbb{C}\setminus\mathbb{Z}$ starting from $\zeta_0 = \gamma(0)$. We fix $R \gg \varepsilon$ so that $\gamma([0,1]) \subset D(0,R)$ and $\mathrm{length}(\gamma) < R$. We will get the analytic continuation of χ along γ by continuously deforming H_0 through an isotopy $H : (s,t) \in \mathbb{R}\times[0,1] \mapsto H_t(s) \in \mathbb{C}\setminus\mathbb{Z}$ which is equal to the identity for $|s|$ large enough. We pick a $\mathscr{C}^1$ function $\eta : \mathbb{C} \to [0,1]$ satisfying $\{\zeta \in \mathbb{C} \mid \eta(\zeta) = 0\} = \mathbb{Z}$. We also set a $\mathscr{C}^1$ function $\rho : \mathbb{C} \to [0,1]$ with compact support so that the conditions $\rho|_{D(0,5R)} = 1$ and $\rho|_{\mathbb{C}\setminus D(0,6R)} = 0$ are fulfilled. In what follows, we see H_0 as an embedded curve $\mathbb{R} \to \mathbb{C}$ and there is no loss of generality in supposing the existence of $s_0 > 0$ so that $H_0(s) \in D(0,3R)$ for $|s| < s_0$, else $H_0(s) \in \mathbb{C}\setminus D(0,3R)$.
One considers the non-autonomous vector field $X(\zeta,t) = \frac{\eta(\zeta)\rho(\zeta)}{\eta(\zeta)+\eta\left(\gamma(t)-\zeta\right)}\gamma'(t)$. We denote by $g : (t_0,t,\zeta_0) \in [0,1]^2 \times \mathbb{C} \mapsto g(t_0,t,\zeta_0) = g^{t_0,t}(\zeta_0) \in \mathbb{C}$ the (well-defined global) flow of the vector field, that is $t \in [0,1] \mapsto \zeta(t) = g^{t_0,t}(\zeta_0)$ is the unique integral curve satisfying both $\frac{\mathrm{d}\zeta}{\mathrm{d}t} = X(\zeta,t)$ and the datum $\zeta(t_0) = \zeta_0$. One finally sets $\phi_t(\zeta) = g^{0,t}(\zeta)$. Notice that any integral curve $\zeta(t)$ of X has length less than $\mathrm{length}(\gamma) < R$, since $|X(\zeta,t)| \leq |\gamma'(t)|$. With this remark and arguments detailed in [MS016], we observe the following properties, for every $t \in [0,1]$:

1. $\phi_t(\gamma(0)) = \gamma(t)$, that is γ is an integral curve. (Notice that $\rho\left(\gamma(t)\right) = 1$ because $\gamma([0,1]) \subset D(0,R)$).
2. $\phi_t(\mathbb{C}\setminus\mathbb{Z}) \subset \mathbb{C}\setminus\mathbb{Z}$. (One has $\phi_t(\omega) = \omega$ for any $\omega \in \mathbb{Z}$ since $\eta(\omega) = 0$).
3. $\phi_t(\zeta) = \zeta$ for any $\zeta \in \mathbb{C}\setminus D(0,6R)$ (since $\rho|_{\mathbb{C}\setminus D(0,6R)} = 0$).
4. for every $\zeta \in D(0,3R)$, $\phi_t(\gamma(0)-\zeta) = \gamma(t) - \phi_t(\zeta)$. Indeed, if $t \mapsto \zeta(t)$ is an integral curve starting from $\zeta(0) \in D(0,3R)$, then $\zeta(t) \in D(0,4R)$ for every $t \in [0,1]$ (the integral curve have length $< R$), thus $\frac{\mathrm{d}\zeta}{\mathrm{d}t} = \frac{\eta(\zeta)}{\eta(\zeta)+\eta\left(\gamma(t)-\zeta\right)}\gamma'(t)$.

Consider $\xi(t)=\gamma(t)-\zeta(t)$; one has $\frac{d\xi}{dt}=\frac{\eta(\xi)\rho(\xi)}{\eta(\xi)+\eta\left(\gamma(t)-\xi\right)}\gamma'(t)$ because $|\xi(t)|<5R$ for every $t\in[0,1]$, thus ξ is an integral curve of X.

5. for every $\zeta\in\mathbb{C}\setminus D(0,3R)$, $|\gamma(t)-\phi_t(\zeta)|>R$. As a matter of fact, observe that if $t\mapsto\zeta(t)$ is an integral curve starting from $\zeta(0)\in\mathbb{C}\setminus D(0,3R)$, then $|\zeta(t)|>2R$ for every $t\in[0,1]$ and therefore $|\gamma(t)-\phi_t(\zeta)|>R$.

We define the isotopy $H:(s,t)\in\mathbb{R}\times[0,1]\mapsto H(s,t)=H_t(s)$ by setting $H_t(s)=\phi_t\big(H_0(s)\big)$. Since H_0 avoids $\mathbb{Z}$, one has $H_t(s)\in\mathbb{C}\setminus\mathbb{Z}$ by property 2. By property 3, we remark that for $|s|$ large enough, H is a constant map. Notice also that $H_0\subset\overset{\bullet}{\mathscr{R}}^{\theta,(0)}$ can be lifted uniquely with respect to $\mathfrak{p}$ on the principal sheet $\mathscr{R}^{\theta,(0)}$ of $\mathscr{R}^{\theta}$. We denote by $\mathscr{H}_0$ this lifting. We use the lifting theorem for homotopies [For81, Ebe007] to get $\mathscr{H}:(s,t)\in\mathbb{R}\times[0,1]\mapsto\mathscr{H}(s,t)=\mathscr{H}_t(s)\in\mathscr{R}^{\theta}$, the continuous mapping which makes commuting the following diagram: $\begin{array}{ccc} & & \mathscr{R}^{\theta} \\ & \mathscr{H}\nearrow & \downarrow\mathfrak{p} \\ \mathbb{R}\times[0,1] & \underset{H}{\longrightarrow} & \mathbb{C}.\end{array}$

We now set $K:(s,t)\in\mathbb{R}\times[0,1]\mapsto K(s,t)=K_t(s)=\gamma(t)-H_t(s)$. We know that $K_0(s)=\gamma(0)-H_0(s)\in\overset{\bullet}{\mathfrak{S}}_{\varepsilon}(I)\subset\overset{\bullet}{\mathscr{R}}^{\theta,(0)}$ for every $s\in\mathbb{R}$. In particular, one can lift K_0 uniquely with respect to $\mathfrak{p}$ into an embedded curve $\mathscr{K}_0$ on the principal sheet $\mathscr{R}^{\theta,(0)}$ of $\mathscr{R}^{\theta}$. Moreover $K_0(s)\in\mathbb{C}\setminus\mathbb{Z}$, for every $s\in\mathbb{R}$. Property 5 ensures that $K_t(s)$ stays in $\overset{\bullet}{\mathfrak{S}}_{\varepsilon}(I)$ for $|s|\geq s_0$, otherwise by property 4, $K_t(s)$ belongs to $\mathbb{C}\setminus\mathbb{Z}$. This implies that K_t can be lifted uniquely with respect to $\mathfrak{p}$ into an embedded curve $\mathscr{K}_t$ which lies on the principal sheet $\mathscr{R}^{\theta,(0)}$ of $\mathscr{R}^{\theta}$ for $|s|\geq s_0$. Applying again the lifting theorem for homotopies, one obtains a continuous mapping $\mathscr{K}:(s,t)\in\mathbb{R}\times[0,1]\mapsto\mathscr{H}(s,t)=\mathscr{H}_t(s)\in\mathscr{R}^{\theta}$ that makes commuting the following diagram: $\begin{array}{ccc} & & \mathscr{R}^{\theta} \\ & \mathscr{K}\nearrow & \downarrow\mathfrak{p} \\ \mathbb{R}\times[0,1] & \underset{K}{\longrightarrow} & \mathbb{C}.\end{array}$

We finally introduce the two holomorphic functions $\Phi,\Psi\in\mathscr{O}(\mathscr{R}^{\theta})$ such that $\Phi(\zeta)=\overset{\vee}{\varphi}\big(\mathfrak{p}(\zeta)\big)$, $\Psi(\zeta)=\overset{\vee}{\psi}\big(\mathfrak{p}(\zeta)\big)$ for $\zeta\in\mathscr{R}^{\theta,(0)}$. With this notation, the germ of holomorphic functions χ at $\zeta_0=\gamma(0)$ reads

$$\chi(\gamma(0)+\xi)=\int_{\mathbb{R}}\Phi\big(\mathscr{H}_0(s)\big)\Psi(\xi+\mathscr{K}_0(s))H_0'(s)ds$$

and its analytic continuation along γ is obtained by

$$\chi(\gamma(t)+\xi)=\int_{\mathbb{R}}\Phi\big(\mathscr{H}_t(s)\big)\Psi(\xi+\mathscr{K}_t(s))H_t'(s)ds. \tag{7.15}$$

Indeed, remark that for $|s|$ large enough, $\Phi\big(\mathscr{H}_t(s)\big)=\overset{\vee}{\varphi}\big(H_t(s)\big)$ and $|\overset{\vee}{\varphi}\big(H_t(s)\big)|\leq\exp(-|H_t(s)|)$. Also, for $|s|\geq s_0$, $\Psi\big(\xi+\mathscr{K}_t(s)\big)=\overset{\vee}{\psi}\big(K_t(s)\big)$ which is bounded since $K_t(s) \in \overset{\bullet}{\mathfrak{S}}_\varepsilon(I)$. Thus the integral (7.15) is well-defined. The fact that (7.15) provides the analytic continuations comes from the Cauchy formula, see analogous arguments in the first volume of this book [MS016]. □

7.7.2.4 Supplements

One often uses other spaces in practice as we now exemplify.

The space $\mathrm{RES}^{(\theta,\alpha)}(L)$ The space $\hat{\mathscr{R}}^{(\theta,\alpha)}(L)$ was introduced by definition 4.24 and we know by proposition 4.6 that $\mathbb{C}\delta \oplus \hat{\mathscr{R}}^{(\theta,\alpha)}(L)$ is a convolution algebra. The following definition thus makes sense.

Definition 7.40. We denote by $\overset{\triangledown}{\mathscr{R}}^{(\theta,\alpha)}(L) \supset \overset{\triangledown}{\mathscr{R}}$ the $\mathbb{C}$-differential commutative and associative convolution algebra made of singularities of the form $\overset{\triangledown}{\varphi} = a\delta + {}^\flat\widehat{\varphi} \in \mathrm{SING}$ with $\widehat{\varphi} \in \hat{\mathscr{R}}^{(\theta,\alpha)}(L)$. The associated space of formal series is denoted by $\widetilde{\mathscr{R}}^{(\theta,\alpha)}(L)$.

By its very definition, any element $\widehat{\varphi} \in \hat{\mathscr{R}}^{(\theta,\alpha)}(L)$ is a germ of holomorphic functions at 0 that can be analytically continued to the Riemann surface $\mathscr{R}^{(\theta,\alpha)}(L)$. This means that any $\overset{\triangledown}{\varphi} \in \overset{\triangledown}{\mathscr{R}}^{(\theta,\alpha)}(L)$ is a simple singularity that has a major $\overset{\vee}{\varphi}$ which can be analytically continued to the universal covering $\widetilde{\mathscr{R}^{(\theta,\alpha)}(L) \setminus \{\underline{0}\}}$ of $\mathscr{R}^{(\theta,\alpha)}(L) \setminus \{\underline{0}\}$. Since $\overset{\triangledown}{\mathscr{R}}^{(\theta,\alpha)}(L)$ is a convolution algebra, we know that for any two singularities $\overset{\triangledown}{\varphi}, \overset{\triangledown}{\psi} \in \overset{\triangledown}{\mathscr{R}}^{(\theta,\alpha)}(L)$, their convolution product $\overset{\triangledown}{\varphi} * \overset{\triangledown}{\psi}$ belongs to $\overset{\triangledown}{\mathscr{R}}^{(\theta,\alpha)}(L)$ as well, thus has a major that can be analytically continued to $\widetilde{\mathscr{R}^{(\theta,\alpha)}(L) \setminus \{\underline{0}\}}$. In substance, this comes from the property that ${}^\flat\widehat{\varphi} * {}^\flat\widehat{\varphi} = {}^\flat(\widehat{\varphi} * \widehat{\varphi})$ for two integrable singularities (proposition 7.6). Now, what about the convolution product $\overset{\triangledown}{\varphi} * \overset{\triangledown}{\psi}$ of two singularities $\overset{\triangledown}{\psi} \in \overset{\triangledown}{\mathscr{R}}^{(\theta,\alpha)}(L)$ and $\overset{\triangledown}{\varphi} \in \mathrm{RES}$? To give the answer, we prefer to shift to a more general case and we introduce a new definition.

Definition 7.41. Let be $\theta \in \{0,\pi\} \subset \mathbb{S}^1$, $\alpha \in]0,\pi/2]$ and $L > 0$. We denote by $\mathrm{RES}^{(\theta,\alpha)}(L)$ the space made of singularities that have majors that can be analytically continued to the Riemann surface $\widetilde{\mathscr{R}^{(\theta,\alpha)}(L) \setminus \{\underline{0}\}}$. The associated space of asymptotic classes is denoted by $\widetilde{\mathrm{RES}}^{(\theta,\alpha)}(L) \subset \mathrm{ASYMP}$.

Proposition 7.16. *The space* $\mathrm{RES}^{(\theta,\alpha)}(L)$ *is a* $\mathbb{C}$*-differential commutative and associative convolution algebra with unit* δ*, contained* RES *and* $\overset{\triangledown}{\mathscr{R}}^{(\theta,\alpha)}(L)$ *as subalgebras.*

Proof. The proof follows that of theorem 7.16 but for the fact that one adds the arguments used at the end of the proof of proposition 4.6.

The spaces $\mathrm{RES}^{(k)}$ The spaces $\widehat{\mathscr{R}}^{(k)}$ were introduced by definition 4.20. They provide new spaces of singularities which are worthy of attention.

Definition 7.42. For $k \in \mathbb{N}^\star$, we denote by $\overset{\triangledown}{\mathscr{R}}^{(k)}$ the space of singularities of the form $\overset{\triangledown}{\varphi} = a\delta + {}^\flat\widehat{\varphi} \in \mathrm{SING}$ with $\widehat{\varphi} \in \widehat{\mathscr{R}}^{(k)}$. The associated space of formal series is denoted by $\widetilde{\mathscr{R}}^{(k)}$.

Remark 7.6. Notice that the set of spaces $(\overset{\triangledown}{\mathscr{R}}^{(k)})_{k\in\mathbb{N}}$ provides an inverse system of spaces whose inverse limit $\lim\limits_{\leftarrow} \overset{\triangledown}{\mathscr{R}}^{(k)} = \bigcap\limits_{k} \overset{\triangledown}{\mathscr{R}}^{(k)}$ is $\overset{\triangledown}{\mathscr{R}}$. This is why we sometimes write $\overset{\triangledown}{\mathscr{R}}^{(\infty)} = \overset{\triangledown}{\mathscr{R}}$.
The space $\overset{\triangledown}{\mathscr{R}}^{(1)}$ is of particular interest since, from propositions 4.1 and 7.6, $\overset{\triangledown}{\mathscr{R}}^{(1)}$ makes a convolution algebra.

The space $\overset{\triangledown}{\mathscr{R}}^{(k)}$ is made of simple singularities that have majors that can be analytically continued to the universal covering $\widetilde{\mathscr{R}^{(k)} \setminus \{\underline{0}\}}$ of $\mathscr{R}^{(k)} \setminus \{\underline{0}\}$. We now consider larger spaces of singularities.

Definition 7.43. Let $k \in \mathbb{N}^\star$ be a positive integer. We denote by $\mathrm{RES}^{(k)}$ the space of singularities that have majors that can be analytically continued to the Riemann surface $\widetilde{\mathscr{R}^{(k)} \setminus \{\underline{0}\}}$. We denote by $\widetilde{\mathrm{RES}}^{(k)} \subset \mathrm{ASYMP}$ the space of asymptotic classes whose formal Borel transform belongs to $\mathrm{RES}^{(k)}$.

Remark 7.7. Notice again that $\lim\limits_{\leftarrow} \mathrm{RES}^{(k)} = \bigcap\limits_{k} \mathrm{RES}^{(k)} = \mathrm{RES}$, and we sometimes write $\mathrm{RES}^{(\infty)} = \mathrm{RES}$.

We will have a special interest in $\mathrm{RES}^{(1)}$ because of the following analogous to proposition 7.16.

Proposition 7.17. *The space* $\mathrm{RES}^{(1)}$ *is a* $\mathbb{C}$*-differential commutative and associative convolution algebra with unit* δ. *It contains* RES *and* $\overset{\triangledown}{\mathscr{R}}^{(1)}$ *as subalgebras.*

We omit the (rather lengthy) proof of this proposition. The main idea is to consider the integral representation (7.14) used in the proof of theorem 7.1 and to adapt the construction made in Sect. 4.3.

Conjecture 7.1. We conjecture that any space $\mathrm{RES}^{(k)}$ makes a convolution algebra as well.

7.8 Alien Operators

Alien operators are powerful tools for analysing the singularities of resurgent functions. These operators are carefully defined and discussed in the first volume of this book [MS016], especially when they operate on the algebra $\mathbb{C}\delta \oplus \widehat{\mathscr{R}}^{\mathrm{simp}}$ of simple resurgent functions. Most of the arguments there can be easily adapted for alien operators acting on $\mathrm{RES}_{\mathbb{Z}}$, once the study of singularities had been made. This is why we introduce the alien operators in a rather sketchy manner in what follows.

7.8.1 Alien Operators Associated with a Triple

7.8.1.1 Mains Definitions

We consider two directions $\theta_1, \theta_2 \in \mathbb{S}^1$, a point $\omega \in \mathbb{Z}$ and a sectorial germ $\overset{\vee}{\varphi} \in \mathscr{O}^0_{\theta_1}$ of direction θ_1. We can think of $\overset{\vee}{\varphi}$ as a sectorial germ on a sector $\overset{\bullet}{\mathit{\Delta}}{}^{R_1}_0(I_1)$ for $0 < R_1 < 1$ and $I_1 \subset \mathbb{S}^1$ an open arc bisected by θ_1, and this is what we do in what follows.
We now assume that $\overset{\vee}{\varphi}$ can be analytically continued to $\widetilde{\mathbb{C} \setminus \mathbb{Z}}$. We consider a path $\gamma : J \to \mathbb{C} \setminus \mathbb{Z}$ starting from $\zeta_1 \in \overset{\bullet}{\mathit{\Delta}}{}^{R_1}_0(I_1)$ and ending at ζ_2 close to ω so that $\zeta_2 - \omega \in \overset{\bullet}{\mathit{\Delta}}{}^{R_2}_0(I_2)$ with $0 < R_2 < 1$ and $I_2 \subset \mathbb{S}^1$ an open arc bisected by θ_2. See Fig. 7.9.
By hypotheses, the analytic continuation $(\mathrm{cont}_\gamma \overset{\vee}{\varphi})$ of $\overset{\vee}{\varphi}$ along γ is a well-defined germ of holomorphic functions at ζ_2 that only depends on the homotopy class of γ (for the relation of homotopy of paths in $\mathbb{C} \setminus \mathbb{Z}$ with fixed extremities). Moreover, if $\overset{\vee}{\psi} \in \mathscr{O}_{\zeta_2 - \omega}$ stands for the germ of holomorphic functions at $\zeta_2 - \omega$ defined by $\overset{\vee}{\psi}(\xi) = \big(\mathrm{cont}_\gamma \overset{\vee}{\varphi}\big)(\omega + \xi)$ then, still by analytic continuations, $\overset{\vee}{\psi}$ determines a unique sectorial germ on $\overset{\bullet}{\mathit{\Delta}}{}^{R_2}_0(I_2)$ and thus, by restriction, a unique sectorial germ $\overset{\vee}{\psi} \in \mathscr{O}^0_{\theta_2}$. This justifies the following definition adapted from the first volume of this book [MS016].

Definition 7.44. Let be $\theta_1, \theta_2 \in \mathbb{S}^1$, $\omega \in \mathbb{Z}$ and $\overset{\vee}{\varphi} \in \mathscr{O}^0_{\theta_1}$ a sectorial germ of direction θ_1 that can be analytically continued to $\widetilde{\mathbb{C} \setminus \mathbb{Z}}$. Let $\gamma : J \to \mathbb{C} \setminus \mathbb{Z}$ be a path starting from a sufficiently small sector $\overset{\bullet}{\mathit{\Delta}}_0(I_1)$ bisected by θ_1 and ending close to ω in a sufficiently small sector of the form $\omega + \overset{\bullet}{\mathit{\Delta}}_0(I_2)$ where I_2 bisects θ_2. Then, one

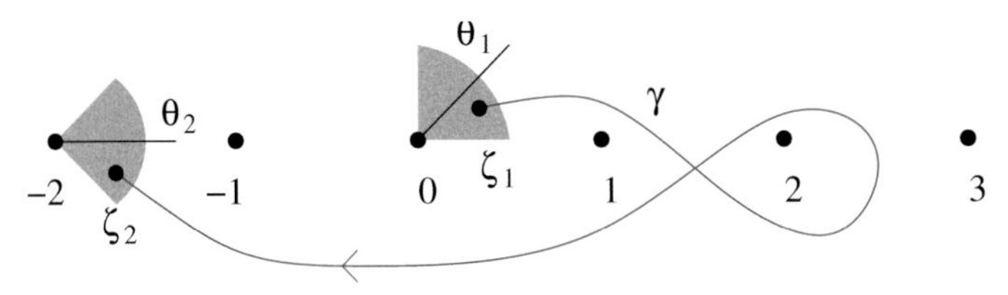

Fig. 7.9 A triple $(\gamma, \theta_1, \theta_2)$ defining the operator $\mathscr{A}^\gamma_\omega(\theta_1, \theta_2)$ at $\omega = -2$.

denotes by $\overset{\bullet}{\mathscr{A}}{}^{\gamma}_{\omega}(\theta_2,\theta_1)\overset{\vee}{\varphi}\in\mathscr{O}^0_{\theta_2}$ the sectorial germ of direction θ_2 represented by $\overset{\vee}{\psi}(\xi)=\big(\mathrm{cont}_\gamma\overset{\vee}{\varphi}\big)(\omega+\xi)$ for $\xi\in\overset{\bullet}{\mathcal{S}}_0(I_2)$.

We now consider two directions $\theta_1,\theta_2\in\widetilde{\mathbb{S}}^1$ and a singularity $\overset{\triangledown}{\varphi}\in\mathrm{RES}_{\mathbb{Z}}$. Thinking of $\overset{\triangledown}{\varphi}$ as a singularity of $\mathrm{SING}_{\theta_1,\alpha_1}$ (for some $\alpha_1>0$), its minor $\widehat{\varphi}$ can be seen as representing a sectorial germ $\widehat{\varphi}\in\mathscr{O}^0_{\overset{\bullet}{\theta}_1}$ of direction $\overset{\bullet}{\theta}_1=\pi(\theta_1)\in\mathbb{S}^1$ which can be analytically continued to $\widetilde{\mathbb{C}\setminus\mathbb{Z}}$. Therefore, under the conditions of definition 7.44, the sectorial germ $\overset{\vee}{\psi}_{\overset{\bullet}{\theta}_2}=\overset{\bullet}{\mathscr{A}}{}^{\gamma}_{\omega}(\overset{\bullet}{\theta}_2,\overset{\bullet}{\theta}_1)\widehat{\varphi}$ of direction $\overset{\bullet}{\theta}_2=\pi(\theta_2)\in\mathbb{S}^1$ is well-defined. Even, by analytic continuations, one can deduce from $\overset{\vee}{\psi}_{\overset{\bullet}{\theta}_2}$ a sectorial germ of direction $I_{\overset{\bullet}{\theta}_2}=]-\pi+\overset{\bullet}{\theta}_2,\overset{\bullet}{\theta}_2+\pi[\subset\mathbb{S}^1$ denoted by $\overset{\vee}{\psi}_{I_{\overset{\bullet}{\theta}_2}}\in\Gamma(I_{\overset{\bullet}{\theta}_2},\mathscr{O}^0)$. By inverse image by π of the sheaf $\mathscr{O}^0$, this sectorial germ $\overset{\vee}{\psi}_{I_{\overset{\bullet}{\theta}_2}}$ determined a uniquely defined sectorial germ of direction $I_{\theta_2}=]-\pi+\theta_2,\theta_2+\pi[\subset\widetilde{\mathbb{S}}^1$ denoted by $\overset{\vee}{\psi}_{I_{\theta_2}}$. Still by analytic continuations, this sectorial germ gives rise to a (multivalued) section on any arc of the form $]-\alpha-2\pi+(\theta_2+\pi),(\theta_2+\pi)+\alpha[\in\widetilde{\mathbb{S}}^1$, $\alpha>0$, that is to an element $\overset{\vee}{\psi}$ of $\mathrm{ANA}=\bigcap_{\alpha>0}\mathrm{ANA}_{(\theta_2+\pi),\alpha}$, whose singularity $\overset{\triangledown}{\psi}$ belongs to $\mathrm{RES}_{\mathbb{Z}}$.

Definition 7.45. Let be $\theta_1,\theta_2\in\widetilde{\mathbb{S}}^1$ and $\omega\in\mathbb{Z}$. Let $\gamma:J\to\mathbb{C}\setminus\mathbb{Z}$ be a path starting from a sufficiently small sector $\overset{\bullet}{\mathcal{S}}_0(I_1)$ bisected by $\overset{\bullet}{\theta}_1=\pi(\theta_1)$ and ending close to ω in a sufficiently small sector of the form $\omega+\overset{\bullet}{\mathcal{S}}_0(I_2)$ where I_2 bisects $\overset{\bullet}{\theta}_2=\pi(\theta_2)$. For any singularity $\overset{\triangledown}{\varphi}\in\mathrm{RES}_{\mathbb{Z}}$, one denotes by $\mathscr{A}^{\gamma}_{\omega}(\theta_2,\theta_1)\overset{\triangledown}{\varphi}$ the singularity $\overset{\triangledown}{\psi}$ which can be represented by a major $\overset{\vee}{\psi}\in\mathrm{ANA}=\Gamma(\widetilde{\mathbb{S}}^1,\mathscr{O}^0)$, whose restriction $\overset{\vee}{\psi}_{\theta_2}\in\mathscr{O}^0_{\theta_2}$ is the sectorial germ of direction θ_2 determined by $\overset{\vee}{\psi}_{\overset{\bullet}{\theta}_2}=\overset{\bullet}{\mathscr{A}}{}^{\gamma}_{\omega}(\overset{\bullet}{\theta}_2,\overset{\bullet}{\theta}_1)\widehat{\varphi}$, where $\widehat{\varphi}$ is the minor of $\overset{\triangledown}{\varphi}$.
The linear operator $\mathscr{A}^{\gamma}_{\omega}(\theta_2,\theta_1):\mathrm{RES}_{\mathbb{Z}}\to\mathrm{RES}_{\mathbb{Z}}$ is called the *alien operator at ω associated with the triple $(\gamma,\theta_1,\theta_2)$.*

The alien operators have their counterparts on asymptotic classes through formal Borel and Laplace transforms.

Definition 7.46. The alien operator $\mathscr{A}^{\gamma}_{\omega}(\theta_2,\theta_1)$ at ω associated with the triple $(\gamma,\theta_1,\theta_2)$ is defined on asymptotic classes by making the following diagram commuting:

$$\begin{array}{ccc}\mathrm{RES} & \overset{\mathscr{A}^{\gamma}_{\omega}(\theta_2,\theta_1)}{\longrightarrow} & \mathrm{RES}\\ \mathscr{L}\downarrow\uparrow\mathscr{B} & & \mathscr{L}\downarrow\uparrow\mathscr{B}\\ \widetilde{\mathrm{RES}} & \overset{\mathscr{A}^{\gamma}_{\omega}(\theta_2,\theta_1)}{\longrightarrow} & \widetilde{\mathrm{RES}}\end{array}.$$

7.8.1.2 The Spaces $\mathrm{RES}^{(\overset{\bullet}{\theta},\alpha)}(L)$ and $\mathrm{RES}^{(k)}$

Alien operators acting on $\mathrm{RES}^{(\overset{\bullet}{\theta},\alpha)}(L)$ We would like to define alien operators acting on the space $\mathrm{RES}^{(\overset{\bullet}{\theta},\alpha)}(L)$. We suppose $\theta \in \{\pi k, k \in \mathbb{Z}\} \subset \widetilde{\mathbb{S}}^1$, $\alpha \in]0,\pi/2]$, $L > 0$ and $m \in \{1,\cdots,\lceil L \rceil\}$. We set $\overset{\bullet}{\theta} = \pi(\theta) \in \{0,\pi\}$ and we consider a singularity $\overset{\triangledown}{\varphi} \in \mathrm{RES}^{(\overset{\bullet}{\theta},\alpha)}(L)$ whose minor is $\widehat{\varphi}$. By the very definition 7.41 of the space $\mathrm{RES}^{(\overset{\bullet}{\theta},\alpha)}(L)$, the sectorial germ $\overset{\vee}{\psi}_{\overset{\bullet}{\theta}_2} = \overset{\bullet}{\mathscr{A}}{}^{\gamma}_{\omega}(\overset{\bullet}{\theta}_2,\overset{\bullet}{\theta})\widehat{\varphi} \in \mathscr{O}^0_{\overset{\bullet}{\theta}_2}$ is well defined under the following conditions:

1. $\omega = me^{\mathrm{i}\overset{\bullet}{\theta}}$ and the path γ is of type $\gamma_{\varepsilon}^{\overset{\bullet}{\theta}}$ with $\varepsilon = (\pm)_{m-1} \in \{+,-\}^{m-1}$. In that case, $\overset{\bullet}{\theta}_2$ should be $\overset{\bullet}{\theta} - \pi$;
2. however, starting from $\overset{\vee}{\psi}_{\overset{\bullet}{\theta}-\pi}$ and be analytic continuations, one can consider as well sectorial germs $\overset{\vee}{\psi}_{\overset{\bullet}{\theta}_2}$ with $\overset{\bullet}{\theta}_2 \in I_{\overset{\bullet}{\theta}} =]-2\pi+\overset{\bullet}{\theta},\overset{\bullet}{\theta}[\subset \mathbb{S}^1$.

By a construction already done, the various sectorial germs $\overset{\vee}{\psi}_{\overset{\bullet}{\theta}_2}$ glue together and provide a sectorial germ $\overset{\vee}{\psi}_{I_{\overset{\bullet}{\theta}}} \in \Gamma(I_{\overset{\bullet}{\theta}},\mathscr{O}^0)$ of direction $I_{\overset{\bullet}{\theta}}$. Still by analytic continuations and moving to multivalued sectorial germs by inverse image by π of the sheaf $\mathscr{O}^0$, one eventually gets an element $\overset{\vee}{\psi}$ of $\mathrm{ANA}_{\theta,\alpha}$ with $\pi(\theta) = \overset{\bullet}{\theta}$. This gives sense to the following definition.

Definition 7.47. Let be $\theta \in \{\pi k, k \in \mathbb{Z}\} \subset \widetilde{\mathbb{S}}^1$, $\alpha \in]0,\pi/2]$ and $L > 0$. We set $\overset{\bullet}{\theta} = \pi(\theta) \in \{0,\pi\} \subset \mathbb{S}^1$. We pick $m \in \{1,\cdots,\lceil L \rceil\}$, we set $\omega = me^{\mathrm{i}\overset{\bullet}{\theta}}$ and we assume that the path γ is of type $\gamma^{\overset{\bullet}{\theta}}_{(\pm)_{m-1}}$. For any singularity $\overset{\triangledown}{\varphi} \in \mathrm{RES}^{(\overset{\bullet}{\theta},\alpha)}(L)$, one denotes by $\mathscr{A}^{\gamma}_{\omega}(\theta,\theta)\overset{\triangledown}{\varphi}$ the singularity $\overset{\triangledown}{\psi} \in \mathrm{SING}_{\theta,\alpha}$ which can be represented by a major $\overset{\vee}{\psi} \in \mathrm{ANA}_{\theta,\alpha}$ whose restriction $\overset{\vee}{\psi}_{\theta-\pi} \in \mathscr{O}^0_{\theta-\pi}$ is the sectorial germ of direction $\theta-\pi$ determined by $\overset{\vee}{\psi}_{\theta-\pi} = \overset{\bullet}{\mathscr{A}}{}^{\gamma}_{\omega}(\overset{\bullet}{\theta}-\pi,\overset{\bullet}{\theta})\widehat{\varphi}$ where $\widehat{\varphi}$ stands for the minor of $\overset{\triangledown}{\varphi}$. This gives rise to a linear operator $\mathscr{A}^{\gamma}_{\omega}(\theta,\theta) : \mathrm{RES}^{(\overset{\bullet}{\theta},\alpha)}(L) \to \mathrm{SING}_{\theta,\alpha}$, still called the alien operator at ω associated with the triple (γ,θ,θ).

Alien operators acting on $\mathrm{RES}^{(k)}$ We now work on the spaces $\mathrm{RES}^{(k)}$ given by definition 7.43. We want to prove that alien operators can be defined on $\mathrm{RES}^{(k)}$, associated with triples of the form (γ,θ,θ) with γ of type $\gamma^{\overset{\bullet}{\theta}}_{(+)_m}$ or $\gamma^{\overset{\bullet}{\theta}}_{(-)_m}$.

We start with $\mathrm{RES}^{(1)}$. Let be $\theta_1 \in \{\pi k, k \in \mathbb{Z}\} \subset \widetilde{\mathbb{S}}^1$ and set $\omega_1 = e^{\mathrm{i}\overset{\bullet}{\theta}_1}$ with $\overset{\bullet}{\theta}_1 = \pi(\theta_1)$. The very definition of $\mathrm{RES}^{(1)}$ and the above reasoning lead straight to the following linear operators, for any integer $m_1 \geq 2$ and any $\varepsilon \in \{-,+\}$:

$$\mathscr{A}_{\omega_1}^{\gamma_{()}^{\overset{\bullet}{\theta}_1}}(\theta_1,\theta_1):\mathrm{RES}^{(1)}\to\mathrm{SING}_{\theta_1,\pi},\quad \mathscr{A}_{m_1\omega_1}^{\gamma_{(\varepsilon)_{m_1-1}}^{\overset{\bullet}{\theta}_1}}(\theta_1,\theta_1):\mathrm{RES}^{(1)}\to\mathrm{SING}_{\theta_1+\pi/2,\pi/2} \tag{7.16}$$

We move to the next case $k=2$, that is we consider the space $\mathrm{RES}^{(2)}\subset\mathrm{RES}^{(1)}$. Of course the above operators (7.16) still act on $\mathrm{RES}^{(2)}$ but, however, their ranges can be made more precise. By the very definition of $\mathrm{RES}^{(2)}$, the minor $\widehat{\varphi}$ of any singularity $\overset{\triangledown}{\varphi}\in\mathrm{RES}^{(2)}$, when considered as a sectorial germ, can be analytically continued along any path γ of type $\gamma_{\varepsilon^{\mathbf{n}_1}}^{\overset{\bullet}{\theta}_1}$ with

$$\varepsilon^{\mathbf{n}_1}\in\{((\pm)^{n_1},(+)_{m_1-1}),((\pm)^{n_1},(-)_{m_1-1})\mid(n_1,m_1)\in(\mathbb{N}^\star)^2\}.$$

Moreover, introducing $\overset{\bullet}{\theta}_2=\overset{\bullet}{\theta}_1+(n-1)\pi$, $\omega_1=\mathrm{e}^{\mathrm{i}\overset{\bullet}{\theta}_1}$, and $\omega_2-\omega_1=\mathrm{e}^{\mathrm{i}\overset{\bullet}{\theta}_2}$, the analytic continuation $\mathrm{cont}_\gamma\,\widehat{\varphi}$ of $\widehat{\varphi}$ along γ is a germ of holomorphic functions whic can be analytically continued onto the simply connected domain $\mathfrak{p}(\mathscr{R}^{\varepsilon^{\mathbf{n}_1},\overset{\bullet}{\theta}})=\mathbb{C}\setminus\{]-\infty,p]\cup[p+1,+\infty[\}$ where $]p,(p+1)[=]\omega_1,\omega_2[$ when $m_1=1$, $]p,(p+1)[=](m_1-1)\omega_2,m_1\omega_2[$ when $m_1\geq 2$. Considering only odd values for n_1 (thus $\overset{\bullet}{\theta}_2=\overset{\bullet}{\theta}_1$ on $\mathbb{S}^1$), one immediately sees that (7.16) becomes:

$$\begin{aligned}
&\mathscr{A}_{\omega_1}^{\gamma_{()}^{\overset{\bullet}{\theta}_1}}(\theta_1,\theta_1):\mathrm{RES}^{(2)}\to\mathrm{RES}^{(1)},\\
&\mathscr{A}_{2\omega_1}^{\gamma_{(\varepsilon)_1}^{\overset{\bullet}{\theta}_1}}(\theta_1,\theta_1):\mathrm{RES}^{(2)}\to\mathrm{SING}_{\theta_1,\pi}\\
&\mathscr{A}_{m_1\omega_1}^{\gamma_{(\varepsilon)_{m_1-1}}^{\overset{\bullet}{\theta}_1}}(\theta_1,\theta_1):\mathrm{RES}^{(2)}\to\mathrm{SING}_{\theta_1+\pi/2,\pi/2},\quad m_1\geq 3.
\end{aligned}\tag{7.17}$$

Notice in particular that the operator $\mathscr{A}_{\omega_1}^{\gamma_{()}^{\overset{\bullet}{\theta}_1}}(\theta_2,\theta_1)$ now acts on $\mathrm{RES}^{(2)}$ as well, for any direction $\theta_2\in\widetilde{\mathbb{S}}^1$.

The reasoning generalizes and we give the result.

Lemma 7.8. *Let be $\theta_1\in\{\pi k,k\in\mathbb{Z}\}\subset\widetilde{\mathbb{S}}^1$. For any integer $k\geq 1$, any $\varepsilon\in\{-,+\}$ and any $m_1\in\mathbb{N}^\star$, setting $\omega_1=\mathrm{e}^{\mathrm{i}\overset{\bullet}{\theta}_1}$, the alien operator $\mathscr{A}_{m_1\omega_1}^{\gamma_{(\varepsilon)_{m_1-1}}^{\overset{\bullet}{\theta}_1}}(\theta_1,\theta_1)$ is well defined on $\mathrm{RES}^{(k)}$ with the range:*

$$\begin{aligned}
&\mathscr{A}_{m_1\omega_1}^{\gamma_{(\varepsilon)_{m_1-1}}^{\overset{\bullet}{\theta}_1}}(\theta_1,\theta_1):\mathrm{RES}^{(k)}\to\mathrm{RES}^{(k-m_1)},\quad 1\leq m_1\leq k-1\\
&\mathscr{A}_{m_1\omega_1}^{\gamma_{(\varepsilon)_{m_1-1}}^{\overset{\bullet}{\theta}_1}}(\theta_1,\theta_1):\mathrm{RES}^{(k)}\to\mathrm{SING}_{\theta_1,\pi},\quad m_1=k\\
&\mathscr{A}_{m_1\omega_1}^{\gamma_{(\varepsilon)_{m_1-1}}^{\overset{\bullet}{\theta}_1}}(\theta_1,\theta_1):\mathrm{RES}^{(k)}\to\mathrm{SING}_{\theta_1+\pi/2,\pi/2},\quad m_1\geq k+1.
\end{aligned}\tag{7.18}$$

7.8.1.3 Miscellaneous Properties

We start with a simple result which is a consequence of the very definitions.

Proposition 7.18. *For any alien operator of the form* $\mathscr{A}^{\gamma}_{\omega}(\theta_2,\theta_1):\mathrm{RES}_{\mathbb{Z}}\to\mathrm{RES}_{\mathbb{Z}}$, *acting on* $\mathrm{RES}_{\mathbb{Z}}$, $\mathrm{RES}^{(\dot{\theta},\alpha)}(L)$ *or* $\mathrm{RES}^{(k)}$, *for any singularity* $\overset{\triangledown}{\varphi}$ *:*

$$\mathscr{A}^{\gamma}_{\omega}(\theta_2,\theta_1)(\overset{\triangledown}{\partial}\overset{\triangledown}{\varphi}) = (\overset{\triangledown}{\partial}-\omega)\mathscr{A}^{\gamma}_{\omega}(\theta_2,\theta_1)\overset{\triangledown}{\varphi}. \tag{7.19}$$

In other words, $[\mathscr{A}^{\gamma}_{\omega}(\theta_2,\theta_1),\overset{\triangledown}{\partial}] = -\omega\mathscr{A}^{\gamma}_{\omega}(\theta_2,\theta_1)$.

We introduce new definitions before keeping on.

Definition 7.48. For any $k\in\mathbb{Z}$, one denotes by $\rho_k\in\mathrm{Aut}(\pi)$ the deck transformation of the cover $(\widetilde{\mathbb{C}},\pi)$, defined by: $\rho_k:\zeta=re^{i\theta}\in\widetilde{\mathbb{C}}\mapsto\rho_k(\zeta)=re^{i\theta+2i\pi k}\in\widetilde{\mathbb{C}}$.
For any singularity of the form $\overset{\triangledown}{\varphi}=\mathrm{sing}_0\overset{\vee}{\varphi}\in\mathrm{SING}$, $\overset{\vee}{\varphi}\in\mathrm{ANA}$, we write $\rho_k.\overset{\triangledown}{\varphi}=\mathrm{sing}_0(\overset{\vee}{\varphi}\circ\rho_k)\in\mathrm{SING}$.
More generally, for any $r\in\mathbb{R}$, one sets $\rho_r:\zeta=re^{i\theta}\in\widetilde{\mathbb{C}}\mapsto\rho_r(\zeta)=re^{i\theta+2i\pi r}\in\widetilde{\mathbb{C}}$ and $\rho_r.\overset{\triangledown}{\varphi}=\mathrm{sing}_0(\overset{\vee}{\varphi}\circ\rho_r)\in\mathrm{SING}$.

Remark 7.8. With this notation, the variation map $\mathrm{var}:\mathrm{SING}\to\mathrm{ANA}$ reads $\mathrm{var}=\mathrm{Id}-\rho_{-1}$.

The alien operators associated with a triple satisfy some identities as can be easily observed:

Proposition 7.19. *For any given alien operator* $\mathscr{A}^{\gamma}_{\omega}(\theta_2,\theta_1):\mathrm{RES}_{\mathbb{Z}}\to\mathrm{RES}_{\mathbb{Z}}$, $\mathscr{A}^{\gamma}_{\omega}(\theta_2,\theta_1+2\pi k)=\mathscr{A}^{\gamma}_{\omega}(\theta_2,\theta_1)\rho_k$ *and* $\mathscr{A}^{\gamma}_{\omega}(\theta_2+2\pi k,\theta_1)=\rho_{-k}.\mathscr{A}^{\gamma}_{\omega}(\theta_2,\theta_1)$, *for any* $k\in\mathbb{Z}$.

Let us consider a point $\omega\in\mathbb{Z}$ and a given triple $(\gamma,\theta_1,\theta_2)$. One can extend the path γ into the path $\gamma\lambda^k_{\omega}$ where λ^k_{ω} is a closed path near ω that surrounds that point like on Fig. 7.10, with winding number $\mathrm{wind}_{\omega}(\lambda^k_{\omega})=k\in\mathbb{Z}$ at that point. One can as well consider the path $\lambda^k_0\gamma$ where λ^k_0 is a closed path surrounding the origin with winding number $\mathrm{wind}_{\omega}(\lambda^k_0)=k\in\mathbb{Z}$. A little thought provides the following result.

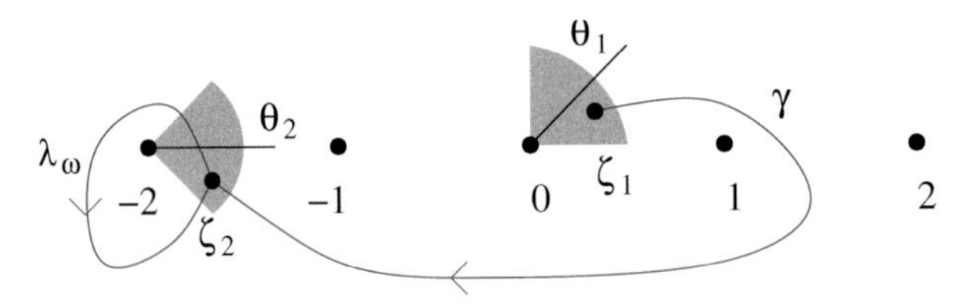

Fig. 7.10 Two triples $(\gamma,\theta_1,\theta_2)$ and $(\gamma\lambda_{\omega},\theta_1,\theta_2)$ for the point $\omega=-2$, with λ_{ω} a closed path of winding number $\mathrm{wind}_{\omega}(\lambda_{\omega})=1$ at ω.

Proposition 7.20. *We consider a triple* $(\gamma,\theta_1,\theta_2)$ *defining alien operator* $\mathscr{A}^{\gamma}_{\omega}(\theta_2,\theta_1)$: $\mathrm{RES}_{\mathbb{Z}}\to\mathrm{RES}_{\mathbb{Z}}$ *at* ω. *We assume that* $\gamma\lambda^k_{\omega}$, *resp.* $\lambda^k_0\gamma$ *is a product of paths so that* λ^k_{ω}, *resp.* λ^k_0, *is a closed path surrounding* ω,*resp.* 0, *and close to that point, with winding number* $\mathrm{wind}_{\omega}(\lambda^k_{\omega})=k$, *resp.* $\mathrm{wind}_0(\lambda^k_0)=k$, $k\in\mathbb{Z}$. *Then,*

$$\mathscr{A}^{\lambda^k_0\gamma}_{\omega}(\theta_2,\theta_1)=\mathscr{A}^{\gamma}_{\omega}(\theta_2,\theta_1)\rho_k.,\qquad \mathscr{A}^{\gamma\lambda^k_{\omega}}_{\omega}(\theta_2,\theta_1)=\rho_k.\mathscr{A}^{\gamma}_{\omega}(\theta_2,\theta_1). \tag{7.20}$$

In particular,

$$\mathscr{A}^{\gamma}_{\omega}(\theta_2,\theta_1+2\pi k)=\mathscr{A}^{\lambda^k_0\gamma}_{\omega}(\theta_2,\theta_1),\qquad \mathscr{A}^{\gamma}_{\omega}(\theta_2+2\pi k,\theta_1)=\mathscr{A}^{\gamma\lambda^{-k}_{\omega}}_{\omega}(\theta_2,\theta_1). \tag{7.21}$$

We end with the following property.

Proposition 7.21. *For any alien operator of the form* $\mathscr{A}^{\gamma}_{\omega}(\theta,\theta)$ *acting on* $\mathrm{RES}_{\mathbb{Z}}$ *or* $\mathrm{RES}^{(\dot{\theta},\alpha)}(L)$, *for any singularity* $\overset{\triangledown}{\varphi}$ *and any resurgent constant* $\overset{\triangledown}{\mathrm{const}}\in\mathrm{CONS}$,

$$\mathscr{A}^{\gamma}_{\omega}(\theta,\theta)\big(\overset{\triangledown}{\mathrm{const}}*\overset{\triangledown}{\varphi}\big)=\overset{\triangledown}{\mathrm{const}}*\big(\mathscr{A}^{\gamma}_{\omega}(\theta,\theta)\overset{\triangledown}{\varphi}\big). \tag{7.22}$$

We stress that in proposition 7.21, only alien operators of the form $\mathscr{A}^{\gamma}_{\omega}(\theta_2,\theta_1)$ with $\theta_1=\theta_2$ are considered. We omit the proof of this proposition which relies on a careful reading of what have been done for showing theorem 7.1.

7.8.2 Composition of Alien Operators

7.8.2.1 Alien Operators on $\mathrm{RES}_{\mathbb{Z}}$

The following definition is adapted from the first volume of this book [MS016].

Definition 7.49. One calls *alien operator at* $\omega\in\mathbb{Z}$ *associated with the couple* (θ^1_1,θ^m_2) any linear combination of composite operators of the form

$$\mathscr{A}^{\gamma_m}_{\omega_m-\omega_{m-1}}(\theta^m_2,\theta^m_1)\circ\cdots\circ\mathscr{A}^{\gamma_2}_{\omega_2-\omega_1}(\theta^2_2,\theta^2_1)\circ\mathscr{A}^{\gamma_1}_{\omega_1}(\theta^1_2,\theta^1_1):\mathrm{RES}_{\mathbb{Z}}\to\mathrm{RES}_{\mathbb{Z}}$$

where $(\omega_1,\cdots,\omega_m)\in\mathbb{Z}^m$, $m\in\mathbb{N}^{\star}$ with $\omega=\omega_m=\sum_{j=1}^m\omega_j-\omega_{j-1}$ and the convention $\omega_0=0$.

Example 7.9. We exemplify the above definition. We set $\omega_1=-2$ and $\omega_2=2$. The alien operator $\mathscr{A}^{\gamma_1}_{\omega_1}(\theta^1_2,\theta^1_1)$ at the point ω_1 is associated with the triple $(\gamma_1,\theta^1_1,\theta^1_2)$ drawn on Fig. 7.11. The alien operator $\mathscr{A}^{\gamma_1}_{\omega_2-\omega_1}(\theta^2_2,\theta^2_1)$ at the point $\omega_2-\omega_1=4$ is associated with the triple $(\gamma_1,\theta^1_1,\theta^1_2)$ drawn on Fig. 7.11. We furthemore assume that $\theta^2_1-\theta^1_2\in[0,2\pi[$ to fix our mind.

From the very definitions of the alien operators and of a minor, one easily checks that the composite alien operator $\mathscr{A}^{\gamma_2}_{\omega_2-\omega_1}(\theta^2_2,\theta^2_1)\circ\mathscr{A}^{\gamma_1}_{\omega_1}(\theta^1_2,\theta^1_1)$ at ω_2, can be written as the difference of two simple alien operators, namely

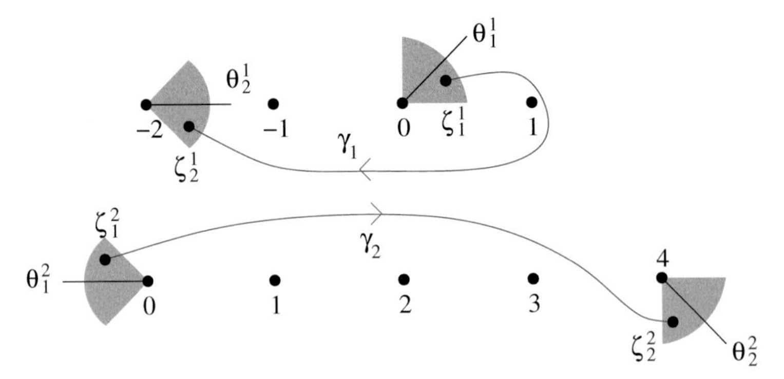

Fig. 7.11 The triple $(\gamma_1, \theta_1^1, \theta_2^1)$ for the point $\omega_1 = -2$, the triple $(\gamma_2, \theta_2^1, \theta_2^2)$ for the point $\omega_2 - \omega_1 = 4$, with $\theta_1^2 = \theta_2^1 + \pi$.

$$\mathscr{A}^{\gamma_2}_{\omega_2-\omega_1}(\theta_2^2, \theta_1^2) \circ \mathscr{A}^{\gamma_1}_{\omega_1}(\theta_2^1, \theta_1^1) = \mathscr{A}^{\Gamma^+}_{\omega_2}(\theta_2^2, \theta_1^1) - \mathscr{A}^{\Gamma^-}_{\omega_2}(\theta_2^2, \theta_1^1).$$

In this equality, Γ^+ and Γ^{-1} stands for the (homotopy class of the) product of paths $\Gamma^+ = \gamma_1 \lambda^+_{\omega_1}(\omega_1 + \gamma_2)$ and $\Gamma^- = \gamma_1 \lambda^-_{\omega_1}(\omega_1 + \gamma_2)$ respectively, where the paths $\lambda^+_{\omega_1}$ and $\lambda^-_{\omega_1}$ drawn on Fig. 7.12, are homotopic to small arcs so that $(\lambda^-_{\omega_1})^{-1}\lambda^+_{\omega_1}$ makes a loop around ω_1 counterclockwise.

Typically, the end point of γ_1 is $\zeta_2^1 = \omega_1 + re^{i\dot{\theta}_1^1}$ while the starting point of γ_2 is $\zeta_1^2 = re^{i\dot{\theta}_1^2}$ with $0 < r \ll 1$. Then, $\lambda^+_{\omega_1} : \dot{\theta} \in [\dot{\theta}_2^1, \dot{\theta}_1^2] \mapsto \omega_1 + re^{i\dot{\theta}}$ while $(\lambda^-_{\omega_1})^{-1} : \dot{\theta} \in [-2\pi + \dot{\theta}_1^2, \dot{\theta}_2^1] \mapsto \omega_1 + re^{i\dot{\theta}}$.

From this result, one deduces from proposition 7.20 that for any $k \in \mathbb{Z}$,

$$\mathscr{A}^{\gamma_2}_{\omega_2-\omega_1}(\theta_2^2, \theta_1^2 + 2\pi k) \circ \mathscr{A}^{\gamma_1}_{\omega_1}(\theta_2^1, \theta_1^1) = \mathscr{A}^{\Gamma_k^+}_{\omega_2}(\theta_2^2, \theta_1^1) - \mathscr{A}^{\Gamma_k^-}_{\omega_2}(\theta_2^2, \theta_1^1).$$

with $\Gamma_k^+ = \gamma_1 \lambda^k_{\omega_1} \lambda^+_{\omega_1}(\omega_1 + \gamma_2)$ and $\Gamma_k^- = \gamma_1 \lambda^k_{\omega_1} \lambda^-_{\omega_1}(\omega_1 + \gamma_2)$ respectively, where $\lambda^k_{\omega_1}$ stands for a closed path around $\omega_1 = -2$ with winding number $\mathrm{wind}_{\omega_1}(\lambda^k_{\omega_1}) = k$ at that point.

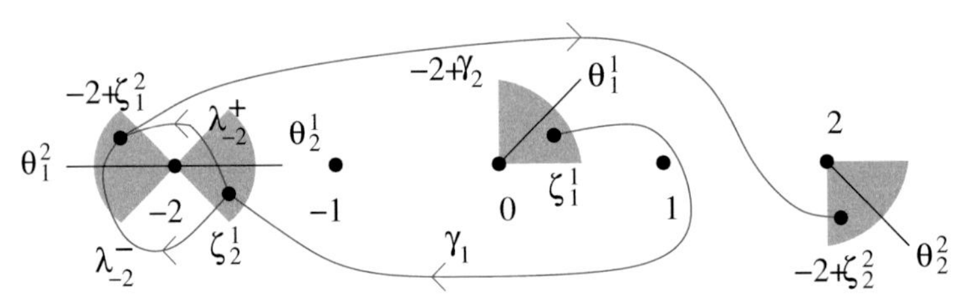

Fig. 7.12 The paths $\Gamma^+ = \gamma_1 \lambda^+_{\omega_1}(\omega_1 + \gamma_2)$ and $\Gamma^+ = \gamma_1 \lambda^-(\omega_1 + \gamma_2)$, $\omega_1 = -2$.

What have been done in the above example can be generalized. This is the matter of the next proposition.

Proposition 7.22. *We consider the two alien operators* $\mathscr{A}^{\gamma_1}_{\omega_1}(\theta_2^1,\theta_1^1)$, $\mathscr{A}^{\gamma_2}_{\omega_2-\omega_1}(\theta_2^2,\theta_1^2)$ *and we assume that* $\theta_1^2-\theta_2^1\in[0,2\pi[$. *Then, for any* $k\in\mathbb{Z}$,

$$\mathscr{A}^{\gamma_2}_{\omega_2-\omega_1}(\theta_2^2,\theta_1^2+2\pi k)\circ\mathscr{A}^{\gamma_1}_{\omega_1}(\theta_2^1,\theta_1^1)=\mathscr{A}^{\Gamma_k^+}_{\omega_2}(\theta_2^2,\theta_1^1)-\mathscr{A}^{\Gamma_k^-}_{\omega_2}(\theta_2^2,\theta_1^1).$$

with $\Gamma_k^+=\gamma_1\lambda^k_{\omega_1}\lambda^+_{\omega_1}(\omega_1+\gamma_2)$ *and* $\Gamma_k^-=\gamma_1\lambda^k_{\omega_1}\lambda^-_{\omega_1}(\omega_1+\gamma_2)$ *respectively, where* $\lambda^k_{\omega_1}$ *stands for a closed path around* ω_1 *with winding number* $\text{wind}_{\omega_1}(\lambda^k_{\omega_1})=k$ *at that point, whereas* $\lambda^+_{\omega_1}$ *and* $\lambda^-_{\omega_1}$ *follows small arcs so that* $(\lambda^-_{\omega_1})^{-1}\lambda^+_{\omega_1}$ *makes a loop around* ω_1 *counterclockwise.*
As a consequence, any alien operator at a point $\omega\in\mathbb{Z}$ *associated with the couple* (θ_1,θ_2) *can be written as a linear combination of alien operators at* ω *associated with triples of the form* $(\gamma,\theta_1,\theta_2)$.

We now focus on paths of type $\gamma^{\dot\theta}_{\varepsilon\mathbf{n}}$. For $m\in\mathbb{N}^\star$, we take a $(m-1)$-tuple of signs $\varepsilon=(\varepsilon_1,\cdots,\varepsilon_{m-1})\in\{+,-\}^{m-1}$ and $\mathbf{n}=(n_1,\cdots,n_{m-1})\in(\mathbb{N}^\star)^{m-1}$. We choose a direction $\theta_1\in\{\pi k,k\in\mathbb{Z}\}$. Following definition 4.7, to a path of type $\gamma^{\dot\theta_1}_{\varepsilon\mathbf{n}}$ one associates a sequence of points and directions defined as follows :

$$\begin{aligned}\dot\theta_{j+1}&=\dot\theta_j+\varepsilon_j(n_j-1)\pi\qquad 1\le j\le m-1\\ \omega_{j+1}-\omega_j&=\mathrm{e}^{\mathrm{i}\dot\theta_{j+1}}\qquad 0\le j\le m-1\\ \omega_0&=0.\end{aligned}\tag{7.23}$$

These data thus provide a uniquely defined alien operator $\mathscr{A}^{\gamma^{\dot\theta_1}_{\varepsilon\mathbf{n}}}_{\omega_m}(\theta_m,\theta_1)$, once the direction $\theta_m\in\widetilde{\mathbb{S}}^1$, $\dot\theta_m=\pi(\theta_m)$ is chosen.

Theorem 7.2. *Let* $m\in\mathbb{N}^\star$ *be a positive integer,* $\varepsilon\in\{+,-\}^{m-1}$, $\mathbf{n}\in(\mathbb{N}^\star)^{m-1}$ *and* $\theta_1\in\{\pi k,k\in\mathbb{Z}\}$. *Let* γ *be a path of type* $\gamma^{\dot\theta_1}_{\varepsilon\mathbf{n}}$, ω_m *and* $\dot\theta_m$ *given by (7.23), and* $\theta_m\in\widetilde{\mathbb{S}}^1$ *so that* $\dot\theta_m=\pi(\theta_m)$. *Then the alien operator* $\mathscr{A}^{\gamma}_{\omega_m}(\theta_m,\theta_1)$ *at* ω_m *associated with the triple* $(\gamma,\theta_1,\theta_m)$ *can be written as a* $\mathbb{Z}$*-linear combination of composite operators of the form* $\mathscr{A}^{\gamma_k}_{\omega'_k-\omega'_{k-1}}(\theta_m,\theta'_k)\circ\cdots\circ\mathscr{A}^{\gamma_2}_{\omega'_2-\omega'_1}(\theta'_2,\theta'_2)\circ\mathscr{A}^{\gamma_1}_{\omega'_1}(\theta'_1,\theta'_1)$ *that satisfy the properties:*

- $(\omega'_1,\cdots,\omega'_k)\in\mathbb{Z}^k$, $k\in\mathbb{N}^\star$ *and* $\omega'_k=\omega_m$;
- $\dot\theta_m=\dot\theta'_k$;
- *for every* $j=1,\cdots,k$, *the path* γ_j *is of type* $\gamma^{\dot\theta'_j}_{(+)_{m_j-1}}$, $m_j\in\mathbb{N}^\star$;
- $\sum_{j=1}^k m_j\le m$.

This theorem is of a purely geometric nature. We omit its proof (see [CNP93-1], Sect. Rés II-2, see also [MS016, Ou012]) and we rather discuss examples which explain the algorithm.

Example 7.10. We consider a path γ of type $\gamma_\varepsilon^{\overset{\bullet}{\theta}_1}$ for $\varepsilon = (+,-,+)$ and we set $\overset{\bullet}{\theta}_1 = 0$, see Fig. 4.2. To the path γ one associates by (7.23) the sequence of points and directions: $\begin{cases} \overset{\bullet}{\theta}_j = 0, & 1 \leq j \leq 4 \\ \omega_0 = 0, & \omega_{j+1} - \omega_j = 1 \quad 0 \leq j \leq 3 \end{cases}$. One sets $\theta_j = \theta' = 0$ for any $j \in [1,4]$. We want to decompose the alien operator $\mathscr{A}^{\gamma}_{\omega_4}(\theta_4,\theta_1)$. From the very definition of the alien operators, one observes that

$$\mathscr{A}^{\gamma^{\overset{\bullet}{\theta}_3}_{(+)}}_{\omega_4-\omega_2}(\theta_4,\theta_3) \circ \mathscr{A}^{\gamma^{\overset{\bullet}{\theta}_1}_{(+)}}_{\omega_2}(\theta_2,\theta_1) = \mathscr{A}^{\gamma^{\overset{\bullet}{\theta}_1}_{(+)3}}_{\omega_4}(\theta_4,\theta_1) - \mathscr{A}^{\gamma^{\overset{\bullet}{\theta}_1}_{(+,-,+)}}_{\omega_4}(\theta_4,\theta_1),$$

and therefore

$$\mathscr{A}^{\gamma^{\overset{\bullet}{\theta}_1}_{(+,-,+)}}_{\omega_4}(\theta_4,\theta_1) = \mathscr{A}^{\gamma^{\overset{\bullet}{\theta}'}_{(+)3}}_{\omega_4}(\theta',\theta') - \mathscr{A}^{\gamma^{\overset{\bullet}{\theta}'}_{(+)}}_{\omega_4-\omega_2}(\theta',\theta') \circ \mathscr{A}^{\gamma^{\overset{\bullet}{\theta}'}_{(+)}}_{\omega_2}(\theta',\theta')$$

Example 7.11. A bit more difficult, we consider a path γ of type $\gamma_{\varepsilon\mathbf{n}}^{\overset{\bullet}{\theta}_1}$ for $\varepsilon = (+,-,+)$, $\mathbf{n} = (1,3,1)$ and $\overset{\bullet}{\theta}_1 = 0$, see Fig. 7.13. The algorithm (7.23) still provides $\begin{cases} \overset{\bullet}{\theta}_j = 0, & 1 \leq j \leq 4 \\ \omega_0 = 0, & \omega_{j+1} - \omega_j = 1 \quad 0 \leq j \leq 3 \end{cases}$. One sets again $\theta_j = \theta' = 0$ for any $j \in [1,4]$. Since

$$\mathscr{A}^{\gamma^{\overset{\bullet}{\theta}_3}_{(+)}}_{\omega_4-\omega_2}(\theta_4,\theta_3 - 2\pi) \circ \mathscr{A}^{\gamma^{\overset{\bullet}{\theta}_1}_{(+)}}_{\omega_2}(\theta_2,\theta_1) = \mathscr{A}^{\gamma^{\overset{\bullet}{\theta}_1}_{(+,-,+)}}_{\omega_4}(\theta_4,\theta_1) - \mathscr{A}^{\gamma^{\overset{\bullet}{\theta}_1}_{(+,-2,+)}}_{\omega_4}(\theta_4,\theta_1),$$

one deduces with the first example that

$$\begin{aligned}
\mathscr{A}^{\gamma^{\overset{\bullet}{\theta}_1}_{(+,-2,+)}}_{\omega_4}(\theta_4,\theta_1) &= \mathscr{A}^{\gamma^{\overset{\bullet}{\theta}'}_{(+,-,+)}}_{\omega_4}(\theta',\theta') - \mathscr{A}^{\gamma^{\overset{\bullet}{\theta}'}_{(+)}}_{\omega_4-\omega_2}(\theta',\theta'-2\pi) \circ \mathscr{A}^{\gamma^{\overset{\bullet}{\theta}'}_{(+)}}_{\omega_2}(\theta',\theta') \\
&= \mathscr{A}^{\gamma^{\overset{\bullet}{\theta}'}_{(+)3}}_{\omega_4}(\theta',\theta') - \mathscr{A}^{\gamma^{\overset{\bullet}{\theta}'}_{(+)}}_{\omega_4-\omega_2}(\theta',\theta') \circ \mathscr{A}^{\gamma^{\overset{\bullet}{\theta}'}_{(+)}}_{\omega_2}(\theta',\theta') \\
&\quad - \mathscr{A}^{\gamma^{\overset{\bullet}{\theta}'}_{(+)}}_{\omega_4-\omega_2}(\theta',\theta'-2\pi) \circ \mathscr{A}^{\gamma^{\overset{\bullet}{\theta}'}_{(+)}}_{\omega_2}(\theta',\theta').
\end{aligned}$$

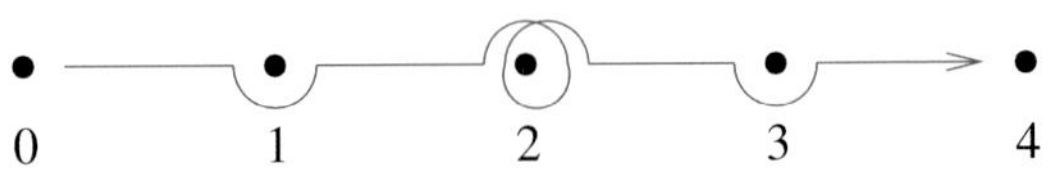

Fig. 7.13 A path of type $\gamma_{\varepsilon\mathbf{n}}^{\overset{\bullet}{\theta}_1}$ for $\varepsilon = (+,-,+)$, $\mathbf{n} = (1,3,1)$ and $\overset{\bullet}{\theta}_1 = 0$.

Example 7.12. A step further, we consider a path γ of type $\gamma_{\varepsilon^{\mathbf{n}}}^{\overset{\bullet}{\theta}_1}$ for $\varepsilon = (-,+,+,+,-)$, $\mathbf{n} = (1,2,1,1,1)$ and take $\theta_1 = 0$, see Fig. 4.3. Using (7.23), we define:

$$\begin{cases} \overset{\bullet}{\theta}_1 = \overset{\bullet}{\theta}_2 = 0 \\ \overset{\bullet}{\theta}_3 = \cdots = \overset{\bullet}{\theta}_6 = \pi \\ \omega_0 = 0,\ \omega_1 - \omega_0 = \omega_2 - \omega_1 = 1 \\ \omega_3 - \omega_2 = \cdots = \omega_6 - \omega_5 = -1. \end{cases}$$

We set $\theta_1 = \theta_2 = \theta_1' = 0$, $\theta_3 = \cdots = \theta_6 = \theta_2' = \pi$. We start with the identity:

$$\mathscr{A}_{\omega_6-\omega_5}^{\gamma_{()}^{\overset{\bullet}{\theta}_6}}(\theta_6,\theta_6) \circ \mathscr{A}_{\omega_5}^{\gamma_{(-,+^2,+,+)}^{\overset{\bullet}{\theta}_1}}(\theta_5,\theta_1) = \mathscr{A}_{\omega_6}^{\gamma_{(-,+^2,+,+,+)}^{\overset{\bullet}{\theta}_1}}(\theta_6,\theta_1) - \mathscr{A}_{\omega_6}^{\gamma}(\theta_6,\theta_1).$$

Next, a little thought yields:

$$\mathscr{A}_{\omega_6-\omega_2}^{\gamma_{(+,+,+)}^{\overset{\bullet}{\theta}_3}}(\theta_6,\theta_3) \circ \mathscr{A}_{\omega_2}^{\gamma_{(-)}^{\overset{\bullet}{\theta}_1}}(\theta_2,\theta_1) = \mathscr{A}_{\omega_6}^{\gamma_{(-,+^2,+,+,+)}^{\overset{\bullet}{\theta}_1}}(\theta_6,\theta_1) - \mathscr{A}_{\omega_6}^{\gamma_{(+)}^{\overset{\bullet}{\theta}_5}}(\theta_6,\theta_5),$$

$$\mathscr{A}_{\omega_5-\omega_2}^{\gamma_{(+,+)}^{\overset{\bullet}{\theta}_3}}(\theta_5,\theta_3) \circ \mathscr{A}_{\omega_2}^{\gamma_{(-)}^{\overset{\bullet}{\theta}_1}}(\theta_2,\theta_1) = \mathscr{A}_{\omega_5}^{\gamma_{(-,+^2,+,+)}^{\overset{\bullet}{\theta}_1}}(\theta_5,\theta_1) - \mathscr{A}_{\omega_5}^{\gamma_{()}^{\overset{\bullet}{\theta}_5}}(\theta_5,\theta_5).$$

Finally, $\mathscr{A}_{\omega_2-\omega_1}^{\gamma_{()}^{\overset{\bullet}{\theta}_2}}(\theta_2,\theta_2) \circ \mathscr{A}_{\omega_1}^{\gamma_{()}^{\overset{\bullet}{\theta}_1}}(\theta_1,\theta_1) = \mathscr{A}_{\omega_2}^{\gamma_{(+)}^{\overset{\bullet}{\theta}_1}}(\theta_2,\theta_1) - \mathscr{A}_{\omega_2}^{\gamma_{(-)}^{\overset{\bullet}{\theta}_1}}(\theta_2,\theta_1)$. Putting things together, one concludes:

$$\begin{aligned} \mathscr{A}_{\omega_6}^{\gamma}(\theta_6,\theta_1) = {} & \mathscr{A}_{\omega_6}^{\gamma_{(+)}^{\overset{\bullet}{\theta}_2'}}(\theta_2',\theta_2') \\ & + \mathscr{A}_{\omega_6-\omega_2}^{\gamma_{(+,+,+)}^{\overset{\bullet}{\theta}_2'}}(\theta_2',\theta_2') \circ \mathscr{A}_{\omega_2}^{\gamma_{(+)}^{\overset{\bullet}{\theta}_1'}}(\theta_1',\theta_1') - \mathscr{A}_{\omega_6-\omega_5}^{\gamma_{()}^{\overset{\bullet}{\theta}_2'}}(\theta_2',\theta_2') \circ \mathscr{A}_{\omega_5}^{\gamma_{()}^{\overset{\bullet}{\theta}_2'}}(\theta_2',\theta_2') \\ & - \mathscr{A}_{\omega_6-\omega_2}^{\gamma_{(+,+,+)}^{\overset{\bullet}{\theta}_2'}}(\theta_2',\theta_2') \circ \mathscr{A}_{\omega_2-\omega_1}^{\gamma_{()}^{\overset{\bullet}{\theta}_1'}}(\theta_1',\theta_1') \circ \mathscr{A}_{\omega_1}^{\gamma_{()}^{\overset{\bullet}{\theta}_1'}}(\theta_1',\theta_1') \\ & - \mathscr{A}_{\omega_6-\omega_5}^{\gamma_{()}^{\overset{\bullet}{\theta}_2'}}(\theta_2',\theta_2') \circ \mathscr{A}_{\omega_5-\omega_2}^{\gamma_{(+,+)}^{\overset{\bullet}{\theta}_2'}}(\theta_2',\theta_2') \circ \mathscr{A}_{\omega_2}^{\gamma_{(+)}^{\overset{\bullet}{\theta}_1'}}(\theta_1',\theta_1') \\ & + \mathscr{A}_{\omega_6-\omega_5}^{\gamma_{()}^{\overset{\bullet}{\theta}_2'}}(\theta_2',\theta_2') \circ \mathscr{A}_{\omega_5-\omega_2}^{\gamma_{(+,+)}^{\overset{\bullet}{\theta}_2'}}(\theta_2',\theta_2') \circ \mathscr{A}_{\omega_2-\omega_1}^{\gamma_{()}^{\overset{\bullet}{\theta}_1'}}(\theta_1',\theta_1') \circ \mathscr{A}_{\omega_1}^{\gamma_{()}^{\overset{\bullet}{\theta}_1'}}(\theta_1',\theta_1'). \end{aligned}$$

7.8.2.2 Alien Operators on $\mathrm{RES}^{(k)}$

We saw with lemma 7.8 that the alien operators associated with triples of the form $(\gamma,\theta_1,\theta_1)$ act on $\mathrm{RES}^{(k)}$ for γ of type $\gamma_{(+)_m}^{\overset{\bullet}{\theta}_1}$ and $\gamma_{(-)_m}^{\overset{\bullet}{\theta}_1}$. We keep on this study according to the guiding line of this section.

We assume $\overset{\bullet}{\theta}_1 \in \{0,\pi\}$ and pick two integers l,k subject to the condition $2 \leq l \leq k$. By the very definition of $\mathrm{RES}^{(k)}$, the minor $\widehat{\varphi}$ of any singularity $\overset{\triangledown}{\varphi} \in \mathrm{RES}^{(k)}$, once considered as a sectorial germ, can be analytically continued along any path γ of type $\gamma_{\varepsilon^{\mathbf{n}_l}}^{\overset{\bullet}{\theta}_1}$ with

$$\varepsilon^{\mathbf{n}_l} \in \{((\pm)_{l-1}^{\mathbf{n}_l},(\varepsilon)_{m_l-1}) \mid \varepsilon \in \{+,-\}, \mathbf{n}_l = (n_1,\cdots,n_{l-1}) \in (\mathbb{N}^\star)^{l-1}, m_l \in \mathbb{N}^\star\}.$$

With the notation of (7.23), the analytic continuation $\mathrm{cont}_\gamma\, \widehat{\varphi}$ of $\widehat{\varphi}$ along γ is a germ of holomorphic functions that can be analytically continued onto the simply connected domain $\mathfrak{p}(\mathscr{R}^{\varepsilon^{\mathbf{n}_l},\overset{\bullet}{\theta}}) = \mathbb{C} \setminus \{]-\infty,p] \cup [p+1,+\infty[\}$ where $]p,(p+1)[=]\omega_{l-1},\omega_l[$ when $m_l = 1$, $]p,(p+1)[=](m_l-1)\omega_l, m_l\omega_l[$ otherwise. These properties translate into the next statement (the details are left to the reader).

Proposition 7.23. *Let be $\theta_1 \in \{\pi k, k \in \mathbb{Z}\} \subset \widetilde{\mathbb{S}}^1$ and $(l,k) \in \mathbb{N}$ with the condition $1 \leq l \leq k$. The following alien operators are well-defined, for any $\varepsilon \in \{-,+\}$, any $\mathbf{n}_l \in \mathbb{N}^{l-1}$ and any $m_l \in \mathbb{N}^\star$. Setting $\overset{\bullet}{\theta}_l, \omega_l$ by (7.23) and $\theta_l \in \widetilde{\mathbb{S}}^1$ with $\overset{\bullet}{\theta}_l = \pi(\theta_l)$,*

$$\begin{aligned}
&\mathscr{A}_{m_l\omega_l}^{\gamma^{\overset{\bullet}{\theta}_1}_{((\pm)^{\mathbf{n}_l}_{l-1}(\varepsilon)_{m_l-1})}}(\theta_l,\theta_1) : \mathrm{RES}^{(k)} \to \mathrm{RES}^{(k-l-m_l+1)}, \quad 1 \leq m_l \leq k-l \\
&\mathscr{A}_{m_l\omega_l}^{\gamma^{\overset{\bullet}{\theta}_1}_{((\pm)^{\mathbf{n}_l}_{l-1}(\varepsilon)_{m_l-1})}}(\theta_l,\theta_1) : \mathrm{RES}^{(k)} \to \mathrm{SING}_{\theta_1,\pi}, \quad m_l = k-l+1 \\
&\mathscr{A}_{m_l\omega_l}^{\gamma^{\overset{\bullet}{\theta}_1}_{((\pm)^{\mathbf{n}_l}_{l-1}(\varepsilon)_{m_l-1})}}(\theta_l,\theta_1) : \mathrm{RES}^{(k)} \to \mathrm{SING}_{\theta_1+\pi/2,\pi/2}, \quad m_1 \geq k-l+2.
\end{aligned} \tag{7.24}$$

Equivalently, $\mathscr{A}_{m_l\omega_l-\omega_{l-1}}^{\gamma^{\overset{\bullet}{\theta}_l}_{(\varepsilon)_{m_l-1}}}(\theta_l,\theta_l) \circ \cdots \mathscr{A}_{\omega_2-\omega_1}^{\gamma^{\overset{\bullet}{\theta}_2}_{()}}(\theta_2,\theta_2) \circ \mathscr{A}_{\omega_1}^{\gamma^{\overset{\bullet}{\theta}_1}_{()}}(\theta_1,\theta_1)$ are well-defined alien operators, with $\overset{\bullet}{\theta}_j, \omega_j$ given by (7.23) and $\theta_j \in \widetilde{\mathbb{S}}^1$ with $\overset{\bullet}{\theta}_j = \pi(\theta_j)$, with the following ranges:

$$\begin{aligned}
&\mathscr{A}_{m_l\omega_l-\omega_{l-1}}^{\gamma^{\overset{\bullet}{\theta}_l}_{(\varepsilon)_{m_l-1}}}(\theta_l,\theta_l) \circ \cdots \circ \mathscr{A}_{\omega_1}^{\gamma^{\overset{\bullet}{\theta}_1}_{()}}(\theta_1,\theta_1) : \mathrm{RES}^{(k)} \to \mathrm{RES}^{(k-l-m_l+1)}, \quad 1 \leq m_l \leq k-l \\
&\mathscr{A}_{m_l\omega_l-\omega_{l-1}}^{\gamma^{\overset{\bullet}{\theta}_l}_{(\varepsilon)_{m_l-1}}}(\theta_l,\theta_l) \circ \cdots \circ \mathscr{A}_{\omega_1}^{\gamma^{\overset{\bullet}{\theta}_1}_{()}}(\theta_1,\theta_1) : \mathrm{RES}^{(k)} \to \mathrm{SING}_{\theta_1,\pi}, \quad m_l = k-l+1 \\
&\mathscr{A}_{m_l\omega_l-\omega_{l-1}}^{\gamma^{\overset{\bullet}{\theta}_l}_{(\varepsilon)_{m_l-1}}}(\theta_l,\theta_l) \circ \cdots \circ \mathscr{A}_{\omega_1}^{\gamma^{\overset{\bullet}{\theta}_1}_{()}}(\theta_1,\theta_1) : \mathrm{RES}^{(k)} \to \mathrm{SING}_{\theta_1+\frac{\pi}{2},\frac{\pi}{2}}, \quad m_1 \geq k-l+2.
\end{aligned} \tag{7.25}$$

We would like now to discuss a kind of converse of proposition 7.23 with the next two propositions.

Proposition 7.24. *Let $k \in \mathbb{N}^\star$ be a positive integer and $\overset{\triangledown}{\varphi} \in \mathrm{RES}^{(k)}$. We suppose that for any $\theta \in \{\pi k, k \in \mathbb{Z}\} \subset \widetilde{\mathbb{S}}^1$ one has $\mathscr{A}_\omega^{\gamma_{()}^{\dot\theta}}(\theta,\theta)\,\overset{\triangledown}{\varphi} \in \mathrm{RES}^{(k)}$, with $\omega = \mathrm{e}^{\mathrm{i}\dot\theta}$, $\dot\theta = \pi(\theta)$. Then $\overset{\triangledown}{\varphi}$ belongs to* $\mathrm{RES}^{(k+1)}$.

Proof. There will be no loss of generality in assuming that $\overset{\triangledown}{\varphi}$ is a simple singularity and this assumption is easier to handle : $\overset{\triangledown}{\varphi} = a\delta + {}^\flat\widehat{\varphi} \in \overset{\triangledown}{\mathscr{R}}{}^{(k)}$ with $\widehat{\varphi} \in \hat{\mathscr{R}}^{(k)}$.
We consider a singularity $\overset{\triangledown}{\mathscr{R}}{}^{(1)}$. Thus, $\widehat{\varphi}$ can be analytically continued to $\mathscr{R}^{(1)}$. Equivalently, for any $\theta_1 \in \{\pi k, k \in \mathbb{Z}\}$, $\widehat{\varphi}$ can be analytically continued along any path γ_1 of type $\gamma^{\dot\theta_1}_{(\varepsilon)_{m-1}}$, $m \in \mathbb{N}^\star$, $\varepsilon \in \{-,+\}$ and $\mathrm{cont}_{\gamma_1}\widehat{\varphi}$ is a germ which can be analytically continued to the star-shaped domain $\mathfrak{p}\big(\mathscr{R}^{(\varepsilon)_{m-1},\dot\theta_1}\big)$.

Let us assume that for any $\theta_1 \in \{\pi k, k \in \mathbb{Z}\}$, $\mathscr{A}_\omega^{\gamma_{()}^{\dot\theta_1}}(\theta_1,\theta_1)\,\overset{\triangledown}{\varphi}$ belongs to $\mathrm{RES}^{(1)}$, where $\omega_1 = \mathrm{e}^{\mathrm{i}\dot\theta_1}$. We claim that $\overset{\triangledown}{\varphi}$ belongs to $\mathrm{RES}^{(2)}$.

Our assumption results in the following property : for any $n_1 \in \mathbb{N}^\star$ and any path γ of type $\gamma^{\dot\theta_1}_{(\pm)_1^{n_1}}$, denoting by $\lambda^-_{\omega_1}$ a clockwise loop around ω_1, the difference $\big(\mathrm{cont}_\gamma - \mathrm{cont}_{\gamma\lambda^-_{\omega_1}}\big)\widehat{\varphi}$ is a sectorial germ which can be analytically continued along any path γ_2 of type $\gamma^{\dot\theta_2}_{(\varepsilon)_{m-1}}$, $m \in \mathbb{N}^\star$, $\varepsilon \in \{-,+\}$, $\dot\theta_2 = \dot\theta_1 + (n_1-1)\pi$. Moreover $\mathrm{cont}_{\gamma_2}\big(\mathrm{cont}_\gamma - \mathrm{cont}_{\gamma\lambda^-_{\omega_1}}\big)\widehat{\varphi}$ is a germ of holomorphic functions which can be analytically continued to the star-shaped domain $\mathfrak{p}\big(\mathscr{R}^{((\pm)_1^{n_1},(\varepsilon)_{m-1}),\dot\theta_1}\big)$.
Start with $n_1 = 1$ and a path γ of type $\gamma^{\dot\theta_1}_{(+)_1}$, *resp.* $\gamma^{\dot\theta_1}_{(-)_1}$. Take a path γ_2 of type $\gamma^{\dot\theta_2}_{(+)_{m-1}}$, $\dot\theta_2 = \dot\theta_1$, *resp.* $\gamma^{\dot\theta_2}_{(-)_{m-1}}$. Notice that $\gamma_1 = \gamma\gamma_2$ is a path of type $\gamma^{\dot\theta_1}_{(\varepsilon)_m}$. Therefore from the above property, $\mathrm{cont}_{\gamma_2}\big(\mathrm{cont}_\gamma\widehat{\varphi}\big) = \mathrm{cont}_{\gamma_1}\widehat{\varphi}$ is well-defined and gives a germ that can be analytically continued to the domain $\mathfrak{p}\big(\mathscr{R}^{(+)_m,\dot\theta_1}\big) = \mathfrak{p}\big(\mathscr{R}^{((+)_1,(+)_{m-1}),\dot\theta_1}\big)$, *resp.* $\mathfrak{p}\big(\mathscr{R}^{(-)_m,\dot\theta_1}\big) = \mathfrak{p}\big(\mathscr{R}^{((-)_1,(-)_{m-1}),\dot\theta_1}\big)$. But this implies that $\mathrm{cont}_{\gamma_2}\big(\mathrm{cont}_{\gamma\lambda^-_{\omega_1}}\widehat{\varphi}\big) = \mathrm{cont}_{\gamma\lambda^-_{\omega_1}\gamma_2}\widehat{\varphi}$ is also well-defined and provides a germ that can be analytically continued to the domain $\mathfrak{p}\big(\mathscr{R}^{(+)_m,\dot\theta_1}\big) = \mathfrak{p}\big(\mathscr{R}^{((-)_1,(+)_{m-1}),\dot\theta_1}\big)$, *resp.* $\mathfrak{p}\big(\mathscr{R}^{(-)_m,\dot\theta_1}\big) = \mathfrak{p}\big(\mathscr{R}^{((-)_1^3,(-)_{m-1}),\dot\theta_1}\big)$. (Notice that the path $\gamma\lambda^-_{\omega_1}\gamma_2$ is a path of type $\gamma^{\dot\theta_1}_{((-)_1,(+)_{m-1})}$, *resp.* $\gamma^{\dot\theta_1}_{((-)_1^3,(-)_{m-1})}$).
Of course, one could have chosen a path γ of type $\gamma^{\dot\theta_1}_{(+)_1}$ and a path γ_2 of type $\gamma^{\dot\theta_1}_{(-)_{m-1}}$. The path $\gamma_1 = \gamma\lambda^-_{\omega_1}\gamma_2$ is a path of type $\gamma^{\dot\theta_1}_{(-)_m}$ and we conclude for the analytic continuation of $\widehat{\varphi}$ along the path $\gamma\gamma_2$ of type $\gamma^{\dot\theta_1}_{((+)_1,(-)_{m-1})}$.

One can pursue this way by induction on n_1 to show our claim. Here, we just add the case $n_1 = 2$ so as to deal with a subtlety. We thus consider a path γ of type $\gamma^{\dot{\theta}_1}_{(+)_1^2}$ and a path γ_2 of type $\gamma^{\dot{\theta}_2}_{(\varepsilon)_{m-1}}$, $\dot{\theta}_2 = \dot{\theta}_1 + \pi$. Notice that the path $\gamma\lambda^-_{\omega_1}\gamma_2$ is homotopic to a path of type $\gamma^{\dot{\theta}_1}_{()}$ when $m = 1$, of type $\gamma^{\dot{\theta}_2}_{(\varepsilon)_{m-2}}$ when $m \geq 2$. Therefore, $\mathrm{cont}_{\gamma_2}\left(\mathrm{cont}_{\gamma\lambda^-_{\omega_1}}\widehat{\varphi}\right)$ is well-defined and one concludes that $\widehat{\varphi}$ can be analytically continued along the path $\gamma_1 = \gamma\gamma_2$ of type $\gamma^{\dot{\theta}_1}_{((+)_1^2,(\varepsilon)_{m-1})}$ and moreover the germ $\mathrm{cont}_{\gamma_1}\widehat{\varphi}$ can be analytically continued to the star-shaped domain $\mathfrak{p}\left(\mathscr{R}^{((+)_1^2,(\varepsilon)_{m-1}),\dot{\theta}_1}\right)$.

The same reasoning can be generalized and gives the proposition. □

A quite similar (and even simpler) reasoning gives the next result.

Proposition 7.25. *Let be $k \in \mathbb{N}^\star$ and $\overset{\triangledown}{\varphi} \in \mathrm{RES}^{(k)}$. We suppose that for any $\theta_1 \in \{\pi k, k \in \mathbb{Z}\} \subset \widetilde{\mathbb{S}}^1$ and any $\mathbf{n} \in \mathbb{N}^{k-1}$, the singularity $\mathscr{A}^{\gamma^{\dot{\theta}_1}_{((\pm)^{\mathbf{n}}_{k-1})}}_{\omega_k}(\theta_k, \theta_1)\overset{\triangledown}{\varphi}$ belongs to $\mathrm{RES}^{(1)}$, where ω_k is given by (7.23). Then $\overset{\triangledown}{\varphi}$ belongs to $\mathrm{RES}^{(k+1)}$.*

We eventually use theorem 7.2 to reformulate proposition 7.25.

Corollary 7.1. *Let $k \in \mathbb{N}^\star$ be a positive integer and $\overset{\triangledown}{\varphi} \in \mathrm{RES}^{(k)}$. We suppose that $\mathscr{A}^{\gamma_k}_{\omega_k - \omega_{k-1}}(\theta_k, \theta_k) \circ \cdots \circ \mathscr{A}^{\gamma_2}_{\omega_2 - \omega_1}(\theta_2, \theta_2) \circ \mathscr{A}^{\gamma_1}_{\omega_1}(\theta_1, \theta_1)\overset{\triangledown}{\varphi}$ belongs to $\mathrm{RES}^{(1)}$ for any composite operator that satisfies the properties:*

- *for every $j = 1, \cdots, k$, the path γ_j is of type $\gamma^{\dot{\theta}_j}_{(+)_{m_j-1}}$, $m_j \in \mathbb{N}^\star$;*
- *$\sum_{j=1}^k m_j = k$.*

Then $\overset{\triangledown}{\varphi}$ belongs to $\mathrm{RES}^{(k+1)}$.

7.8.3 Alien Derivations

We now specialize our analysis to some particularly interesting alien operators.

7.8.3.1 Definitions

Definition 7.50. Let be $\theta \in \{\pi k, k \in \mathbb{Z}\} \subset \widetilde{\mathbb{S}}^1$, $\alpha \in]0, \pi/2]$ and $L > 0$. We set $\dot{\theta} = \pi(\theta) \in \{0, \pi\} \subset \mathbb{S}^1$. Let be $\omega = m e^{\mathrm{i}\theta} \in \widetilde{\mathbb{C}}$ for $m \in \{1, \cdots, \lceil L \rceil\}$, *resp.* $m \in \mathbb{N}^\star$. The alien operators Δ^+_ω and Δ_ω at ω,

$$\Delta^+_{\omega}, \Delta_{\omega} : \mathrm{RES}^{(\overset{\bullet}{\theta},\alpha)}(L) \to \mathrm{SING}_{\theta,\alpha}, \quad \textit{resp.} \quad \Delta^+_{\omega}, \Delta_{\omega} : \mathrm{RES} \to \mathrm{RES},$$

are defined as follows:

$$\Delta^+_{\omega} \overset{\triangledown}{\varphi} = \mathscr{A}^{\overset{\bullet}{\gamma}{}^{\theta}_{(+)^{m-1}}}_{\overset{\bullet}{\omega}}(\theta,\theta)\, \overset{\triangledown}{\varphi} \tag{7.26}$$

$$\Delta_{\omega} \overset{\triangledown}{\varphi} = \sum_{\varepsilon=(\varepsilon_1,\cdots,\varepsilon_{m-1})\in\{+,-\}^{m-1}} \frac{p(\varepsilon)!q(\varepsilon)!}{m!} \mathscr{A}^{\overset{\bullet}{\gamma}{}^{\theta}_{\varepsilon}}_{\overset{\bullet}{\omega}}(\theta,\theta)\, \overset{\triangledown}{\varphi},$$

where $p(\varepsilon)$, *resp.* $q(\varepsilon) = m-1-p(\varepsilon)$, denotes the number of "+" signs, *resp.* "−" signs in the sequence ε.

Definition 7.51. The alien operators $\Delta^+_{\omega}, \Delta_{\omega} : \widetilde{\mathrm{RES}} \to \widetilde{\mathrm{RES}}$ for asymptotic classes are defined by making the following diagrams commuting:

$$\begin{array}{ccc} \mathrm{RES} & \overset{\Delta^+_{\omega},\Delta_{\omega}}{\longrightarrow} & \mathrm{RES} \\ \mathscr{L} \downarrow\uparrow \mathscr{B} & & \mathscr{L} \downarrow\uparrow \mathscr{B} \\ \widetilde{\mathrm{RES}} & \overset{\Delta^+_{\omega},\Delta_{\omega}}{\longrightarrow} & \widetilde{\mathrm{RES}} \end{array}.$$

7.8.3.2 Properties

Theorem 7.3. *Under the hypotheses of definition 7.50, the alien operators* $\Delta^+_{\omega} : \mathrm{RES}^{(\overset{\bullet}{\theta},\alpha)}(L) \to \mathrm{SING}_{\theta,\alpha}$, *resp.* $\Delta^+_{\omega} : \mathrm{RES} \to \mathrm{RES}$, *satisfy the identity:*

$$\Delta^+_{\omega}(\overset{\triangledown}{\varphi} * \overset{\triangledown}{\psi}) = (\Delta^+_{\omega} \overset{\triangledown}{\varphi}) * \overset{\triangledown}{\psi} + \sum_{\overset{\bullet}{\omega}_1+\overset{\bullet}{\omega}_2=\overset{\bullet}{\omega}} (\Delta^+_{\omega_1} \overset{\triangledown}{\varphi}) * (\Delta^+_{\omega_2} \overset{\triangledown}{\psi}) + \overset{\triangledown}{\varphi} * (\Delta^+_{\omega} \overset{\triangledown}{\psi}) \tag{7.27}$$

where the sum runs over all $\omega_1 = m_1 e^{i\theta}$, $\omega_2 = m_2 e^{i\theta}$, *with* $m_1, m_2 \in \mathbb{N}^\star$ *such that* $m_1 + m_2 = m$.

The alien operators $\Delta_{\omega} : \mathrm{RES}^{(\overset{\bullet}{\theta},\alpha)}(L) \to \mathrm{SING}_{\theta,\alpha}$, *resp.* $\Delta_{\omega} : \mathrm{RES} \to \mathrm{RES}$, *satisfy the Leibniz rule,* $\Delta_{\omega}(\overset{\triangledown}{\varphi} * \overset{\triangledown}{\psi}) = (\Delta_{\omega} \overset{\triangledown}{\varphi}) * \overset{\triangledown}{\psi} + \overset{\triangledown}{\varphi} * (\Delta_{\omega} \overset{\triangledown}{\psi})$. *Moreover,* $\Delta_{\omega}(\overset{\triangledown\triangledown}{\partial\varphi}) = (\overset{\triangledown}{\partial} - \overset{\bullet}{\omega})(\Delta_{\omega} \overset{\triangledown}{\varphi})$. *Eventually,* $\Delta^+_{\omega} \overset{\triangledown}{\mathrm{cons}} = \Delta_{\omega} \overset{\triangledown}{\mathrm{cons}} = 0$ *for any resurgent constant* $\overset{\triangledown}{\mathrm{cons}}$.

Proof. We give the proof for the identity (7.27) only, so as to exemplify the use of singularities. Moreover we work on the space $\overset{\triangledown}{\mathscr{R}}{}^{(\overset{\bullet}{\theta},\alpha)}(L)$.

The reader is invited to compare with the proof made in the first volume of this book [MS016] for simple resurgent functions.

There is no loss of generality in assuming that $\overset{\triangledown}{\varphi} = {}^{\flat}\widehat{\varphi}$, $\overset{\triangledown}{\psi} = {}^{\flat}\widehat{\psi}$ with $\widehat{\varphi}, \widehat{\psi} \in \widehat{\mathscr{R}}^{(\theta,\alpha)}(L)$. By proposition 7.6 one has ${}^{\flat}\widehat{\varphi} * {}^{\flat}\widehat{\varphi} = {}^{\flat}(\widehat{\varphi} * \widehat{\varphi})$, therefore we can use arguments developed in chapter 4 (see in particular the proof of theorem 4.1), which allow us some abuse of notation.

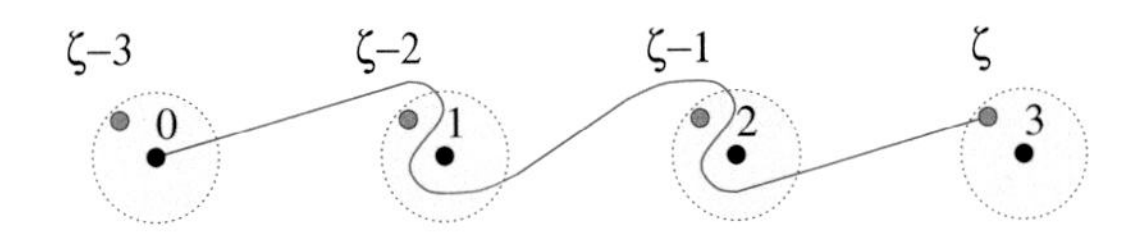

Fig. 7.14 Symmetrically contractile path H_1 and contributions to $\Delta^+_\omega(\widehat{\varphi}*\widehat{\psi})$ for $\omega=3$. Pinchings occur between 1 and $\zeta-2$, and between 2 and $\zeta-1$.

The analytic continuation of the convolution product $\widehat{\varphi}*\widehat{\psi}$ along a path γ of type $\overset{\bullet}{\gamma}{}^{\theta}_{(+)_{m-1}}=\pi(\gamma^{\theta}_{(+)_{m-1}})$, ending at $\zeta=\omega+\overset{\bullet}{\xi}_0$ near $](m-1)e^{i\overset{\bullet}{\theta}},me^{i\overset{\bullet}{\theta}}[$, is the germ of holomorphic functions defined as follows:

$$(\mathrm{cont}_\gamma\,\widehat{\varphi}*\widehat{\psi})(\omega+\overset{\bullet}{\xi})=\int_{H_1}\widehat{\varphi}(\eta_1)\widehat{\psi}(\eta_2+\overset{\bullet}{\xi}-\overset{\bullet}{\xi}_0)+\int_0^{\overset{\bullet}{\xi}-\overset{\bullet}{\xi}_0}\widehat{\varphi}(\zeta+\eta)\widehat{\psi}(\overset{\bullet}{\xi}-\overset{\bullet}{\xi}_0-\eta)d\eta.$$

Here H_1 $(=\mathfrak{p}\circ\mathscr{H}_1)$ is a symmatrically contractile path deduced from γ, $\widehat{\varphi}(\eta_1)=\mathrm{cont}_{H_1|[0,s]}\,\widehat{\varphi}\big(H_1(s)\big)$, $\widehat{\psi}(\eta_2+\overset{\bullet}{\xi}-\overset{\bullet}{\xi}_0)=\mathrm{cont}_{H_1^{-1}|[0,s]}\,\widehat{\psi}\big(H_1^{-1}(s)+\overset{\bullet}{\xi}-\overset{\bullet}{\xi}_0\big)$ and $\widehat{\varphi}(\zeta+\eta)=\mathrm{cont}_{H_1}\,\widehat{\varphi}\big(H_1(1)+\eta\big)$. To get the associated singularity, that is $\Delta^+_\omega(\overset{\triangledown}{\varphi}*\overset{\triangledown}{\psi})$, one only needs to consider the restrictions:

1. of the first integral near the "pinching points" (see Fig. 7.14), where one easily recognizes convolution products for majors and these provide the contribution $\sum_{\overset{\bullet}{\omega}_1+\overset{\bullet}{\omega}_2=\overset{\bullet}{\omega}}\big(\Delta^+_{\omega_1}\overset{\triangledown}{\varphi}\big)*\big(\Delta^+_{\omega_2}\overset{\triangledown}{\psi}\big)$ to the singularity $\Delta^+_\omega(\overset{\triangledown}{\varphi}*\overset{\triangledown}{\psi})$;
2. of the two integrals near the end points, which provide the missing contributions (use proposition 7.2).

This ends the proof. □

Definition 7.52. The linear operators Δ_ω are called alien derivations and RES is called a resurgent algebra (since stable under alien derivations).

We refer to the first volume of this book [MS016] (see also [CNP93-1, Eca81-1, Eca85]) for the proof of the next statements.

Theorem 7.4. *For any* $\theta\in\{\,k\pi,k\in\mathbb{Z}\}$, $\omega\in\widetilde{\mathbb{C}}$ *with* $\arg(\omega)=\theta$,

$$\Delta_\omega=\sum_{s\in\mathbb{N}^\star}\frac{(-1)^{s-1}}{s}\sum_{\substack{\arg(\omega_1)=\cdots=\arg(\omega_{s-1})=\theta\\ 0\prec\overset{\bullet}{\omega}_1\prec\cdots\prec\overset{\bullet}{\omega}_s\prec\overset{\bullet}{\omega}}}\Delta^+_{\omega-\omega_{s-1}}\circ\cdots\circ\Delta^+_{\omega_2-\omega_1}\circ\Delta^+_{\omega_1},\qquad(7.28)$$

$$\Delta^+_\omega=\sum_{s\in\mathbb{N}^\star}\frac{1}{s!}\sum_{\substack{\arg(\omega_1)=\cdots=\arg(\omega_{s-1})=\theta\\ 0\prec\overset{\bullet}{\omega}_1\prec\cdots\prec\overset{\bullet}{\omega}_s\prec\overset{\bullet}{\omega}}}\Delta_{\omega-\omega_{s-1}}\circ\cdots\circ\Delta_{\omega_2-\omega_1}\circ\Delta_{\omega_1},\qquad(7.29)$$

In the above theorem, $\prec$ stands for the total order on $[0,\omega]$ induced by $t\in[0,1]\mapsto t\omega\in[0,\omega]$.

The alien derivations own the property of generating the whole set of alien operators. We precise this claim with the following upshot from theorem 7.2 and theorem 7.4.

Theorem 7.5. *Let* $m \in \mathbb{N}^\star$ *be a positive integer,* $\varepsilon \in \{+,-\}^{m-1}$, $\mathbf{n} \in (\mathbb{N}^\star)^{m-1}$ *and* $\theta_1 \in \{\pi k, k \in \mathbb{Z}\}$. *Let* γ *be a path of type* $\gamma_{\varepsilon\mathbf{n}}^{\overset{\bullet}{\theta}_1}$, $\overset{\bullet}{\omega}_m$ *and* $\overset{\bullet}{\theta}_m$ *given by (7.23), and* $\theta_m \in \widetilde{\mathbb{S}}^1$ *so that* $\overset{\bullet}{\theta}_m = \pi(\theta_m)$. *Then the alien operator* $\mathscr{A}^{\gamma}_{\overset{\bullet}{\omega}_m}(\theta_m, \theta_1)$ *at* $\overset{\bullet}{\omega}_m$ *associated with the triple* $(\gamma, \theta_1, \theta_m)$ *can be written as a* $\mathbb{Z}$*-linear, resp.* $\mathbb{Q}$*-linear combination of composite operators of the form*

$$\rho_{k_n}.\big(\Delta^+_{\omega_n} \circ \cdots \circ \Delta^+_{\omega_2} \circ \Delta^+_{\omega_1}\big), \quad \textit{resp.} \quad \rho_{k_n}.\big(\Delta_{\omega_n} \circ \cdots \circ \Delta_{\omega_2} \circ \Delta_{\omega_1}\big),$$

that satisfy the properties:

- $(\overset{\bullet}{\omega}_1, \cdots, \overset{\bullet}{\omega}_n) \in (\mathbb{Z}^\star)^n$, $n \in \mathbb{N}^\star$ *and* $\pi\Big(\sum_{j=1}^n \omega_j\Big) = \overset{\bullet}{\omega}_m$;
- $\overset{\bullet}{\theta}_m = \arg(\omega_n) + 2\pi k_n$, $k_n \in \mathbb{Z}$;
- $\sum_{j=1}^n |\omega_j| \leq m$.

Example 7.13. We continue the example 7.10. The path γ is of type γ^0_ε for $\varepsilon = (+,-,+)$ and we know that $\mathscr{A}_4^{\gamma^0_{(+,-,+)}}(0,0) = \Delta_4^+ - \Delta_2^+ \circ \Delta_2^+$. (On the right-hand side of the equality, $(4,2)$ stands for $(4\mathrm{e}^{\mathrm{i}0}, 2\mathrm{e}^{\mathrm{i}0})$). Using theorem 7.4, one gets:

$$\begin{aligned} \mathscr{A}_4^{\gamma^0_{(+,-,+)}}(0,0) = \Delta_4 + \tfrac{1}{2!}\big(\Delta_3 \circ \Delta_1 + \Delta_2 \circ \Delta_2 + \Delta_1 \circ \Delta_3\big) \\ + \tfrac{1}{3!}\big(\Delta_2 \circ \Delta_1 \circ \Delta_1 + \Delta_1 \circ \Delta_2 \circ \Delta_1 + \Delta_1 \circ \Delta_1 \circ \Delta_1\big) + \tfrac{1}{4!}\Delta_1 \circ \Delta_1 \circ \Delta_1 \circ \Delta_1 \\ -\big(\Delta_2 + \tfrac{1}{2!}\Delta_1 \circ \Delta_1\big) \circ \big(\Delta_2 + \tfrac{1}{2!}\Delta_1 \circ \Delta_1\big). \end{aligned}$$

Example 7.14. We continue the example 7.11. The path γ is of type $\gamma_{\varepsilon\mathbf{n}}^{\overset{\bullet}{\theta}_1}$ for $\varepsilon = (+,-,+)$, $\mathbf{n} = (1,3,1)$ and we have shown the identity:

$$\mathscr{A}_4^{\gamma^0_{(+,-2,+)}}(0,0) = \Delta_4^+ - \Delta_2^+ \circ \Delta_2^+ - \rho_{-1}.\Delta^+_{2\mathrm{e}^{-2\mathrm{i}\pi}} \circ \Delta_2^+.$$

This can be expressed in term of alien derivatives as well.

We end with an observation. By its very definition, any singularity $\overset{\triangledown}{\varphi} \in \overset{\triangledown}{\mathscr{R}}^{(\theta,\alpha)}(L)$ has a *regular* minor. This property involves the following relationships for the action of the alien operators. (These are essentially consequences of propositions 7.19 and 7.20).

Proposition 7.26. *We suppose* $\alpha \in]0, \pi/2]$, $L > 0$ *and* $m \in \{1, \cdots, \lceil L \rceil\}$. *The following equalities hold for any* $k \in \mathbb{Z}$:

- *for any* $\overset{\triangledown}{\varphi} \in \overset{\triangledown}{\mathscr{R}}^{(0,\alpha)}(L)$, $\Delta^+_{m\mathrm{e}^{\mathrm{i}\pi 2k}} \overset{\triangledown}{\varphi} = \rho_{-k}.\big(\Delta^+_{m\mathrm{e}^{\mathrm{i}\pi 0}} \overset{\triangledown}{\varphi}\big)$, $\Delta_{m\mathrm{e}^{\mathrm{i}\pi 2k}} \overset{\triangledown}{\varphi} = \rho_{-k}.\big(\Delta_{m\mathrm{e}^{\mathrm{i}\pi 0}} \overset{\triangledown}{\varphi}\big)$;
- *for any* $\overset{\triangledown}{\varphi} \in \overset{\triangledown}{\mathscr{R}}^{(\pi,\alpha)}(L)$, $\Delta^+_{m\mathrm{e}^{\mathrm{i}\pi(2k+1)}} \overset{\triangledown}{\varphi} = \rho_{-k}.\big(\Delta^+_{m\mathrm{e}^{\mathrm{i}\pi}} \overset{\triangledown}{\varphi}\big)$, $\Delta_{m\mathrm{e}^{\mathrm{i}\pi(2k+1)}} \overset{\triangledown}{\varphi} = \rho_{-k}.\big(\Delta_{m\mathrm{e}^{\mathrm{i}\pi}} \overset{\triangledown}{\varphi}\big)$;

- *moreover, if* $\overset{\triangledown}{\varphi} \in \overset{\triangledown}{\mathscr{R}}^{(0,\alpha)}(L) \cap \overset{\triangledown}{\mathscr{R}}^{(\pi,\alpha)}(L)$ *and if its minor* $\widehat{\varphi}$ *is even, then* $\Delta^+_{e^{i\pi}} \overset{\triangledown}{\varphi} = \rho_{-1/2}.\left(\Delta^+_1 \overset{\triangledown}{\varphi}\right)$, $\Delta_{e^{i\pi}} \overset{\triangledown}{\varphi} = \rho_{-1/2}.\left(\Delta_1 \overset{\triangledown}{\varphi}\right)$ *with* $1 = e^{i0}$, *while if* $\widehat{\varphi}$ *is odd, then* $\Delta^+_{e^{i\pi}} \overset{\triangledown}{\varphi} = -\rho_{-1/2}.\left(\Delta^+_1 \overset{\triangledown}{\varphi}\right)$, $\Delta_{e^{i\pi}} \overset{\triangledown}{\varphi} = -\rho_{-1/2}.\left(\Delta_1 \overset{\triangledown}{\varphi}\right)$.

Example 7.15. We consider $\widehat{\varphi}(\zeta) = \dfrac{\zeta}{e^{2i\pi\zeta} - 1} \in \hat{\mathscr{R}}$. This is a meromorphic function with simple poles at $\mathbb{Z}^\star$ whose residue at $m \in \mathbb{Z}^\star$ is $\mathrm{res}_m \widehat{\varphi} = m$. Introducing the singularity $\overset{\triangledown}{\varphi} = {}^\flat\widehat{\varphi}$, one easily deduces that for every $k \in \mathbb{Z}$ and every $m \in \mathbb{N}^\star$,

$$\Delta_{me^{i\pi k}} \overset{\triangledown}{\varphi} = \Delta^+_{me^{i\pi k}} \overset{\triangledown}{\varphi} = (-1)^k m\delta. \tag{7.30}$$

The formal Laplace transform $\mathscr{L} \overset{\triangledown}{\varphi}$ is an asymptotic class $\overset{\triangle}{\varphi} = {}^\natural\widetilde{\varphi}$ that can be represented by a $\mathbb{Z}$-resurgent series $\widetilde{\varphi} \in \widetilde{\mathscr{R}}_{\mathbb{Z}}$ and (7.30) translates into

$$\Delta_{me^{i\pi k}} \overset{\triangle}{\varphi} = \Delta^+_{me^{i\pi k}} \overset{\triangle}{\varphi} = (-1)^k m. \tag{7.31}$$

We now look at the singularity $\overset{\triangledown}{\psi}_{\sigma,n} = \overset{\triangledown}{J}_{\sigma,n} * \overset{\triangledown}{\varphi}$ for $(\sigma,n) \in \mathbb{C} \times \mathbb{N}$. By the Leibniz rule and since $\overset{\triangledown}{J}_{\sigma,n}$ is a resurgent constant,

$$\Delta_{me^{i\pi k}} \overset{\triangledown}{\psi}_{\sigma,n} = \overset{\triangledown}{J}_{\sigma,n} * \Delta_{me^{i\pi k}} \overset{\triangledown}{\varphi} = (-1)^k m \overset{\triangledown}{J}_{\sigma,n} \in \bigcap_{\alpha>0} \mathrm{SING}_{\pi k,\alpha}. \tag{7.32}$$

The asymptotic class associated to $\overset{\triangledown}{\psi}_{\sigma,n}$ by formal Laplace transform is $\overset{\triangle}{\psi}_{\sigma,n} = \overset{\triangle}{J}_{\sigma,n} \overset{\triangle}{\varphi} \in \widetilde{\mathrm{RES}}$. The identity (7.32) provides:

$$\Delta_{me^{i\pi k}} \overset{\triangle}{\psi}_{\sigma,n} = (-1)^k m \overset{\triangle}{J}_{\sigma,n} \in \bigcap_{\alpha>0} \mathrm{ASYMP}_{\pi k,\alpha}. \tag{7.33}$$

7.8.3.3 The Spaces $\mathrm{RES}^{(k)}$

We have already describe the action of the alien operators on the spaces $\mathrm{RES}^{(k)}$. We can draw some consequences from theorem 7.3.

Corollary 7.2. *Let be* $k \in \mathbb{N}^\star$ *and* $\omega \in \widetilde{\mathbb{C}}$ *such that* $\overset{\bullet}{\omega}$ *is an integer and* $|\omega| \leq k$. *The alien operator* Δ_ω *acts on* $\mathrm{RES}^{(k)}$ *and*

$$\begin{aligned} &\Delta_\omega : \mathrm{RES}^{(k)} \to \mathrm{RES}^{(k-|\omega|)}, \quad \textit{when} \quad 1 \leq |\omega| \leq k-1 \\ &\Delta_\omega : \mathrm{RES}^{(k)} \to \mathrm{SING}_{\arg(\omega),\pi}, \quad \textit{when} \quad |\omega| = k. \end{aligned} \tag{7.34}$$

Moreover for any $\overset{\triangledown}{\varphi}, \overset{\triangledown}{\psi} \in \mathrm{RES}^{(k)}$ *:*

- $\Delta_\omega(\overset{\triangledown}{\partial}\overset{\triangledown}{\varphi}) = (\overset{\triangledown}{\partial} - \overset{\bullet}{\omega})(\Delta_\omega \overset{\triangledown}{\varphi})$;
- $\Delta_\omega\left(\overset{\triangledown}{\varphi} * \overset{\triangledown}{\psi}\right)$ *belongs to* $\mathrm{RES}^{(1)}$ *when* $1 \leq |\omega| \leq k-1$ *and to* $\mathrm{SING}_{\arg(\omega),\pi}$ *when* $|\overset{\bullet}{\omega}| = k$ *and furthermore* $\Delta_\omega\left(\overset{\triangledown}{\varphi} * \overset{\triangledown}{\psi}\right) = \left(\Delta_\omega \overset{\triangledown}{\varphi}\right) * \overset{\triangledown}{\psi} + \overset{\triangledown}{\varphi} * \left(\Delta_\omega \overset{\triangledown}{\psi}\right)$ *(Leibniz rule).*

Proof. The identity (7.34) is a consequence of proposition 7.23. The commutation formula $[\Delta_\omega, \overset{\triangledown}{\partial}] = -\overset{\bullet}{\omega}\,\Delta_\omega$ ensues from proposition 7.18. Notice now that for any $k \in \mathbb{N}^\star$, any $L \in]k-1,k]$ and any $\alpha \in]0,\pi/2]$, one has $\mathrm{RES}^{(\overset{\bullet}{\theta},\alpha)}(L) \supset \mathrm{RES}^{(k)}$. Pick two singularities $\overset{\triangledown}{\varphi}, \overset{\triangledown}{\psi} \in \mathrm{RES}^{(k)}$ and consider them as belonging to $\mathrm{RES}^{(\overset{\bullet}{\theta},\alpha)}(L)$. One can apply theorem 7.3 to get: $\Delta_\omega\left(\overset{\triangledown}{\varphi} * \overset{\triangledown}{\psi}\right) = \left(\Delta_\omega \overset{\triangledown}{\varphi}\right) * \overset{\triangledown}{\psi} + \overset{\triangledown}{\varphi} * \left(\Delta_\omega \overset{\triangledown}{\psi}\right) \in \mathrm{SING}_{\theta,\alpha}$. Also, we know that $\Delta_\omega \overset{\triangledown}{\varphi}$ and $\Delta_\omega \overset{\triangledown}{\psi}$ belong to $\mathrm{RES}^{(k-m)}$ or $\mathrm{SING}_{\theta,\pi}$ depending on $|\omega|$. Finally when $1 \leq |\omega| \leq k-1$, one can work in $\mathrm{RES}^{(1)} \supset \mathrm{RES}^{(k-m)}$ which is a convolution algebra by proposition 7.17 and this provides the conclusion. □

Definition 7.53. The alien operators $\Delta^+_\omega, \Delta_\omega : \widetilde{\mathrm{RES}}^{(k)} \to \widetilde{\mathrm{RES}}^{(k-|\omega|)}$ for $1 \leq |\omega| \leq k-1$, *resp.* $\Delta^+_\omega, \Delta_\omega : \widetilde{\mathrm{RES}}^{(k)} \to \mathrm{ASYMP}_{\arg(\omega),\pi}$, for $|\omega| = k$, for asymptotic classes are defined by making the following diagrams commuting:

$$\begin{array}{ccc} \mathrm{RES}^{(k)} & \overset{\Delta^+_\omega,\Delta_\omega}{\longrightarrow} & \mathrm{RES}^{(k-|\omega|)} \\ \mathscr{L}\downarrow\uparrow\mathscr{B} & & \mathscr{L}\downarrow\uparrow\mathscr{B} \\ \widetilde{\mathrm{RES}}^{(k)} & \overset{\Delta^+_\omega,\Delta_\omega}{\longrightarrow} & \widetilde{\mathrm{RES}}^{(k-|\omega|)} \end{array}, \text{ resp. } \begin{array}{ccc} \mathrm{RES}^{(k)} & \overset{\Delta^+_\omega,\Delta_\omega}{\longrightarrow} & \mathrm{SING}_{\arg(\omega),\pi} \\ \mathscr{L}\downarrow\uparrow\mathscr{B} & & \mathscr{L}\downarrow\uparrow\mathscr{B} \\ \widetilde{\mathrm{RES}}^{(k)} & \overset{\Delta^+_\omega,\Delta_\omega}{\longrightarrow} & \mathrm{ASYMP}_{\arg(\omega),\pi} \end{array}.$$

We add a property that will be useful in the sequel.

Corollary 7.3. *Let $k \in \mathbb{N}^\star$ be a positive integer and $\overset{\triangledown}{\varphi} \in \mathrm{RES}^{(k)}$. We suppose that for any $n \in \mathbb{N}^\star$, $\Delta_{\omega_n} \circ \cdots \circ \Delta_{\omega_2} \circ \Delta_{\omega_1} \overset{\triangledown}{\varphi}$ belongs to $\mathrm{RES}^{(1)}$ for any composite operator that satisfies the properties: $(\overset{\bullet}{\omega}_1, \cdots, \overset{\bullet}{\omega}_n) \in (\mathbb{Z}^\star)^n$ and $\sum_{j=1}^n |\omega_j| = k$. Then $\overset{\triangledown}{\varphi}$ belongs to $\mathrm{RES}^{(k+1)}$.*

Proof. This is a direct consequence of both corollary 7.1 and theorem 7.2. □

7.9 Ramified Resurgent Functions

As already said, one uses various spaces of resurgent functions, accordingly to the problem under consideration. We introduce some of them.

7.9.1 Simple and Simply Ramified Resurgent Functions

We start with the resurgent algebra of simple resurgent singularities, much discussed in the first volume of this book [MS016] (see also [CNP93-1, Eca81-1]) and we make use of proposition 7.6.

Definition 7.54. A $\mathbb{Z}$-resurgent singularity $\overset{\triangledown}{\varphi} \in \mathrm{RES}$ is said to be a *simple resurgent* singularity when $\overset{\triangledown}{\varphi} = a\delta + {}^\flat\widehat{\varphi} \in \mathrm{SING}^{\mathrm{simp}}$ and, for any alien operator $\mathscr{A}^{\gamma}_{\omega}(\theta_2,\theta_1)$, $\mathscr{A}^{\gamma}_{\omega}(\theta_2,\theta_1)\ \overset{\triangledown}{\varphi}$ belongs to $\mathrm{SING}^{\mathrm{simp}}$. The minor $\widehat{\varphi}$, *resp.* the 1-Gevrey series $\widetilde{\varphi} = a + \mathscr{L}\widehat{\varphi}$, associated with a simple $\mathbb{Z}$-resurgent singularity is a *simple resurgent function*, *resp.* a *simple resurgent series*, and one denotes by $\widehat{\mathscr{R}_{\mathbb{Z}}}^{\mathrm{simp}}$, *resp.* $\widetilde{\mathscr{R}_{\mathbb{Z}}}^{\mathrm{simp}}$ the space of simple $\mathbb{Z}$-resurgent functions, *resp.* series. The resurgent subalgebra made of simple resurgent singularities is denoted by $\mathrm{RES}^{\mathrm{simp}}_{\mathbb{Z}}$ and the corresponding space of asymptotic classes is denoted by $\widetilde{\mathrm{RES}}_{\mathbb{Z}}^{\mathrm{simp}}$.

As usual in this course, we use abridged notation. One can make acting the alien operators on the space $\widetilde{\mathscr{R}}^{\mathrm{simp}}$.

Definition 7.55. The alien operators $\Delta^{+}_{\omega}, \Delta_{\omega} : \widetilde{\mathscr{R}}^{\mathrm{simp}} \to \widetilde{\mathscr{R}}^{\mathrm{simp}}$ are defined by making the following diagrams commuting:

$$\begin{array}{ccc} \widetilde{\mathrm{RES}}^{\mathrm{simp}} & \overset{\Delta^{+}_{\omega},\Delta_{\omega}}{\longrightarrow} & \widetilde{\mathrm{RES}}^{\mathrm{simp}} \\ T_1 \downarrow\uparrow^{\natural} & & T_1 \downarrow\uparrow^{\natural} \\ \widetilde{\mathscr{R}}^{\mathrm{simp}} & \overset{\Delta^{+}_{\omega},\Delta_{\omega}}{\longrightarrow} & \widetilde{\mathscr{R}}^{\mathrm{simp}} \end{array}.$$

Obviously (from proposition 7.26), for any $\widetilde{\varphi} \in \widetilde{\mathscr{R}}^{\mathrm{simp}}$, the alien derivative $\Delta_{\omega}\widetilde{\varphi}$ only depends on $\overset{\bullet}{\omega}$, thus one could define $\Delta^{+}_{\omega}, \Delta_{\omega} : \widetilde{\mathscr{R}}^{\mathrm{simp}} \to \widetilde{\mathscr{R}}^{\mathrm{simp}}$ for $\omega \in \mathbb{Z}^{\star}$.

Before introducing the simply ramified resurgent functions, we need to state the following straightforward consequence of proposition 7.13.

Lemma 7.9. *The space* $\mathrm{SING}^{\mathrm{s.ram}}$ *of simply ramified singularities* $\overset{\triangledown}{\varphi} = \sum_{n=0}^{N} a_n \overset{\triangledown}{I}_{-n} + {}^\flat\widehat{\varphi}$, $\widehat{\varphi} \in \mathscr{O}_0$, *is a convolution subalgebra.*

Definition 7.56. One denotes by $\mathrm{ASYMP}^{\mathrm{s.ram}}$ the space of asymptotic classes associated with $\mathrm{SING}^{\mathrm{s.ram}}$. The restriction of the Taylor map to $\mathrm{ASYMP}^{\mathrm{s.ram}}$ is denoted by $T_1^{\mathrm{s.ram}}$. One denotes by $\natural^{\mathrm{s.ram}}$ its composition inverse, that is the natural extension of the mapping $\natural$ to $\mathbb{C}[z] \oplus \mathbb{C}[[z^{-1}]]_1$.

Definition 7.57. A $\mathbb{Z}$-resurgent singularity $\overset{\triangledown}{\varphi} \in \mathrm{RES}$ is a *simply ramified resurgent* singularity if $\overset{\triangledown}{\varphi} = \sum_{n=0}^{N} a_{-n} \overset{\triangledown}{I}_{-n} + {}^\flat\widehat{\varphi} \in \mathrm{SING}^{\mathrm{s.ram}}$ and if, for any alien operator $\mathscr{A}^{\gamma}_{\omega}(\theta_2,\theta_1)$, $\mathscr{A}^{\gamma}_{\omega}(\theta_2,\theta_1)\ \overset{\triangledown}{\varphi}$ belongs to $\mathrm{SING}^{\mathrm{s.ram}}$. The resurgent subalgebra made of simply ramified resurgent singularities is denoted by $\mathrm{RES}^{\mathrm{s.ram}}_{\mathbb{Z}}$ to which corresponds the space of asymptotic classes $\widetilde{\mathrm{RES}}^{\mathrm{s.ram}}$. The space of the associated formal series $\widetilde{\varphi}(z) = \sum_{n=-N}^{\infty} a_n z^{-n}$ is denoted by $\widetilde{\mathscr{R}_{\mathbb{Z}}}^{\mathrm{s.ram}}$

One can define the alien operators $\Delta_\omega^+, \Delta_\omega : \widetilde{\mathscr{R}}^{\text{s.ram}} \to \widetilde{\mathscr{R}}^{\text{s.ram}}$ in the same manner than in definition 7.55 and, again, for any $\widetilde{\varphi} \in \widetilde{\mathscr{R}}^{\text{s.ram}}$, the alien derivative $\Delta_\omega \widetilde{\varphi}$ only depends on $\overset{\bullet}{\omega}$.

7.9.2 Ramified Resurgent Functions

The following definition makes sense by propositions 7.6 and 7.13.

Definition 7.58. We denote by $\text{SING}^{\text{ram}} \subset \text{SING}$ the convolution subalgebra generated by the simple singularities and the set of singularities $\{\overset{\triangledown}{J}_{\sigma,m}, (\sigma,m) \in \mathbb{C} \times \mathbb{N}\}$. An element of $\overset{\triangledown}{\varphi} \in \text{SING}^{\text{ram}}$ is called a *ramified singularity* and reads as a finite sum $\overset{\triangledown}{\varphi} = \sum_{(\sigma,m)} \overset{\triangledown}{J}_{\sigma,m} * \overset{\triangledown}{\varphi}_{(\sigma,m)}$ with $\overset{\triangledown}{\varphi}_{(\sigma,m)} \in \text{SING}^{\text{simp}}$. The associated space of asymptotic classes is denoted by $\text{ASYMP}^{\text{ram}} \subset \text{ASYMP}$.

To a ramified singularity $\overset{\triangledown}{\varphi} = \sum_{(\sigma,m)} \overset{\triangledown}{J}_{\sigma,m} * \overset{\triangledown}{\varphi}_{(\sigma,m)}$ is associated, by formal Laplace transform, an asymptotic class $\overset{\vartriangle}{\varphi} \in \text{ASYMP}^{\text{ram}}$ of the form $\overset{\vartriangle}{\varphi} = \sum_{(\sigma,m)} \overset{\vartriangle}{J}_{\sigma,m} \overset{\vartriangle}{\varphi}_{(\sigma,m)}$ with $\overset{\vartriangle}{\varphi}_{(\sigma,m)} = {}^\natural\widetilde{\varphi}_{(\sigma,m)} \in \text{ASYMP}^{\text{simp}}$. This asymptotic class provides a formal expansion of the type

$$\widetilde{\varphi}(z) = \sum_{(\sigma,m)} (-1)^m \frac{\log^m(z)}{z^\sigma} \widetilde{\varphi}_{(\sigma,m)} \in \bigoplus_{(\sigma,m)} \frac{\log^m(z)}{z^\sigma} \mathbb{C}[[z^{-1}]]_1$$

through the Taylor map, for any given arc of $\widetilde{\mathbb{S}}^1$.

We have encountered such formal expansions when we considered the formal integral for Painlevé I (theorem 5.1).

In the same way that $\mathbb{C}[[z^{-1}]]_1$ can be thought of as a constant sheaf on $\mathbb{S}^1$, the space $\bigoplus_{(\sigma,m)} \frac{\log^m(z)}{z^\sigma} \mathbb{C}[[z^{-1}]]_1$ can be seen as a constant sheaf on $\widetilde{\mathbb{S}}^1$. This justifies the following definition.

Definition 7.59. Let be $\theta \in \widetilde{\mathbb{S}}^1$ and $\alpha > 0$. We denote by $\widetilde{\text{Nils}}_1$, *resp.* $\widetilde{\text{Nils}}_{1,(\theta,\alpha)}$. the space of global sections of the sheaf $\bigoplus_{(\sigma,m)} \frac{\log^m(z)}{z^\sigma} \mathbb{C}[[z^{-1}]]_1$, *resp.* section on $J^\star =]-\pi/2-\alpha-\theta, -\theta+\alpha+\pi/2[$. We call $\widetilde{\text{Nils}}_1$ the differential algebra of *1-Gevrey Nilsson series.*

The restriction of the Taylor map to $\text{ASYMP}^{\text{ram}}$ is denoted by T_1^{ram}. One denotes by

$$^{\natural\text{ram}} : \begin{array}{lcl} \widetilde{\text{Nils}}_1 & \to & \text{ASYMP}^{\text{ram}} \\ \widetilde{\varphi} = \sum_{(\sigma,m)} \widetilde{J}_{\sigma,m} \widetilde{\varphi}_{(\sigma,m)} & \to & {}^{\natural\text{ram}}\widetilde{\varphi} = \sum_{(\sigma,m)} \overset{\vartriangle}{J}_{\sigma,m} {}^\natural\widetilde{\varphi}_{(\sigma,m)} \end{array}$$

its composition inverse, where $\widetilde{J}_{\sigma,m}(z) = (-1)^m \frac{\log^m(z)}{z^\sigma}$.

One can define the space $\widetilde{\text{Nils}}$ as well, made of formal expansions of the form $\widetilde{\varphi} = \sum_{(\sigma,m)} \widetilde{J}_{\sigma,m} \widetilde{\varphi}_{(\sigma,m)}$ with $\widetilde{\varphi}_{(\sigma,m)} \in \mathbb{C}[[z^{-1}]]$. Let us consider an element $\widetilde{\varphi} \in \widetilde{\text{Nils}}$ under the form $\widetilde{\varphi} = \sum_{i=1}^n \frac{\widetilde{\varphi}_i}{z^{\sigma_i}}$, $\widetilde{\varphi}_i \in \mathbb{C}[[z^{-1}]]$. We can of course assume that for any $i \neq j$, $\sigma_i - \sigma_j \notin \mathbb{Z}$. We denote $\omega_i = \mathrm{e}^{-2\mathrm{i}\pi\sigma_i}$ and we remark that $\omega_i - \omega_j \neq 0$ for any $i \neq j$. We set $\rho.\widetilde{\varphi}(z) = \widetilde{\varphi}(z\mathrm{e}^{2\mathrm{i}\pi})$ and more generally $\rho_k.\widetilde{\varphi}(z) = \widetilde{\varphi}(z\mathrm{e}^{2\mathrm{i}\pi k})$ for any $k \in \mathbb{Z}$. We notice that $\rho_k.\widetilde{\varphi} = \sum_{i=1}^n \omega_i^k \frac{\widetilde{\varphi}_i}{z^{\sigma_i}}$. Therefore, ${}^t(\widetilde{\varphi}, \rho_1.\widetilde{\varphi}, \cdots, \rho_n.\widetilde{\varphi}) = A^t \left(\frac{\widetilde{\varphi}_1}{z^{\sigma_1}}, \frac{\widetilde{\varphi}_2}{z^{\sigma_2}}, \cdots, \frac{\widetilde{\varphi}_n}{z^{\sigma_n}} \right)$ where A stands for the $n \times n$ invertible Vandermonde matrix $A = \begin{pmatrix} 1 & \cdots & 1 \\ \omega_1 & \cdots & \omega_n \\ \vdots & & \vdots \\ \omega_1^n & \cdots & \omega_n^n \end{pmatrix}$. This implies that for each integer $i \in [1,n]$, $\frac{\widetilde{\varphi}_i}{z^{\sigma_i}}$ is a linear combination of $\widetilde{\varphi}, \rho.\widetilde{\varphi}, \cdots, \rho_n.\widetilde{\varphi}$. This observation can be generalized:

Lemma 7.10. *Let $\widetilde{\varphi} = \sum_i \sum_{m=0}^{r_i-1} \widetilde{J}_{\sigma_i,m} \widetilde{\varphi}_{(\sigma_i,m)}$ be an element of $\widetilde{\text{Nils}}$. Then the series $\widetilde{\varphi}_{(\sigma_i,m)} \in \mathbb{C}[[z^{-1}]]$ are uniquely determined by $\widetilde{\varphi}$ and its monodromy (that is $\rho.\widetilde{\varphi}$, $\rho_2.\widetilde{\varphi}$, etc.) once one imposes that $\sigma_i - \sigma_j \notin \mathbb{Z}$ whenever $\widetilde{\varphi}_{(\sigma_i,m)}.\widetilde{\varphi}_{(\sigma_j,m)} \neq 0$.*

Proof. This is a well-known fact and we follow a reasoning from [Nar010]. We only show how $\widetilde{\varphi}$ determines the series $\widetilde{\varphi}_{(\sigma_i,m)}$ since we will use this result in a moment. If $\omega = \mathrm{e}^{-2\mathrm{i}\pi\sigma}$, observe that $(\rho - \omega)\left(\frac{\log^m(z)}{z^\sigma}\right) = \omega \sum_{l=0}^{m-1} \binom{m}{l} (2\mathrm{i}\pi)^{m-l} \frac{\log^l(z)}{z^\sigma}$ and therefore $(\rho - \omega)^m \left(\frac{\log^m(z)}{z^\sigma}\right) = m! \frac{\omega^m}{z^\sigma}$ while $(\rho - \omega)^{m+1} \left(\frac{\log^m(z)}{z^\sigma}\right) = 0$. As a consequence, for any $\widetilde{\varphi} \in \widetilde{\text{Nils}}$ one has $P(\rho)\widetilde{\varphi} \in \widetilde{\text{Nils}}$ for any polynomial $P \in \mathbb{C}[X]$, and there exists a polynomial $P \in \mathbb{C}[X]$ such that $P(\rho)\widetilde{\varphi} = 0$. We denote by $d(\widetilde{\varphi})$ the degree of the minimal polynomial of the action of ρ on $\widetilde{\varphi}$. We then make a reasoning by induction on $d(\widetilde{\varphi})$.
Suppose that $d(\widetilde{\varphi}) = 1$. This means that there exists $\omega = \mathrm{e}^{-2\mathrm{i}\pi\sigma} \in \mathbb{C}$ such that $(\rho - \omega)\widetilde{\varphi} = 0$, thus $\rho\left(z^\sigma \widetilde{\varphi}\right) = z^\sigma \widetilde{\varphi}$. Therefore $\widetilde{\varphi}$ is of the form $\widetilde{\varphi} = \frac{\widetilde{\varphi}_{(\sigma_1,0)}}{z^{\sigma_1}}$ with $\widetilde{\varphi}_{(\sigma_1,0)} \in \mathbb{C}[[z^{-1}]]$ and a convenient choice of $\sigma_1 \in \mathbb{C}$ so that $\sigma_1 - \sigma \in \mathbb{Z}$. (Thus $\widetilde{\varphi}_{(\sigma_1,0)} = \rho\left(z^{\sigma_1}\widetilde{\varphi}\right)$).
Suppose now that for any $\widetilde{\varphi} \in \widetilde{\text{Nils}}$ such that $d(\widetilde{\varphi}) \leq d$, its decomposition is (uniquely) determined by $\widetilde{\varphi}, \rho.\widetilde{\varphi}, \cdots, \rho_d.\widetilde{\varphi}$.
Let be $\widetilde{\varphi} \in \widetilde{\text{Nils}}$ with $d(\widetilde{\varphi}) = d+1$. The minimal polynomial of the action of ρ on $\widetilde{\varphi}$ is $P(X) = \prod_i (X - \omega_i)^{r_i}$ with $\sum_i r_i = d+1$. Write:

$$\widetilde{P}(X) = (X - \omega_1)^{r_1 - 1} \prod_{i \neq 1} (X - \omega_i)^{r_i} = (X - \omega_i)^{r_1 - 1} Q(X).$$

From the fact that $(\rho-\omega_1)\widetilde{P}(\rho)\widetilde{\varphi}=0$, we deduce the identity $\widetilde{P}(\rho)\widetilde{\varphi}=\dfrac{\widetilde{\phi}}{z^{\sigma_1}}$ with $\widetilde{\phi}\in\mathbb{C}[[z^{-1}]]_1$ and a convenient $\sigma_1\in\mathbb{C}$ such that $\omega_1=\mathrm{e}^{-2\mathrm{i}\pi\sigma_1}$. Since

$$\widetilde{P}(\rho)\left(\frac{\log^{r_1-1}(z)}{z^{\sigma_1}}\right)=Q(\rho)\left((r_1-1)!\frac{\omega_1^{r_1-1}}{z^{\sigma_1}}\right)=Q(\omega_1)(r_1-1)!\frac{\omega_1^{r_1-1}}{z^{\sigma_1}},$$

we see that necessarily $\widetilde{P}(\rho)\left(\widetilde{J}_{\sigma_1,r_1-1}\widetilde{\varphi}_{\sigma_1,r_1-1}\right)=(-1)^{r_1-1}\dfrac{\widetilde{\phi}}{z^{\sigma_1}}$ and $\widetilde{\varphi}_{\sigma_1,r_1-1}=(-1)^{r_1-1}\dfrac{\widetilde{\phi}}{(r_1-1)!\omega_1^{r_1-1}Q(\omega_1)}$.

We finally observe that $\widetilde{P}(\rho)\left(\widetilde{\phi}-\widetilde{J}_{\sigma_1,r_1-1}\widetilde{\varphi}_{\sigma_1,r_1-1}\right)=0$ and we can apply the induction hypothesis on $\widetilde{\phi}-\widetilde{J}_{\sigma_1,r_1-1}\widetilde{\varphi}_{\sigma_1,r_1-1}$. This ends the proof. □

We are in good position to define the ramified resurgent functions [Sau006, Eca81-1, Eca85], see also [DH002, Del005, Del010, LR011].

Definition 7.60. A $\mathbb{Z}$-resurgent singularity $\overset{\triangledown}{\varphi}\in\mathrm{RES}_{\mathbb{Z}}$ is a *ramified resurgent* singularity when $\overset{\triangledown}{\varphi}\in\mathrm{SING}^{\mathrm{ram}}$ whereas, any alien operator $\mathscr{A}^{\gamma}_{\omega}(\theta_2,\theta_1)$, $\mathscr{A}^{\gamma}_{\omega}(\theta_2,\theta_1)\ \overset{\triangledown}{\varphi}$ belongs to $\mathrm{SING}^{\mathrm{ram}}$. The space of ramified resurgent singularities makes a resurgent subalgebra denoted by $\mathrm{RES}^{\mathrm{ram}}_{\mathbb{Z}}$. The corresponding space of asymptotic classes, *resp.* formal expansions, is denoted by $\widetilde{\mathrm{RES}}_{\mathbb{Z}}^{\mathrm{ram}}$, *resp.* $\widetilde{\mathscr{R}}_{\mathbb{Z}}^{\mathrm{ram}}$.

We state a result that derives directly from lemma 7.10

Proposition 7.27. *The formal expansion* $\widetilde{\varphi}=\sum_{(\sigma,m)}\widetilde{J}_{\sigma,m}\widetilde{\varphi}_{(\sigma,m)}\in\widetilde{\mathrm{Nils}}$ *belongs to* $\widetilde{\mathscr{R}}^{\mathrm{ram}}$ *if and only if each of its components* $\widetilde{\varphi}_{(\sigma,m)}$ *belongs to* $\widetilde{\mathscr{R}}^{\mathrm{ram}}$.

Definition 7.61. The alien operators $\Delta^{+}_{\omega},\Delta_{\omega}:\widetilde{\mathscr{R}}^{\mathrm{ram}}\to\widetilde{\mathscr{R}}^{\mathrm{ram}}$ are defined by making the following diagrams commuting:

$$\begin{array}{ccc}\widetilde{\mathrm{RES}}^{\mathrm{ram}} & \overset{\Delta^{+}_{\omega},\Delta_{\omega}}{\longrightarrow} & \widetilde{\mathrm{RES}}^{\mathrm{ram}}\\ T_1^{\mathrm{ram}}\downarrow\uparrow{}^{\natural\mathrm{ram}} & & T_1^{\mathrm{ram}}\downarrow\uparrow{}^{\natural\mathrm{ram}}\\ \widetilde{\mathscr{R}}^{\mathrm{ram}} & \overset{\Delta^{+}_{\omega},\Delta_{\omega}}{\longrightarrow} & \widetilde{\mathscr{R}}^{\mathrm{ram}}\end{array}.$$

We eventually lay down a direct consequence of proposition 7.19. (We warn to the change of sign).

Proposition 7.28. *Let* $\widetilde{\varphi}$ *be an element of* $\widetilde{\mathscr{R}}^{\mathrm{ram}}$. *Then, for any* $\omega\in\widetilde{\mathbb{C}}$ *with* $\overset{\bullet}{\omega}\in\mathbb{Z}^{\star}$, *for any* $k\in\mathbb{Z}$,

$$\Delta_{\omega\mathrm{e}^{2\mathrm{i}\pi k}}\widetilde{\varphi}=\rho_k.\left(\Delta_{\omega}\,\rho_{-k}.\widetilde{\varphi}\right),\qquad\Delta_{\omega\mathrm{e}^{\mathrm{i}\pi}}\widetilde{\varphi}=\rho_{1/2}.\left(\Delta_{\omega}\,\rho_{-1/2}.\widetilde{\varphi}\right).$$

Example 7.16. Suppose that $\widetilde{\varphi}\in\mathbb{C}[[z^{-1}]]_1$ belongs to $\widetilde{\mathscr{R}}^{\mathrm{ram}}$ with $\Delta_{\omega}\widetilde{\varphi}=\frac{\log(z)}{z^{\sigma}}\widetilde{\psi}$, $\widetilde{\psi}\in\mathbb{C}[[z^{-1}]]$. For $k\in\mathbb{Z}$, $\rho_{-k}.\widetilde{\varphi}(z)=\widetilde{\varphi}(z)$, then $\Delta_{\omega\mathrm{e}^{2\mathrm{i}\pi k}}\widetilde{\varphi}(z)=\frac{\log(z+2\pi k)}{z^{\sigma}\mathrm{e}^{2\mathrm{i}\pi k\sigma}}\widetilde{\psi}(z)$.

Suppose furthermore that $\widetilde{\varphi}$ is even, so that $\rho_{-1/2}.\widetilde{\varphi}(z) = \widetilde{\varphi}(z)$. On deduces that $\Delta_{\omega e^{i\pi}} \widetilde{\varphi}(z) = \frac{\log(z+\pi)}{z^{\sigma} e^{i\pi\sigma}} \widetilde{\psi}(-z)$.

7.10 Comments

We mentioned in Sect. 4.7 the generalisation of the resurgence theory with the notion of "endlessly continuable functions" [CNP93-1, Eca85]. The whole constructions made in this chapter can be extended as well to this context.

We of course owe the main ideas presented here from the work of Ecalle, who started his theory in the 1970's [Eca78]. We have borrowed most of the materials to Pham *et al.* [CNP93-1], in particular the microfunctions and the sheaf approach. To compare with other written papers devoted to resurgence theory, we have paid more attention to the sheaf and associated spaces of asymptotic classes. Finally, the responsability for possible mistakes is ours.

References

CNP93-1. B. Candelpergher, C. Nosmas, F. Pham, *Approche de la résurgence*, Actualités mathématiques, Hermann, Paris (1993).

Cos009. O. Costin, *Asymptotics and Borel summability*, Chapman & Hall/CRC Monographs and Surveys in Pure and Applied Mathematics, 141. CRC Press, Boca Raton, FL, 2009.

Del005. E. Delabaere, *Addendum to the hyperasymptotics for multidimensional Laplace integrals.* Analyzable functions and applications, 177-190, Contemp. Math., 373, Amer. Math. Soc., Providence, RI, 2005.

Del010. E. Delabaere, *Singular integrals and the stationary phase methods.* Algebraic approach to differential equations, 136-209, World Sci. Publ., Hackensack, NJ, 2010.

DH002. E. Delabaere, C. J. Howls, *Global asymptotics for multiple integrals with boundaries.* Duke Math. J. **112** (2002), 2, 199-264.

DP99. E. Delabaere, F. Pham, *Resurgent methods in semi-classical asymptotics*, Ann. Inst. Henri Poincaré, Sect. A **71** (1999), no 1, 1-94.

Ebe007. W. Ebeling, *Functions of several complex variables and their singularities.* Translated from the 2001 German original by Philip G. Spain. Graduate Studies in Mathematics, 83. American Mathematical Society, Providence, RI, 2007.

Eca78. J. Écalle, *Un analogue des fonctions automorphes: les fonctions résurgentes.* Séminaire Choquet, 17e année (1977/78), Initiation à l'analyse, Fasc. 1, Exp. No. 11, 9 pp., Secrétariat Math., Paris, 1978.

Eca81-1. J. Écalle, *Les algèbres de fonctions résurgentes*, Publ. Math. d'Orsay, Université Paris-Sud, 1981.05 (1981).

Eca85. J. Ecalle, *Les fonctions résurgentes. Tome III : l'équation du pont et la classification analytique des objets locaux.* Publ. Math. d'Orsay, Université Paris-Sud, 1985.05 (1985).

Eca92. J. Écalle, *Fonctions analysables et preuve constructive de la conjecture de Dulac.* Actualités mathématiques, Hermann, Paris (1992).

Eca93-1. J. Écalle, *Six lectures on transseries, Analysable functions and the Constructive proof of Dulac's conjecture*, Bifurcations and periodic orbits of vector fields (Montreal, PQ, 1992), 75-184, NATO Adv. Sci. Inst. Ser. C Math. Phys. Sci., 408, Kluwer Acad. Publ., Dordrecht, 1993.

For81. O. Forster, *Lectures on Riemann Surfaces*, Graduate texts in mathematics; 81, Springer, New York (1981).

God73. R. Godement, *Topologie algébrique et théorie des faisceaux.* Publications de l'Institut de Mathématique de l'Université de Strasbourg, XIII. Actualités Scientifiques et Industrielles, No. 1252. Hermann, Paris, 1973.

KKK86. M. Kashiwara, T. Kawai, T. Kimura, *Foundations of algebraic analysis.* Princeton Mathematical Series, **37**. Princeton University Press, Princeton, NJ, 1986.

Lod016. M. Loday-Richaud, *Divergent Series, Summability and Resurgence II. Simple and Multiple Summability.* Lecture Notes in Mathematics, **2154**. Springer, Heidelberg, 2016.

LR011. M. Loday-Richaud, P. Remy, *Resurgence, Stokes phenomenon and alien derivatives for level-one linear differential systems.* J. Differential Equations **250** (2011), no. 3, 1591-1630.

Mal91. B. Malgrange, *Equations différentielles à coefficients polynomiaux.* Progress in Mathematics, 96. Birkhäuser Boston, Inc., Boston, MA, 1991.

Mal95. B. Malgrange, *Sommation des séries divergentes.* Expositiones Mathematicae **13**, 163-222 (1995).

MS016. C. Mitschi, D. Sauzin, *Divergent Series, Summability and Resurgence I. Monodromy and Resurgence.* Lecture Notes in Mathematics, **2153**. Springer, Heidelberg, 2016.

Mor93. M. Morimoto, *An introduction to Sato's hyperfunctions.* Translated and revised from the 1976 Japanese original by the author. Translations of Mathematical Monographs, 129. American Mathematical Society, Providence, RI, 1993.

Nar010. L. Narváez Macarro, *D-modules in dimension 1.* Algebraic approach to differential equations, 1-51, World Sci. Publ., Hackensack, NJ, 2010.

OSS003. C. Olivé, D. Sauzin, T. Seara, *Resurgence in a Hamilton-Jacobi equation.* Proceedings of the International Conference in Honor of Frédéric Pham (Nice, 2002). Ann. Inst. Fourier (Grenoble) **53** (2003), no. 4, 1185-1235.

Ou012. Y. Ou, *Sur la stabilité par produit de convolution d'algèbres de résurgence.* PhD thesis, Université d'Angers (2012).

Pha005. F. Pham, *Intégrales singulières.* Savoirs Actuels (Les Ulis). EDP Sciences, Les Ulis; CNRS Editions, Paris, 2005

Sau006. D. Sauzin, *Resurgent functions and splitting problems*, RIMS Kokyuroku 1493 (31/05/2006) 48-117

Sau013. D. Sauzin, *On the stability under convolution of resurgent functions.* Funkcial. Ekvac. **56** (2013), no. 3, 397-413.

Sau015. D. Sauzin, *Nonlinear analysis with resurgent functions.* Ann. Sci. Ecole Norm. Sup. (6) **48** (2015), no. 3, 667-702.

SS96. B. Sternin, V. Shatalov, *Borel-Laplace transform and asymptotic theory. Introduction to resurgent analysis.* CRC Press, Boca Raton, FL, 1996.

Chapter 8
Resurgent Structure For The First Painlevé Equation

Abstract We show the resurgence property for the formal series solution of the prepared form associated with the first Painlevé equation. The detailed resurgent structure is given in Sect. 8.1. Its proof is given using the so-called bridge equation (Sect. 8.4), after some preliminaries (Sect. 8.3). The nonlinear Stokes phenomena are briefly analyzed in Sect. 8.2.

8.1 The Main Theorem

8.1.1 Reminder

The formal integral of the prepared ODE (3.6) associated with the first Painlevé equation was described with theorem 5.1 and its corollary 5.1. It can be written under the following equivalent form:

$$\widetilde{w}(z,\mathbf{U}) = \widetilde{W}_{\mathbf{0}}(z) + \sum_{n=0}^{\infty} \sum_{\mathbf{k}\in \Xi_{n+1,0}\setminus \Xi_{n,0}} \mathbf{U}^{\mathbf{k}} e^{-\boldsymbol{\lambda}.\mathbf{k}z} \widetilde{W}_{\mathbf{k}}(z), \tag{8.1}$$

where $\widetilde{W}_{\mathbf{0}} = \widetilde{w}_{\mathbf{0}} = \widetilde{w}_{\mathbf{0}}^{[0]}$ and for any $n \in \mathbb{N}$ and any $\mathbf{k} \in \Xi_{n+1,0} \setminus \Xi_{n,0}$,

$$\widetilde{W}_{\mathbf{k}} = \sum_{l=0}^{n} \frac{1}{l!} (\boldsymbol{\varkappa}.\mathbf{k})^l \log^l(z) \widetilde{W}_{\mathbf{k}-1}^{[0]}, \quad \widetilde{W}_{\mathbf{k}}^{[0]} = z^{-\boldsymbol{\tau}.\mathbf{k}} \widetilde{w}_{\mathbf{k}}^{[0]}. \tag{8.2}$$

The formal series $\widetilde{w}_{\mathbf{0}} \in \mathbb{C}[[z^{-1}]]$ solves (3.6), namely

$$P(\partial)\widetilde{w}_{\mathbf{0}} + \frac{1}{z} Q(\partial)\widetilde{w}_{\mathbf{0}} = F(z,\widetilde{w}_{\mathbf{0}}) = f_0 + f_1 \widetilde{w}_{\mathbf{0}} + f_2 \widetilde{w}_{\mathbf{0}}^2, \tag{8.3}$$

E. Delabaere, *Divergent Series, Summability and Resurgence III*,
Lecture Notes in Mathematics 2155, DOI 10.1007/978-3-319-29000-3_8

$P(\partial) = \partial^2 - 1$, $Q(\partial) = -3\partial$, $f_0(z) = \dfrac{392}{625} z^{-2}$, $f_1(z) = -4z^{-2}$, $f_2(z) = \dfrac{1}{2} z^{-2}$, while the $\widetilde{W}_{\mathbf{k}}$ satisfy a hierarchy of equations given by lemma 5.3 that we recall:

$$\mathfrak{P}_{\mathbf{k}} \widetilde{W}_{\mathbf{k}} = \sum_{\substack{\mathbf{k}_1+\mathbf{k}_2=\mathbf{k} \\ |\mathbf{k}_i|\geq 1}} \frac{\widetilde{W}_{\mathbf{k}_1} \widetilde{W}_{\mathbf{k}_2}}{2!} \frac{\partial^2 F(z,\widetilde{w}_{\mathbf{0}})}{\partial w^2}, \tag{8.4}$$

$$\mathfrak{P}_{\mathbf{k}} = \mathfrak{P}_{\mathbf{k}}(\widetilde{w}_{\mathbf{0}}) = P(-\boldsymbol{\lambda}.\mathbf{k}+\partial) + \frac{1}{z} Q(-\boldsymbol{\lambda}.\mathbf{k}+\partial) - \frac{\partial F(z,\widetilde{w}_{\mathbf{0}})}{\partial w}.$$

To what concerns the $\mathbf{k}$-th series $\widetilde{w}_{\mathbf{k}}^{[0]} \in \mathbb{C}[[z^{-1}]]$, we have a result that ensues directly from theorem 6.1:

Proposition 8.1. *For any $\mathbf{k} \in \mathbb{N}^2$, the $\mathbf{k}$-th series $\widetilde{w}_{\mathbf{k}}^{[0]}$ belongs to $\widetilde{\mathscr{R}}^{(1)}$, the asymptotic class $\overset{\triangle}{W}_{\mathbf{k}} = {}^{\natural\mathrm{ram}}\widetilde{W}_{\mathbf{k}}$ belongs to $\widetilde{\mathrm{RES}}^{(1)}$ and the singularity $\overset{\triangledown}{W}_{\mathbf{k}} = \mathscr{B} \overset{\triangle}{W}_{\mathbf{k}}$ belongs to $\mathrm{RES}^{(1)}$.*

Notice that $\widetilde{I}_{-\boldsymbol{\tau}.\mathbf{k}} \widetilde{W}_{\mathbf{k}} = \sum_{l=0}^{n} \frac{1}{l!} (-\boldsymbol{\varkappa}.\mathbf{k})^l \widetilde{J}_{-\boldsymbol{\tau}.\mathbf{1},l} \widetilde{w}_{\mathbf{k}-\mathbf{1}}^{[0]}$. for any $n \in \mathbb{N}$ and any $\mathbf{k} \in \Xi_{n+1,0} \setminus \Xi_{n,0}$. Therefore, $\overset{\triangledown}{W}_{\mathbf{k}} = \sum_{l=0}^{n} \frac{1}{l!} (-\boldsymbol{\varkappa}.\mathbf{k})^l \overset{\triangledown}{J}_{-\boldsymbol{\tau}.\mathbf{1},l} * \overset{\triangledown}{w}_{\mathbf{k}-\mathbf{1}}^{[0]}$ where $\overset{\triangledown}{w}_{\mathbf{e}_i}^{[0]} = \delta + {}^{\flat}\widehat{w}_{\mathbf{e}_i}$ for $i =, 1, 2$, otherwise $\overset{\triangledown}{w}_{\mathbf{k}}^{[0]} = {}^{\flat}\widehat{w}_{\mathbf{k}}^{[0]}$.

8.1.2 The Main Theorem

We formulate the main result of this chapter.

Theorem 8.1. *The formal integral $\widetilde{w}(z,\mathbf{U})$ of the prepared form (3.6) associated with the first Painlevé equation, is resurgent. More precisely, for any $\mathbf{k} \in \mathbb{N}^2$, $\widetilde{W}_{\mathbf{k}}$ belongs to the space $\widetilde{\mathscr{R}}_{\mathbb{Z}}^{\mathrm{ram}}$ of ramified resurgent formal expansions.*
We set $\omega_1^j = \mathrm{e}^{2i\pi j}$ ($\overset{\bullet}{\omega}{}_1^j = \lambda_1$) and $\omega_2^j = \mathrm{e}^{2i\pi(j+1/2)}$ ($\overset{\bullet}{\omega}{}_2^j = \lambda_2$) for any $j \in \mathbb{Z}$. Then, for every $\omega \in \widetilde{\mathbb{C}}$ of the form $\omega = k_0 \omega_1^j$, resp. $\omega = k_0 \omega_2^j$, with $k_0 \in \mathbb{N}^\star$, there exist two sequences of complex numbers $\big(A_n(\omega)\big)_{n\in\mathbb{N}}$ and $\big(B_n(\omega)\big)_{n\in\mathbb{N}}$, uniquely determined by ω such that, for any $\mathbf{k} = (k_1,k_2) \in \mathbb{N}^2$ and any $n \in \mathbb{N}$,

$$\Delta_\omega \widetilde{W}_{\mathbf{k}+\mathbf{n}} = \sum_{m=-1}^{n} \Big((k_1+m+k_0) A_{n-m}(\omega) + (k_2+m) B_{n-m}(\omega) \Big) \widetilde{W}_{\mathbf{k}+\mathbf{m}+k_0\mathbf{e}_1},$$

resp. (8.5)

$$\Delta_\omega \widetilde{W}_{\mathbf{k}+\mathbf{n}} = \sum_{m=-1}^{n} \Big((k_2+m+k_0) A_{n-m}(\omega) + (k_1+m) B_{n-m}(\omega) \Big) \widetilde{W}_{\mathbf{k}+\mathbf{m}+k_0\mathbf{e}_2},$$

where by convention $\widetilde{W}_{(k_1,k_2)} = 0$ if $k_1 < 0$ or $k_2 < 0$.
The sequences $\big(A_n(\omega)\big)_{n\in\mathbb{N}}$ and $\big(B_n(\omega)\big)_{n\in\mathbb{N}}$ are subject to the conditions:

$A_n(\omega)=0$ when $|\omega|\geq n+2$ and $B_n(\omega)=0$ when $|\omega|\geq n+1$. Also, $\big(A_n(\omega)\big)_{n\in\mathbb{N}}$ and $\big(B_n(\omega)\big)_{n\in\mathbb{N}}$ are known for every $\omega\in\widetilde{\mathbb{C}}$ once they are known for $\arg\omega=0$ only. In particular, $A_0(\omega_i^j)=(-1)^jA_0(\omega_i^0)$ while $A_0(\omega_2^j)=-\mathrm{i}A_0(\omega_1^j)$.

The proof of this theorem will be given in Sect. 8.4.

8.1.2.1 Remarks

We detail (8.5) for $n=0$. For any $j\in\mathbb{Z}$ and any $k_0\in\mathbb{N}^\star$,

$$\begin{aligned}\Delta_{k_0\omega_1^j}\widetilde{w}_{\mathbf{0}} &= A_0(k_0\omega_1^j)\widetilde{W}_{\mathbf{e}_1}=A_0(k_0\omega_1^j)z^{3/2}\widetilde{w}_{\mathbf{e}_1} \\ \Delta_{k_0\omega_2^j}\widetilde{w}_{\mathbf{0}} &= A_0(k_0\omega_2^j)\widetilde{W}_{\mathbf{e}_2}=A_0(k_0\omega_2^j)z^{3/2}\widetilde{w}_{\mathbf{e}_2}\end{aligned} \tag{8.6}$$

and $A_0(k_0\omega_i^j)=0$ when $k_0\geq 2$.

When $\mathbf{k}\in\Xi_{1,0}$, we use abridged notation $\widetilde{w}_{\mathbf{k}}=\widetilde{w}_{\mathbf{k}}^{[0]}$.

By proposition 7.28, $\Delta_{\omega_i^j}\widetilde{w}_{\mathbf{0}}=\rho_j.\Big(\Delta_{\omega_i^0}\rho_{-j}.\widetilde{w}_{\mathbf{0}}\Big)$, $i=1,2$. Therefore, $A_0(\omega_i^j)=(-1)^jA_0(\omega_i^0)$. Remember that $\widetilde{w}_{\mathbf{0}}$ is even, thus $\widetilde{w}_{\mathbf{0}}=\rho_{-1/2}.\widetilde{w}_{\mathbf{0}}$, while $\widetilde{w}_{\mathbf{e}_2}=\rho_{1/2}.\widetilde{w}_{\mathbf{e}_1}$. By proposition 7.28 again, $\Delta_{\omega_2^j}\widetilde{w}_{\mathbf{0}}=\rho_{1/2}.\Big(\Delta_{\omega_1^j}\rho_{-1/2}.\widetilde{w}_{\mathbf{0}}\Big)$ and we deduce that $A_0(\omega_2^j)=-\mathrm{i}A_0(\omega_1^j)$.

Now for any $k_1\in\mathbb{N}^\star$,

$$\begin{aligned}\Delta_{k_0\omega_1^j}\widetilde{W}_{k_1\mathbf{e}_1} &= (k_1+k_0)A_0(k_0\omega_1^j)\widetilde{W}_{(k_1+k_0)\mathbf{e}_1} \\ \Delta_{k_0\omega_2^j}\widetilde{W}_{k_1\mathbf{e}_1} &= k_0A_0(k_0\omega_2^j)\widetilde{W}_{k_1\mathbf{e}_1+k_0\mathbf{e}_2}+(k_1-1)B_1(k_0\omega_2^j)\widetilde{W}_{k_1\mathbf{e}_1+k_0\mathbf{e}_2-1}.\end{aligned} \tag{8.7}$$

and $B_1(k_0\omega_2^j)=0$ when $k_0\geq 2$. We have in particular $\Delta_{\omega_2^j}\widetilde{W}_{\mathbf{e}_1}=A_0(\omega_2^j)\widetilde{W}_1$, thus $\Delta_{\omega_2^j}\widetilde{w}_{\mathbf{e}_1}=A_0(\omega_2^j)z^{3/2}\widetilde{w}_1$. Also, for $k_1\geq 2$,

$$\Delta_{\omega_2^j}\widetilde{W}_{k_1\mathbf{e}_1}=A_0(\omega_2^j)\widetilde{W}_{(k_1-1)\mathbf{e}_1+1}+(k_1-1)B_1(\omega_2^j)\widetilde{W}_{(k_1-1)\mathbf{e}_1}$$

and using (8.2),

$$\begin{aligned}\Delta_{\omega_2^j}\widetilde{w}_{k_1\mathbf{e}_1} &= A_0(\omega_2^j)\Big((k_1-1)\varkappa_1\log(z)z^{-3/2}\widetilde{w}_{(k_1-1)\mathbf{e}_1}+z^{3/2}\widetilde{w}^{[0]}_{(k_1-1)\mathbf{e}_1+1}\Big) \\ &+ (k_1-1)B_1(\omega_2^j)z^{-3/2}\widetilde{w}_{(k_1-1)\mathbf{e}_1}.\end{aligned}$$

By proposition 7.28, $\Delta_{\omega_2^j}\widetilde{w}_{2\mathbf{e}_1}=\rho_j.\Big(\Delta_{\omega_i^0}\rho_{-j}.\widetilde{w}_{2\mathbf{e}_1}\Big)$, therefore

$$\begin{aligned}\Delta_{\omega_2^j}\widetilde{w}_{2\mathbf{e}_1} &= (-1)^j A_0(\omega_2^0)\Big(\varkappa_1 \log(z+2\mathrm{i}\pi j) z^{-3/2}\widetilde{w}_{\mathbf{e}_1} + z^{3/2}\widetilde{w}^{[0]}_{\mathbf{e}_1+1}\Big)\\ &+ (-1)^j B_1(\omega_2^0) z^{-3/2}\widetilde{w}_{\mathbf{e}_1}\end{aligned}$$

and one deduces: $B_1(\omega_2^j) = (-1)^j\Big(B_1(\omega_2^0) + 2\mathrm{i}\pi j \varkappa_1 A_0(\omega_2^0)\Big)$. Of course, by symmetry: $B_1(\omega_1^j) = (-1)^j\Big(B_1(\omega_1^0) + 2\mathrm{i}\pi j \varkappa_2 A_0(\omega_1^0)\Big)$.
In the same way, $\Delta_{\omega_2^j}\widetilde{w}_{2\mathbf{e}_1} = \rho_{1/2}.\Big(\Delta_{\omega_1^j}\rho_{-1/2}.\widetilde{w}_{2\mathbf{e}_1}\Big)$ and we know that $\rho_{-1/2}.\widetilde{w}_{2\mathbf{e}_1} = \widetilde{w}_{2\mathbf{e}_2}$, $\rho_{1/2}.\widetilde{w}_{\mathbf{e}_2} = \widetilde{w}_{\mathbf{e}_1}$, $\rho_{1/2}.\widetilde{w}^{[0]}_{\mathbf{e}_2+1} = \widetilde{w}^{[0]}_{\mathbf{e}_1+1}$. Thus,

$$\begin{aligned}\Delta_{\omega_2^j}\widetilde{w}_{2\mathbf{e}_1} &= -\mathrm{i}A_0(\omega_1^j)\Big(-\varkappa_2 \log(z+\mathrm{i}\pi) z^{-3/2}\widetilde{w}_{\mathbf{e}_1} + z^{3/2}\widetilde{w}^{[0]}_{\mathbf{e}_1+1}\Big)\\ &+ \mathrm{i}B_1(\omega_1^j) z^{-3/2}\widetilde{w}_{\mathbf{e}_1}\end{aligned}$$

and $B_1(\omega_2^j) = \mathrm{i}\Big(B_1(\omega_1^j) + \mathrm{i}\pi\varkappa_2 A_0(\omega_1^j)\Big)$.

8.1.2.2 Resurgence Coefficients and Analytic Classification

Definition 8.1. The coefficents $A_n(\omega)$ and $B_n(\omega)$ given by theorem 8.1 are called the *resurgence cofficients* for the first Painlevé equation. The coefficient $A_0(\omega_1^0)$ and $A_0(\omega_2^0)$ are the *Stokes cofficients*.

The resurgence coefficients are also called *higher order Stokes cofficients* in exponential asymptotics.

As a rule and apart from some integrable equations, the resurgence coefficients are seldom known by closed formulas but can be calculated numerically : see for instance [DH002, Del005, Del010] and specifically [Old005] for hyperasymptotic methods, see also [ASV012]. For the first Painlevé equation and its Stokes cofficients, an explicit expression has been obtained by Kapaev [Kap88, KK93, Kap004] using isomonodromic methods, see also [Tak000, KT005] for an exact WKB offspring. This result has also been founded by Costin *et al.* [CHT014] by means of resurgent analysis and we give this expression.

Proposition 8.2. *In theorem 8.1, the Stokes coefficients are* $A_0(\omega_1^0) = -\mathrm{i}\sqrt{\dfrac{6}{5\pi}}$ *and* $A_0(\omega_2^0) = -\sqrt{\dfrac{6}{5\pi}}$.

The Stokes coefficients are also known for the second Painlevé equation, see [FIKN006] and references therein. It is likely that the method of Costin *et al.* [CHT014] can be used to get the other resurgence cofficients for the first Painlevé equation.

We saw with corollary 5.2 that the formal integral can be interpreted as a formal transformation $\widetilde{w} = \widetilde{\Phi}(z,\mathbf{u})$, $\widetilde{\Phi}(z,\mathbf{u}) = \sum_{\mathbf{k}\in\mathbb{N}^2}\mathbf{u}^{\mathbf{k}}\widetilde{w}^{[0]}_{\mathbf{k}}(z) \in \mathbb{C}[[z^{-1},\mathbf{u}]]$ that formally conjugates the prepared equation (3.6) to its normal form (5.66). We mentioned (without proof) in Sect. 6.3 that this formal transformation gives rise to analytic transformations through Borel-Laplace summation. In other words, equation (3.6) and the normal form (5.66) are *analytically conjugated.*

It can be shown (see for instance the arguments given in [CNP93-1]) that for any two differential equations that are formally conjugated to (5.66), then these differential equations are analytically conjugated if and only if their resurgence coefficients are the same. Therefore in this way, the resurgence coefficients are also called the *holomorphic invariants* of Ecalle. See [Eca85] for further details.

8.2 Stokes Phenomenon

Knowing the Stokes coefficients $A_0(\omega)$ provides a complete description for the lower order Stokes phenomenon. In what follows, we use the notation of theorem 8.1 and we denote $\theta_i^j = \arg(\omega_i^j)$, $i = 1,2$, $j \in \mathbb{Z}$. We simply refer to the first volume of this book [MS016] for the notion of "symbolic Stokes automorphism" $\Delta^+_{\theta_i^j}$ and of "symbolic Stokes infinitesimal generator" $\Delta_{\theta_i^j}$, for a given direction θ_i^j (see also [DH99, DH002, Del005, Del010]). We only recall their expressions and relationships, in our frame:

$$\begin{aligned}
&\Delta^+_{\theta_i^j}, \Delta_{\theta_i^j} : \bigoplus_{k\in\mathbb{N}} \mathrm{e}^{-k\lambda_i z} \widetilde{\mathscr{R}}_{\mathbb{Z}}^{\mathrm{ram}} \to \bigoplus_{k\in\mathbb{N}} \mathrm{e}^{-k\lambda_i z} \widetilde{\mathscr{R}}_{\mathbb{Z}}^{\mathrm{ram}},\\
&\Delta^+_{\theta_i^j} = \mathrm{Id} + \sum_{k_0\in\mathbb{N}^\star} \dot{\Delta}^+_{k_0\omega_i^j}, \quad \Delta_{\theta_i^j} = \sum_{k_0\in\mathbb{N}^\star} \dot{\Delta}_{k_0\omega_i^j}\\
&\Delta^+_{\theta_i^j} = \exp\left(\Delta_{\theta_i^j}\right) = \mathrm{Id} + \sum_{\ell\in\mathbb{N}^\star} \frac{\mathrm{e}^{-k_0\lambda_i z}}{\ell!} \sum_{\substack{k_1+\cdots+k_\ell=k_0\\ k_i\geq 1}} \Delta_{k_\ell\omega_i^j} \circ\cdots\circ \Delta_{k_1\omega_i^j}.
\end{aligned} \tag{8.8}$$

Let us consider the formal series $\widetilde{w}_{\mathbf{0}}$. From theorem 8.1, one sees that

$$\Delta^+_{\theta_i^j} \widetilde{w}_{\mathbf{0}} = \widetilde{w}_{\mathbf{0}} + \sum_{k\in\mathbb{N}^\star} A_0(\omega_i^j)^k \mathrm{e}^{-k\lambda_i z} \widetilde{W}_{k\mathbf{e}_i} \tag{8.9}$$

where, on the right-hand side, one recognizes the transseries solutions. The action of the symbolic Stokes automorphism allows to compare left and right Borel-Laplace summation: in their intersection domain of convergence,

$$\mathscr{S}^{\theta_i^j-} \widetilde{w}_{\mathbf{0}} = \mathscr{S}^{\theta_i^j+} \widetilde{w}_{\mathbf{0}} + \sum_{k\in\mathbb{N}^\star} A_0(\omega_i^j)^k \mathrm{e}^{-k\lambda_i z} \mathscr{S}^{\theta_i^j+} \widetilde{W}_{k\mathbf{e}_i}. \tag{8.10}$$

This allows in particular to analytically continue the sum $\mathscr{S}^{\theta_i^j-} \widetilde{w}_{\mathbf{0}}$, thus the tritruncated solutions, onto a wider domain.

The same calculation can be made for the (convenient) transseries as well, and one easily gets, for $i = 1,2$:

$$\mathscr{S}^{\theta_i^{j-}}\left(\widetilde{w}_{\mathbf{0}} + \sum_{k\in\mathbb{N}^\star} U_i^k \mathrm{e}^{-k\lambda_i z}\widetilde{W}_{k\mathbf{e}_i}\right) = \mathscr{S}^{\theta_i^{j+}}\left(\widetilde{w}_{\mathbf{0}} + \sum_{k\in\mathbb{N}^\star} \left(U_i + A_0(\omega_i^j)\right)^k \mathrm{e}^{-k\lambda_i z}\widetilde{W}_{k\mathbf{e}_i}\right). \tag{8.11}$$

Once again, this provides analytic continuations of the truncated solutions onto a wider domain.

It is a good place to mention medianization, since the $k\mathbf{e}_i$-th series $\widetilde{w}_{k\mathbf{e}_i}$ are all *real* formal series. For instance, since $\widetilde{w}_{\mathbf{0}}$ belongs to $\mathbb{R}[[z^{-1}]]$, its left and right Borel-Laplace sum are complex conjugate: $\overline{\mathscr{S}^{\theta_1^{0+}}\,\widetilde{w}_{\mathbf{0}}(z)} = \mathscr{S}^{\theta_1^{0-}}\,\widetilde{w}_{\mathbf{0}}(\overline{z})$. However, neither $\mathscr{S}^{\theta_1^{0+}}\,\widetilde{w}_{\mathbf{0}}$ nor $\mathscr{S}^{\theta_1^{0-}}\,\widetilde{w}_{\mathbf{0}}$ are real analytic functions, because of the Stokes phenomenon. The question is therefore the following one : can we construct from $\widetilde{w}_{\mathbf{0}}$ a real analytic function by a suitable morphism of differential algebras ?

The naive idea of taking their mean does not work (why ?).

The answer is "yes", by medianization or good averages, and is not unique. We refer to [Men99, Eve004] for this question and its subtleties, see also [Cos009].

Remark 8.1. The fact that the Stokes coefficient $A_0(\omega_1^0)$ is nonzero can be deduced from the identity (8.10) : if $A_0(\omega_1^0) = 0$, then necessary the associated trituncated solution would be holomorphic on $\mathbb{C}\setminus K$ where K is a compact domain. This would mean that this trituncated solution has only a finite number of poles and that contradicts theorem 2.2. The fact that $A_0(\omega_1^0)$ is pure imaginary can be seen also from (8.10) and from the realness of $\widetilde{w}_{\mathbf{0}}$. For $\arg(z) = 0$ and $|z|$ large enough, one can write

$$\overline{\mathscr{S}^{\theta_1^{0+}}\,\widetilde{w}_{\mathbf{0}}(z)} = \mathscr{S}^{\theta_1^{0+}}\,\widetilde{w}_{\mathbf{0}}(z) + \sum_{k\in\mathbb{N}^\star} A_0(\omega_1^0)^k \mathrm{e}^{-k\lambda_i z}\mathscr{S}^{\theta_1^{0+}}\,\widetilde{W}_{k\mathbf{e}_i}(z), \tag{8.12}$$

and $\overline{A_0(\omega_1^0)} = -A_0(\omega_1^0)$ comes as an upshot.

As already said, the resurgent coefficients can be numerically calculated by hyperasymptotic methods. In return, resurgence coefficients and higher order Stokes phenomena play a crucial role in the hyperasymptotic approximations to Borel-Laplace sums, see for instance [DH002, Old005] and references therein.

8.3 The Alien Derivatives for the Seen Singularities

The idea that leads to theorem 8.1 relies on the following observations. We know by proposition 8.1 that the singularity $\overset{\triangledown}{W}_{\mathbf{k}}$ belongs to $\mathrm{RES}^{(1)}$, for any $\mathbf{k}\in\mathbb{N}^2$, and we can apply corollary 7.2 : for any $\omega\in\widetilde{\mathbb{C}}$ so that $\overset{\bullet}{\omega} = \pm 1$ (the so-called seen singularities), the alien derivative $\Delta_\omega \overset{\triangledown}{W}_{\mathbf{k}}$ is well-defined. If these alien derivatives belong to $\mathrm{RES}^{(1)}$, then we see with corollary 7.3 that the singularities $\overset{\triangledown}{W}_{\mathbf{k}}$ belongs to $\mathrm{RES}^{(2)}$. A reasoning by induction allows to conclude.

In this section, we explain how to calculate these alien derivatives with various methods and we direct our efforts towards $\widetilde{w}_{\mathbf{0}}$.

8.3.1 Preparations

The formal series $\widetilde{w}_{\mathbf{0}}$ being solution of the equation (8.3), we introduce by proposition 7.4 the singularities $\overset{\triangledown}{w}_{\mathbf{0}}= {}^{\flat}\widehat{w}_{\mathbf{0}}$, $\overset{\triangledown}{f}_0= {}^{\flat}\widehat{f}_0$, $\overset{\triangledown}{f}_1= {}^{\flat}\widehat{f}_1$ and $\overset{\triangledown}{f}_2= {}^{\flat}\widehat{f}_2$. Notice that $\overset{\triangledown}{f}_0$, $\overset{\triangledown}{f}_1$ and $\overset{\triangledown}{f}_2$ obviously belong to $\overset{\triangledown}{\mathrm{CONS}}$.
Equation (8.3) translates into the fact that $\overset{\triangledown}{w}_{\mathbf{0}}$ satisfies the following convolution equation:

$$
\begin{aligned}
P(\overset{\triangledown}{\partial})\, \overset{\triangledown}{w}_{\mathbf{0}} + \overset{\triangledown}{I}_1 * \big[Q(\overset{\triangledown}{\partial})\, \overset{\triangledown}{w}_{\mathbf{0}}\,\big] &= \overset{\triangledown}{F}(\zeta, \overset{\triangledown}{w}_{\mathbf{0}}) \\
&= \overset{\triangledown}{f}_0 + \overset{\triangledown}{f}_1 * \overset{\triangledown}{w}_{\mathbf{0}} + \overset{\triangledown}{f}_2 * \overset{\triangledown}{w}_{\mathbf{0}}^{*2}\,.
\end{aligned}
\tag{8.13}
$$

One can rather introduce the asymptotic class $\overset{\vartriangle}{w}_{\mathbf{0}}= {}^{\natural}\widetilde{w}_{\mathbf{0}} \in \mathrm{ASYMP}^{\mathrm{simp}}$ (cf. definition 7.27) and equation (8.3) becomes:

$$
\begin{aligned}
P(\partial)\, \overset{\vartriangle}{w}_{\mathbf{0}} + \frac{1}{z} Q(\partial)\, \overset{\vartriangle}{w}_{\mathbf{0}} &= F(z, \overset{\vartriangle}{w}_{\mathbf{0}}) \\
&= f_0 + f_1\, \overset{\vartriangle}{w}_{\mathbf{0}} + f_2\, \overset{\vartriangle}{w}_{\mathbf{0}}^{2}
\end{aligned}
\tag{8.14}
$$

As already said, we know that $\overset{\triangledown}{w}_{\mathbf{0}}$ belongs to $\mathrm{RES}^{(1)}$, *resp.* $\overset{\vartriangle}{w}_{\mathbf{0}}$ belongs to $\widetilde{\mathrm{RES}}^{(1)}$, and corollary 7.2 can be applied : with the notation of theorem 8.1, $\overset{\triangledown}{W}= \Delta_{\omega_1^0}\, \overset{\triangledown}{w}_{\mathbf{0}}$ is a well-defined singularity of $\mathrm{SING}_{0,\pi}$, *resp.* $\overset{\vartriangle}{W}= \Delta_{\omega_1^0}\, \overset{\vartriangle}{w}_{\mathbf{0}}$ is a well-defined asymptotic class of $\mathrm{ASYMP}_{0,\pi}$.

The singularities $\overset{\triangledown}{f}_0$, $\overset{\triangledown}{f}_1$, $\overset{\triangledown}{f}_2$ and $\overset{\triangledown}{I}_1$ are all constant of resurgence. Therefore, they vanish under the action of any alien derivation. Adding to this remark the fact that $\Delta_{\omega_1^0}$ satisfies the Leibniz rule and the commutation rule $[\Delta_{\omega_1^0}, \overset{\triangledown}{\partial}] = -\Delta_{\omega_1^0}$ (corollary 7.2 and remember that $\overset{\bullet}{\omega}{}_1^0 = 1$), one deduces from (8.13) that $\overset{\triangledown}{W}$ solves in $\mathrm{SING}_{0,\pi}$ the following associated linear convolution equation:

$$
\begin{aligned}
P(\overset{\triangledown}{\partial} - 1)\, \overset{\triangledown}{W} + \overset{\triangledown}{I}_1 * \big[Q(\overset{\triangledown}{\partial} - 1)\, \overset{\triangledown}{W}\,\big] &= \frac{\partial \overset{\triangledown}{F}(\zeta, \overset{\vartriangle}{w}_{\mathbf{0}})}{\partial w} * \overset{\triangledown}{W} \\
&= \left(\overset{\triangledown}{f}_1 + 2\, \overset{\triangledown}{f}_2 * \overset{\triangledown}{w}_{\mathbf{0}}\right) * \overset{\triangledown}{W}\,.
\end{aligned}
\tag{8.15}
$$

For the same reasons, the asymptotic class $\overset{\triangle}{W}$ is solution in $\mathrm{ASYMP}_{0,\pi}$ of a linear ODE:

$$P(\partial-1)\overset{\triangle}{W}+\frac{1}{z}Q(\partial-1)\overset{\triangle}{W}=\frac{\partial F(z,\overset{\triangle}{w}_{\mathbf{0}})}{\partial w}\overset{\triangle}{W}\,. \tag{8.16}$$

Of course, (8.16) can be deduced also from (8.15) by formal Laplace transform (definition 7.25 and proposition 7.10).

The differential equation (8.16) is nothing but the equation

$$\mathfrak{P}_{\mathbf{e}_1}(\overset{\triangle}{w}_{\mathbf{0}})\overset{\triangle}{W}=0 \tag{8.17}$$

where $\mathfrak{P}_{\mathbf{e}_1}$ is the linear operator recalled in (8.4). We know by lemma 5.4 that the differential equation $\mathfrak{P}_{\mathbf{e}_1}(\widetilde{w}_{\mathbf{0}})\widetilde{\mathscr{W}}=0$, that is (8.17) through the Taylor map, has its general formal solution which belongs to the direct sum $\widetilde{\mathrm{Nils}}_1\oplus e^{2z}\widetilde{\mathrm{Nils}}_1$, under the form

$$\begin{aligned}\widetilde{\mathscr{W}}(z)&=C_1z^{\frac{3}{2}}\widetilde{w}_{\mathbf{e}_1}(z)+C_2\mathrm{e}^{2z}z^{\frac{3}{2}}\widetilde{w}_{\mathbf{e}_2}(z)\\&=C_1\widetilde{W}_{\mathbf{e}_1}(z)+C_2\mathrm{e}^{2z}\widetilde{W}_{\mathbf{e}_2}(z),\end{aligned} \tag{8.18}$$

where $\widetilde{W}_{\mathbf{e}_1}$ and $\widetilde{W}_{\mathbf{e}_2}$ belong to the space $\widetilde{\mathrm{Nils}}_1$ of 1-Gevrey Nilsson series.
One should precise what we mean by "general formal solution". The linear operator $\mathfrak{P}_{\mathbf{e}_1}$ is of order 2 in z and the particular solutions $\widetilde{W}_{\mathbf{e}_1}$ and $\mathrm{e}^{2z}\widetilde{W}_{\mathbf{e}_2}$ are two independent formal solutions : their wronskian is $\begin{vmatrix}\widetilde{W}_{\mathbf{e}_1} & \mathrm{e}^{2z}\widetilde{W}_{\mathbf{e}_2}\\ \partial\widetilde{W}_{\mathbf{e}_1} & \partial(\mathrm{e}^{2z}\widetilde{W}_{\mathbf{e}_2})\end{vmatrix}=2z^3\mathrm{e}^{2z}$. Thus, if $\widetilde{\mathscr{W}}$ belongs to a differential algebra containing $\widetilde{\mathrm{Nils}}_1\oplus e^{2z}\widetilde{\mathrm{Nils}}_1$ as sub-vector space, for instance the direct sum $\prod\limits_{k\in\mathbb{Z}}e^{-kz}\widetilde{\mathrm{Nils}}_1$ and if $\mathfrak{P}_{\mathbf{e}_1}(\widetilde{w}_{\mathbf{0}})\widetilde{W}=0$, then $\widetilde{W}$ is of the form (8.18) with $C_1,C_2\in\mathbb{C}$ given by the Kramer's formulas: $C_2=-\frac{z^{-3}\mathrm{e}^{-2z}}{2}\begin{vmatrix}\widetilde{W} & \widetilde{W}_{\mathbf{e}_1}\\ \partial\widetilde{W} & \partial\widetilde{W}_{\mathbf{e}_1}\end{vmatrix}$,
$C_1=\frac{z^{-3}\mathrm{e}^{-2z}}{2}\begin{vmatrix}\widetilde{W} & \mathrm{e}^{2z}\widetilde{W}_{\mathbf{e}_2}\\ \partial\widetilde{W} & \partial(\mathrm{e}^{2z}\widetilde{W}_{\mathbf{e}_2})\end{vmatrix}$.
We claim that the general solution of equation (8.17) in $\prod_{k\in\mathbb{Z}}e^{-kz}\mathrm{ASYMP}_{0,\pi}$ is a linear combination of $\overset{\triangle}{W}_{\mathbf{e}_1}\in\mathrm{ASYMP}^{\mathrm{ram}}$ and $\mathrm{e}^{2z}\,\overset{\triangle}{W}_{\mathbf{e}_2}\in\mathrm{e}^{2z}\mathrm{ASYMP}^{\mathrm{ram}}$ with $\overset{\triangle}{W}_{\mathbf{e}_i}={\natural}^{\mathrm{ram}}\widetilde{W}_{\mathbf{e}_i}$. Consequently:

Lemma 8.1. *There exists $A_0(\omega_1^0)\in\mathbb{C}$ such that the singularity $\Delta_{\omega_1^0}\overset{\triangledown}{w}_{\mathbf{0}}\in\mathrm{SING}_{0,\pi}$ is of the form*

$$\Delta_{\omega_1^0}\overset{\triangledown}{w}_{\mathbf{0}}=A_0(\omega_1^0)\,\overset{\triangledown}{I}_{-\frac{3}{2}}*\overset{\triangledown}{w}_{\mathbf{e}_1}=A_0(\omega_1^0)\,\overset{\triangledown}{W}_{\mathbf{e}_1},$$

thus can be extended uniquely to an element of SING. *In other equivalent words,* $\Delta_{\omega_1^0}\overset{\triangle}{w}_{\mathbf{0}}=A_0(\omega_1^0)\,\overset{\triangle}{W}_{\mathbf{e}_1}\in\mathrm{ASYMP}^{\mathrm{ram}}$, $\Delta_{\omega_1^0}\widetilde{w}_{\mathbf{0}}=A_0(\omega_1^0)\widetilde{W}_{\mathbf{e}_1}\in\widetilde{\mathrm{Nils}}_1$.

As promised, we show lemma 8.1 by two different approaches in the sequel.

8.3.2 Alien Derivations, First Approach

We follow here ideas developed in [GS001, OSS003, Sau006, Ras010, LR011] and we start with results coming from general nonsense in 1-Gevrey theory (the proof is saved for an exercise).

Lemma 8.2. *Let $\widetilde{w} \in z^{-1}\mathbb{C}[[z^{-1}]]_1$ be a 1-Gevrey series with vanishing constant term, and $\widehat{w} \in \mathscr{O}_0$ its minor. The following properties are satisfied.*

1. *The formal series $(1+\widetilde{w}) \in \mathbb{C}[[z^{-1}]]_1$ is invertible. Its inverse $(1+\widetilde{w})^{-1}$ is 1-Gevrey and has a formal Borel transform $\mathscr{B}(1+\widetilde{w})^{-1} \in \mathbb{C}\delta \oplus \mathscr{O}_0$ of the form $(\delta+\widehat{w})^{*-1} = \delta + \sum_{n\geq 1}(-1)^n \widehat{w}^{*n}$.*
2. *The formal series $\log(1+\widetilde{w}) = \sum_{n\geq 1} \frac{(-1)^{n+1}}{n}\widetilde{w}^n$ is a 1-Gevrey with vanishing constant term, whose minor is given by $\log_*(\delta+\widehat{w}) := \sum_{n\geq 1} \frac{(-1)^{n+1}}{n}\widehat{w}^{*n}$.*
3. *The formal series $\widetilde{w}$ is exponentiable in the sense that its exponential $\mathrm{e}^{\widetilde{w}} = \sum_{n\geq 1} \frac{1}{n!}\widetilde{w}^n$ is a 1-Gevrey series, whose minor is of the form $\exp_*(\widehat{w}) := \delta + \sum_{n\geq 1} \frac{1}{n!}\widehat{w}^{*n}$.*
4. *Moreover, $\log \circ \exp \widetilde{w} = \widetilde{w}$ and $\exp \circ \log(1+\widetilde{w}) = 1+\widetilde{w}$.*

Remark 8.2. More general results along that line in resurgence theory can be obtained, see [CNP93-1, MS016] and specially [Sau015].

We are now ready to calculate the alien derivative $\overset{\triangledown}{W} = \Delta_{\omega_1^0} \overset{\triangledown}{w}_{\mathbf{0}} \in \mathrm{SING}_{0,\pi}$. We consider the 1-Gevrey Nilsson series $\widetilde{W}_{\mathbf{e}_1} = z^{3/2}\widetilde{w}_{\mathbf{e}_1} \in \widetilde{\mathrm{Nils}_1}$ solution of (8.16) (more precisely its transform through the Taylor map), and its associated singularity $\overset{\triangledown}{W}_{\mathbf{e}_1} = \overset{\triangledown}{I}_{-\frac{3}{2}} * \overset{\triangledown}{w}_{\mathbf{e}_1} \in \mathrm{SING}$, where $\overset{\triangledown}{w}_{\mathbf{e}_1} = \delta + {}^{\flat}\widehat{w}_{\mathbf{e}_1}$. (Remember that $\widetilde{w}_{\mathbf{e}_1}$ has 1 for its constant term). Since $\widetilde{w}_{\mathbf{e}_1}$ is invertible in $\mathbb{C}[[z^{-1}]]$, so does $\overset{\triangledown}{w}_{\mathbf{e}_1}$ in SING, its inverse being given by $\overset{\triangledown}{w}_{\mathbf{e}_1}^{*-1} = \delta + {}^{\flat}\left(\sum_{n\geq 1}(-1)^n \widehat{w}_{\mathbf{e}_1}^{*n}\right)$. Accordingly, $\overset{\triangledown}{W}_{\mathbf{e}_1}$ is invertible in SING and $\overset{\triangledown}{W}_{\mathbf{e}_1}^{*-1} = \overset{\triangledown}{I}_{\frac{3}{2}} * \overset{\triangledown}{w}_{\mathbf{e}_1}^{*-1}$. We now introduce the singularity $\overset{\triangledown}{S} \in \mathrm{SING}_{0,\pi}$ defined by

$$\overset{\triangledown}{W} = \overset{\triangledown}{S} * \overset{\triangledown}{W}_{\mathbf{e}_1} \tag{8.19}$$

and we want to show that $\overset{\triangledown}{S} = A_0(\omega_1^0)\delta$ for some $A_0(\omega_1^0) \in \mathbb{C}$. Plugging (8.19) into (8.15), using the property that $\overset{\triangledown}{\partial}$ is a derivation in $\mathrm{SING}_{0,\pi}$ (cf. proposition 7.6) and that $\overset{\triangledown}{W}_{\mathbf{e}_1}$ solves (8.15), one easily gets for $\overset{\triangledown}{S}$ the following equation:

$$((\overset{\triangledown}{\partial}^2 - \overset{\triangledown}{\partial})\overset{\triangledown}{S}) * \overset{\triangledown}{W}_{\mathbf{e}_1} + 2(\overset{\triangledown}{\partial}\overset{\triangledown}{S}) * (\partial \overset{\triangledown}{W}_{\mathbf{e}_1}) - 3\overset{\triangledown}{I}_1 * (\overset{\triangledown}{\partial}\overset{\triangledown}{S}) * \overset{\triangledown}{W}_{\mathbf{e}_1} = 0. \tag{8.20}$$

Since $\overset{\triangledown}{\partial}\overset{\triangledown}{W}_{\mathbf{e}_1} = \frac{3}{2}\overset{\triangledown}{I}_{-\frac{1}{2}} * \overset{\triangledown}{w}_{\mathbf{e}_1} + \overset{\triangledown}{I}_{-\frac{3}{2}} * (\overset{\triangledown}{\partial}\overset{\triangledown}{w}_{\mathbf{e}_1})$, equation (8.20) reduces to the equation

$$\overset{\triangledown}{\partial}^2 \overset{\triangledown}{S} = \left[\delta - 2\overset{\triangledown}{\chi}\right] * \overset{\triangledown}{\partial}\overset{\triangledown}{S}, \qquad \overset{\triangledown}{\chi} = \overset{\triangledown}{w}_{\mathbf{e}_1}^{*-1} * (\overset{\triangledown}{\partial}\overset{\triangledown}{w}_{\mathbf{e}_1}), \tag{8.21}$$

where $\overset{\triangledown}{\chi} = {}^{\flat}\widehat{\chi}$ is the singularity associated with the minor $\widehat{\chi}(\zeta)$ of

$$\widetilde{\chi}(z) = \frac{\partial \widetilde{w}_{\mathbf{e}_1}}{\widetilde{w}_{\mathbf{e}_1}} \in z^{-2}\mathbb{C}[[z^{-1}]]_1.$$

The formal series $\widetilde{\chi}$ has a unique primitive $\widetilde{\chi}_0(z) = \partial^{-1}\widetilde{\chi}(z) = \log\left(\widetilde{w}_{\mathbf{e}_1}(z)\right)$ in the maximal ideal $z^{-1}\mathbb{C}[[z^{-1}]]_1$ of $\mathbb{C}[[z^{-1}]]_1$ and, thus, $\widehat{\chi}_0$ as well as its associated singularity $\overset{\triangledown}{\chi}_0$ is exponentiable in SING. (Lemma 8.2)

More simply, $\exp_*(\overset{\triangledown}{\chi}_0) = \delta + {}^{\flat}\widehat{w}_{\mathbf{e}_1}$, thus $\exp_*(2\overset{\triangledown}{\chi}_0) = \delta + {}^{\flat}(2\widehat{w}_{\mathbf{e}_1} + \widehat{w}_{\mathbf{e}_1}^{*2})$.

We introduce $\overset{\triangledown}{S}_0 \in \mathrm{SING}_{0,\pi}$ given by the identity:

$$\overset{\triangledown\triangledown}{\partial S} = \overset{\triangledown}{S}_0 * \exp_*(-2\overset{\triangledown}{\chi}_0). \tag{8.22}$$

By construction, $\overset{\triangledown}{\partial} \exp_*(-2\overset{\triangledown}{\chi}_0) = -2\overset{\triangledown}{\chi} * \exp_*(-2\overset{\triangledown}{\chi}_0)$. One deduces from (8.21) that $\overset{\triangledown}{S}_0$ solves the convolution equation $\partial \overset{\triangledown}{S}_0 - \overset{\triangledown}{S}_0 = 0$. This translates into the fact that $(\zeta + 1)\overset{\vee}{S}_0$ is holomorphic near $\zeta = 0$, where $\overset{\vee}{S}_0$ stands for any major of $\overset{\triangledown}{S}_0$. Therefore $\overset{\vee}{S}_0$ is holomorphic as well near $\zeta = 0$, thus $\overset{\triangledown}{S}_0 = 0$. From (8.22), this means that $\overset{\triangledown\triangledown}{\partial S} = 0$, that is $\zeta\overset{\vee}{S}(\zeta)$ is holomorphic near $\zeta = 0$ for any major $\overset{\vee}{S}$ of $\overset{\triangledown}{S}$. This allows to conclude that there exists a constant $A_0(\omega_1^0) \in \mathbb{C}$ such that $\overset{\triangledown}{S} = A_0(\omega_1^0)\delta$. Thus, $\Delta_{\omega_1^0}\overset{\triangledown}{w}_{\mathbf{0}} = A_0(\omega_1^0)\overset{\triangledown}{W}_{\mathbf{e}_1}$ which implies that $\Delta_{\omega_1^0}\overset{\triangledown}{w}_{\mathbf{0}}$ can be continued to an element of SING. This ends the proof of lemma 8.1 with the first approach.

8.3.3 Alien Derivations, Second Approach

The second approach We now propose another approach, based on the notion of asymptotic classes, that uses tools akin to Gevrey and 1-summability theories, see also [DDP93, DDP97, DH99, DH002, Del005, Del008].

We know that $\overset{\vartriangle}{W} = \Delta_{\omega_1^0}\overset{\vartriangle}{w}_{\mathbf{0}} \in \mathrm{ASYMP}_{0,\pi}$ satisfies the condition $\mathfrak{P}_{\mathbf{e}_1}(\overset{\vartriangle}{w}_{\mathbf{0}})\overset{\vartriangle}{W} = 0$. We look at the equation $\mathfrak{P}_{\mathbf{e}_1}(\widetilde{w}_{\mathbf{0}})\widetilde{W} = 0$. The operator $\mathfrak{P}_{\mathbf{e}_1}(\widetilde{w}_{\mathbf{0}})$ is of order two in z and has two linearly independent formal solutions $\widetilde{W}_{\mathbf{e}_1} = z^{\frac{3}{2}}\widetilde{w}_{\mathbf{e}_1} \in \widetilde{\mathrm{Nils}}_1$ and $\mathrm{e}^{2z}\widetilde{W}_{\mathbf{e}_2} = \mathrm{e}^{2z}z^{\frac{3}{2}}\widetilde{w}_{\mathbf{e}_2} \in \mathrm{e}^{2z}\widetilde{\mathrm{Nils}}_1$.

Let us represent the asymptotic classes $\overset{\vartriangle}{w}_{\mathbf{0}} = {}^{\natural}\widetilde{w}_{\mathbf{0}}$, $\overset{\vartriangle}{w}_{\mathbf{e}_1} = {}^{\natural}\widetilde{w}_{\mathbf{e}_1}$ and $\overset{\vartriangle}{w}_{\mathbf{e}_2} = {}^{\natural}\widetilde{w}_{\mathbf{e}_2}$ on restriction to $\mathrm{ASYMP}_{0,\pi}$. We pick a (good) open covering (I_i) of $J^\star =]-3\pi/2, 3\pi/2[$ with open arcs I_i of aperture less than π. We use the Borel-Ritt theorem for 1-Gevrey asymptotics to get, for each subscript i: $w_{\mathbf{0},i}$, $w_{\mathbf{e}_1,i}$, $w_{\mathbf{e}_2,i}, \in \overline{\mathscr{A}}_1(I_i)$ whose 1-Gevrey asymptotics is given by $\widetilde{w}_{\mathbf{0}}, \widetilde{w}_{\mathbf{e}_1}, \widetilde{w}_{\mathbf{e}_2}$ respectively.

We know that each of these 1-Gevrey germ is uniquely defined up to 1-exponentially flat germs, that is up to elements of $\overline{\mathscr{A}}^{\leq -1}(I_i)$. As a consequence, the collections $(w_{\mathbf{0},i})$, $(w_{\mathbf{e}_1,i})$, $(w_{\mathbf{e}_2,i})$ represent the asymptotic classes we have in mind.
For each subscript i, observe that

$$T_1(I_i)\Big(\mathfrak{D}_{\mathbf{e}_1}(w_{\mathbf{0},i})w_{\mathbf{e}_1,i}\Big) = \mathfrak{D}_{\mathbf{e}_1}(\widetilde{w}_{\mathbf{0}})\widetilde{w}_{\mathbf{e}_1} = 0$$

with $\mathfrak{D}_{\mathbf{e}_1}$ the linear operator given by definition 5.5, because the 1-Gevrey Taylor map $T_1(I_i)$ is a morphism of differential algebras. This ensures that $\mathfrak{D}_{\mathbf{e}_1}(w_{\mathbf{0},i})w_{\mathbf{e}_1,i}$ belongs to $\overline{\mathscr{A}}^{\leq -1}(I_i)$.

We draw a first conclusion : $\mathfrak{D}_{\mathbf{e}_1}(\overset{\triangle}{w}_{\mathbf{0}})\,\overset{\triangle}{w}_{\mathbf{e}_1} = 0$ in $\mathrm{ASYMP}_{0,\pi}$ and thus, $\mathfrak{P}_{\mathbf{e}_1}(\overset{\triangle}{w}_{\mathbf{0}})\,\overset{\triangle}{W}_{\mathbf{e}_1} = 0$ as well with $\overset{\triangle}{W}_{\mathbf{e}_1} = z^{3/2}\,\overset{\triangle}{w}_{\mathbf{e}_1} \in \mathrm{ASYMP}_{0,\pi}$.

We add a property that ensues from an analogue of the M.A.E.T. (theorem 3.1) and for which we refer to [Lod016, Mal91]: one can even find $h_{i,\mathbf{e}_1} \in \overline{\mathscr{A}}^{\leq -1}(I_i)$ so that $\mathfrak{D}_{\mathbf{e}_1}(w_{\mathbf{0},i})(w_{\mathbf{e}_1,i} - h_{\mathbf{e}_1,i})$ vanishes exactly, for each subscript i. Thus, one can find a representative $w_{\mathbf{e}_1,i} \in \overline{\mathscr{A}}_1(I_i)$ of $\overset{\triangle}{w}_{\mathbf{e}_1}$ so that $\mathfrak{D}_{\mathbf{e}_1}(w_{\mathbf{0},i})w_{\mathbf{e}_1,i} = 0$ and thus, $\mathfrak{P}_{\mathbf{e}_1}(w_{\mathbf{0},i})W_{\mathbf{e}_1,i} = 0$ as well with $W_{\mathbf{e}_1,i} = z^{\frac{3}{2}}w_{\mathbf{e}_1,i}$.

The same reasoning yields: one can find a representative $w_{\mathbf{e}_2,i} \in \overline{\mathscr{A}}_1(I_i)$ of $\overset{\triangle}{w}_{\mathbf{e}_2}$ so that $\mathfrak{D}_{\mathbf{e}_2}(w_{i,\mathbf{0}})w_{\mathbf{e}_2,i} = 0$, thus $\mathfrak{P}_{\mathbf{e}_1}(w_{\mathbf{0},i})\mathrm{e}^{2z}w_{\mathbf{e}_2,i} = 0$ with $W_{\mathbf{e}_2,i} = z^{\frac{3}{2}}w_{\mathbf{e}_2,i}$.

Therefore $\mathfrak{D}_{\mathbf{e}_2}(\overset{\triangle}{w}_{\mathbf{0}})\,\overset{\triangle}{w}_{\mathbf{e}_2} = 0$ in $\mathrm{ASYMP}_{0,\pi}$ and thus $\mathfrak{P}_{\mathbf{e}_1}(\overset{\triangle}{w}_{\mathbf{0}})\mathrm{e}^{2z}\,\overset{\triangle}{W}_{\mathbf{e}_2} = 0$. The key point if that $\mathrm{e}^{2z}\,\overset{\triangle}{W}_{\mathbf{e}_2}$ belongs to $\mathrm{e}^{2z}\mathrm{ASYMP}_{0,\pi}$ which is a vector space in direct sum with $\mathrm{ASYMP}_{0,\pi}$.

Putting things together, keeping the same notation, we see that the kernel of the linear differential operator $\mathfrak{P}_{\mathbf{e}_1}(w_{i,\mathbf{0}})$ in the space of sectorial germs of direction I_i is spanned by $W_{\mathbf{e}_1,i}$ and $\mathrm{e}^{2z}W_{\mathbf{e}_2,i}$.

We now go back to the asymptotic class $\overset{\triangle}{W} \in \mathrm{ASYMP}_{0,\pi}$ that satisfies $\mathfrak{P}_{\mathbf{e}_1}(\overset{\triangle}{w}_{\mathbf{0}})\,\overset{\triangle}{W} = 0$. Considering a refinement of (I_i) if necessary, one can find for each subscript i a representative $W_i \in \overline{\mathscr{A}}^{\leq 0}(I_i)$ of $\overset{\triangle}{W}$ and a 1-exponentially flat germ $b_i \in \overline{\mathscr{A}}^{\leq -1}(I_i)$ such that $\mathfrak{P}_{\mathbf{e}_1}(w_{\mathbf{0},i})W_i = b_i$. To get W_i, we apply the usual variation of constants method. One gets W_i under the form

$$\begin{aligned} W_i &= B_i(z) + C_1 W_{\mathbf{e}_1,i} + C_2\mathrm{e}^{2z}W_{\mathbf{e}_2,i}, \quad C_1, C_2 \in \mathbb{C}, \qquad (8.23)\\ 2B_i(z) &= W_{\mathbf{e}_2,i}\int z^{-3}W_{\mathbf{e}_1,i}.b_i - W_{\mathbf{e}_i,i}\int z^{-3}W_{\mathbf{e}_2,i}.b_i. \end{aligned}$$

It is a simple exercise to show that B_i belongs to $\overline{\mathscr{A}}^{\leq -1}(I_i)$ and one easily concludes that W_i has to be equal to $C_1 W_{\mathbf{e}_1,i}$ modulo $\overline{\mathscr{A}}^{\leq -1}(I_i)$.

Depending on the arc, the term $C_2\mathrm{e}^{2z}W_{\mathbf{e}_2,i}$ either belongs to $\overline{\mathscr{A}}^{\leq -1}(I_i)$ (so one can take $C_2 = 0$) or escapes from $W_i \in \overline{\mathscr{A}}^{\leq 0}(I_i)$ (thus one has to impose $C_2 = 0$).

This ends the second proof of lemma 8.1: the general solution of the linear equation $\mathfrak{P}_{\mathbf{e}_1}(\overset{\triangle}{w}_{\mathbf{0}})\overset{\triangle}{W}=0$ in $\mathrm{ASYMP}_{0,\pi}$ is $C_1\overset{\triangle}{W}_{\mathbf{e}_1}$ and, consequently, there exists a constant $A_0(\omega_1^0)\in\mathbb{C}$ so that $\Delta_{\omega_1^0}\overset{\triangle}{w}_{\mathbf{0}}=A_0(\omega_1^0)\overset{\triangle}{W}_{\mathbf{e}_1}$ in $\mathrm{ASYMP}_{0,\pi}$. Thus, $\Delta_{\omega_1^0}\overset{\triangle}{w}_{\mathbf{0}}$ can be uniquely continued to an element of ASYMP.

Conclusion What we have shown amounts to the following upshot. The solutions of the equation $\mathfrak{P}_{\mathbf{e}_1}(\widetilde{w}_{\mathbf{0}})\widetilde{\mathscr{W}}=0$ in the differential algebra $\prod_{k\in\mathbb{Z}}e^{-kz}\widetilde{\mathrm{Nils}}_1$ are spanned by the independent solutions $\widetilde{W}_{\mathbf{e}_1}\in\widetilde{\mathrm{Nils}}_1$ and $\mathrm{e}^{2z}\widetilde{W}_{\mathbf{e}_2}\in\mathrm{e}^{2z}\widetilde{\mathrm{Nils}}_1$. This implies that the solutions of the equation $\mathfrak{P}_{\mathbf{e}_1}(\overset{\triangle}{w}_{\mathbf{0}})\overset{\triangle}{\mathscr{W}}=0$ in the differential algebra $\prod_{k\in\mathbb{Z}}e^{-kz}\mathrm{ASYMP}$, *resp.* $\prod_{k\in\mathbb{Z}}e^{-kz}\mathrm{ASYMP}_{0,\pi}$, are spanned by the independent solutions $\overset{\triangle}{W}_{\mathbf{e}_1}\in\mathrm{ASYMP}$ and $\mathrm{e}^{2z}\overset{\triangle}{W}_{\mathbf{e}_2}\in\mathrm{e}^{2z}\mathrm{ASYMP}$, *resp.* their restrictions in $\mathrm{ASYMP}_{0,\pi}$ and $\mathrm{e}^{2z}\mathrm{ASYMP}_{0,\pi}$ respectively. This result can be generalized as follows.

Lemma 8.3. *For* $\mathbf{k}\in\mathbb{N}^2$, *we denote by* $\overset{\triangle}{W}_{\mathbf{k}}\in\mathrm{ASYMP}^{\mathrm{ram}}$ *the asympotic class defined by* $\overset{\triangle}{W}_{\mathbf{k}}=\natural^{\mathrm{ram}}\widetilde{W}_{\mathbf{k}}$ *where* $\widetilde{W}_{\mathbf{k}}\in\widetilde{\mathrm{Nils}}_1$ *satisfies (8.4). Let* $\theta\in\widetilde{\mathbb{S}}^1$ *be any direction,* $\alpha>0$ *and* $\mathbf{k}\in\mathbb{N}^2\setminus\{\mathbf{0}\}$. *If* $\overset{\triangle}{\mathscr{W}}\in\prod_{\ell\in\mathbb{Z}}e^{-\ell z}\mathrm{ASYMP}_{\theta,\alpha}$ *solves the linear differential equation*

$$\mathfrak{P}_{\mathbf{k}}\overset{\triangle}{\mathscr{W}}=\sum_{\substack{\mathbf{k}_1+\mathbf{k}_2=\mathbf{k}\\|\mathbf{k}_i|\geq 1}}\frac{\overset{\triangle}{W}_{\mathbf{k}_1}\overset{\triangle}{W}_{\mathbf{k}_2}}{2!}\frac{\partial^2F(z,\overset{\triangle}{w}_{\mathbf{0}})}{\partial w^2},\quad \mathfrak{P}_{\mathbf{k}}=\mathfrak{P}_{\mathbf{k}}(\overset{\triangle}{w}_{\mathbf{0}}),\tag{8.24}$$

then there exist uniquely determined constants $C_1,C_2\in\mathbb{C}$ *so that*

$$\overset{\triangle}{\mathscr{W}}=\overset{\triangle}{W}_{\mathbf{k}}+\mathrm{e}^{\boldsymbol{\lambda}.\mathbf{k}z}\left(C_1\mathrm{e}^{-\lambda_1 z}\overset{\triangle}{W}_{\mathbf{e}_1}+C_2\mathrm{e}^{-\lambda_2 z}\overset{\triangle}{W}_{\mathbf{e}_2}\right).\tag{8.25}$$

Proof. The general formal solution for the equation (8.4) is of the form $\widetilde{\mathscr{W}}=\widetilde{W}_{\mathbf{k}}+\mathrm{e}^{\boldsymbol{\lambda}.\mathbf{k}z}\left(C_1\mathrm{e}^{-\lambda_1 z}\widetilde{W}_{\mathbf{e}_1}+C_2\mathrm{e}^{-\lambda_2 z}\widetilde{W}_{\mathbf{e}_2}\right)$. We already know that $\mathrm{e}^{\boldsymbol{\lambda}.\mathbf{k}z}\left(C_1\mathrm{e}^{-\lambda_1 z}\overset{\triangle}{W}_{\mathbf{e}_1}+C_2\mathrm{e}^{-\lambda_2 z}\overset{\triangle}{W}_{\mathbf{e}_2}\right)$ provides the general solution for the homogeneous equation $\mathfrak{P}_{\mathbf{k}}(\overset{\triangle}{w}_{\mathbf{0}})\overset{\triangle}{\mathscr{W}}=0$ in $\prod_{\ell\in\mathbb{Z}}e^{-\ell z}\mathrm{ASYMP}_{\theta,\alpha}$. This asymptotic class $\overset{\triangle}{W}_{\mathbf{k}}$ is of the form (8.2), namely

$$\overset{\triangle}{W}_{\mathbf{k}}=\sum_{l}\frac{1}{l!}(\boldsymbol{\varkappa}.\mathbf{k})^l\log^l(z)z^{-\boldsymbol{\tau}.\mathbf{k}}\overset{\triangle}{w}{}_{\mathbf{k}}^{[0]},\quad \overset{\triangle}{w}{}_{\mathbf{k}}^{[0]}=\natural\,\widetilde{w}_{\mathbf{k}}^{[0]},$$

with $\widetilde{w}_{\mathbf{k}}^{[0]}\in\mathbb{C}[[z^{-1}]]_1$ satisfying as linear differential equation given in corollary 5.1. This allows to conclude that $\overset{\triangle}{W}_{\mathbf{k}}$ is a particular solution for the equation (8.24) and one ends the proof with the arguments of the above second approach.

8.3.4 A Step Further

What have been previously done works as well for the other alien derivatives $\Delta_{\omega_i^j} \overset{\triangledown}{w}_{\mathbf{0}} = A_0(\omega_i^j)\, \overset{\triangledown}{W}_{\mathbf{e}_i}$, *resp.* $\Delta_{\omega_i^j} \overset{\triangle}{w}_{\mathbf{0}} = A_0(\omega_i^j)\, \overset{\triangle}{W}_{\mathbf{e}_i}$, for any $i = 1, 2$ and $j \in \mathbb{Z}$. Since $\overset{\triangle}{W}_{\mathbf{e}_1}$ and $\overset{\triangle}{W}_{\mathbf{e}_2}$ belong to $\widetilde{\mathrm{RES}}^{(1)}$ (proposition 8.1), one infers from corollary 7.3 that $\widetilde{w}_{\mathbf{0}}$ belongs to $\widetilde{\mathscr{R}}^{(2)}$. In particular, the alien derivatives $\Delta_{2\omega_1^j} \overset{\triangledown}{w}_{\mathbf{0}} \in \mathrm{SING}_{2\pi j, \pi}$, *resp.* $\Delta_{2\omega_1^j} \overset{\triangle}{w}_{\mathbf{0}} \in \mathrm{ASYMP}_{2\pi j, \pi}$ and $\Delta_{2\omega_2^j} \overset{\triangledown}{w}_{\mathbf{0}} \in \mathrm{SING}_{2\pi(j+1/2), \pi}$, *resp.* $\Delta_{2\omega_2^j} \overset{\triangle}{w}_{\mathbf{0}} \in \mathrm{ASYMP}_{2\pi(j+1/2), \pi}$, are well-defined. As a matter of fact, these alien derivatives are quite simple !

Lemma 8.4. *For any $\omega \in \widetilde{\mathbb{C}}$ so that $\overset{\bullet}{\omega} = \pm 2$, one has $\Delta_\omega \overset{\triangledown}{w}_{\mathbf{0}} = 0$. Equivalently, $\Delta_\omega \overset{\triangle}{w}_{\mathbf{0}} = 0$, $\Delta_\omega \widetilde{w}_{\mathbf{0}} = 0$.*

Proof. We only calculate $\overset{\triangle}{W} = \Delta_{2\omega_1^j} \overset{\triangle}{w}_{\mathbf{0}}$. Through the alien derivation $\Delta_{2\omega_1^j}$, equation (8.3) is transformed into the linear ODE

$$P(\partial - 2)\, \overset{\triangle}{W} + \frac{1}{z} Q(\partial - 2)\, \overset{\triangle}{W} = \frac{\partial F(z, \overset{\triangle}{w}_{\mathbf{0}})}{\partial w}\, \overset{\triangle}{W} \tag{8.26}$$

as a consequence of corollary 7.2. We recognize the equation $\mathfrak{P}_{2\mathbf{e}_1}(\overset{\triangle}{w}_{\mathbf{0}})\, \overset{\triangle}{W} = 0$. By lemma 5.4, the general formal solution for the linear equation $\mathfrak{P}_{2\mathbf{e}_1}(\widetilde{w}_{\mathbf{0}}) \widetilde{\mathscr{W}} = 0$ is of the form $C_1 \mathrm{e}^z \widetilde{W}_{\mathbf{e}_1} + C_2 \mathrm{e}^{3z} \widetilde{W}_{\mathbf{e}_2}$ and we either conclude with the reasoning made in Sect. 8.3.2 (still write $\overset{\triangledown}{W}$ under the form $\overset{\triangledown}{W} = \overset{\triangledown}{S} * \overset{\triangledown}{W}_{\mathbf{e}_1}$ and show that $\overset{\triangledown}{S} = 0$) or rather directly with lemma 8.3 : the solutions of the equation $\mathfrak{P}_{2\mathbf{e}_1}(\overset{\triangle}{w}_{\mathbf{0}})\, \overset{\triangle}{\mathscr{W}} = 0$ in $\prod_{k \in \mathbb{Z}} e^{-kz} \mathrm{ASYMP}$ are $C_1 \mathrm{e}^z\, \overset{\triangledown}{W}_{\mathbf{e}_1} + C_2 \mathrm{e}^{3z}\, \overset{\triangledown}{W}_{\mathbf{e}_2}$ and one concludes that $\Delta_{2\omega_1^j} \overset{\triangle}{w}_{\mathbf{0}} = 0$ since the alien derivative belongs to $\mathrm{ASYMP}_{2\pi j, \pi}$. □

We can keep on that way to get the complete resurgent structure for $\widetilde{w}_{\mathbf{0}}$ and, at the same time, to analytically continue its minor $\widehat{w}_{\mathbf{0}}$. Let us see what happens a step further.

To show that $\widetilde{w}_{\mathbf{0}}$ belongs to $\widetilde{\mathscr{R}}^{(3)}$, we have to complete the informations given by lemma 8.4. Following corollary 7.3, we would like to show that $\Delta_{\omega_2} \circ \Delta_{\omega_1} \overset{\triangle}{w}_{\mathbf{0}}$ belongs to $\widetilde{\mathrm{RES}}^{(1)}$ for any $\omega_1, \omega_2 \in \widetilde{\mathbb{C}}$ so that $\overset{\bullet}{\omega}_1 = \pm 1$ and $\overset{\bullet}{\omega}_2 = \pm 1$. From what we know, this amount to showing that the alien derivatives $\Delta_{\omega_2} \overset{\triangle}{W}_{\mathbf{e}_i}$ belong to $\widetilde{\mathrm{RES}}^{(1)}$. Let us look at $\overset{\triangle}{W} = \Delta_{\omega_1^0} \overset{\triangle}{W}_{\mathbf{e}_1} \in \mathrm{ASYMP}_{0,\pi}$. From the identity $\mathfrak{P}_{\mathbf{e}_1}(\overset{\triangle}{w}_{\mathbf{0}})\, \overset{\triangle}{W}_{\mathbf{e}_1} = 0$ (equation (8.16)) and corollary 7.2, we draw:

$$P(\partial-2)\overset{\triangle}{W}+\frac{1}{z}Q(\partial-2)\overset{\triangle}{W}=\frac{\partial F(z,\overset{\triangle}{w}_{\mathbf{0}})}{\partial w}\overset{\triangle}{W}+\overset{\triangle}{W}_{\mathbf{e}_1}\Delta_{\omega_1^0}\overset{\triangle}{w}_{\mathbf{0}}\frac{\partial^2 F(z,\overset{\triangle}{w}_{\mathbf{0}})}{\partial w^2},$$

that is

$$\mathfrak{P}_{2\mathbf{e}_1}(\overset{\triangle}{w}_{\mathbf{0}})\overset{\triangle}{W}=A_0(\omega_1^0)\overset{\triangle}{W}{}^2_{\mathbf{e}_1}\frac{\partial^2 F(z,\overset{\triangle}{w}_{\mathbf{0}})}{\partial w^2}. \tag{8.27}$$

where $A_0(\omega_1^0)$ is the resurgent constant given in lemma 8.1. Observe that the general formal solution for the equation $\mathfrak{P}_{2\mathbf{e}_1}(\widetilde{w}_{\mathbf{0}})\widetilde{\mathscr{W}}=A_0(\omega_1^0)\widetilde{W}^2_{\mathbf{e}_1}\frac{\partial^2 F(z,\widetilde{w}_{\mathbf{0}})}{\partial w^2}$, deduced from (8.27) through the Taylor map, reads:

$$\widetilde{\mathscr{W}}=2A_0(\omega_1^0)\widetilde{W}_{2\mathbf{e}_1}+C_1\mathrm{e}^{z}\widetilde{W}_{\mathbf{e}_1}+C_2\mathrm{e}^{3z}\widetilde{W}_{\mathbf{e}_2}\in\prod_{k\in\mathbb{Z}}e^{-kz}\widetilde{\mathrm{Nils}}_1$$

with $C_1,C_2\in\mathbb{C}$. By lemma 8.3 one gets $\Delta_{\omega_1^0}\overset{\triangle}{W}_{\mathbf{e}_1}=2A_0(\omega_1^0)\overset{\triangle}{W}_{2\mathbf{e}_1}$, which thus belongs to $\widetilde{\mathrm{RES}}^{(1)}$ by proposition 8.1.

Of course, one can keep on that way, by induction. However, a lesson has to be learned from what precedes : the resurgent structure is closely coupled to the formal integral and it is much time to introduce the bridge equation.

8.4 The Bridge Equation and Proof of the Main Theorem

We go back to the formal integral

$$\widetilde{w}(z,\mathbf{U})=\sum_{\mathbf{k}\in\mathbb{N}^2}\mathbf{U}^{\mathbf{k}}\mathrm{e}^{-\boldsymbol{\lambda}.\mathbf{k}z}\widetilde{W}_{\mathbf{k}}\quad\in\prod_{k\in\mathbb{Z}}e^{-kz}\widetilde{\mathrm{Nils}}_1[[\mathbf{U}]] \tag{8.28}$$

and we consider its derivatives with respect to the indeterminate U_i, $i=1,2$:

$$\begin{aligned}\frac{\partial\widetilde{w}}{\partial U_i}(z,\mathbf{U})&=\sum_{\mathbf{k}\in\mathbb{N}^2}\mathbf{k}.\mathbf{e}_i\mathbf{U}^{\mathbf{k}-\mathbf{e}_i}\mathrm{e}^{-\lambda.\mathbf{k}z}\widetilde{W}_{\mathbf{k}}\quad\in\prod_{k\in\mathbb{Z}}e^{-kz}\widetilde{\mathrm{Nils}}_1[[\mathbf{U}]]\\&=\widetilde{W}_{\mathbf{e}_i}+O(U_1,U_2).\end{aligned} \tag{8.29}$$

Since the formal integral $\widetilde{w}$ solves the differential equation $P(\partial)\widetilde{w}+\frac{1}{z}Q(\partial)\widetilde{w}=F(z,\widetilde{w})$, one deduces that the following identity holds for $i=1,2$:

$$\left(P(\partial)+\frac{1}{z}Q(\partial)-\frac{\partial F(z,\widetilde{w})}{\partial w}\right)\frac{\partial\widetilde{w}}{\partial U_i}=0,\ \text{ i.e. }\mathfrak{P}_{\mathbf{0}}(\widetilde{w})\frac{\partial\widetilde{w}}{\partial U_i}=0. \tag{8.30}$$

The formal solutions for the equation $\mathfrak{P}_{\mathbf{0}}(\widetilde{w}_{\mathbf{0}})\widetilde{\mathscr{W}}=0$ is spanned by $\mathrm{e}^{-\lambda_1 z}\widetilde{W}_{\mathbf{e}_1}$ and $\mathrm{e}^{-\lambda_2 z}\widetilde{W}_{\mathbf{e}_2}$. Therefore, $\dfrac{\partial\widetilde{w}}{\partial U_1}$ and $\dfrac{\partial\widetilde{w}}{\partial U_2}$ are two linearly independent solutions for the order two linear differential equation $\mathfrak{P}_{\mathbf{0}}(\widetilde{w})\widetilde{\mathscr{W}} = 0$, explicitly (wronsk stands for the wronskian):

$$\text{wronsk}\left(\frac{\partial \widetilde{w}}{\partial U_1}, \frac{\partial \widetilde{w}}{\partial U_2}\right) = \text{wronsk}\left(\mathrm{e}^{-\lambda_1 z}\widetilde{W}_{\mathbf{e}_1}, \mathrm{e}^{-\lambda_2 z}\widetilde{W}_{\mathbf{e}_2}\right) = 2z^3.$$

Lemma 8.3 translates into the fact that for any series of the form

$$\overset{\triangle}{\mathscr{W}}(z,\mathbf{U}) = \sum_{\mathbf{k}\in\mathbb{N}^2} \mathbf{U}^{\mathbf{k}} \overset{\triangle}{\mathscr{W}}_{\mathbf{k}}, \quad \overset{\triangle}{\mathscr{W}}_{\mathbf{k}} \in \prod_{k\in\mathbb{Z}} e^{-kz}\text{ASYMP}_{\theta,\alpha},$$

which satisfies the second order equation $\mathfrak{P}_{\mathbf{0}}(\overset{\triangle}{w})\overset{\triangle}{W} = 0$, there exist uniquely determined constants $A(\omega,\mathbf{U}) \in \mathbb{C}[[\mathbf{U}]]$ and $B(\omega,\mathbf{U}) \in \mathbb{C}[[\mathbf{U}]]$ such that

$$\overset{\triangle}{\mathscr{W}}(z,\mathbf{U}) = A(\omega,\mathbf{U})\frac{\partial \overset{\triangle}{w}}{\partial U_1} + B(\omega,\mathbf{U})\frac{\partial \overset{\triangle}{w}}{\partial U_2}, \quad \frac{\partial \overset{\triangle}{w}}{\partial U_i} = ^{\natural\text{ram}}\frac{\partial \widetilde{w}}{\partial U_i} \tag{8.31}$$

To the formal integral $\widetilde{w}(z,\mathbf{U})$, one associates its analogue through the mapping $^{\natural\text{ram}}$:

$$\overset{\triangle}{w}(z,\mathbf{U}) = \sum_{\mathbf{k}\in\mathbb{N}^2} \mathbf{U}^{\mathbf{k}}\mathrm{e}^{-\boldsymbol{\lambda}.\mathbf{k}z}\overset{\triangle}{W}_{\mathbf{k}}, \quad \overset{\triangle}{W}_{\mathbf{k}} = ^{\natural\text{ram}}\widetilde{W}_{\mathbf{k}}. \tag{8.32}$$

We pick $\omega \in \widetilde{\mathbb{C}}$ and we assume for the moment that $\dot{\omega} = \pm 1$. By proposition 8.1 and corollary 7.2, the alien derivation Δ_ω acts on the formal integral $\overset{\triangle}{w}(z,\mathbf{U})$. As a matter of fact, it will be easier to use the *dotted alien derivation*, $\dot{\Delta}_\omega = \mathrm{e}^{-\omega z}\Delta_\omega$ which has the virtue of commuting with the derivation ∂. Therefore,

$$\dot{\Delta}_\omega \overset{\triangle}{w}(z,\mathbf{U}) = \sum_{\mathbf{k}\in\mathbb{N}^2} \mathbf{U}^{\mathbf{k}}\mathrm{e}^{-\boldsymbol{\lambda}.\mathbf{k}z}\dot{\Delta}_\omega \overset{\triangle}{W}_{\mathbf{k}}, \quad \dot{\Delta}_\omega \overset{\triangle}{W}_{\mathbf{k}} \in \mathrm{e}^{-\omega z}\text{ASYMP}_{\arg(\omega),\pi}$$

and

$$\mathfrak{P}_{\mathbf{0}}(\overset{\triangle}{w})\dot{\Delta}_\omega \overset{\triangle}{w} = 0.$$

We deduce that the decomposition (8.31) holds for $\dot{\Delta}_\omega \overset{\triangle}{w}$. This decomposition $\dot{\Delta}_\omega \overset{\triangle}{w} = A(\omega,\mathbf{U})\frac{\partial \overset{\triangle}{w}}{\partial U_1} + B(\omega,\mathbf{U})\frac{\partial \overset{\triangle}{w}}{\partial U_2}$ is the so-called *bridge equation* of Ecalle, that is a link between alien derivatives and the usual partial derivatives.
Let $\Xi \subset \mathbb{N}^2$ be the set defined by $\Xi = \Xi_0 = \{k\mathbf{e}_1, k\mathbf{e}_2 \,|\, k \in \mathbb{N}\}$ and set $\Xi_n = \mathbf{n} + \Xi$ for any $n \in \mathbb{N}^\star$. With this notation, the formal integral can be written as follows:

$$\widetilde{w}(z,\mathbf{U}) = \sum_{n=0}^{\infty}\sum_{\mathbf{k}\in\Xi_n} \mathbf{U}^{\mathbf{k}}\mathrm{e}^{-\boldsymbol{\lambda}.\mathbf{k}z}\widetilde{W}_{\mathbf{k}}(z) = \sum_{n=0}^{\infty}\sum_{\mathbf{k}\in\Xi} \mathbf{U}^{\mathbf{k}+\mathbf{n}}\mathrm{e}^{-\boldsymbol{\lambda}.\mathbf{k}z}\widetilde{W}_{\mathbf{k}+\mathbf{n}}(z) \tag{8.33}$$

To fix our mind, suppose that $\overset{\bullet}{\omega} = k_0\lambda_1$ with $k_0 = 1$ at the moment. We get from the decomposition (8.31) the identity:

$$\sum_{n=0}^{\infty}\sum_{\mathbf{k}\in\Xi}\mathbf{U}^{\mathbf{k}+\mathbf{n}}e^{-\boldsymbol{\lambda}.(\mathbf{k}+k_0\mathbf{e}_1)z}\Delta_\omega \overset{\triangle}{W}_{\mathbf{k}+\mathbf{n}}= \\ A(\omega,\mathbf{U})\sum_{n=0}^{\infty}\sum_{\mathbf{k}\in\Xi}(\mathbf{k}+\mathbf{n}).\mathbf{e}_1\mathbf{U}^{\mathbf{k}+\mathbf{n}-\mathbf{e}_1}e^{-\boldsymbol{\lambda}.\mathbf{k}z}\overset{\triangle}{W}_{\mathbf{k}+\mathbf{n}} \\ +B(\omega,\mathbf{U})\sum_{n=0}^{\infty}\sum_{\mathbf{k}\in\Xi}(\mathbf{k}+\mathbf{n}).\mathbf{e}_2\mathbf{U}^{\mathbf{k}+\mathbf{n}-\mathbf{e}_2}e^{-\boldsymbol{\lambda}.\mathbf{k}z}\overset{\triangle}{W}_{\mathbf{k}+\mathbf{n}} \tag{8.34}$$

Each component $\mathbf{U}^{\mathbf{k}+\mathbf{n}}e^{-\boldsymbol{\lambda}.(\mathbf{k}+k_0\mathbf{e}_1)z}\Delta_\omega \overset{\triangle}{W}_{\mathbf{k}+\mathbf{n}}\in e^{-\boldsymbol{\lambda}.(\mathbf{k}+k_0\mathbf{e}_1)z}\mathrm{ASYMP}_{\arg(\omega),\pi}$ has its counterpart on the right-hand side of the equality. Necessarily,

$$A(\omega,\mathbf{U}) = \mathbf{U}^{(1-k_0)\mathbf{e}_1}\sum_{n\geq 0}A_n(\omega)\mathbf{U}^{\mathbf{n}} \\ B(\omega,\mathbf{U}) = \mathbf{U}^{\mathbf{e}_2-k_0\mathbf{e}_1}\sum_{n\geq 0}B_n(\omega)\mathbf{U}^{\mathbf{n}}. \tag{8.35}$$

This implies on the one hand hand that $A_n(\omega) = 0$ when $|\omega| \geq n+2$ while $B_n(\omega) = 0$ when $|\omega| \geq n+1$. On the other hand,

$$\Delta_\omega \overset{\triangle}{W}_{\mathbf{k}+\mathbf{n}} = \sum_{m=-1}^{n} A_{n-m}(\omega)(\mathbf{k}+\mathbf{m}+k_0\mathbf{e}_1).\mathbf{e}_1\overset{\triangle}{W}_{\mathbf{k}+\mathbf{m}+k_0\mathbf{e}_1} \\ + \sum_{m=-1}^{n} B_{n-m}(\omega)(\mathbf{k}+\mathbf{m}+k_0\mathbf{e}_1).\mathbf{e}_2\overset{\triangle}{W}_{\mathbf{k}+\mathbf{m}+k_0\mathbf{e}_1} \tag{8.36}$$

with the convention used in theorem 8.1. The case $\overset{\bullet}{\omega} = k_0\lambda_2$ with $k_0 = 1$ is obtained by symmetry.

This result implies that the asymptotic class $\overset{\triangle}{W}_{\mathbf{k}} = {}^{\natural\mathrm{ram}}\widetilde{W}_{\mathbf{k}}$ belongs to $\widetilde{\mathrm{RES}}^{(2)}$, as a consequence of corollary 7.3. An easy induction on $k_0 \in \mathbb{N}^\star$ allows then to conclude that the $\widetilde{W}_{\mathbf{k}}$ belong to $\widetilde{\mathscr{R}}_{\mathbb{Z}}^{\mathrm{ram}}$. The rest of the theorem is shown by arguments used in remark 8.1.2.1. This ends the proof of theorem 8.1.

8.5 Comments

For differential systems of level 1 of the type (5.67), the resurgent study of the Stokes phenomenon and of the action of the symbolic Stokes automorphism Δ_θ^+ on transseries solutions have been obtained by Costin [Cos98], under some conditions. This has been later extended to more general differential equations (with no resonance), and also for difference equations of the type (5.68), in particular by Braaksma and his students (see [Bra001, Kui003]). These works make use of (so-called) "staircase distributions" [Cos98, Cos009] and do not make appeal to alien derivations. The method explained in this chapter is closer to the ideas of Ecalle, leading to the bridge equation and the full resurgent structure. Also, as we saw on the particular example of the first Painlevé equation, this method provides (theoretically) the whole set of Ecalle's holomophic invariants and passes the resonance cases under some conditions (no quasi-resonance, no nihilence [Eca85]).

References

ASV012. I. Aniceto, R. Schiappa, M. Vonk, *The resurgence of instantons in string theory.* Commun. Number Theory Phys. **6** (2012), no. 2, 339-496.

Bra001. B.L.J. Braaksma, *Transseries for a class of nonlinear difference equations. In memory of W. A. Harris, Jr.* J. Differ. Equations Appl. **7** (2001), no. 5, 717-750.

CNP93-1. B. Candelpergher, C. Nosmas, F. Pham, *Approche de la résurgence*, Actualités mathématiques, Hermann, Paris (1993).

Cos98. O. Costin, *On Borel summation and Stokes phenomena for rank-*1 *nonlinear systems of ordinary differential equations.* Duke Math. J. **93** (1998), no. 2, 289-344.

Cos99. O. Costin, *Correlation between pole location and asymptotic behavior for Painlevé I solutions.* Comm. Pure Appl. Math. **52** (1999), no. 4, 461-478.

Cos009. O. Costin, *Asymptotics and Borel summability*, Chapman & Hall/CRC Monographs and Surveys in Pure and Applied Mathematics, 141. CRC Press, Boca Raton, FL, 2009.

CC001. O. Costin, R. Costin, *On the formation of singularities of solutions of nonlinear differential systems in antistokes directions.* Invent. Math. **145** (2001), no. 3, 425-485.

CHT014. O. Costin, M. Huang, S. Tanveer, *Proof of the Dubrovin conjecture and analysis of the tritronquée solutions of* P_I, Duke Math. J. **163** (2014), no. 4, 665-704.

Del005. E. Delabaere, *Addendum to the hyperasymptotics for multidimensional Laplace integrals.* Analyzable functions and applications, 177-190, Contemp. Math., 373, Amer. Math. Soc., Providence, RI, 2005.

Del008. E. Delabaere, *Exact WKB analysis near a simple turning point.* Algebraic analysis of differential equations from microlocal analysis to exponential asymptotics, 101-117, Springer, Tokyo, 2008.

Del010. E. Delabaere, *Singular integrals and the stationary phase methods.* Algebraic approach to differential equations, 136-209, World Sci. Publ., Hackensack, NJ, 2010.

DDP93. E. Delabaere, H. Dillinger, F. Pham, *Résurgence de Voros et périodes des courbes hyperelliptiques*, Annales de l'Institut Fourier **43** (1993), no. 1, 163-199.

DDP97. E. Delabaere, H. Dillinger, F. Pham, *Exact semi-classical expansions for one dimensional quantum oscillators*, Journal Math. Phys. **38** (1997), 12, 6126-6184.

DH002. E. Delabaere, C. J. Howls, *Global asymptotics for multiple integrals with boundaries.* Duke Math. J. **112** (2002), 2, 199-264.

DH99. E. Delabaere, F. Pham, *Resurgent methods in semi-classical asymptotics*, Ann. Inst. Henri Poincaré, Sect. A **71** (1999), no 1, 1-94.

Eca81-1. J. Écalle, *Les algèbres de fonctions résurgentes*, Publ. Math. d'Orsay, Université Paris-Sud, 1981.05 (1981).

Eca85. J. Ecalle, *Les fonctions résurgentes. Tome III : l'équation du pont et la classification analytique des objets locaux.* Publ. Math. d'Orsay, Université Paris-Sud, 1985.05 (1985).

Eca92. J. Écalle, *Fonctions analysables et preuve constructive de la conjecture de Dulac.* Actualités mathématiques, Hermann, Paris (1992).

Eca93-1. J. Écalle, *Six lectures on transseries, Analysable functions and the Constructive proof of Dulac's conjecture*, Bifurcations and periodic orbits of vector fields (Montreal, PQ, 1992), 75-184, NATO Adv. Sci. Inst. Ser. C Math. Phys. Sci., 408, Kluwer Acad. Publ., Dordrecht, 1993.

Eve004. C. Even, *Degrés de liberté des moyennes de convolution préservant la réalité.* Ann. Fac. Sci. Toulouse Math. (6) 13 (2004), no. 3, 377-420.

FIKN006. A.S. Fokas, A.R. Its, A.A. Kapaev, V.Y. Novokshenov, *Painlevé transcendents. The Riemann-Hilbert approach.* Mathematical Surveys and Monographs, 128. American Mathematical Society, Providence, RI, 2006.

GS001. V. Gelfreich, D. Sauzin, *Borel summation and splitting of separatrices for the Hénon map*, Ann. Inst. Fourier (Grenoble) **51** (2001), no 2, 513-567.

Kap88. A. Kapaev, *Asymptotic behavior of the solutions of the Painlevé equation of the first kind.* Differentsial'nye Uravneniya **24** (1988), no. 10, 1684-1695, 1835; translation in Differential Equations **24** (1988), no. 10, 1107-1115 (1989).

Kap004. A.A. Kapaev, *Quasi-linear stokes phenomenon for the Painlevé first equation.* J. Phys. A **37** (2004), no. 46, 11149-11167.

KK93. A. Kapaev, A.V. Kitaev, *Connection formulae for the first Painlevé transcendent in the complex domain.* Lett. Math. Phys. **27** (1993), no. 4, 243-252.

KT005. T. Kawai, Y. Takei, *Algebraic analysis of singular perturbation theory.* Translated from the 1998 Japanese original by Goro Kato. Translations of Mathematical Monographs, 227. Iwanami Series in Modern Mathematics. American Mathematical Society, Providence, RI, 2005.

Kui003. G.R. Kuik, *Transseries in Difference and Differential Equations.* PhD thesis, Rijksuniversiteit Groningen (2003).

Lod016. M. Loday-Richaud, *Divergent Series, Summability and Resurgence II. Simple and Multiple Summability.* Lecture Notes in Mathematics, **2154**. Springer, Heidelberg, 2016.

LR011. M. Loday-Richaud, P. Remy, *Resurgence, Stokes phenomenon and alien derivatives for level-one linear differential systems.* J. Differential Equations **250** (2011), no. 3, 1591-1630.

Men99. F. Menous, *Les bonnes moyennes uniformisantes et une application à la resommation réelle.* Ann. Fac. Sci. Toulouse Math. (6) 8 (1999), no. 4, 579-628.

MS016. C. Mitschi, D. Sauzin, *Divergent Series, Summability and Resurgence I. Monodromy and Resurgence.* Lecture Notes in Mathematics, **2153**. Springer, Heidelberg, 2016.

Old005. A.B. Olde Daalhuis, *Hyperasymptotics for nonlinear ODEs. II. The first Painlevé equation and a second-order Riccati equation.* Proc. R. Soc. Lond. Ser. A Math. Phys. Eng. Sci. **461** (2005), no. 2062, 3005-3021.

OSS003. C. Olivé, D. Sauzin, T. Seara, *Resurgence in a Hamilton-Jacobi equation.* Proceedings of the International Conference in Honor of Frédéric Pham (Nice, 2002). Ann. Inst. Fourier (Grenoble) **53** (2003), no. 4, 1185-1235.

Ou012. Y. Ou, *Sur la stabilité par produit de convolution d'algèbres de résurgence.* PhD thesis, Université d'Angers (2012).

Ras010. J.-M. Rasoamanana, *Résurgence-sommabilité de séries formelles ramifiées dépendant d'un paramètre et solutions d'équations différentielles linéaires.* Ann. Fac. Sci. Toulouse Math. **(6)** 19 (2010), no. 2, 303-343.

Sau006. D. Sauzin, *Resurgent functions and splitting problems*, RIMS Kokyuroku 1493 (31/05/2006) 48-117

Sau013. D. Sauzin, *On the stability under convolution of resurgent functions.* Funkcial. Ekvac. **56** (2013), no. 3, 397-413.

Sau015. D. Sauzin, *Nonlinear analysis with resurgent functions.* Ann. Sci. Ecole Norm. Sup. (6) **48** (2015), no. 3, 667-702.

Tak000. Y. Takei, *An explicit description of the connection formula for the first Painlevé equation.* In Toward the exact WKB analysis of differential equations, linear or non-linear (Kyoto, 1998), 204, 271-296, Kyoto Univ. Press, Kyoto, 2000.

Index

E. Delabaere, *Divergent Series, Summability and Resurgence III*,
Lecture Notes in Mathematics 2155, DOI 10.1007/978-3-319-29000-3